Biomedical Research Developments for Improved Healthcare

Pranav Kumar Prabhakar
Lovely Professional University, India

IGI Global
PUBLISHER of TIMELY KNOWLEDGE

A volume in the Advances in Bioinformatics and Biomedical Engineering (ABBE) Book Series

Published in the United States of America by
IGI Global
Medical Information Science Reference (an imprint of IGI Global)
701 E. Chocolate Avenue
Hershey PA, USA 17033
Tel: 717-533-8845
Fax: 717-533-8661
E-mail: cust@igi-global.com
Web site: http://www.igi-global.com

Library of Congress Cataloging-in-Publication Data

Names: Kumar Prabhakar, Pranav, editor.
Title: Biomedical research developments for improved healthcare / edited by Pranav Prabhakar.
Description: Hershey, PA : Medical Information Science Reference, [2024] | Includes bibliographical references and index. | Summary: "The primary objective of the book is to inform, educate, and inspire readers about the exciting and rapidly evolving field of biomedical research and its profound impact on healthcare"-- Provided by publisher.
Identifiers: LCCN 2023051293 (print) | LCCN 2023051294 (ebook) | ISBN 9798369319222 (hardcover) | ISBN 9798369319239 (ebook)
Subjects: MESH: Biomedical Engineering | Artificial Intelligence | Biomedical Technology
Classification: LCC R857.B5 (print) | LCC R857.B5 (ebook) | NLM QT 36 | DDC 610.28--dc23/eng/20240102
LC record available at https://lccn.loc.gov/2023051293
LC ebook record available at https://lccn.loc.gov/2023051294

This book is published in the IGI Global book series Advances in Bioinformatics and Biomedical Engineering (ABBE) (ISSN: 2327-7033; eISSN: 2327-7041)

British Cataloguing in Publication Data
A Cataloguing in Publication record for this book is available from the British Library.

For electronic access to this publication, please contact: eresources@igi-global.com.

Advances in Bioinformatics and Biomedical Engineering (ABBE) Book Series

Ahmad Taher Azar
College of Computer & Information Sciences, Prince Sultan University, Riyadh, Saudi Arabia & Faculty of Computers and Artificial Intelligence, Benha University, Benha, Egypt

ISSN:2327-7033
EISSN:2327-7041

Mission

The fields of biology and medicine are constantly changing as research evolves and novel engineering applications and methods of data analysis are developed. Continued research in the areas of bioinformatics and biomedical engineering is essential to continuing to advance the available knowledge and tools available to medical and healthcare professionals.

The **Advances in Bioinformatics and Biomedical Engineering (ABBE) Book Series** publishes research on all areas of bioinformatics and bioengineering including the development and testing of new computational methods, the management and analysis of biological data, and the implementation of novel engineering applications in all areas of medicine and biology. Through showcasing the latest in bioinformatics and biomedical engineering research, ABBE aims to be an essential resource for healthcare and medical professionals.

Coverage

- Applied Biosystems
- Genetic Engineering
- Computational Biology
- Neural Engineering
- Drug Design
- Data Mining
- Protein Structure
- Biostatistics
- Algorithms
- Artificial Organs

IGI Global is currently accepting manuscripts for publication within this series. To submit a proposal for a volume in this series, please contact our Acquisition Editors at Acquisitions@igi-global.com or visit: http://www.igi-global.com/publish/.

The Advances in Bioinformatics and Biomedical Engineering (ABBE) Book Series (ISSN 2327-7033) is published by IGI Global, 701 E. Chocolate Avenue, Hershey, PA 17033-1240, USA, www.igi-global.com. This series is composed of titles available for purchase individually; each title is edited to be contextually exclusive from any other title within the series. For pricing and ordering information please visit http://www.igi-global.com/book-series/advances-bioinformatics-biomedical-engineering/73671. Postmaster: Send all address changes to above address.

Titles in this Series

For a list of additional titles in this series, please visit: http://www.igi-global.com/book-series/advances-bioinformatics-biomedical-engineering/73671

Quantum Innovations at the Nexus of Biomedical Intelligence
Vishal Dutt (AVN Innovations Pvt. Ltd., India) Abhishek Kumar (Department of CSE, UIE, Chandigarh University, Punjab, India) Sachin Ahuja (Chandigarh University, India) Anupam Baliyan (Geeta University, India) and Narayan Vyas (AVN Innovations Pvt. Ltd, India)
Medical Information Science Reference • © 2024 • 287pp • H/C (ISBN: 9798369314791) • US $330.00

Physiology and Function of Glial Cells in Health and Disease
Bilal El-Mansoury (Faculty of Sciences, Chouaib Doukkali University, Morocco) Omar El Hiba (Chouaib Doukkali University, Morocco) and Arumugam Radhakrishnan Jayakumar (University of Mimi, USA)
Engineering Science Reference • © 2024 • 589pp • H/C (ISBN: 9781668496756) • US $325.00

Applications of Synthetic Biology in Health, Energy, and Environment
Muhammad Arshad (International Islamic University, Islamabad, Pakistan)
Medical Information Science Reference • © 2023 • 454pp • H/C (ISBN: 9781668465776) • US $325.00

Isolation, Characterization, and Therapeutic Applications of Natural Bioactive Compounds
Ajeet Singh (ICAR-Indian Institute of Wheat and Barley Research, India)
Medical Information Science Reference • © 2022 • 401pp • H/C (ISBN: 9781668473375) • US $345.00

AI-Enabled Smart Healthcare Using Biomedical Signals
Rahul Kumar Chaurasiya (Maulana Azad National Institute of Technology, Bhopal, India) Dheeraj Agrawal (Maulana Azad National Institute of Technology, Bhopal, India) and Ram Bilas Pachori (Indian Institute of Technology, Indore, India)
Medical Information Science Reference • © 2022 • 322pp • H/C (ISBN: 9781668439470) • US $435.00

Advancements in Controlled Drug Delivery Systems
Shekhar Verma (University College of Pharmacy, Pandit Deendayal Upadhyay Memorial Health Science, India & Ayush University of Chattisgarh, India) and Santosh Kumar Verma (Yulin University, China)
Medical Information Science Reference • © 2022 • 395pp • H/C (ISBN: 9781799889083) • US $360.00

Emerging Developments and Applications of Low Temperature Plasma
Aamir Shahzad (Government College University, Faisalabad, Pakistan) and Maogang He (Xi'an Jiaotong University, China)
Engineering Science Reference • © 2022 • 239pp • H/C (ISBN: 9781799883982) • US $250.00

IGI Global
PUBLISHER of TIMELY KNOWLEDGE

701 East Chocolate Avenue, Hershey, PA 17033, USA
Tel: 717-533-8845 x100 • Fax: 717-533-8661
E-Mail: cust@igi-global.com • www.igi-global.com

Table of Contents

Detailed Table of Contents

Light harvesting applications have proven to be highly valuable in various fields such as photovoltaics, photo catalysis, and photo polymerization. When bacteriochlorophyll is manipulated at the molecular level, it opens up numerous possibilities for improving its physical and chemical properties for harvesting light. . In their research study, the authors suggest using molecular graph representations from the Computational Material Repository (CMR) dataset to forecast the energy gaps of bacteriochlorophyll. The electro-topological-state index is the description for the molecular network that the authors choose expressly. The physical configuration of carbon in methane bonds associated to the attached group, as found by the authors, has a considerable impact on the energy gaps of bacteriochlorophyll. This observation is consistent with the three-dimensional diffusion patterns of the molecule's leading-edge molecular orbitals, as predicted by quantum mechanical calculations.

This chapter investigates the application of artificial intelligence (AI) to the upkeep of safety equipment for employees across several industries. Wear and tear, exposure to dangerous materials, and other factors can reduce the effectiveness of protective equipment, making routine maintenance essential to ensuring that it stays functional and offers sufficient protection. It is explained how AI can be used for preventative maintenance, checks and repairs, and extending the lifespan of safety equipment. AI is capable of identifying tears or other damage in personal protective equipment (PPE), guiding repairs, and extending the lifespan of equipment. Incorporating AI into protective equipment maintenance can improve safety, save time and money, and as technology advances, it has the potential to completely change how protective equipment is maintained and managed. The chapter concludes that in order to preserve worker safety and wellbeing, enterprises should integrate AI into their strategy for maintaining protective equipment.

Metabolomics is a fascinating area that specializes in analyzing the small molecules in our frame, like hormones and enzymes, that play an important role in our metabolism. With the aid of reading these molecules, scientists can gain insights into our metabolic fitness and the way our body processes nutrients. It entails different factors such as energy production, nutrient usage, and waste elimination. Through metabolomics, researchers can become aware of precise metabolic biomarkers that imply the state of our fitness. Through studying these markers, they are able to hit upon early symptoms of metabolic disorders like diabetes, obesity, and cardiovascular ailment. This knowledge can then be used to broaden customized interventions and treatments to improve metabolic health. It is able to assist in the development of centered treatment plans, personalized nutrition plans, and preventive strategies. In this chapter, we will attempt towards achieving and maintaining optimal health with the study of metabolomics biomarkers.

People who are at higher risk for cardiovascular disease can be found using biomarkers. Acute coronary syndrome must be diagnosed and treated promptly and accurately, and cardiac biomarkers are crucial for both. For therapy to start as soon as possible, and possibly, a diagnosis within the golden period, is crucial. Among the numerous biomarkers, the myocardial enzymes, a number of myocardial proteins, peptides, and many other compounds were the first ones to be studied. The stable molecules known as microRNAs are the newest addition to the repertoire and can be found in circulation. Many cardiological conditions are still incurable, and traditional treatments frequently have unfavourable side effects or only slightly improve patient outcomes. Hence, personalised therapy of cardiovascular diseases (CVD), also known as personalised or precision cardiology, is the process of choosing the most appropriate course of action for a given patient.

high surface-to-volume ratio and superior penetration across cell membranes, and their nanostructure provides flexibility in designing effective strategies to tackle biofilm-associated infection. This chapter summarizes the various research endeavours in which nanoparticles have been explored for the prevention and or/treatment of bacterial biofilms on orthopedic implants.

The evolutionary foundations of brain development represent point of departure. It delves into the depths of evolutionary history, unveiling the intricate journey that has shaped the human brain. Comparative neuroanatomy and phylogeny provide the canvas upon which we paint the portrait of our cognitive evolution. It contemplates the constraints, both genetic and epigenetic, that have sculpted the human brain's architecture. These insights serve as the bedrock upon which Evolutionary Global Neuroscience stands. In the quest to unlock the enigmatic complexities of the human brain, Evolutionary Global Neuroscience emerges as a transformative paradigm that transcends disciplinary boundaries. This expansive field explores the profound interplay between our evolutionary heritage and contemporary challenges in cognition, brain health, and the ever-evolving landscape of neurotechnology. Cognition, that quintessential hallmark of human existence, is next to claim our attention. It interrogate the adaptive value of cognition across epochs, deciphering its role in survival and prosperity. Yet, in this age of rapid societal transformation, the concept of "evolutionary mismatch" comes to the fore. It explore how this mismatch underlies a burgeoning epidemic of brain disorders, underscoring the urgent need for a new paradigm in addressing cognitive health. This paper illuminates the grim landscape of modernity's brain disorders. Alzheimer's disease and autism spectrum disorders serve as poignant exemplars of the evolutionary puzzle that engulfs us. From genetic predispositions to environmental influences, it contemplates the multifaceted origins of these disorders, ever mindful of the shadows cast by evolutionary history. The promise of Evolutionary Global Neuroscience, however, is not confined to diagnosis and treatment. It extends to innovative approaches that safeguard cognitive well-being. Drawing inspiration from our evolutionary past, it uncovers preventive strategies, therapeutic interventions, and personalized medicine approaches. Lifestyle modifications and environmental adaptations emerge as potent tools in preserving cognitive resilience. Amidst this transformative journey, neurotechnology emerges as the catalyst of innovation. Neuroimaging, brain-machine interfaces, neuromodulation, and artificial intelligence converge to redefine the boundaries of human brain exploration. The fusion of biology and technology holds the key to unlocking new frontiers in brain science, offering hope to individuals grappling with brain disorders. Yet, as it propels ourselve toward these novel discoveries, it confront an array of ethical challenges. Questions of cognitive enhancement, informed consent, privacy, and equitable access to neurotechnological advancements loom large. Our abstract engages with these ethical considerations, advocating for responsible innovation that harmonizes with our ethical and societal values. Evolutionary Global Neuroscience is not merely a scientific endeavor; it is a call to action. It beckons us to unite disciplines, fostering interdisciplinary collaboration that transcends borders and cultures. It urges us to invest in longitudinal and cross-cultural research that expands our understanding of the human brain's exquisite diversity. It demands a commitment to brain health literacy and thoughtful policy recommendations to ensure that the fruits of neurotechnological advancement are accessible to all. This paper journeyed through the realm of Evolutionary Global Neuroscience, a journey that begins in our evolutionary past and charts a course toward an enlightened, innovative, and equitable future for brain science, brain health, and the human mind.

Preface

In an era where the confluence of science and technology is propelling healthcare into uncharted territories, *Biomedical Research Developments for Improved Healthcare* stands as a testament to the transformative power of innovation in medicine. As the editor of this comprehensive reference book, it is my pleasure to present a collection that delves deep into the dynamic world of biomedical research, illuminating the groundbreaking advancements shaping the future of healthcare.

The overarching objective of this book is to serve as a beacon, guiding readers through the intricacies of biomedical research while fostering a profound appreciation for its far-reaching impact on healthcare. We aim to achieve multiple objectives, encompassing education, innovation, ethical considerations, and a forward-looking perspective on the evolving landscape of biomedical research.

Education and Awareness are at the forefront of our intentions. By providing a comprehensive understanding of the latest advancements in biomedical research, we aspire to equip readers with the knowledge needed to navigate the complex and ever-evolving realm of healthcare. The book explores a myriad of topics, ranging from medical imaging, genomics, and therapies to artificial intelligence and biomedical engineering.

This collection is not just about the science; it is about humanizing healthcare. By sharing patient stories and testimonials, we aim to create a connection between the readers and the individuals whose lives have been transformed by biomedical research. Beyond the scientific and technological facets, this approach emphasizes the real-world impact of our collective efforts in the pursuit of improved healthcare.

Chapter 1 explores the transformative impact of Artificial Intelligence (AI) and Machine Learning (ML) in healthcare, focusing on clinical decision support systems. It delves into real-time recommendations and guidelines, demonstrating how AI and ML enhance patient outcomes, especially in diagnostics through accurate interpretation of medical images.

Addressing AI systems designed by humans, Chapter 2 discusses their ability to process and analyze datasets swiftly. It explores the recent renaissance in AI, driven by deep learning, and examines its potential applications in clinical practice. The focus is on AI's impact on diagnostic accuracy and future research directions.

Chapter 3 delves into the manipulation of bacteriochlorophyll at the molecular level for light harvesting applications. Using the Computational Material Repository dataset, the authors predict energy gaps and emphasize the impact of the physical configuration of carbon in methane bonds on the properties of bacteriochlorophyll.

Examining AI's role in maintaining safety equipment across industries, Chapter 4 discusses how AI can identify damage, guide repairs, and extend the lifespan of protective equipment. It highlights the potential for AI to revolutionize how safety equipment is managed, emphasizing the importance of integrating AI for worker safety.

Focusing on the Smart Mask, Chapter 5 explores an IoT platform designed to proactively regulate and prevent the spread of illnesses. By integrating sophisticated sensor technologies, smart algorithms, and wireless connectivity, the Smart Mask analyzes real-time health data to enable accurate diagnoses, personalized treatment programs, and health trajectory predictions.

Chapter 6 provides an overview of the interdisciplinary field of artificial intelligence, covering disciplines such as computer science, data analytics, neuroscience, and more. It discusses the role of AI in governmental agendas, addressing social welfare, legal frameworks, economic consequences, global security implications, and equity issues.

Exploring the field of metabolomics, Chapter 7 highlights the analysis of small molecules to gain insights into metabolic health. It discusses the identification of metabolic biomarkers and their role in developing personalized interventions, treatment plans, and preventive strategies for metabolic disorders.

Chapter 8 focuses on biomarkers for cardiovascular diseases, emphasizing the role of myocardial enzymes, proteins, peptides, and microRNAs. It discusses the importance of personalized therapy in cardiovascular diseases, addressing the need for prompt and accurate diagnoses for conditions like acute coronary syndrome.

Examining Computational Fluid Dynamics (CFD), Chapter 9 explores its versatile applications in understanding physiological systems and guiding medical design. It discusses the transformative role of CFD in drug delivery, physiological flow analysis, surgical planning, and medical device development.

Chapter 10 analyzes the transformative role of AI in predicting and prognosing ovarian cancer. It critically examines existing research, highlighting AI's potential to offer personalized and optimized treatment strategies, thereby improving patient outcomes and overall survival rates.

Focusing on orthopedic implants, Chapter 11 explores the application of nanoparticles in preventing and treating bacterial biofilm infections. It discusses the challenges associated with biofilm-associated infections and highlights how nanoparticles offer effective strategies to tackle these challenges.

Chapter 12 delves into the evolutionary foundations of brain development, exploring the journey that has shaped the human brain. It contemplates the genetic and epigenetic constraints, shedding light on modern brain disorders. The chapter extends its focus to innovative approaches for safeguarding cognitive well-being.

Chapter 13 explores Attention-Deficit/Hyperactivity Disorder (ADHD), examining its multifactorial origins involving genetic, environmental, and neurological factors. It discusses the impact of family history, environmental risk factors, and altered brain structure and activity revealed through neuroimaging studies.

Analyzing the multifactorial nature of diabetes, Chapter 14 explores precision medicine by examining genetic and environmental determinants. It emphasizes the importance of accuracy in therapeutic approaches and discusses the role of epigenetics in tailoring therapies based on an individual's unique epigenetic profile.

Chapter 15 provides a comprehensive overview of precision medicine, exploring its foundations, applications in disease diagnosis and treatment, integration of big data and artificial intelligence, and ethical considerations. The chapter emphasizes precision medicine's transformative role in tailoring medical interventions to the unique characteristics of each individual.

Ethical considerations loom large in the discussion, addressing pressing issues such as privacy concerns with AI in healthcare and the ethical dilemmas surrounding gene editing and other advanced therapies. Our exploration of these topics seeks to stimulate thoughtful reflection on the responsibilities that accompany scientific progress.

Looking toward the future, the book provides insights into emerging trends and challenges, encouraging readers to contemplate the evolving landscape of healthcare and the potential impact of cutting-edge technologies. It is our hope that this forward-looking perspective will inspire curiosity and a sense of anticipation among our readers.

The intended audience for *Biomedical Research Developments for Improved Healthcare* is as diverse as the subject matter it covers. From healthcare professionals and researchers to students, policymakers, and the general public, the book is tailored to cater to a broad spectrum of interests and expertise. Each chapter is crafted with the aim of engaging and informing readers at various levels of familiarity with biomedical research and healthcare.

The recommended topics offer a roadmap for an in-depth exploration of the field, covering historical developments, medical imaging advances, genomics and precision medicine, immunotherapy and gene therapy, artificial intelligence in healthcare, biomedical engineering, ethical considerations, patient stories, future trends, healthcare policy, collaboration in research, global healthcare disparities, and the broader impact of biomedical research on industries beyond healthcare.

As you embark on this journey through the pages of *Biomedical Research Developments for Improved Healthcare*, I invite you to embrace the spirit of curiosity and discovery. May this book serve as a source of knowledge, inspiration, and appreciation for the remarkable strides being made in biomedical research to usher in a brighter and healthier future for all.

Editor,

Pranav Kumar Prabhakar
Lovely Professional University, India

Chapter 1
Artificial Intelligence in Healthcare: A Revolutionary Paradigm Shift

Keshav Singh
https://orcid.org/0009-0008-1862-9268
Raja Balwant Singh Engineering Technical Campus, Agra, India

Vivek Kumar Srivastava
https://orcid.org/0000-0003-0295-3991
Raja Balwant Singh Engineering Technical Campus, Agra, India

Ashish Shukla
Raja Balwant Singh Engineering Technical Campus, Agra, India

ABSTRACT

Healthcare practitioners can access real-time recommendations and guidelines through the incorporation of AI into clinical decision support systems, improving patient care and safety. The chapter offers a thorough examination of AI and ML applications in healthcare, with a focus on its function in enhancing patient outcomes in terms of diagnosis and treatment. Healthcare has undergone a transformation because of AI and ML, which efficiently analyse massive volumes of medical data. They particularly excel in diagnostics. In interpreting medical pictures like X-rays and MRIs, AI-powered systems have shown extraordinary accuracy, resulting in early illness diagnosis and fewer diagnostic errors.

INTRODUCTION

The relationship between medical research and Artificial Intelligence (AI) has transformed the breadth of diagnosis, treatment, and total patient care in the modern healthcare environment. The healthcare sector is experiencing a paradigm shift as a result of AI's revolutionary power, bringing in a new era of precision, efficiency, and personalised therapy. In order to show the enormous potential and hitherto

DOI: 10.4018/979-8-3693-1922-2.ch001

unheard-of prospects that result from the integration of AI and healthcare, this book tries to study the complex interaction between the two fields. (Sharma *et.al.,*2022)

The field of artificial intelligence, which is defined by the creation of intelligent computer systems that are able to carry out tasks that ordinarily need human intelligence, has developed into a key component of innovation in a number of different industries. Artificial intelligence (AI) has surpassed traditional boundaries in the healthcare industry, providing professionals with cutting-edge tools and methods to optimise resource allocation, expedite operations, and deliver customised medical solutions. (Shuaib *et.al,*2020) Artificial intelligence (AI) technologies are transforming the healthcare industry, offering a future that is not only data-driven but also fundamentally patient-centric. These technologies range from machine learning and natural language processing to predictive analytics and pattern identification.

AI's importance in healthcare stems from its capacity to enhance human talents, allowing medical personnel to make well-informed judgements quickly and precisely. AI algorithms can quickly spot patterns, forecast results, and provide individualised treatment plans by assimilating large datasets. (Miller *et.al.,*2018) This improves the standard of care and patient outcomes. Furthermore, healthcare personnel can concentrate more on the intricate and subtle facets of patient care because to AI's ability to automate repetitive chores and administrative procedures, which promotes a compassionate and all-encompassing approach to patient care.

This chapter delves into the complexities of AI applications in healthcare, examining how they affect administration, diagnosis, and treatment. We also want to talk about the ethical, legal, and social ramifications of incorporating AI into the healthcare system, highlighting the necessity of a well-rounded strategy that puts patient safety, data privacy, and legal compliance first. (Morley *et.al.,*2020)

This chapter seeks to offer a thorough grasp of the transformative role that AI will play in reshaping healthcare through a close assessment of case studies and a critical analysis of opportunities and challenges. It is crucial to acknowledge the enormous potential that artificial intelligence (AI) offers as we set out on this revolutionary path, all the while making sure that the human element is kept at the core of healthcare delivery.

Fundamentals of Healthcare and AI

A long history of technological discoveries, social changes, and medical advancements has formed the cornerstones of contemporary healthcare systems. Historical turning points in the history of healthcare include the discovery of antibiotics, the creation of immunisation programmes, and the implementation of extensive public health initiatives. These developments occurred from the earliest days of empirical medicine to the organised healthcare institutions of the 20th century. The healthcare industry has embraced digitization due to the quick advancement of technology, which has resulted in the integration of telemedicine, remote patient monitoring systems, and electronic health records. The way healthcare systems have developed has made it easier for artificial intelligence (AI) to be assimilated as a revolutionary force that is changing how patients are cared for and managed. (Bughin *et.al.,*2017)

Fundamentally, artificial intelligence is the development of intelligent systems that can mimic cognitive processes in humans, including learning, reasoning, solving problems, and making decisions. The core ideas of artificial intelligence (AI) span several fields, such as computer vision, robotics, natural language processing, and machine learning. Natural language processing (NLP) enables human-machine interaction through speech recognition and language interpretation, while machine learning methods allow AI systems to identify patterns in large datasets and provide data-driven predictions.(Chadha *et.al.*,2015) Moreover, AI can now do jobs like image analysis and surgery thanks to advancements in robotics and computer vision technology, which improves the precision of medical interventions and expands the capacities of healthcare workers.

The application of AI in healthcare is a major step towards data-driven diagnostics, precision medicine, and efficient healthcare administration. Healthcare organisations may obtain valuable insights from intricate datasets and improve disease detection, prognosis, and treatment planning by utilising AI-driven analytics. Systems with AI capabilities also make it easier to automate administrative duties, improve resource allocation, and boost operational effectiveness. Additionally, the application of AI in healthcare promotes the creation of individualised treatment plans based on each patient's unique profile and medical background, which enhances patient satisfaction and outcomes overall. AI's integration promises to completely change the way healthcare is administered, provided, and experienced as it continues to permeate different parts of the healthcare ecosystem. This signals the beginning of a time when compassion and innovation will come together to improve global health. (Tian *et.al.,*2019)

Applications of AI in Diagnosis and Treatment

AI has completely changed many facets of healthcare, including diagnosis and therapy. Its use in these fields has greatly improved medical procedures' accuracy, efficiency, and speed. In particular, AI has aided in the development of robotics and AI in surgery, precision medicine and tailored treatment, and AI-powered diagnostic systems. Improvements in surgical outcomes, customised treatments, and more precise diagnostics have all been made possible by these developments. (Andras *et.al.*,2020)

AI-Powered Diagnostic Tools

1. Medical Imaging Analysis: Patterns and anomalies that might not be immediately apparent to the human eye can be found by using AI algorithms to examine complicated medical imaging data, including MRI, CT, and X-ray scans. As a result, diagnoses for diseases like tumours, fractures, and anomalies in organs have become more accurate. (Yoon *et.al.*,2019)
2. Pathology and Histopathology Assistance: Artificial intelligence (AI)-enabled instruments can help pathologists analyse tissue samples and spot cellular abnormalities, which can help in the early detection of diseases like cancer. Large volumes of data can be processed rapidly by these instruments, which enables quicker and more precise diagnosis. (Serag *et.al.,*2019)
3. Virtual Health Assistants and Symptom Checkers: Chatbots and virtual health assistants with AI capabilities can converse with patients, get their medical history, and provide preliminary diagnostic recommendations. They can assist patients in making sense of their symptoms and in determining if they require emergency medical care.

Precision Medicine and Personalized Treatment

1. Genomic Analysis and Drug Development: Large-scale genomic data can be analysed by AI to find genetic variants that might affect how a patient reacts to a particular medication. This makes it possible to create individualised treatment programmes and find more precisely targeted medications, which improves results and lowers side effects. (Malik *et.al.,*2012)

2. Predictive Analytics for Treatment Response: Using information from a patient's genetic composition, medical history, and lifestyle, AI systems can forecast how the patient will react to a specific medication. This aids doctors in personalising treatment regimens, maximising effectiveness and lowering the possibility of negative effects. (Ahuja and Abhimanyu S, 2019)

Robotics and AI in Surgery

1. Assisted and Robotic Surgery: Robotic devices driven by artificial intelligence (AI) can help surgeons with intricate procedures by giving them access to real-time data, improving accuracy, and lowering the possibility of human error. By enabling minimally invasive surgery, these technologies shorten recovery periods and lower the risk of problems after surgery. (Wright & J. D., 2017)
2. Surgical Training Simulators: AI-powered surgical simulators give aspiring surgeons a risk-free virtual environment in which to hone their craft. By offering performance evaluations and feedback, these simulators shorten the learning curve and guarantee safer, more effective surgical procedures. (Dye & Neelam C, 2023)

AI has revolutionised the healthcare industry by bringing cutting-edge diagnostic tools, enabling individualised treatment plans, and improving surgical accuracy and safety. These advancements have lowered medical errors, greatly enhanced patient outcomes, and cleared the path for a more effective and customised healthcare system.

AI-Driven Healthcare Management and Administration

AI-driven solutions have played a key role in transforming healthcare administration and management by enabling workflow optimisation, resource allocation, and more effective operations. Better patient outcomes and more efficient healthcare delivery are the ultimate results of these developments, which have also revolutionised the management of health data, expedited resource allocation, and optimised healthcare workflows.

Health Data Management With AI

1. Data Integration and Analysis: Large volumes of patient data from many sources, including as wearable technology, medical imaging, and electronic health records, can be integrated and analysed using AI. Artificial intelligence (AI) can make more accurate diagnoses and individualised treatment regimens by using machine learning algorithms to recognise patterns, forecast the course of diseases, and give clinicians insightful information. (Chen *et.al.*,2021)
2. Predictive Analytics for Healthcare Trends: Based on population health data, AI-powered analytics can forecast disease outbreaks and trends in healthcare, giving administrators and providers of healthcare the ability to proactively plan interventions and allocate resources to reduce risks and improve patient care. (Agbehadji *et.al.*,2020)

AI-enabled Healthcare Resource Allocation

1. Optimizing Staff Allocation: AI systems are capable of analysing staff performance indicators, historical patterns, and patient data to optimise the distribution of medical professionals. AI can improve patient care and staff efficiency by anticipating patient influx and acuity levels and ensuring that the appropriate staff members are present in the correct departments at the right time. (Robertson et.al.,2016)
2. Efficient Inventory Management: AI is able to forecast future demand for pharmaceuticals, equipment, and medical supplies by analysing consumption trends. Healthcare institutions can ensure the availability of essential resources, minimise waste, and lower operating costs by optimising inventory management, which will ultimately result in a more sustainable healthcare system.

AI in Healthcare Workflow Optimization

1. Streamlining Administrative Processes: Administrative chores like making appointments, handling bills and insurance claims, and processing paperwork may all be automated with the help of AI-powered solutions. Healthcare practitioners can concentrate more time and resources towards patient care by simplifying these processes, which will improve patient satisfaction and overall operational efficiency. (Barriga *et.al.*,2022)
2. Enhancing Care Coordination and Communication: Healthcare teams may collaborate more easily and effectively by using AI-enabled communication platforms and care coordination technologies. This allows for prompt interventions, real-time information exchange, and coordinated care delivery. This enhances patient safety and results, streamlines processes, and lowers medical errors.

The way healthcare organisations gather, analyse, and use data, allocate resources, and streamline workflows has been profoundly changed by AI-driven healthcare management and administration. These innovations could result in better patient outcomes and experiences by raising the general standard and efficiency of healthcare delivery. (Letterie and Gerard 2021)

Ethical and Legal Implications of AI in Healthcare

Numerous advantages have resulted from the use of AI in healthcare, including enhanced diagnosis, individualised treatment plans, and efficient operations. However, in order to guarantee patient safety, privacy, and fair access to care, these developments also bring up important ethical and legal issues that need to be properly handled. Privacy and data security, legal frameworks, and ethical issues in AI-based healthcare decision-making are a few major areas of concern.

Privacy and Data Security Concerns

1. Data Protection and Patient Privacy: AI in healthcare depends on gathering and analysing private patient data, such as genetic data and medical records. To protect patient privacy and stop unauthorised access or breaches, it is essential to deploy strong security measures and ensure strict adherence to data protection standards.(Johnson *et.al.,*2021)

2. Informed Consent and Transparency: Patients' informed consent must be obtained before using patient data for AI-driven analysis, and healthcare practitioners and AI developers are required to maintain transparency in this regard. Trust and patient confidentiality are reliant on open communication on data usage, storage, and any hazards.

Regulatory Frameworks for AI in Healthcare

1. Quality and Safety Standards: It is essential to create thorough regulatory frameworks that specify the safety and quality requirements for AI technology used in healthcare. These frameworks ought to cover the assessment, verification, and oversight of AI algorithms to guarantee that they fulfil the essential safety and performance standards. (Floridi *et.al.,*2021)
2. Accountability and Liability: It is critical to define legal obligations and liabilities for AI-driven healthcare interventions. It is imperative to establish protocols for accountability in the event of AI-related errors or malfunctions in order to safeguard patients' rights and guarantee that healthcare practitioners and AI developers bear responsibility for any unfavourable consequences.

Ethical Considerations in AI-Based Healthcare Decision-Making

1. Fairness and Bias Mitigation: Artificial intelligence (AI) algorithms used in healthcare decision-making need to be created without prejudices that could result in unfair treatment. In order to reduce biases and advance equitable healthcare delivery, it is important to make sure AI systems are trained on a variety of representative datasets. (Hoffman et.al.,2019)
2. Transparency in Decision-making Processes: Transparency in AI-driven decision-making processes is a must for healthcare practitioners, especially when AI systems suggest particular procedures or treatments. To enable patients to make knowledgeable decisions about their care, they should have access to clear explanations of how AI algorithms arrived at specific recommendations.
3. Human Oversight and Responsibility: It's crucial to strike a balance between human monitoring and AI-driven decision-making. It is imperative that healthcare personnel maintain ultimate accountability for patient care, with AI functioning as a supplementary instrument rather than supplanting human discernment and compassion.

Healthcare practitioners, legislators, AI developers, and patients must work together to address the ethical and legal ramifications of AI in healthcare. The healthcare sector can fully utilise AI while guaranteeing the protection of patients' rights and welfare by placing a high priority on patient privacy, putting strong regulatory frameworks in place, and maintaining ethical standards in decision-making processes.

Challenges and Future Prospects

The integration of AI in healthcare presents a promising future with the potential to transform the way medical services are delivered and improve patient outcomes. However, this transition is not without its challenges. Understanding and addressing these technical hurdles, socio-economic impacts, and anticipating future trends are crucial in maximizing the benefits of AI in healthcare.

Technical Challenges in Implementing AI in Healthcare

1. Data Quality and Accessibility: One of the biggest challenges still facing AI algorithms is ensuring that high-quality, standardised, and easily accessible data is available. Keeping data accurate, guaranteeing interoperability, and integrating data from several sources are critical to the effective application of AI-driven healthcare solutions. (Tsamados *et.al.,*2021)
2. Algorithm Bias and Interpretability: Gaining the trust of patients and healthcare providers requires addressing algorithm biases and guaranteeing the interpretability of AI-driven choices. To address any potential biases in the decision-making process and to comprehend the rationale behind AI-driven suggestions, transparent and explicable AI models are required.

Socio-Economic Impacts of AI in Healthcare

1. Workforce Displacement and Reskilling: Concerns regarding possible job displacement for specific healthcare roles may arise from the integration of AI in healthcare. To minimise workforce disruption and encourage the successful application of AI in healthcare settings, it is imperative to guarantee that healthcare personnel possess the requisite training and competencies to collaborate with AI technology. (Cohen *et.al.,*1985)
2. Healthcare Disparities and Access: Although AI can increase healthcare equity and accessibility, there is a chance that it will exacerbate already-existing socioeconomic gaps. To stop growing inequalities in healthcare delivery, it is crucial to guarantee that AI-driven healthcare solutions are available to all populations, particularly those in underprivileged places and with low financial means.(Cannon & Yael, 2021)

Future Trends and Potential Developments in AI-Driven Healthcare

1. AI-enabled Preventive Healthcare: Future advancements might concentrate on using AI to change the focus of healthcare from treatment to prevention. Predictive analytics powered by AI can assist in identifying groups that are at-risk, allowing for preventative interventions and individualised preventive healthcare plans.
2. Integration of AI with Internet of Medical Things (IoMT): It is anticipated that the convergence of AI and IoMT would make real-time patient health data monitoring and analysis possible, enabling prompt interventions and early health issue detection. Enhancing patient involvement in their own healthcare management and revolutionising remote patient monitoring are two possible outcomes of this integration. (Razdan *et.al.,*2022)
3. Advancements in Drug Discovery and Personalized Medicine: AI-driven technologies are probably going to be highly important in helping to build personalised medicine and in accelerating the drug discovery process. Large-scale datasets can be analysed by AI to find possible therapeutic targets, forecast drug efficacy, and customise therapies based on the unique characteristics of each patient, resulting in more focused and efficient therapies.

By addressing technical challenges, mitigating socio-economic impacts, and staying attuned to future trends, the healthcare industry can effectively harness the transformative potential of AI, ultimately leading to improved healthcare accessibility, quality, and patient outcomes.

Case Studies and Success Stories

Real-world Examples of AI Implementation in Healthcare

1. IBM Watson for Oncology: IBM Watson for Oncology is an AI-powered system that assists oncologists in making evidence-based treatment decisions. By analyzing vast amounts of medical literature, patient data, and clinical trial information, Watson provides personalized treatment recommendations, helping oncologists devise more targeted and effective cancer treatment plans. (Hamilton *et.al.,*2019)
2. Google's DeepMind Health: DeepMind Health, a subsidiary of Google's DeepMind, has collaborated with healthcare institutions to develop AI algorithms for various healthcare applications, including the analysis of medical images, early detection of eye diseases, and predicting patient deterioration. These AI-driven solutions have demonstrated improved diagnostic accuracy and enhanced patient outcomes. (Powles *et.al.*,2017)

Impact Assessment of AI in Healthcare Institutions

1. Enhanced Diagnostic Accuracy and Efficiency: The use of AI in healthcare facilities has increased diagnostic efficiency and accuracy, allowing for the quicker and more accurate diagnosis of complicated medical diseases. As a result, there have been fewer diagnostic mistakes and quicker therapy adjustments, which has improved patient outcomes in the end.
2. Streamlined Operational Processes and Resource Optimization: Streamlining operational operations like patient scheduling, resource allocation, and inventory management has been made easier by AI-driven solutions. Healthcare organisations have seen increased total resource utilisation, lower operating costs, and better workflow efficiency as a result of these process optimisations.

Lessons Learned and Best Practices

1. Collaboration between Healthcare Professionals and AI Developers: Successful AI implementation in healthcare institutions often involves close collaboration between healthcare professionals and AI developers. Engaging clinicians in the development and validation of AI algorithms ensures that these technologies align with clinical workflows and meet the specific needs of healthcare settings.
2. Continuous Evaluation and Validation of AI Algorithms: In order to guarantee AI algorithms' dependability, security, and efficacy in actual healthcare settings, ongoing assessment and validation are crucial. For AI-driven solutions to be optimised and remain relevant over time, regular

evaluations and updates based on real-time input from patients and healthcare professionals are essential.

3. Ethical Considerations and Patient-Centric Approaches: Prioritizing ethical considerations and adopting patient-centric approaches are integral to successful AI implementation in healthcare. Healthcare institutions should prioritize patient privacy, consent, and transparency throughout the AI implementation process, fostering trust and promoting patient engagement in their healthcare journey.

Through a thorough analysis of these case studies and success stories, thorough impact assessments, and adoption of insightful lessons and best practises, healthcare organisations can effectively incorporate AI-driven solutions, enhance patient care, and maximise operational efficiency—all of which will eventually contribute to the improvement of healthcare outcomes and delivery.

CONCLUSION

The revolutionary impact of AI in healthcare is evident across various domains, including diagnostics, treatment, management, and administration. AI-powered diagnostic tools have improved accuracy and speed in identifying complex medical conditions, while precision medicine has enabled personalized treatment plans tailored to individual patients. Robotics and AI in surgery have enhanced surgical precision and safety, leading to better patient outcomes. Moreover, AI has streamlined healthcare management, optimized resource allocation, and improved workflow efficiency, fundamentally transforming the healthcare landscape.

Key takeaways from the integration of AI in healthcare highlight the importance of data security, transparency, and ethical considerations in decision-making. Future research endeavours should focus on addressing technical challenges such as data quality, algorithm bias, and interpretability, while also considering the socio-economic impacts of AI implementation, including workforce reskilling and healthcare accessibility. Continued collaboration between healthcare professionals, AI developers, and policymakers is crucial for establishing robust regulatory frameworks and implementing patient-centric AI solutions in healthcare settings.

The future of AI-driven healthcare holds tremendous promise, with ongoing advancements in predictive analytics, preventive healthcare, and personalized medicine. Integration with the Internet of Medical Things (IoMT) is expected to revolutionize remote patient monitoring and improve patient engagement. However, it is imperative to approach these developments with a careful balance between technological innovation and ethical responsibility. The future of AI-driven healthcare will depend on continued interdisciplinary collaboration, adherence to ethical guidelines, and a patient-centered approach, ultimately leading to a more efficient, accessible, and equitable healthcare system for all.

REFERENCES

Agbehadji, I. E., Awuzie, B. O., Ngowi, A. B., & Millham, R. C. (2020). Review of big data analytics, artificial intelligence and nature-inspired computing models towards accurate detection of COVID-19 pandemic cases and contact tracing. *International Journal of Environmental Research and Public Health*, *17*(15), 5330. doi:10.3390/ijerph17155330 PMID:32722154

Ahuja, A. S. (2019). The impact of artificial intelligence in medicine on the future role of the physician. *PeerJ*, *7*, e7702. doi:10.7717/peerj.7702 PMID:31592346

Andras, I., Mazzone, E., van Leeuwen, F. W. B., De Naeyer, G., van Oosterom, M. N., Beato, S., Buckle, T., O'Sullivan, S., van Leeuwen, P. J., Beulens, A., Crisan, N., D'Hondt, F., Schatteman, P., van Der Poel, H., Dell'Oglio, P., & Mottrie, A. (2020). Artificial intelligence and robotics: A combination that is changing the operating room. *World Journal of Urology*, *38*(10), 2359–2366. doi:10.1007/s00345-019-03037-6 PMID:31776737

Barriga, A., Rutle, A., & Heldal, R. (2022). AI-powered model repair: An experience report—lessons learned, challenges, and opportunities. *Software & Systems Modeling*, *21*(3), 1135–1157. doi:10.1007/s10270-022-00983-5

Cannon, Y. (2021). Closing the health justice gap: Access to justice in furtherance of health equity. *Colum. Hum. Rts. L. Rev.*, *53*, 517.

Chadha, N. (2015). R. C. Gangwar, and Rajeev Bedi. "Current Challenges and Application of Speech Recognition Process using Natural Language Processing: A Survey.". *Int. J. Comput. Appl*, *131*(11), 28–31.

Chen, Z.-H., Lin, L., Wu, C.-F., Li, C.-F., Xu, R.-H., & Sun, Y. (2021). Artificial intelligence for assisting cancer diagnosis and treatment in the era of precision medicine. *Cancer Communications*, *41*(11), 1100–1115. doi:10.1002/cac2.12215 PMID:34613667

Dey, N. C. (2023). Innovative Digital Teaching and Learning Practices in Society in View of Nursing Education/Profession in India: A Comprehensive Review. *Profession in India: A Comprehensive Review*.

Ghaheri, A., Shoar, S., Naderan, M., & Hoseini, S. S. (2015). The applications of genetic algorithms in medicine. *Oman Medical Journal*, *30*(6), 406–416. doi:10.5001/omj.2015.82 PMID:26676060

Hamilton, J. G., Garzon, M. G., Westerman, J. S., Shuk, E., Hay, J. L., Walters, C., Elkin, E., & … . (2019). "A tool, not a crutch": Patient perspectives about IBM Watson for oncology trained by Memorial Sloan Kettering. *Journal of Oncology Practice / American Society of Clinical Oncology*, *15*(4), e277–e288. doi:10.1200/JOP.18.00417 PMID:30689492

Hoffman, S., & Podgurski, A. (2019). Artificial intelligence and discrimination in health care. *Yale J. Health Pol'y L. & Ethics*, *19*, 1.

Johnson, K. B., Wei, W.-Q., Weeraratne, D., Frisse, M. E., Misulis, K., Rhee, K., Zhao, J., & Snowdon, J. L. (2021). Precision medicine, AI, and the future of personalized health care. *Clinical and Translational Science*, *14*(1), 86–93. doi:10.1111/cts.12884 PMID:32961010

Letterie, G. (2021). Three ways of knowing: The integration of clinical expertise, evidence-based medicine, and artificial intelligence in assisted reproductive technologies. *Journal of Assisted Reproduction and Genetics*, *38*(7), 1617–1625. doi:10.1007/s10815-021-02159-4 PMID:33870475

Malik, A. I., Zai, C. C., Abu, Z., Nowrouzi, B., & Beitchman, J. H. (2012). The role of oxytocin and oxytocin receptor gene variants in childhood-onset aggression. *Genes Brain & Behavior*, *11*(5), 545–551. doi:10.1111/j.1601-183X.2012.00776.x PMID:22372486

Miller, D. D., & Brown, E. W. (2018). Artificial intelligence in medical practice: The question to the answer? *The American Journal of Medicine*, *131*(2), 129–133. doi:10.1016/j.amjmed.2017.10.035 PMID:29126825

Morley, J., & Caio, C. V. (2020). The ethics of AI in health care: A mapping review. *Social Science & Medicine*, *260*, 113172. doi:10.1016/j.socscimed.2020.113172 PMID:32702587

Powles, J., & Hodson, H. (2017). Google DeepMind and healthcare in an age of algorithms. *Health and Technology*, *7*(4), 351–367. doi:10.1007/s12553-017-0179-1 PMID:29308344

Razdan, S., & Sharma, S. (2022). Internet of medical things (IoMT): Overview, emerging technologies, and case studies. *IETE Technical Review*, *39*(4), 775–788. doi:10.1080/02564602.2021.1927863

Robertson, S., Back, N., & Bartlett, J. D. (2016). Explaining match outcome in elite Australian Rules football using team performance indicators. *Journal of Sports Sciences*, *34*(7), 637–644. doi:10.1080/02640414.2015.1066026 PMID:26176890

Serag, A., Ion-Margineanu, A., Qureshi, H., McMillan, R., Saint Martin, M.-J., Diamond, J., O'Reilly, P., & Hamilton, P. (2019). Translational AI and deep learning in diagnostic pathology. *Frontiers in Medicine*, *6*, 185. doi:10.3389/fmed.2019.00185 PMID:31632973

Sharma, A., Singh, A., Gupta, V., & Arya, S. (2022). Advancements and future prospects of wearable sensing technology for healthcare applications. *Sensors & Diagnostics*, *1*(3), 387–404. doi:10.1039/D2SD00005A

Shuaib, A., Arian, H., & Shuaib, A. (2020). The increasing role of artificial intelligence in health care: Will robots replace doctors in the future? *International Journal of General Medicine*, *13*, 891–896. doi:10.2147/IJGM.S268093 PMID:33116781

Tian, S., Yang, W., Le Grange, J. M., Wang, P., Huang, W., & Ye, Z. (2019). Smart healthcare: Making medical care more intelligent. *Global Health Journal (Amsterdam, Netherlands)*, *3*(3), 62–65. doi:10.1016/j.glohj.2019.07.001

Tsamados, A., Aggarwal, N., Cowls, J., Morley, J., Roberts, H., Taddeo, M., & Floridi, L. (2021). *The ethics of algorithms: key problems and solutions*. Ethics, Governance, and Policies in Artificial Intelligence.

Wright, J. D. (2017). Robotic-assisted surgery: Balancing evidence and implementation. *Journal of the American Medical Association*, *318*(16), 1545–1547. doi:10.1001/jama.2017.13696 PMID:29067404

Yoon, H. J., Jeong, Y. J., Kang, H., Jeong, J. E., & Kang, D.-Y. (2019). Medical image analysis using artificial intelligence. *Progress in Medical Physics*, *30*(2), 49–58. doi:10.14316/pmp.2019.30.2.49

Chapter 2
Artificial Intelligence and Machine Learning in Healthcare

Nilanjana Sarkar
https://orcid.org/0000-0002-4948-8700
Purvanchal Institute of Dental Sciences, Gorakhpur, India

Sumit Goel
https://orcid.org/0000-0002-0543-2000
Purvanchal Institute of Dental Sciences, Gorakhpur, India

ABSTRACT

Artificial intelligence (AI) systems are systems designed by humans that, given a complex goal, act in the physical or digital dimension by perceiving their environment through data acquisition, interpreting the collected structured or unstructured data, reasoning on the knowledge, or processing the information, derived from this data and deciding the best action(s) to take to achieve the given goal. It is precisely AI's ability to carry out speedy processing and analysis of datasets that is one of its key strengths. The recent renaissance in AI largely has been driven by the successful application of deep learning — which involves training an artificial neural network with many layers (that is, a 'deep' neural network) on huge datasets. The rise and dissemination of AI in clinical medicine will refine our diagnostic accuracy and rule-out capabilities. In this Book Chapter, we focus on the AI applications that could augment or change clinical practice, identify the impact arising from the development of AI diagnostic systems and suggest future research directions.

INTRODUCTION

Artificial Intelligence-based technologies are advancing healthcare in the fields of telemedicine, assessment, and biomedical research. AI-driven optimization decreases errors and improves workflow, given the technical and organizational challenges are addressed, as for example, robust infrastructures to support responsible innovation and effective post-market surveillance (Baumgartner et al., 2023). The operational potentials of AI are different from human intelligence and intuitive capacities and expand

DOI: 10.4018/979-8-3693-1922-2.ch002

the classical evidence and experience-based medical knowledge, among other things, by insights based on statistical correlations. Complex relationships within large heterogeneous data sets such as published research outcomes or biomedical datasets can be identified with AI. This can open up new opportunities by discovering novel treatments with reduced research and development time and costs. Especially, precision medicine, made possible by AI, promise a tailored medical approach to healthcare according to each person's specific genotype and phenotype. This would that help understand drug interactions as well as group-specific risks and individual differences (Baumgartner, 2021).

No innovation is free of its shortcomings. It is believed by some that AI may reproduce and exacerbate existing inequalities. Various forms of bias, axes of discrimination and foundational flaws within practical medicine should be addressed and equity through AI should be promoted. The AI commune must build an integrated best practice method for execution and safeguarding by incorporating active best practices of principled inclusivity, software growth, implementation science, and individual–workstation interaction, as pointed out by Kumar et al (2023).

One of the maximum burdens that a practising physician faces today is the association and performance of organizational tasks. Fukuda et al (2020) explains that by automating these tasks, healthcare institutions could help resolve the trouble and allow physicians to do their best, i.e., spend more time with patients.

In this chapter, we are going to explore how Artificial Intelligence and Machine Learning technologies along with biomedical research are impacting healthcare and what promise does this hold for the future.

INTEGRATING AI IN BIOMEDICAL RESEARCH TO ENHANCE HEALTHCARE

The primary aim of biomedical research is to supplement healthcare data with other resources by conducting required research, applying the research findings in medicine, regenerative medicine and associated areas and developing better products and services, by combining engineering, virology, and healthcare data (Javaid et al., 2020)

Be it conceptualisation and evaluation of new technologies, development of novel patient care techniques or the research of biological processes, all these require bioengineers. According to Javaid et al (2023), the primary supportive technologies of biomedical engineering are Additive Manufacturing, Artificial Intelligence, Internet of Things, Virtual Reality, Holography and other medical imaging technologies.

Several innovations and progressive journeys are observed in the broad biomedical engineering domain in the last few years as depicted in the Figure 1.

Figure 1. The journey of biomedical engineering-related development through the ages

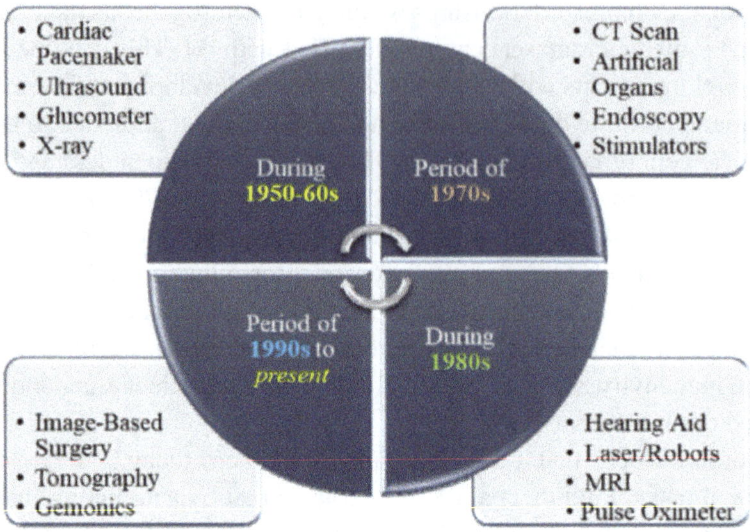

The application of interdisciplinary research for the integration of mechanical components, electronics and biological organism parts including areas of neurology and robotics fall under Biomechatronics. Interdisciplinary collaboration led to designing of essential tools and gadgets, including Magnetic Resonance Imaging (MRI) machines, dialysis machines, diagnostic instruments, and ultrasound as depicted in Figure 2.

Figure 2. Smart integration of big data, artificial intelligence, digital methodologies and production steps

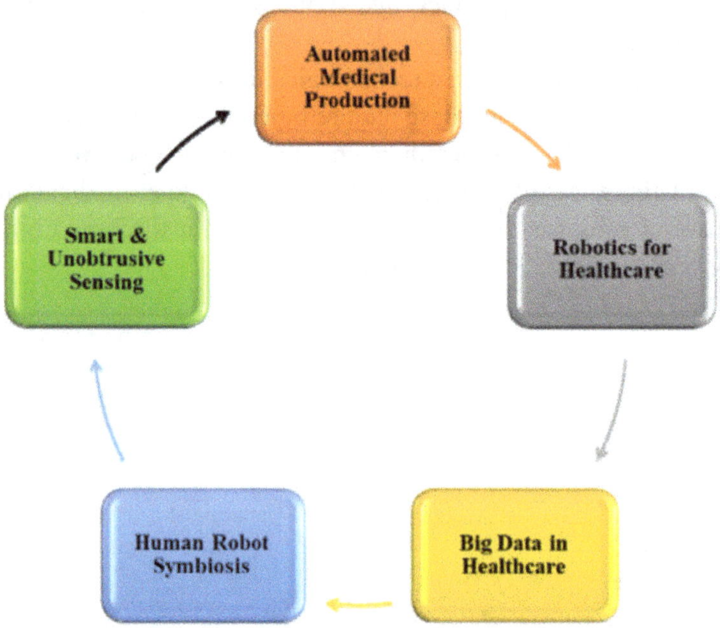

Table 1. Biomedical Research Applications In Healthcare

Applications	Description
Develop solutions for healthcare problems	This helps change how patients are treated and lower their healthcare costs with better expertise in physiology, biology, healthcare and health information technology, mechanics and engineering
Design and maintenance of the equipment	Biomedical electronics is the bioengineering industry for creating, designing and maintaining the equipment used for the health sector, such as hospitals and clinics. Due to the development and launch of today's equipment, such as intensive unit monitoring systems, CT imaging systems, dialysis machines and surgical lasers, bio-medical electronics as a discipline significantly improved medical business.
Improve treatment of patients	Biomedical engineers have been engaged in developing medical treatments, systems, equipment and products, ranging from improving 3D medical imaging machines to improved biomaterials for the replacement of knees to accelerating rehabilitation of physical impairments and increase in life expectancy.
Technological developments	Biomedical engineering is an area of study that uses engineering concepts. It integrates fundamental science with engineering to create, enhance and generate medical devices or processes.
Treatment facilities	Biomaterials use both natural and laboratory materials which are employed in medical equipment or which come in touch with biological systems. For example, artificial hearts, hip articulations and sutures disintegrate by themselves. Rehabilitation engineering develops through engineering, biology and computer science instruments to support individuals with physical and cognitive disabilities.
Health care management	Biomedical engineering impacts healthcare management and facilities worldwide, so professionals need much training. Hospitals and treatment centres require skilled biomedical engineers to employ technology efficiently. Therefore, the demand for biomedical engineers is essential and evident in a global health crisis
Analyse patients	Biomechanics applied to biological or medicinal challenges using classical mechanics. The study included motion, deformation of materials, movement in and through biological and synthetic media and membranes in the body and the devices and transport of chemical elements. Tissue engineering and genetic engineering attempt to address biomedical issues microscopically. These fields use cellular and subcellular anatomy, biochemistry and mechanics to understand illness's mechanisms and interfere in exceptional locations.
Diagnosis	The method of medicine practised is changed by revolutionary developments in medical imaging and diagnostics. How doctors treat sickness and trauma has changed entirely new medical systems, which arise in research laboratories by biomedical engineers across the world, improving the quality and life of humans.

Biomedical Research Applications in Healthcare

The recent achievements in next-generation artificial intelligence and deep learning holds a great promise as a biomedical research tool with many applications.

An immense amount of data is generated in the healthcare industry through clinical records, health records of patients, results of medical examinations, and data collected by using diagnostic and health management devices (Cremin et al., 2022). A substantial amount of data is also generated by the field of biomedical research. This collective data is called the Big data.

Healthcare and medical research improve disease outcomes by finding hidden associations or patterns within data derived from these wealth of available data resources. Such data can be used in prediction studies to determine highly probable outcomes of disease as outlined by Schwalbe et al (2020). These investigations are leading to advances in delivering personalized medicine tailored to individual patients' needs and the development of automated analysis to aid in the diagnosis and treatment of diseases (Cirillo and Valencia, 2019). Big data has also contributed to better understanding of epidemiology within populations for a variety of pathogens. Immunological surveillance centers routinely process big data to identify pathogens at high risk of becoming endemic within populations.

Also, Hassan et al., (2022) added that the big data from genomic libraries, such as whole genome sequencing and whole-exome sequencing libraries, also play key roles in accelerating data-driven biomedical discovery.

Recent advances in genomics, including single-cell genome and transcriptome sequencing of circulating tumor DNA in liquid biopsies, and metagenomics, are already markedly affecting medicine and are steadily being integrated into standard medical practice for early-stage diagnosis and various stages of disease treatment as described by Rong et al (2022) in a review.

Omics technology generates big data that require sophisticated bioinformatic analysis and has enabled the detailed study of many cellular components in their roles and behaviors across many cells and tissues. The availability of high-quality omics data has substantially contributed to understanding across the many tiers of biological conditions in health and disease as explained by Cremin et al (2022). Integration of big data across different branches through multi-omics is one of the breakthroughs in current biomedical research; this integration has paved the way for pathogen detection, disease diagnosis and timely therapeutic intervention (Fig. 3).

Figure 3. Data sources and applications of multi-omics. Multi-omics data generated from various experimental pipelines (genomics, epigenomics, transcriptomics, proteomics and metabolomics) in preclinical and clinical investigations provide useful biological information at multiple levels (genotype to phenotype), thus aiding in clarifying the mechanisms and factors that contribute to disease pathology.

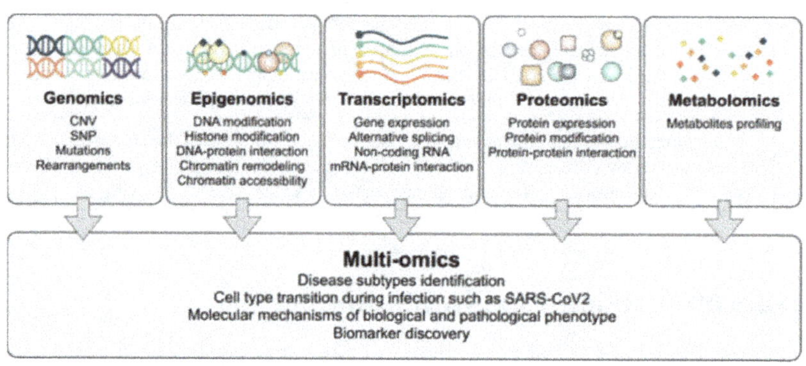

WHAT ARE ARTIFICIAL INTELLIGENCE SYSTEMS?

As stated in the Communication from the Commission to the European Parliament, the European Council, the European Economic and Social Committee and the Committee of the Regions on Artificial Intelligence for Europe (2018), Artificial intelligence (AI) systems are software (and possibly also hardware) systems designed by humans that, given a complex goal, act in the physical or digital dimension by perceiving their environment through data acquisition, interpreting the collected structured or unstructured data, reasoning on the knowledge, or processing the information, derived from this data and deciding the best action(s) to take to achieve the given goal. AI systems can either use symbolic rules or learn a numeric model, and they can also adapt their behaviour by analysing how the environment is affected by their previous actions.

Parry and Aneja (2020) pointed out that Artificial intelligence for health includes machine learning (ML), natural language processing (NLP), speech recognition (text-to-speech and speech-to-text), image recognition and machine vision, expert systems (a computer system that emulates the decision-making ability of a human expert), robotics, and systems for planning, scheduling and optimization. ML is a core component of AI that allows systems to automatically learn and improve without being explicitly programmed. Deep learning (DL), a type of ML, is inspired by the human brain and uses multi-layered neural networks to find complex patterns and relationships in large datasets that traditional ML may miss. NLP is a subfield of AI that helps computers understand, interpret and manipulate human language. It is precisely AI's ability to carry out speedy processing and analysis of datasets that is one of its key strengths.

AI is a set of versatile capabilities that can be applied in various contexts. It is an umbrella term encompassing a wide range of evolving tools and techniques, enhanced through iterative cycles of social interaction, technical development, utilization and reinvention by users (Williams, 2019).

The period of 1940s–1970s saw the rise of AI, i.e., the development of the term AI, through the historical work of Turing on computation. The period from 1970s witnessed knowledge-based or expert systems, which tried to mimic the human reasoning process by formulation of robust decision rules. In the 1990s with the re-introduced neural network systems along with the availability of big data, affordable computer data and cloud storage, there was a paradigm shift to machine learning systems, i.e., the induction of rules ("models") from large (training) data sets instead of relying on explicit rules programmed by humans, as described by Delipetrev et al (2020).

Application of AI in healthcare technology has revolutionised diagnosis, application and care delivery in many aspects. One of them is the collaboration between Google and Aravind Eye Care System in India to develop an algorithm for retinal screening system which gives physician-level specificity and sensitivity for detecting diabetic retinopathy (Simonite, 2015). It also identifies previously unrecognized associations between image patterns in the fundus photograph and cardiovascular risk factors (Poplin et al., 2018). Such an application is beneficial in areas where patients suffer vision loss before the disease is detected due to shortage of specialist doctors or due to low awareness. Based on a historical trial by Abràmoff et al (2018), US Food and Drug Administration (FDA) approved this application for detecting moderate-to-severe diabetic retinopathy and is in clinical practice in eye hospitals in India.

A recently FDA-approved AI-assisted colonoscopy from GI Genius™ system is in clinical use and has generated mixed reviews from endoscopists, according to a study conducted by Tham et al (2023).

AI has made remarkable progress in radiology tasks such as breast cancer screening; an AI system developed by McKinney et al (2020) that outperformed all the human readers. Brot and Mango (2024) discussed a few FDA-approved AI-powered decision support applications for breast ultrasound in clinical usage like BU-CAD, Koios DS and S-Detect. Perl et al (2021) showed in a comparative study that deep 3D convolutional neural network (CNN) CAD software has showed higher specificity while detecting pulmonary solid nodules in comparison with conventional CAD softwares. Apart from radiology, there is also considerable advancement in cardiac function assessment; Ouyang et al (2020) presented the first video-based deep learning algorithm, EchoNet-Dynamic, that surpasses human expert performance in the critical tasks of segmenting the left ventricle, estimating ejection fraction, and assessing cardiomyopathy. Trained on echocardiogram videos, this model accurately segments the left ventricle, predicts ejection fraction and reliably classifies heart failure with reduced ejection fraction.

A pioneering contribution from Google Health in integrating AI in healthcare is 'Care Studio' that provides a time-saving platform for clinicians to spend more time with the patients for their care rather than searching information for diseases such as Diabetes, Hypertension, Hyperlipidemia, Hypertrophic Cardiomyopathy etc. Google Health's 'DermAssist' app can help find skin conditions but the Melanoma Project, the largest available pigmented lesion dataset, relies heavily on images collected from patients with fair skin from the United Kingdom, United States, and Australia, AI may offer less accurate diagnoses for conditions encountered in patients with darker skin due to this skewed ethnic and geographic variability of the datasets. To ensure fairness, dermatologists should critically assess for bias within the AI tools used by patients, as pointed out by Dave et al (2023). AlDubayan et al (2020) conducted a cross-sectional study comparing Germline testing using deep learning with standard genetic testing. Germline genetic testing identifies a class of inherited genetic changes called pathogenic variants, which are associated with an increased risk of developing cancer and other diseases. This can identify patients and families with inherited cancer susceptibility in whom established gene-specific screening recommendations can be implemented. It was found that Germline genetic testing using deep learning, compared with the current standard genetic testing method, was associated with higher sensitivity and specificity for detection of pathogenic variants.

APPLICATIONS OF AI IN HEALTHCARE

The power of machine learning and the neural networks underlying it are frequently demonstrated by AI's growing power to win complex games, as shown by Ye et al (2020). Deep reinforcement learning (DRL) has been widely used for building agents to learn complex control in competitive environments like the complex games. This suggests that neural networks are evolving in ways that not only mimic but also exceed the human brain (Kirkpatrick et al., 2017).

Sarker (2021a) stated that according to the types of tasks that they intend to solve, basic machine-learning algorithms fall roughly into four categories: Supervised learning, Unsupervised learning, Semi-supervised learning, and Reinforcement learning as shown in the Fig. 4.

Figure 4. Machine learning types

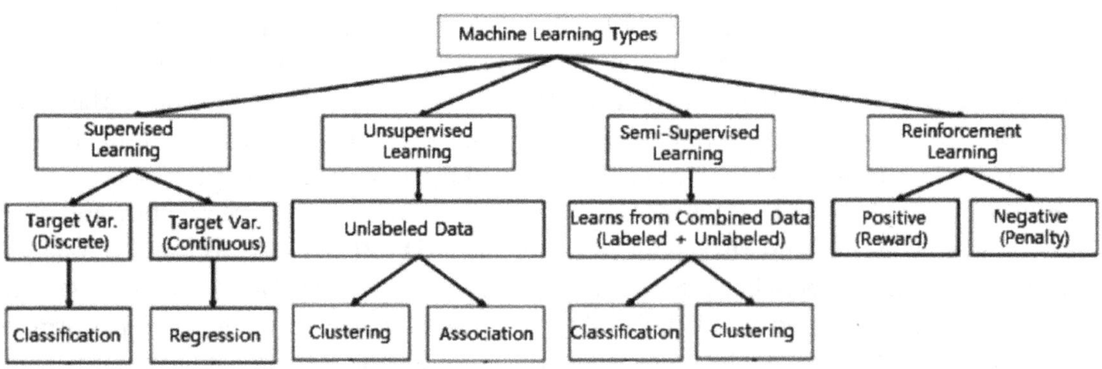

Machine-learning methods enable the development of AI applications that facilitate the discovery of previously unrecognized patterns in the data without the need to specify decision rules for each specific task or to account for complex interactions among input features. Machine learning has thus become the preferred framework for building AI utilities (Krizhevsky et al., 2012).

Adoption of EHR/EMR Systems

The recent renaissance in AI has to a large extent been driven by the successful application of deep learning — which involves training an artificial neural network with many layers (that is, a 'deep' neural network) on huge datasets — to large sources of labelled data. The basic architecture of deep neural networks consists of an input layer and an output layer, and several hidden layers in between (Sarker 2021b).

Many modern neural networks have more than 100 layers. Neural networks with many layers can model complex relations between the input and output but may require more data, computation time or advanced architecture designs to achieve optimal performance. Large sources of medical data that can be fed into deep-learning data-hungry algorithms have become widely available, owing to the establishment of many large-scale studies; in particular, the Cancer Genome Atlas (Tomczak et al.,2015) and the UK Biobank (Sudlow et al., 2015).

It is precisely AI's ability to carry out speedy processing and analysis of datasets that is one of its key strengths. With the increasing adoption of health informatics and electronic medical records (EMR), AI will become increasingly useful, as better AI system integration will improve care delivery, permit research to be conducted and informed health planning. In Kenya, a cloud-based open EMR platform has contributed to improving child and maternal health and HIV/AIDS treatments in rural areas by helping to achieve a more complete data collection (Haskew et al.,2015).

Decision Support Systems

Decision support systems that collect medically relevant information and provide suggestions to clinicians have been developed. There are several ways to integrate decision support systems into the clinical workflow; for example, for actively gathering information from patients and EHRs, present suggestions to clinicians and store the system outputs in EHRs. In many proposed fully automated clinical systems, the autonomous tools collect information from the patients, make decisions and output the results into EHRs. Data from EHR systems provide detailed information about the patients, including clinical notes and laboratory values, enabling the application of natural-language-processing methods to extract codified vocabularies (Jha and Topol, 2016).

Recently, applications of medical-image diagnostic systems have expanded the frontiers of AI into areas that were previously the domain of human experts. This frontier continues to expand into other areas of medicine, such as clinical practice (Golub et al., 1999), translational medical research (Wang et al., 2005, Yu et al., 2016, Yu et al., 2018, Check Hayden, 2014) and basic biomedical research (Shademan et al., 2016). The recent confluence of large-scale annotated clinical data acquisition, advancement in machine-learning methods, opensource machine-learning packages, and affordable and rapidly growing computational power and cloud storage has fuelled the recent exponential growth in AI. AI systems have specialist-level performance in many diagnostic tasks, can better predict patient prognosis than clinicians, and can assist in surgical interventions.

Table 2. Current and potential AI applications in medicine

Basic biomedical research	Translational research	Clinical practice
Automated experiments	Biomarker discovery	Disease diagnosis
Automated data collection	Drug–target Prioritization	Interpretation of patient genomes
Gene function annotation	Drug discovery	Treatment selection
Prediction of transcription factor binding sites	Drug repurposing	Automated surgery
Simulation of molecular dynamics	Prediction of chemical toxicity	Patient monitoring
Literature mining	Genetic variant annotation	Patient risk stratification for primary prevention

NLP Technologies

The use of NLP (Natural Language Processing) technologies allows machines to identify keywords and phrases and enables them to determine the meaning of text. NLP algorithms are used, for example, to simplify clinical documentation and enable voice-to-text dictation, as explained by Kaufman et al (2016). These technologies allow them to interact with patients and produce accurate records of consultations without having to type at the same time. This reduces the electronic health record (EHR) overload. Once converted into structured data, health systems can classify patients and summarize their condition on arrival.

Breakthroughs have been made in natural language processing for biomedical applications. In the area of biomedical question answering (BioQA), the aim is to find fast and accurate answers to user-formulated questions from a reservoir of documents and datasets (Abacha & Zweigenbaum, 2015). Therefore, natural language-processing techniques can be expected to search for informative answers. To begin with, the biomedical questions must be classified into different categories in order to extract appropriate information from the answer. ML can categorize biomedical questions into four basic types with an accuracy of nearly 90% (Sarrouti, 2017). Next, an intelligent biomedical document retrieval system can efficiently retrieve sections of the documents that are most likely to contain the answers to the biomedical questions, one of which systems was designed by Noh and Kavuluru (2018).

A natural language processing chatbot is a software program that can understand and respond to human speech. Bots powered by NLP allow people to communicate with computers in a way that feels natural and human-like — mimicking person-to-person conversations (Pollock, 2023).

MH (Mental Health) chatbots are AI-powered chatbots that provide MH support, guidance, and resources through a conversational interface (Dekker et al, 2020). These chatbots replicate human interactions, respond to user inputs, and deliver tailored MH care (Inkster et al, 2018). MH chatbots can target a range of MH concerns, including anxiety, depression, and stress (Abd-Alrazaq et al, 2019). These can provide coping strategies, mindfulness exercises, and information about MH conditions and treatments and, in some cases, connect users to MH professionals. Haque and Rubya (2023) conducted a study regarding popularly used MH apps using built-in chatbots and concluded that chatbots in MH apps have considerable potential in terms of being conversational companions, virtual friends, and immediate helpers. The chatbot's ability to be present 24/7 and to create a judgment-free zone enabled users to talk comfortably about their issues and concerns.

Precision research platform Deep 6 AI employs artificial intelligence (AI) and natural language processing (NLP) to sift through vast amounts of electronic medical record (EMR) data. Deep 6 along with Graticule has developed a new algorithm designed to expedite patient screening and recruitment for clinical trials. It identifies eligible patients by extracting critical information from both structured and unstructured data sources, such as clinical notes and radiology reports. Typically, it takes up to two years for life sciences companies to get access to EMR data to develop novel patient matching algorithms. Also, it overburdens an already overworked hospital to perform clinical data pulling for clinical trials (Clinical Trials Arena, 2024). The screening algorithm qualifies and prioritises eligible subjects using the insights to give a precise picture of the individuals who would benefit from the intervention.

Deriving Health Status Through Wearable Devices

Modern wearable devices record biomedical signals of our body, including heart rate, voice, tremor, and limb movement and give real-time feedback of physiological system (Loncar-Turukalo et al., 2019) These biological signals could be useful for detecting diseases and inferring health conditions, especially useful in tracking chronic disease progression and in disabled persons (Majumder et al., 2017). The wearable sensors are been widely used mainly due to their inconspicuousness, small footprint and lower cost. Remote patient monitoring became an important tool throughout the COVID19 pandemic (Jouffroy et al.,2020). Several wearable devices used in the healthcare environment to quantify disease progression are wearable cutaneous haptic interface (WCHI), Smart electro-clothing systems, 5DT data glove, Flex Sensor data glove, VICON System, Polysomnography sensors, Pulse Oximetry, etc (Vijayan et al, 2021).

Caregiving

The Stanford Thailand Research Consortium was established in 2018 to find out an integrated solution for the remote monitoring, assessment and support of seniors living independently at home. It involves the use of multiple sensors for the detection and recording of daily activities, lifestyle patterns, emotions, and vital signs, as well as the development of intelligent and rapid response towards any deviation from normal (Senior Care, Stanford Partnership in AI-assisted care, 2018).

A fall-detection system based on radar Doppler time–frequency signatures and a sparse Bayesian classifier can reduce fall risks and complications for seniors. An "ambient-assisted living" (AAL) has been developed to allow AI processing information to be gathered from different communication channels or technologies (Wu et al., 2015).

In this scenario, sensors collect data about the ambient environment and human behavior, which is then analyzed by cloud computing or edge intelligence. A decision is then made regarding what actions are necessary, and this decision is used to activate alarms or preventative measures (García-Vázquez et al., 2010).

Research and Education

Further research and education is possible with AI to accelerate drug development, primarily by searching for patterns in clinical data (Ehrenstein et al., 2019). Routinely collected health data such as EHR data that was not initially collected for research purposes, together with data collected from wearables, is being used to train research AI (Askin et al., 2023).

Clinical Data Analytics

Whether used for research or to train clinical AI, data analytics is increasingly important in healthcare (Ahmed et al., 2020). As the analytics engines become more powerful their roles, they will evolve from descriptive to diagnostic to predictive. At some point the analytics likely will turn prescriptive, with the AI itself deciding to execute a task. As the amount of data that is fed into AI increases, so generally does its predictive abilities (Rajkomar et al., 2018). Many of the most important breakthroughs therefore depend on feeding AI. For example, one proof of concept study used almost one quarter of a million electronic patient records, including clinical notes from two different hospitals and a total of almost 47 billion data points, and achieved high accuracy in predicting in-hospital mortality, unplanned readmissions, prolonged stay, and final discharge diagnoses (Wirtz et al., 2023).

Imaging, Pathology, and Radiology

By harnessing computation power, large datasets and 'convolutional' neural networks (CNNs), deep learning has transformed not just medical-image analysis, but the whole field of computer vision. CNNs use a special type of layer (that is, a convolutional layer) to summarize and transform clusters of pixels in images to extract high-level features. CNNs have proven crucial to the success of deep learning in image analysis and are also responsible for the subsequent revolution in medical imaging (Yamashita et al., 2018). AI is particularly adept at pattern recognition, making it a natural fit for reading scans. For example, it has been used to detect skin cancer (Ahmadi et al., 2022), colon cancer (Kavitha et al., 2022), evidence of stroke (Ozaltin et al., 2022), and pneumonia (Lamia et al., 2022). With the help of modern machine-learning methods, many radiology applications of AI, such as the detection of lung nodules using computed tomography images (Kang et al.,2017), the diagnosis of pulmonary tuberculosis and common lung diseases with chest radiography (Wang, 2022, Thanoon et al., 2023) and breast-mass identification using mammography scans (Nasser and Yusof, 2023), have reached expert-level diagnostic accuracies. AI-based image recognition and machine learning can see far more detail in MRI and X-ray images than human eyes can register. For example, different types of glioblastomas have distinct genetic abnormalities, and doctors treat each one based on those abnormalities (Buchlak et al., 2021). But radiologists cannot identify genetic abnormalities of these brain cancers from images alone. The machine learning program at the Mayo Clinic can quickly and reliably identify the abnormalities (Romero-Brufau et al., 2021). These studies employed a technique known as transfer learning, where well-established deep neural networks trained on millions of natural, non-medical images are borrowed, and then the neural network connections are fine-tuned by using thousands of biomedical images. The deep-learning algorithm outperformed the average dermatologist in a comparison of the algorithm predictions (Hee et al., 2022)

A group has recently shown the possibility of developing a low-cost point of care for lymphoma diagnosis based on basic imaging and deep learning (Im et al., 2018).

Prediction Diagnosis

Properly trained AI has the potential to dramatically improve diagnosis. Machine-learning algorithms can predict the health of embryos for in vitro fertilization treatment or do the analysis of emergency calls to detect cardiac arrest (Goyal et al., 2020, Alamgir et al., 2021). A controversial application, now discontinued, has been Streams, an algorithm running on Google's DeepMind that has been trained on more than a million patient records held by the NHS Trust, that alerts clinicians about acute kidney disease in their patients (Yao et al., 2021).

Ningrum et al (2024) developed an artificial intelligence model with spatiotemporal data to predict Dengue outbreak and Dengue incidence case with 89.25% accuracy. They used meteorological, climatological, and Dengue epidemiology surveillance data during 2014-2021 from many government institutions and open access data sources and utilised a machine learning approach.

Experts from IISc (Indian Institute of Science, Bangalore) have developed an AI-based model that will provide dengue predictive analysis four weeks in advance. The mobile application will help get an accurate ground-level picture of prevention activities and incidence. This model can predict other diseases too.

Prediction Diagnosis in Children

Numerous diseases of childhood may benefit from machine learning including chronic, complex conditions with high intra-disease heterogeneity, rheumatological and dermatological disorders where immunosuppressive therapy is important (Momtazmanesh et al., 2022). Pairing highly dimensional data such as continuous observations collected during intensive care stays, with outcomes, admission durations and complications, may be useful in highlighting patients at an early stage who require escalation of care, or who can be discharged to the ward earlier (King et al., 2019). Even common seasonal childhood illness such as bronchiolitis, viral induced wheeze and gastroenteritis may benefit from machine learning research (Ashton et al., 2023). Utilising clinical parameters and discharge outcomes may help to identify factors associated with morbidity or patients who can be safely discharged.

Infectious Diseases

Recent reports have shown the added value of machine learning for image processing where classical tools could not identify early signs of diseases (Chen et al., 2017).

The Table 3 outlines the series of machine learning tools used in the field of infectious diseases and the aspects they target, as described by Agrebi and Larbi (2020). Expected outcomes from the contribution of AI are presented.

Table 3. AI tools, their use, and potential outcomes

Artificial intelligence	Infectious diseases	Outcomes
• Bayesian networks • Weak signal detection • Artificial neural network • Fuzzy clustering • Support vector machine • Artificial immune recognition system • k-Nearest neighbor • Decision tree • Random forest • ARIMA • 3-step floating catchment area • Unsupervised learning • Super learner	• Pathogen mutation • Diagnosis • Zoonosis • Outbreak • Source of infection • Epidemic prediction • Pandemic prediction • Resistance prediction • Drug discovery • Host genetic • Host-pathogen interaction • Adherence to therapy • Missing data	• Decision support • Reducing time for diagnosis, epidemic prediction, drug discovery • Identification of strategies for blocking transmission • Enabling low-income countries • Improving health • Saving life • Saving costs • Better be prepared • Personalized medicine • Forensic approach

The Utilization of AI in Infectious Diseases

- **Improved diagnosis and blocking transmission** - respiration rate, heart rate, and facial temperature were used to successfully classify individuals at higher risk for influenza using neural network and fuzzy clustering method (Amendolara et al., 2023). This demonstrates the ability to develop effective methods for identifying populations at risk. This triage is necessary and part of the process, even in the case of emergent infectious diseases, where efforts must be prioritized. Machine learning algorithms were developed to detect red blood cells (RBCs) infected with malaria from digital in-line holographic microscopy data (Go et al, 2018).
- **Epidemiology and transmission** - Epidemiological studies should be acquired along a specific timeline, with infection-related data collected in a longitudinal manner. When done properly, mathematical models can predict the size of emerging infectious diseases. Many of the machine learning methods besides giving indications on size and location, which is often the communicated information, are primarily used by mathematicians to estimate variables related to infection (e.g., incubation time, transmission mode, symptoms, resistance to treatments) (Wang et al., 2023).
- **Treatments and antimicrobial drug resistance -** Despite a good ability to diagnose malaria and probably with improved diagnosis in the near future, there is a strong problem with antibacterial and antiparasitic drugs resistance (Blasco et al., 2017). Antibiotic resistance can be better tackled with the existence of databases reflecting this phenomenon (Shafranskaya et al., 2021).

Ophthalmology

Fundus photography is a non-invasive procedure that uses retinal cameras to capture images of the retina, optic disc and macula. It can detect and monitor diseases such as Diabetic Retinopathy (DR), glaucoma, neoplasms of the retina and age-related macular degeneration, and plays a vital role in identifying causes of preventable blindness (Lee et al., 2015). In particular, clinical guidelines from the American Diabetes Association recommend DR screening for diabetic patients with minimal or no retinopathy every year, and more frequent examination for patients with progressing DR. Traditionally, fundus photographs are examined and interpreted by ophthalmologists, which is difficult to scale to the millions of diabetic patients at risk of developing sight-threatening DR.

Deep learning can extract previously unrecognized associations between retinal image patterns and age, gender, systolic blood pressure and smoking status, as well as major adverse cardiac events, illustrating the utility of machine learning in eliciting new knowledge from the raw data. The performance of a convolutional neural network exceeded their pre-specified sensitivity and specificity and was authorized by the FDA for use by healthcare providers to detect diabetic macular oedema and moderate-to-severe DR (Abràmoff et al., 2018).

Machine Learning for Biomarker Discovery and Precision Medicine

Biomarker discovery relies on identifying previously unrecognized correlations between thousands of measurements and phenotypes. Omics technologies have enabled the high-throughput measurement of thousands of genes and proteins, and of millions of genomic and epigenomic aberrations (Yu and Snyder, 2016). However, it is almost impossible for researchers to manually analyse and interpret the vast amount of data gathered from omics approaches. Machine-learning methods can identify the molecular patterns associated with disease status and disease subtypes, account for the high-level interactions among the measurements, and derive omics signatures to predict disease phenotypes (Babu and Snyder, 2023). Gene expression, protein abundance levels and DNA methylation profiles can predict the status of a number of diseases, including cancers, infectious diseases and the risk of Down's syndrome. Many of the biomarker panels derived from machine learning have outperformed those selected by experts or by conventional statistical methods. A number of them have been approved by the FDA and can be routinely used to guide treatment selection (McDermott et al., 2013). With the recent establishment of nationwide biobanks, standardized high-throughput profiling methods and advanced machine-learning methods, more robust and accurate biomarkers are expected to arise.

The application of big data in genomics is rapidly changing therapeutic development. With ongoing improvements in the field of bioinformatics and the combination of basic and clinical science with machine learning and high-power computing, major advances have been made in disease diagnosis and the monitoring of molecular changes in organ regions, tissues and circulating body fluids, such as the blood, plasma/serum, urine and cerebrospinal fluid Collectively, the use of big data in biomedicine involves extracting the essential information from large amounts of research and clinical data to assist in decisions advancing the field toward effective patient-centric healthcare and disease management (Guo et al., 2023).

Standardized treatments do not work for every patient, given the complexity of each person's history and genetic makeup, so researchers are using advanced analytics to personalize regimens. Decisions can be based on data analysis and patient monitoring with use of remote diagnostic devices. A startup called Turbine uses AI to design personalized cancer-treatment regimens. The technology models cell biology on the molecular level and seeks to identify the best drug to use for specific tumors. It can also identify complex biomarkers and search for combination therapies by performing millions of simulated experiments each day (Liao et al., 2023).

Customized medical treatments or Precision Medicine entails knowing the health outcomes of millions of other people with similar symptoms, prognoses, and ages is invaluable to providers that promise one-of-a-kind pharmaceuticals, physiotherapies, and other actions designed to deliver the greatest benefit with the fewest side effects. The National Institutes of Health (NIH) defines precision medicine as the "approach for disease treatment and prevention that takes into account individual variability in genes, environment, and lifestyle for each person" (Akhoon, 2021).

The emphasis is placed on tailored prevention, diagnosis and treatment for each individual based on genetics, epigenetics, and other lifestyle considerations (Prosperi et al., 2018).

Several companies are already using machine learning or other AI technologies to tailor their treatments to individual patients. Mindmaze uses machine learning to optimize rehabilitation activities for stroke patients (Choo and Chang, 2022).

Clinical Outcome Prediction and Patient Monitoring

In addition to identifying biomarkers related to clinical phenotypes, the use of EHRs to predict clinical outcomes shows great promise. Bayesian networks can predict mortality, readmission and length of hospital stay by using EHRs from the emergency department (Wong et al., 2018). Data from health insurance claims can be used to predict mortality in elderly patients (Wei et al., 2011), patient attributes in the medical notes can be employed to classify cancer patients with different responses to chemotherapy, and clinical predictors for the prognosis of patients can be identified (Krishnan et al., 2023). These could be used to help guide patients and their physicians in choosing an individualized treatment strategy.

Patient monitoring is crucial in intensive care units, operating rooms, emergency rooms and cardiac wards where timeliness in clinical decision-making can be measured in seconds. Routine monitoring devices in these high-acuity contexts generate a large amount of data and thus represent a great opportunity for AI-assisted alert systems (Ndichu et al., 2023). By using vital signs and the Modified Early Warning Score, a prediction model for cardiac arrest was established (Nishijima et al., 2016). Demographics, laboratory results and vital signs can also be used to predict cardiac arrest, transfer into the intensive care unit, or death (Aggarwal et al., 2020). In addition, an interpretable machine-learning model can assist anaesthesiologists in predicting hypoxaemia events during surgery (Lee et al., 2023). This suggests that, with deep-learning algorithms, raw patient-monitoring data could be better used to avoid information overload and alert overload while enabling more accurate clinical prediction and timely decision-making.

MEDICAL ROBOTICS

Robotic technologies have shown to have clear potential for providing innovation in treatments and treatment modalities for a wide variety of diseases and disorders. This includes improving both the standard and accessibility of care, enhancing patient health outcomes, supporting caregivers, and aiding healthcare workers. Thus, the spectrum of robotic system niches in medicine and healthcare currently spans a wide range of environments, user populations, and interaction modalities (Ienca 2022).

Table 4 summarizes some of the most recent advances in the healthcare robot and AI field[159]. These robots and AI technologies are deployed in the organized provision of medical care to individuals, including efforts to maintain, treat, or restore physical, mental, or emotional wellbeing (Ienca 2022).

Table 4. Robots and AI applications for healthcare

Embodied robots	
Type	**Category**
Surgical robots	Cardiac surgery, Neurosurgery, Plastic and reconstructive surgery, Dental surgery, Endocrine surgery, Endoscopic Surgery, Gastrointestinal surgery, Gynecology.
Assistive robots	Socially assistive robots – Therapeutic robots, Care robots Physically Assistive robots – Support, Performance, Rehabilitation
Healthcare Service robots	Routine task, Tele-presence, Disinfectant
Non-embodied AI	
Type	Category
Basic biomedical Research	Automated experiments - Automated data collection - Gene function annotation - Prediction of transcription factor binding sites - Simulation of molecular dynamics
Translational Research	Biomarker discovery - Drug–target prioritization - Drug discovery - Drug repurposing - Prediction of chemical toxicity - Genetic variant annotation
Clinical Practice	Disease diagnosis - Treatment design - Interpretation of patient genomes - treatment selection - Patient monitoring - Precision medicine (as genetics and genomics look for mutations and links to diseases from the information present in DNA) - patient risk stratification for primary prevention

ROBOTIC SURGERIES: PROCEDURAL AI AND AUTONOMOUS ROBOTIC SURGERY, ROBOTICS IN BARIATRIC SURGERY

Procedural AI

Procedural aspects of medicine, particularly surgery, will likely remain in the human domain longer than other branches of medicine. The development of new cancer treatments has become a crucial aspect of medical advancements. Nanobots, as one of the most promising applications of nanomedicines, are at the forefront of multidisciplinary research. With the progress of nanotechnology, nanobots enable the assembly and deployment of functional molecular/nanosized machines and are increasingly being utilized in cancer diagnosis and therapeutic treatment (Lundberg et al., 2018). In recent years, various practical applications of nanobots for cancer treatments have transitioned from theory to practice, from in vitro experiments to in vivo applications.

Autonomous Robotic Surgery

Robotic systems controlled by AI are routinely used in assembly lines in industry and in many biomedical laboratories. However, the development and adoption of autonomous robots in medical interventions has been considerably slower. For many decades, robotic surgery has been synonymous with robotically assisted surgery — systems that facilitate surgical procedures and enable motions smoother than those achievable by human hands, but that still require a surgeon for movement control. For instance, in the FDA-approved da Vinci surgical system for minimally invasive operations, surgeons operate the robot from a console (Kong et al., 2023). Such systems are designed to translate the surgeon's hand movements into the movements of instruments inside the patient and are therefore not autonomous.

With the continuing development of pre-programmed, image-guided and teleoperated surgical robots, more robot-assisted or automated intervention methods are expected to be incorporated in surgical practice (Leung and Vyas, 2014).

Current clinical practice is going to change in the foreseeable future with the AI applications. AI systems can enhance clinical decision-making, help in disease diagnosis, identify previously unrecognized imaging or genomic patterns associated with patient phenotypes, and assist in surgical interventions. AI applications have the advantage of bringing clinical expertise to remote regions where specialists are scarce or not available (Fiorini et al., 2022, Pakkasjärvi et al., 2023).

Robotics in Bariatric Surgery

Bariatric surgery is witnessing an ever-increasing demand. There is always a challenge of operating on patients with high body mass index (BMI), revisional bariatric procedures (RBP), managing complications and reoperative abdomens. Surgeons find bariatric surgery technically demanding because of large patients, large livers, thick abdominal walls and substantial visceral fat making exposure, dissection, and reconstruction difficult.

Robotic surgery has provided the surgeons with the advantage of three-dimensional vision, increased dexterity and precision (Wilson and Sudan, 2013), improved ergonomics and lesser operator fatigue. Ergonomics in laparoscopic surgery can be very challenging with big patients and uncomfortable postures, which lead to surgeon fatigue and work-related musculoskeletal symptoms (Nakadi et al., 2006).

The main limitation with robotic surgery is the perceived higher cost and set-up time compared with laparoscopy. But with increased experience, it is seen that set-up times reduce, and costs may also come down as material prices reduce (Talamini et al., 2003).

Robotics provides augmented and higher quality inputs from the patient to the surgeon, and the output is refined to a superior quality before reaching back to the target (Esposito et al., 2013) With the advent of newer technologies in robotics like fluorescence, integration of images, virtual and augmented reality, telesurgery, single site platforms, natural orifice surgery and haptic feedback, these are like empowering tool to the surgeons which can potentially change the way surgery is practiced today (Ma et al., 2020).

THE INTERNET OF THINGS

Internet of Things (IoT) based in-home health monitoring applications are one of the key mobile health (mHealth) applications that provide proactive and preventive digital health interventions (Mohanta et al.,

2019). The world is facing the aging population and the inherent need for assisted-living environments for senior citizens. Such advances include improvements in optimized network architecture, indoor networks coverage, increased device reliability and performance, ultra-low device cost, low device power consumption, and improved device and network security and privacy. Current studies of in-home health monitoring systems presented many benefits including improved safety, quality of life and reduction in hospitalization and cost (Philip et al., 2021).

Aminizadeh et al classified medical data processing platforms used for healthcare delivery into five categories: Internet of Things (IoT), cloud, edge, fog, and hybrid platforms (2023). Several challenges remain throughout the setting of remotely monitoring health and behavior, such as data collecting, modeling, and processing (Wang et al., 2023). The aim of healthcare systems, which is to restore patients' health, involves the application of effective medical data enabling evidence-based intervention (Hasasneh et al., 2023, Karar et al., 2022). Due to records management, compliance and regulatory obligations, and patient care, the healthcare industry produces a lot of data (Ahmed et al., 2023, Kc et al., 2021). Also, the current tendency is for these massive volumes of data to be rapidly digitalized and assembled with data gathered from individual and mobility devices (Gupta et al., 2020, Juyal et al., 2021).

The (IoT) enables sensors and smart objects to communicate without the direct involvement of human agents, necessitating near real-time processing. (Guo et al., 2021) In smart hospital infrastructures, a huge amount of sensitive data is exchanged among nodes throughout radio interfaces. These data have to be protected for accurate event detection and decision making. Anomalies are generally data that deviate from usual patterns from breach in data. In the context of smart hospitals, there are two types of anomalies: e-health related data anomalies and IoT network anomalies (Ukil et al., 2016). E-health related data anomalies are generated by events, such as fires (unusual temperature and humidity data), patient state change (unusual heart beat rates), etc. These anomalies are very critical and they must be detected in real-time with no false alerts, especially for hospital infrastructures (Said et al., 2021).

KNOWLEDGE EMBODIMENT IN AI

Understanding the effects of AI robotic systems on people and their connections among them requires clarity in the concept of knowledge embodiment, which has largely remained obscure.

Knowledge embodiment refers to the transformation of knowledge into a form in which its value becomes evident. Knowledge embodiment in robotic systems with artificial intelligence has been successfully accomplished in a collaboration between Anhui Provincial Hospital and iFlytek (Pee et al., 2019). Three AI robotic systems were implemented in 2016:

- A receptionist robot named Xiaoyi,
- Speech-based electronic health records (EHR) system, and
- An imaging diagnostic system

"Xiaoyi" answers non-emergency patients' and visitors' inquiries about healthcare services in standard Mandarin as well as 22 dialects of Chinese (Sun, 2021). It can refer a patient visitor to the relevant physician and room based on the patient's verbal description of symptoms and availability of medical facilities and staff. For patients with scheduled appointment, Xiaoyi will direct them to the self-registration kiosks.

The speech-based EHR system is an AI robotic system that integrates microphone and speech-recognition software to process natural languages. It is used to update EHR during clinical examination, ultrasound to enter data during examination using speech recognition (Pee et al., 2019).

The imaging diagnostic system integrates computer hardware and image processing software to visually analyze CT images. It has highly sophisticated diagnostic abilities.

The medical diagnostic and treatment assistant robot "Yizhi" also suggests diagnosis and treatment to physicians by analyzing information collected from various sources, including laboratory diagnosis, patient medical history, past medical cases, and conversation between the physician and patient during clinical examination (Sun, 2021).

Cloud and Edge Computing

The revolution in information technologies, and the spread of the Internet of Things (IoT) and smart city industrial systems, have developed widespread use of smart systems. Healthcare is a complex, 24/7 service that requires efficient and reliable follow-up on daily operations, service and resources. Patient data collection in healthcare systems requires extensive resources to collect, input, share and analyze information to be used for certain medical services. Traditionally, patient information management has been a slow, erroneous process where real-time accessibility was not possible. Cloud computing process is able to run on multiple tenants and is the best approach to enhance medical services owing to its flexibility and remote delivery (Gholamhosseini et al., 2023). A study by Dubey and Vishwakarma (2016) highlights the application of cloud computing in healthcare in a smart city. Mobile cloud computing and wireless body area networks can also be enhanced by the help of smart healthcare applications (Oueida et al., 2018).

In cloud and edge computing, the workflow and resource pools of the proposed smart healthcare system are defined in the cloud (Saini and Raj, 2022) where the process is executed, and resources are assigned. Each edge node in every resource reports the completion of an assigned task. Then, the resource is reassigned by the scheduling algorithm in the cloud. Research has shown that cloud computing is more reliable, efficient and faster than regular client-server computing.

The main problem faced by healthcare systems over the last decade is the overcrowding of emergency departments (ED), exponential rise in chronic diseases and this means that resources and workflow must be optimized. A Petri net workflow model proposed by Dotoli et al. (2010) improved the structure and dynamics of an ED at the general hospital of Bari, Italy. The model defined complete patient flow management and proposed an optimized solution with new resource dimensions to guarantee maximum patient flow. A study by Mahulea et al. (2014) proves that synchronization and concurrency are the powerful tools that make Petri nets ideal for modeling and analyzing healthcare systems.

The deployment of IoT into healthcare have been seen to improve the system leading to a better allocation of resources and an enhancement in performance measures.

Rolim et al. (2010) proposed automating the cloud and edge computing process using sensors attached to medical devices. The sensors are connected to external gateways through wireless networks, the information is stored in the cloud and is accessible to all medical staff. This proposal ensures a cost-effective, reliable cloud computing system that is integrated with medical devices. Doukas et al. (2010) developed a mobile system that provides storage, updating and retrieval of electronic healthcare data through cloud computing. This greatly improves the management of patient health records and images.

Fog Computing

Fog computing is an extended version of the cloud computing paradigm that includes new applications and services based on characteristics, such as low latency, location awareness, mobility, strong presence of streaming and real-time applications (Das and Inuwa, 2023).

The edge technology is the first framework to store, process, and compute immediate data closer to the healthcare workflow while cloud is the second stage of data storage and processing like past patients' histories. The fog/edge computing framework include the cloud, resources at the edge of the network, and healthcare system devices or sensors to execute certain related health tasks, all along maintaining privacy of patient data. (Butpheng et al.,2020).

HEALTHCARE AI REGULATION

A new ethical framework for AI has been suggested comprising of two new "rights;" a right to choose "meaningful human contact" over robot contact and "the right to refuse being profiled, tracked, measured, analysed, coached or manipulated" (Dubber et al., 2020).

Trustworthy AI should be lawful, have an ethical purpose, and should be technically and socially robust (to better avoid unintentional harms) (Jha et al., 2023). The guidance expresses the concepts of ethical purpose and human-centric development as based on four principles or values; respect for human autonomy, prevention of harm, fairness and explicability (Dlugatch et al., 2023).

Increasingly, AI personal assistants are being tuned to better understand the context of their interactions with humans. In part, this is achieved by analyzing non-verbal sounds rather than just decoding the language (Díaz-Rodríguez et al., 2023).

As these technologies evolve, there may be questions about imposing limits on artificial empathy, a question of particular relevance to caregiver bots or even end-of-life comfort bots. Like other health information technologies, healthcare AI is projected to improve access, reduce cost, and improve quality. (World Health Organization, 2021).

In everyday language an object or a technology is referred as being trustworthy, as an (indirect) reference to the belief that the people and processes involved in making and deploying that object or technology are themselves trustworthy. It seems that it is in this more indirect understanding that the term 'trustworthy AI' is used in documents such as the guidelines published by European Commission's High-level Expert Group on Artificial Intelligence (Iniesta, 2023). Trust and trustworthiness can be an important enabling factor when it comes to the acceptability and use of AI (Zhang and Zhang, 2023).

THE RIGHT TO HEALTHCARE AND THE IMPLICATIONS FOR THE USE OF ROBOTICS

The extent to which a new technology should be incorporated in the healthcare realm and the ethics of using them is one of the most important discussions once that AI and Robotics are here to stay.

The use and uptake of medical robots in healthcare is a complicated matter partly because of the technology's complex nature. In the case of robots for healthcare, the innovation encounters a vast, complex, and somewhat fragmented field governed by multiple systems of undefined norms. The existing networks of AI and Robotics governance encapsulate many norms such as autonomy and consent and non-discrimination, and the function of human rights always has to be discussed first (Riek, 2016).

Furthermore, this process of technological innovation and integration can also benefit from prospective regard for human rights considerations (Panch et al, 2019).

One of the important issues delaying the incorporation of AI and Robotics in the healthcare sector is the endless discussion on right to health under AI and medical robots; for which there are some core considerations put forward.

These core considerations are derived from the core obligations of States, as identified by the World Health Organization as being core components of the right to health and selected based on their acute relevance in the context of robotics for health care (Table 5). WHO deems these as essential elements that should serve as guiding principles for "governments as they make decisions about the goals, resources, focus and scale of public health law reform activities" (WHO, 2017) (Ienca et al., 2022).

Table 5. Some core obligations under the right to health for medical robots

Focus	Core considerations
Privacy and data Protection	Data-driven technologies such as medical robots that process personal data need to comply with the regulations on privacy and data protection, including the General Data Protection Regulation.
Non-discrimination	Equality and non-discrimination require that there be no distinction in the ability to exercise rights and freedoms based on race, colour, sex, language, religion, political or other opinions, national or social origin, property, birth, or another status.
Availability, accessibility, acceptability, and quality	Enough functioning robots scientifically and medically approved and of good quality, and accessible (physically accessible, affordable, and accessible information) to everyone without discrimination. All robots and AI must be respectful of medical ethics and culturally appropriate, sensitive to gender and age, and designed to respect confidentiality and improve users' health status.
Dignity	Human dignity acknowledges human beings' primacy, entitling them with human rights and should be one of the lenses through which AI development for healthcare should realize the potential benefits of this technology while preserving the individuals, or society at large, rights.

Core Obligations Under the Right to Health for Medical Robots

The right to health and healthcare can play a significant role in policy approaches to the translation and integration of innovative technologies (Ienca et al., 2022).

For the responsible integration of robotics, the groundwork is laid by stating the requirement of medical robots once they are in use and the adoption of the technology. To fully realise a future of widespread robot adoption, the necessary infrastructure must be developed (Fosch-Villaronga, 2015).

FUTURE EXPECTATIONS WITH AI AND ML IN HEALTHCARE

Inclusion of AI in workflows would be able to make healthcare cheaper, efficient, more personalized, and more equitable. As AI technologies are data driven, including diverse data in workflow is imperative to creating meaningful technology. Also, how meticulously the ML algorithms are applied to real life situations would affect the accuracy of technologies available to public in general.

A truly intelligent health system would be possible when AI data has been judiciously applied to all sectors of research and development.

CONCLUSION

Over the last decade, healthcare systems have been considered too complex to manage effectively, which has prompted decision makers to simply follow-up on daily operations and maintain a stable flow of patients. With IoT, smart cities and other new technologies, smart healthcare can solve the major ED problem of resource assignment efficiency. Due to reliability, cloud and edge computing are used to accommodate a proposed framework. When considering the application of machine learning techniques, the question to be answered is all important. Machine learning is not a panacea for the limitations of conventional statistics, nor will machine learning methodologies magically discover perfectly accurate disease prediction models. However, machine learning, as a concept, does hold great promise and can complement conventional statistics and study design. The utilisation of machine learning in highly dimensional data coupled with vast patient numbers can produce excellent results with real impact for patients in areas such as image interpretation and drug discovery. Acknowledging the challenges and moving towards transparency, standardised reporting and improved understanding of machine learning research, is important to take the next steps towards clinical impact.

REFERENCES

Abacha, A. B., & Zweigenbaum, P. (2015). MEANS: A medical question-answering system combining NLP techniques and semantic Web technologies. *Information Processing & Management, 51*(5), 570–594. doi:10.1016/j.ipm.2015.04.006

Abd-Alrazaq, A. A., Alajlani, M., Alalwan, A. A., Bewick, B. M., Gardner, P., & Househ, M. (2019). An overview of the features of chatbots in mental health: A scoping review. *International Journal of Medical Informatics, 132*, 103978. doi:10.1016/j.ijmedinf.2019.103978 PMID:31622850

Abràmoff, M. D., Lavin, P. T., Birch, M., Shah, N., & Folk, J. C. (2018). Pivotal trial of an autonomous AI-based diagnostic system for detection of diabetic retinopathy in primary care offices. *NPJ Digital Medicine, 1*(1), 39. doi:10.1038/s41746-018-0040-6 PMID:31304320

Aggarwal, N., Ahmed, M., Basu, S., Curtin, J. J., Evans, B. J., Matheny, M. E., Nundy, S., Sendak, M. P., Shachar, C., Shah, R. U., & Thadaney-Israni, S. (2020). (2020). Advancing Artificial Intelligence in Health Settings Outside the Hospital and Clinic. *NAM Perspectives*. doi:10.31478/202011f PMID:35291747

Agrebi, S., & Larbi, A. (2020). Use of artificial intelligence in infectious diseases. In *Artificial intelligence in precision health* (pp. 415–438). Academic Press. doi:10.1016/B978-0-12-817133-2.00018-5

Ahmadi, M. R., & Ameri, A. (2022). Skin Cancer Detection Based on Deep Learning. *Journal of Biomedical Physics & Engineering, 12*(6), 559–568. doi:10.31661/jbpe.v0i0.2207-1517 PMID:36569567

Ahmed, I., Jeon, G., & Chehri, A. (2023). An IoT-enabled smart health care system for screening of COVID-19 with multi layers features fusion and selection. *Computing*, *105*(4), 743–760. doi:10.1007/s00607-021-00992-0

Ahmed, Z., Mohamed, K., Zeeshan, S., & Dong, X. (2020). Artificial intelligence with multi-functional machine learning platform development for better healthcare and precision medicine. *Database (Oxford)*, *2020*, baaa010. doi:10.1093/database/baaa010 PMID:32185396

Akhoon, N. (2021). Precision Medicine: A New Paradigm in Therapeutics. *International Journal of Preventive Medicine*, *12*(1), 12. PMID:34084309

Alamgir, A., Mousa, O., & Shah, Z. (2021). Artificial Intelligence in Predicting Cardiac Arrest: Scoping Review. *JMIR Medical Informatics*, *9*(12), e30798. doi:10.2196/30798 PMID:34927595

AlDubayan, S. H., Conway, J. R., Camp, S. Y., Witkowski, L., Kofman, E., Reardon, B., Han, S., Moore, N., Elmarakeby, H., Salari, K., Choudhry, H., Al-Rubaish, A. M., Al-Sulaiman, A. A., Al-Ali, A. K., Taylor-Weiner, A., & Van Allen, E. M. (2020). Detection of Pathogenic Variants With Germline Genetic Testing Using Deep Learning vs Standard Methods in Patients With Prostate Cancer and Melanoma. *Journal of the American Medical Association*, *324*(19), 1957–1969. doi:10.1001/jama.2020.20457 PMID:33201204

Amendolara, A. B., Sant, D., Rotstein, H. G., & Fortune, E. (2023). LSTM-based recurrent neural network provides effective short term flu forecasting. *BMC Public Health*, *23*(1), 1788. doi:10.1186/s12889-023-16720-6 PMID:37710241

Aminizadeh, S., Heidari, A., Toumaj, S., Darbandi, M., Navimipour, N. J., Rezaei, M., Talebi, S., Azad, P., & Unal, M. (2023). The applications of machine learning techniques in medical data processing based on distributed computing and the Internet of Things. *Computer Methods and Programs in Biomedicine*, *241*, 107745. doi:10.1016/j.cmpb.2023.107745 PMID:37579550

Anggraini Ningrum, D. N., Li, Y. J., Hsu, C. Y., Solihuddin Muhtar, M., & Pandu Suhito, H. (2024). Artificial Intelligence Approach for Severe Dengue Early Warning System. *Studies in Health Technology and Informatics*, *310*, 881–885. doi:10.3233/SHTI231091 PMID:38269935

Ashton, J. J., Young, A., Johnson, M. J., & Beattie, R. M. (2023). Using machine learning to impact on long-term clinical care: Principles, challenges, and practicalities. *Pediatric Research*, *93*(2), 324–333. doi:10.1038/s41390-022-02194-6 PMID:35906306

Askin, S., Burkhalter, D., Calado, G., & El Dakrouni, S. (2023). Artificial Intelligence Applied to clinical trials: Opportunities and challenges. *Health and Technology*, *13*(2), 203–213. doi:10.1007/s12553-023-00738-2 PMID:36923325

Babu, M., & Snyder, M. (2023). Multi-omics profiling for health. *Molecular & Cellular Proteomics*, *22*(6), 100561. doi:10.1016/j.mcpro.2023.100561 PMID:37119971

Barragán-Montero, A., Javaid, U., Valdés, G., Nguyen, D., Desbordes, P., Macq, B., Willems, S., Vandewinckele, L., Holmström, M., Löfman, F., Michiels, S., Souris, K., Sterpin, E., & Lee, J. A. (2021). Artificial intelligence and machine learning for medical imaging: A technology review. *Physica medica: PM: an international journal devoted to the applications of physics to medicine and biology: official journal of the Italian Association of Biomedical Physics (AIFB), 83*, 242–256. doi:10.1016/j.ejmp.2021.04.016

Baumgartner, R. (2021). Precision medicine and digital phenotyping: Digital medicine's way from more data to better health. *Big Data & Society, 8*(2), 20539517211066452. doi:10.1177/20539517211066452

Baumgartner, R., Arora, P., Bath, C., Burljaev, D., Ciereszko, K., Custers, B., Ding, J., Ernst, W., Fosch-Villaronga, E., Galanos, V., Gremsl, T., Hendl, T., Kropp, C., Lenk, C., Martin, P., Mbelu, S., Morais Dos Santos Bruss, S., Napiwodzka, K., Nowak, E., & Williams, R. (2023). Fair and equitable AI in biomedical research and healthcare: Social science perspectives. *Artificial Intelligence in Medicine, 144*, 102658. doi:10.1016/j.artmed.2023.102658 PMID:37783540

Blasco, B., Leroy, D., & Fidock, D. A. (2017). Antimalarial drug resistance: Linking Plasmodium falciparum parasite biology to the clinic. *Nature Medicine, 23*(8), 917–928. doi:10.1038/nm.4381 PMID:28777791

Brot, H. F., & Mango, V. L. (2024). Artificial intelligence in breast ultrasound: Application in clinical practice. *Ultrasonography (Seoul, Korea), 43*(1), 3–14. doi:10.14366/usg.23116 PMID:38109894

Buchlak, Q. D., Esmaili, N., Leveque, J. C., Bennett, C., Farrokhi, F., & Piccardi, M. (2021). Machine learning applications to neuroimaging for glioma detection and classification: An artificial intelligence augmented systematic review. *Journal of clinical neuroscience: official journal of the Neurosurgical Society of Australasia, 89*, 177–198. doi:10.1016/j.jocn.2021.04.043

Butpheng, C., Yeh, K. H., & Xiong, H. (2020). Security and privacy in IoT-cloud-based e-health systems—A comprehensive review. *Symmetry, 12*(7), 1191. doi:10.3390/sym12071191

Care, S. (2018). Stanford Partnership in AI-assisted care. *Stanford Medicine*. https://med.stanford.edu/pacresearch/research/senior-care.html

Check Hayden, E. (2014). The automated lab. *Nature, 516*(7529), 131–132. doi:10.1038/516131a PMID:25471888

Chen, J. H., & Asch, S. M. (2017). Machine learning and prediction in medicine—Beyond the peak of inflated expectations. *The New England Journal of Medicine, 376*(26), 2507–2509. doi:10.1056/NEJMp1702071 PMID:28657867

Choo, Y. J., & Chang, M. C. (2022). Use of Machine Learning in Stroke Rehabilitation: A Narrative Review. *Brain & Neurorehabilitation, 15*(3), e26. doi:10.12786/bn.2022.15.e26 PMID:36742082

Cirillo, D., & Valencia, A. (2019). Big data analytics for personalized medicine. *Current Opinion in Biotechnology, 58*, 161–167. doi:10.1016/j.copbio.2019.03.004 PMID:30965188

Clinical Trials Arena. (2024). *Deep 6 AI and Graticule develop clinical trial recruitment algorithm.* Clinical Trials Arena. https://www.clinicaltrialsarena.com/news/deep-6-ai-graticule-algorithm/?cf-view

Cremin, C. J., Dash, S., & Huang, X. (2022). Big data: Historic advances and emerging trends in biomedical research. *Current Research in Biotechnology*, *4*, 138–151. doi:10.1016/j.crbiot.2022.02.004

Das, R., & Inuwa, M. M. (2023). A review on fog computing: issues, characteristics, challenges, and potential applications. *Telematics and Informatics Reports*, 100049.

Dave, P., Nambudiri, V., & Grant-Kels, J. M. (2023). The introduction of "Dr AI": What dermatologists should consider. *Journal of the American Academy of Dermatology*, *88*(6), 1401–1402. doi:10.1016/j.jaad.2022.01.014 PMID:35074441

Dekker, I., De Jong, E. M., Schippers, M. C., De Bruijn-Smolders, M., Alexiou, A., & Giesbers, B. (2020). Optimizing Students' Mental Health and Academic Performance: AI-Enhanced Life Crafting. *Frontiers in Psychology*, *11*, 1063. doi:10.3389/fpsyg.2020.01063 PMID:32581935

Delipetrev, B., Tsinaraki, C., & Kostic, U. (2020). *Historical evolution of artificial intelligence.*

Díaz-Rodríguez, N., Del Ser, J., Coeckelbergh, M., de Prado, M. L., Herrera-Viedma, E., & Herrera, F. (2023). Connecting the dots in trustworthy Artificial Intelligence: From AI principles, ethics, and key requirements to responsible AI systems and regulation. *Information Fusion*, *99*, 101896. doi:10.1016/j.inffus.2023.101896

Dlugatch, R., Georgieva, A., & Kerasidou, A. (2023). Trustworthy artificial intelligence and ethical design: Public perceptions of trustworthiness of an AI-based decision-support tool in the context of intrapartum care. *BMC Medical Ethics*, *24*(1), 1–16. doi:10.1186/s12910-023-00917-w PMID:37340408

Dotoli, M., Fanti, M. P., Iacobellis, G., Martino, L., Moretti, A. M., & Ukovich, W. (2010, February). Modeling and management of a hospital department via petri nets. In *2010 IEEE Workshop on Health Care Management (WHCM)* (pp. 1-6). IEEE. 10.1109/WHCM.2010.5441248

Doukas, C., Pliakas, T., & Maglogiannis, I. (2010, August). Mobile healthcare information management utilizing Cloud Computing and Android OS. In *2010 Annual International Conference of the IEEE Engineering in Medicine and Biology* (pp. 1037-1040). IEEE. 10.1109/IEMBS.2010.5628061

Dubber, M. D., Pasquale, F., & Das, S. (Eds.). (2020). *The Oxford handbook of ethics of AI.* Oxford Handbooks. doi:10.1093/oxfordhb/9780190067397.001.0001

Dubey, N., & Vishwakarma, S. (2016). Cloud computing in healthcare. *Int. J. Curr. Trends Eng. Res*, *2*(5), 211–216.

Ehrenstein, V., Kharrazi, H., & Lehmann, H. (2019) Obtaining Data From Electronic Health Records. In: *Gliklich RE, Leavy MB, Dreyer NA, editors. Tools and Technologies for Registry Interoperability, Registries for Evaluating Patient Outcomes: A User's Guide,* 3rd Edition, Addendum 2 [Internet]. Rockville (MD): Agency for Healthcare Research and Quality (US). https://www.ncbi.nlm.nih.gov/books/NBK551878/

Esposito, C., El Ghoneimi, A., Yamataka, A., Rothenberg, S., Bailez, M., Ferro, M., Gamba, P., Castagnetti, M., Mattioli, G., Delagausie, P., Antoniou, D., Montupet, P., Marte, A., Saxena, A., Bertozzi, M., Philippe, P., Varlet, F., Lardy, H., Caldamone, A., & Schier, F. (2013). Work-related upper limb musculoskeletal disorders in paediatric laparoscopic surgery. A multicenter survey. *Journal of Pediatric Surgery*, *48*(8), 1750–1756. doi:10.1016/j.jpedsurg.2013.01.054 PMID:23932617

Fiorini, P., Goldberg, K. Y., Liu, Y., & Taylor, R. H. (2022). Concepts and Trends n Autonomy for Robot-Assisted Surgery. *Proceedings of the IEEE. Institute of Electrical and Electronics Engineers, 110*(7), 993–1011. 10.1109/JPROC.2022.3176828

Fosch-Villaronga, E. (2015). Creation of a care robot impact assessment. *Int J Soc Behav Edu Econ Bus Ind Eng*, *9*, 1913–1917.

Fukuda, M., Inamoto, K., Shibata, N., Ariji, Y., Yanashita, Y., Kutsuna, S., Nakata, K., Katsumata, A., Fujita, H., & Ariji, E. (2020). Evaluation of an artificial intelligence system for detecting vertical root fracture on panoramic radiography. *Oral Radiology*, *36*(4), 337–343. doi:10.1007/s11282-019-00409-x PMID:31535278

García-Vázquez, J. P., Rodríguez, M. D., Tentori, M. E., Saldaña, D., Andrade, Á. G., & Espinoza, A. N. (2010). An agent-based architecture for developing activity-aware systems for assisting elderly. *Journal of Universal Computer Science*, *16*(12), 1500–1520.

Gholamhosseini, L., Behmanesh, A., Nasiri, S., Ehsanzadeh, S. J., & Sadoughi, F. (2023). Cloud-Based Internet of Things in Healthcare Applications: A Systematic Literature Review. *Frontiers in Health Informatics*, *12*, 145. doi:10.30699/fhi.v12i0.451

Go, T., Kim, J. H., Byeon, H., & Lee, S. J. (2018). Machine learning-based in-line holographic sensing of unstained malaria-infected red blood cells. *Journal of Biophotonics*, *11*(9), e201800101. doi:10.1002/jbio.201800101 PMID:29676064

Golub, T. R., Slonim, D. K., Tamayo, P., Huard, C., Gaasenbeek, M., Mesirov, J. P., Coller, H., Loh, M. L., Downing, J. R., Caligiuri, M. A., Bloomfield, C. D., & Lander, E. S. (1999). Molecular classification of cancer: Class discovery and class prediction by gene expression monitoring. *Science*, *286*(5439), 531–537. doi:10.1126/science.286.5439.531 PMID:10521349

Goyal, A., Kuchana, M., & Ayyagari, K. P. R. (2020). Machine learning predicts live-birth occurrence before in-vitro fertilization treatment. *Scientific Reports*, *10*(1), 20925. doi:10.1038/s41598-020-76928-z PMID:33262383

Guo, K., Wu, M., Soo, Z., Yang, Y., Zhang, Y., Zhang, Q., Lin, H., Grosser, M., Venter, D., Zhang, G., & Lu, J. (2023). Artificial intelligence-driven biomedical genomics. *Knowledge-Based Systems*, *279*, 110937. doi:10.1016/j.knosys.2023.110937

Guo, Z., Shen, Y., Wan, S., Shang, W. L., & Yu, K. (2021). Hybrid intelligence-driven medical image recognition for remote patient diagnosis in internet of medical things. *IEEE Journal of Biomedical and Health Informatics*, *26*(12), 5817–5828. doi:10.1109/JBHI.2021.3139541 PMID:34971545

Gupta, S., Johnson, E. M., Peacock, J. G., Jiang, L., McBee, M. P., Sneider, M. B., & Krupinski, E. A. (2020). Radiology, mobile devices, and internet of things (IoT). *Journal of Digital Imaging*, *33*(3), 735–746. doi:10.1007/s10278-019-00311-2 PMID:31898039

Haque, M. D. R., & Rubya, S. (2023). An Overview of Chatbot-Based Mobile Mental Health Apps: Insights From App Description and User Reviews. *JMIR mHealth and uHealth*, *11*, e44838. doi:10.2196/44838 PMID:37213181

Hasasneh, A., Hijazi, H., Talib, M. A., Afadar, Y., Nassif, A. B., & Nasir, Q. (2023). Wearable Devices and Explainable Unsupervised Learning for COVID-19 Detection and Monitoring. *Diagnostics (Basel)*, *13*(19), 3071. doi:10.3390/diagnostics13193071 PMID:37835814

Haskew, J., Rø, G., Turner, K., Kimanga, D., Sirengo, M., & Sharif, S. (2015). Implementation of a cloud-based electronic medical record to reduce gaps in the HIV treatment continuum in rural Kenya. *PLoS One*, *10*(8), e0135361. doi:10.1371/journal.pone.0135361 PMID:26252212

Hassan, M., Awan, F. M., Naz, A., deAndrés-Galiana, E. J., Alvarez, O., Cernea, A., Fernández-Brillet, L., Fernández-Martínez, J. L., & Kloczkowski, A. (2022). Innovations in Genomics and Big Data Analytics for Personalized Medicine and Health Care: A Review. *International Journal of Molecular Sciences*, *23*(9), 4645. doi:10.3390/ijms23094645 PMID:35563034

Hee, K., Cosa, A., Santhanam, N., Jannesari, M., Maros, M., & Ganslandt, T. (2022). Transfer learning for medical image classification: A literature review. *BMC Medical Imaging*, *22*(1), 69. Advance online publication. doi:10.1186/s12880-022-00793-7 PMID:35418051

Hleg, A. (2019). *A Definition of AI: Main Capabilities and Disciplines–EU High-Level Expert Group on Artificial Intelligence.* [online] European Commission. https://www.aepd.es/sites/default/files/2019-09/ai-definition.pdf

Ienca, M., Pollicino, O., Liguori, L., Stefanini, E., & Andorno, R. (Eds.). (2022). *The Cambridge handbook of information technology, life sciences and human rights*. Cambridge University Press. doi:10.1017/9781108775038

Im, H., Pathania, D., McFarland, P. J., Sohani, A. R., Degani, I., Allen, M., Coble, B., Kilcoyne, A., Hong, S., Rohrer, L., Abramson, J. S., Dryden-Peterson, S., Fexon, L., Pivovarov, M., Chabner, B., Lee, H., Castro, C. M., & Weissleder, R. (2018). Design and clinical validation of a point-of-care device for the diagnosis of lymphoma via contrast-enhanced microholography and machine learning. *Nature Biomedical Engineering*, *2*(9), 666–674. doi:10.1038/s41551-018-0265-3 PMID:30555750

Iniesta, R. (2023). The human role to guarantee an ethical AI in healthcare: A five-facts approach. *AI and Ethics*, 1–13. doi:10.1007/s43681-023-00353-x

Inkster, B., Sarda, S., & Subramanian, V. (2018). An Empathy-Driven, Conversational Artificial Intelligence Agent (Wysa) for Digital Mental Well-Being: Real-World Data Evaluation Mixed-Methods Study. *JMIR mHealth and uHealth*, *6*(11), e12106. doi:10.2196/12106 PMID:30470676

Iqbal, J., Jahangir, K., Mashkoor, Y., Sultana, N., Mehmood, D., Ashraf, M., Iqbal, A., & Hafeez, M. H. (2022). The future of artificial intelligence in neurosurgery: A narrative review. *Surgical Neurology International*, *13*, 13. doi:10.25259/SNI_877_2022 PMID:36447868

Javaid, M., Haleem, A., Singh, R. P., & Suman, R. (2023). Sustaining the healthcare systems through the conceptual of biomedical engineering: A study with recent and future potentials. *Biomedical Technology*, *1*, 39–47. doi:10.1016/j.bmt.2022.11.004

Jha, D., Rauniyar, A., Srivastava, A., Hagos, D. H., Tomar, N. K., Sharma, V., Keles, E., Zhang, Z., Demir, U., Topcu, A., Yazidi, A., Haakegard, J. E., & Bagci, U. (2023). Ensuring Trustworthy Medical Artificial Intelligencethrough Ethical and Philosophical Principles. *arXiv preprint arXiv:2304.11530.*

Jha, S., & Topol, E. J. (2016). Adapting to artificial intelligence: Radiologists and pathologists as information specialists. *Journal of the American Medical Association*, *316*(22), 2353–2354. doi:10.1001/jama.2016.17438 PMID:27898975

Jouffroy, R., Jost, D., & Prunet, B. (2020). Prehospital pulse oximetry: A red flag for early detection of silent hypoxemia in COVID-19 patients. *Critical Care (London, England)*, *24*(1), 313. doi:10.1186/s13054-020-03036-9 PMID:32513249

Juyal, S., Sharma, S., & Shukla, A. S. (2021). Smart skin health monitoring using AI-enabled cloud-based IoT. *Materials Today: Proceedings*, *46*, 10539–10545. doi:10.1016/j.matpr.2021.01.074

Kang, G., Liu, K., Hou, B., & Zhang, N. (2017). 3D multi-view convolutional neural networks for lung nodule classification. *PLoS One*, *12*(11), e0188290. doi:10.1371/journal.pone.0188290 PMID:29145492

Karar, M. E., Alotaibi, B., & Alotaibi, M. (2022). Intelligent medical IoT-enabled automated microscopic image diagnosis of acute blood cancers. *Sensors (Basel)*, *22*(6), 2348. doi:10.3390/s22062348 PMID:35336523

Kaufman, D. R., Sheehan, B., Stetson, P., Bhatt, A. R., Field, A. I., Patel, C., & Maisel, J. M. (2016). Natural Language Processing-Enabled and Conventional Data Capture Methods for Input to Electronic Health Records: A Comparative Usability Study. *JMIR Medical Informatics*, *4*(4), e35. doi:10.2196/medinform.5544 PMID:27793791

Kavitha, M. S., Gangadaran, P., Jackson, A., Venmathi, M. B. A., Kurita, T., & Ahn, B. C. (2022). Deep Neural Network Models for Colon Cancer Screening. Cancers (Basel).14(15):3707. *Published*, *2022*(Jul), 29. doi:10.3390/cancers14153707 PMID:35954370

Kc, K., Yin, Z., Wu, M., & Wu, Z. (2021). Evaluation of deep learning-based approaches for COVID-19 classification based on chest X-ray images. *Signal, Image and Video Processing*, *15*(5), 959–966. doi:10.1007/s11760-020-01820-2 PMID:33432267

King, J., O'Neill, B., Ramsay, P., Linden, M. A., Darweish Medniuk, A., Outtrim, J., & Blackwood, B. (2019). Identifying patients' support needs following critical illness: A scoping review of the qualitative literature. *Critical Care*, *23*(1), 1–12. doi:10.1186/s13054-019-2441-6 PMID:31126335

Kirkpatrick, J., Pascanu, R., Rabinowitz, N., Veness, J., Desjardins, G., Rusu, A. A., Milan, K., Quan, J., Ramalho, T., Grabska-Barwinska, A., Hassabis, D., Clopath, C., Kumaran, D., & Hadsell, R. (2017). Overcoming catastrophic forgetting in neural networks. *Proceedings of the National Academy of Sciences of the United States of America*, *114*(13), 3521–3526. doi:10.1073/pnas.1611835114 PMID:28292907

Kong, X., Gao, P., Wang, J., Fang, Y., & Hwang, K. C. (2023). Advances of medical nanorobots for future cancer treatments. *Journal of Hematology & Oncology*, *16*(1), 74. doi:10.1186/s13045-023-01463-z PMID:37452423

Krishnan, G., Singh, S., Pathania, M., Gosavi, S., Abhishek, S., Parchani, A., & Dhar, M. (2023). Artificial intelligence in clinical medicine: Catalyzing a sustainable global healthcare paradigm. *Frontiers in Artificial Intelligence*, *6*, 6. doi:10.3389/frai.2023.1227091 PMID:37705603

Krizhevsky, A., Sutskever, I., & Hinton, G. E. (2012). Imagenet classification with deep convolutional neural networks. *Advances in Neural Information Processing Systems*, 25.

Kumar, Y., Koul, A., Singla, R., & Ijaz, M. F. (2023). Artificial intelligence in disease diagnosis: A systematic literature review, synthesizing framework and future research agenda. *Journal of Ambient Intelligence and Humanized Computing*, *14*(7), 8459–8486. doi:10.1007/s12652-021-03612-z PMID:35039756

Lamia, A., & Fawaz, A. (2022). Detection of Pneumonia Infection by Using Deep Learning on a Mobile Platform. *Computational Intelligence and Neuroscience*, *7925668*, 1–9. doi:10.1155/2022/7925668 PMID:35942467

Lee, H., Yang, H. L., Ryu, H. G., Jung, C. W., Cho, Y. J., Yoon, S. B., Yoon, H.-K., & Lee, H. C. (2023). Real-time machine learning model to predict in-hospital cardiac arrest using heart rate variability in ICU. *NPJ Digital Medicine*, *6*(1), 215. doi:10.1038/s41746-023-00960-2 PMID:37993540

Lee, R., Wong, T. Y., & Sabanayagam, C. (2015). Epidemiology of diabetic retinopathy, diabetic macular edema and related vision loss. *Eye and Vision (London, England)*, *2*(1), 1–25. doi:10.1186/s40662-015-0026-2 PMID:26605370

Leung, T., & Vyas, D. (2014). Robotic surgery: Applications. *American Journal of Robotic Surgery*, *1*(1), 38–41. doi:10.1166/ajrs.2014.1004 PMID:26501128

Liao, J., Li, X., Gan, Y., Han, S., Rong, P., Wang, W., Li, W., & Zhou, L. (2023). Artificial intelligence assists precision medicine in cancer treatment. *Frontiers in Oncology*, *12*, 998222. doi:10.3389/fonc.2022.998222 PMID:36686757

Loncar-Turukalo, T., Zdravevski, E., Machado da Silva, J., Chouvarda, I., & Trajkovik, V. (2019). Literature on Wearable Technology for Connected Health: Scoping Review of Research Trends, Advances, and Barriers. *Journal of Medical Internet Research*, *21*(9), e14017. doi:10.2196/14017 PMID:31489843

Lundberg, S. M., Nair, B., Vavilala, M. S., Horibe, M., Eisses, M. J., Adams, T., Liston, D. E., Low, D. K., Newman, S. F., Kim, J., & Lee, S. I. (2018). Explainable machine-learning predictions for the prevention of hypoxaemia during surgery. *Nature Biomedical Engineering*, *2*(10), 749–760. doi:10.1038/s41551-018-0304-0 PMID:31001455

Ma, R., Vanstrum, E. B., Lee, R., Chen, J., & Hung, A. J. (2020). Machine learning in the optimization of robotics in the operative field. *Current Opinion in Urology*, *30*(6), 808–816. doi:10.1097/MOU.0000000000000816 PMID:32925312

Mahulea, C., Mahulea, L., García-Soriano, J. M., & Colom, J. M. (2014, October). Petri nets with resources for modeling primary healthcare systems. In *2014 18th International Conference on System Theory, Control and Computing (ICSTCC)* (pp. 639-644). IEEE. 10.1109/ICSTCC.2014.6982489

Majumder, S., Mondal, T., & Deen, M. J. (2017). Wearable Sensors for Remote Health Monitoring. *Sensors (Basel)*, *17*(1), 130. doi:10.3390/s17010130 PMID:28085085

Marr, B. (2022). How AI and machine learning will impact the future of healthcare. *Forbes*. https://www.forbes.com/sites/bernardmarr/2022/09/14/how-ai-and-machine-learning-will-impact-the-future-of-healthcare/?sh=3e9c257047e5

McDermott, J. E., Wang, J., Mitchell, H., Webb-Robertson, B. J., Hafen, R., Ramey, J., & Rodland, K. D. (2013). Challenges in biomarker discovery: Combining expert insights with statistical analysis of complex omics data. *Expert Opinion on Medical Diagnostics*, *7*(1), 37–51. doi:10.1517/17530059.2012.718329 PMID:23335946

McKinney, S. M., Sieniek, M., Godbole, V., Godwin, J., Antropova, N., Ashrafian, H., Back, T., Chesus, M., Corrado, G. S., Darzi, A., Etemadi, M., Garcia-Vicente, F., Gilbert, F. J., Halling-Brown, M., Hassabis, D., Jansen, S., Karthikesalingam, A., Kelly, C. J., King, D., & Shetty, S. (2020). International evaluation of an AI system for breast cancer screening. *Nature*, *577*(7788), 89–94. doi:10.1038/s41586-019-1799-6 PMID:31894144

Mohanta, B., Das, P., & Patnaik, S. (2019, May). Healthcare 5.0: A paradigm shift in digital healthcare system using artificial intelligence, IOT and 5G communication. In *2019 International Conference on Applied Machine Learning (ICAML)* (pp. 191-196). IEEE. 10.1109/ICAML48257.2019.00044

Momtazmanesh, S., Nowroozi, A., & Rezaei, N. (2022). Artificial intelligence in rheumatoid arthritis: current status and future perspectives: a state-of-the-art review. *Rheumatology and Therapy*, *9*(5), 1249–1304. doi:10.1007/s40744-022-00475-4 PMID:35849321

Nakadi, I. E., Mélot, C., Closset, J., De Moor, V., Bétroune, K., Feron, P., Lingier, P., & Gelin, M. (2006). Evaluation of da Vinci Nissen fundoplication clinical results and cost minimization. *World Journal of Surgery*, *30*(6), 1050–1054. doi:10.1007/s00268-005-7950-6 PMID:16736336

Nasser, M., & Yusof, U. K. (2023). Deep Learning Based Methods for Breast Cancer Diagnosis: A Systematic Review and Future Direction. *Diagnostics (Basel)*, *13*(1), 161. doi:10.3390/diagnostics13010161 PMID:36611453

Ndichu, S., Ban, T., Takahashi, T., & Inoue, D. (2023). AI-Assisted Security Alert Data Analysis with Imbalanced Learning Methods. *Applied Sciences (Basel, Switzerland)*, *13*(3), 1977. doi:10.3390/app13031977

Nishijima, I., Oyadomari, S., Maedomari, S., Toma, R., Igei, C., Kobata, S., Koyama, J., Tomori, R., Kawamitsu, N., Yamamoto, Y., Tsuchida, M., Tokeshi, Y., Ikemura, R., Miyagi, K., Okiyama, K., & Iha, K. (2016). Use of a modified early warning score system to reduce the rate of in-hospital cardiac arrest. *Journal of Intensive Care*, *4*(1), 1–6. doi:10.1186/s40560-016-0134-7 PMID:26865981

Noh, J., & Kavuluru, R. (2018). Document Retrieval for Biomedical Question Answering with Neural Sentence Matching. *Proceedings of the ... International Conference on Machine Learning and Applications. International Conference on Machine Learning and Applications*. IEEE. 10.1109/ICMLA.2018.00036

Oueida, S., Kotb, Y., Aloqaily, M., Jararweh, Y., & Baker, T. (2018). An edge computing based smart healthcare framework for resource management. *Sensors (Basel), 18*(12), 4307. doi:10.3390/s18124307 PMID:30563267

Ouyang, D., He, B., Ghorbani, A., Yuan, N., Ebinger, J., Langlotz, C. P., Heidenreich, P. A., Harrington, R. A., Liang, D. H., Ashley, E. A., & Zou, J. Y. (2020). Video-based AI for beat-to-beat assessment of cardiac function. *Nature, 580*(7802), 252–256. doi:10.1038/s41586-020-2145-8 PMID:32269341

Ozaltin, O., Coskun, O., Yeniay, O., & Subasi, A. (2022). A Deep Learning Approach for Detecting Stroke from Brain CT Images Using OzNet. *Bioengineering (Basel, Switzerland), 9*(12), 783. doi:10.3390/bioengineering9120783 PMID:36550989

Pakkasjärvi, N., Luthra, T., & Anand, S. (2023). Artificial Intelligence in Surgical Learning. *Surgeries, 4*(1), 86–97. doi:10.3390/surgeries4010010

Panch, T., Mattie, H., & Celi, L. A. (2019). The "inconvenient truth" about AI in healthcare. *NPJ Digital Medicine, 2*(1), 77. doi:10.1038/s41746-019-0155-4 PMID:31453372

Parry, C. M., & Aneja, U. (2020). *Artificial Intelligence for Healthcare: Insights from India*. Chatham House. https://www.chathamhouse.org/2020/07/artificial-intelligence-healthcare-insights-india/3-ai-healthcare-india-applications

Pee, L. G., Pan, S. L., & Cui, L. (2019). Artificial intelligence in healthcare robots: A social informatics study of knowledge embodiment. *Journal of the Association for Information Science and Technology, 70*(4), 351–369. doi:10.1002/asi.24145

Perl, R. M., Grimmer, R., Hepp, T., & Horger, M. S. (2021). Can a novel deep neural network improve the computer-aided detection of solid pulmonary nodules and the rate of false-positive findings in comparison to an established machine learning computer-aided detection? *Investigative Radiology, 56*(2), 103–108. doi:10.1097/RLI.0000000000000713 PMID:32796198

Philip, N. Y., Rodrigues, J. J., Wang, H., Fong, S. J., & Chen, J. (2021). Internet of Things for in-home health monitoring systems: Current advances, challenges and future directions. *IEEE Journal on Selected Areas in Communications, 39*(2), 300–310. doi:10.1109/JSAC.2020.3042421

Pollock, L. (2023, August 22). *What Is an NLP Chatbot — And How Do NLP-Powered Bots Work?* Ultimate.ai. https://www.ultimate.ai/blog/ai-automation/how-nlp-text-based-chatbots-work

Poplin, R., Varadarajan, A. V., Blumer, K., Liu, Y., McConnell, M. V., Corrado, G. S., Peng, L., & Webster, D. R. (2018). Prediction of cardiovascular risk factors from retinal fundus photographs via deep learning. *Nature Biomedical Engineering, 2*(3), 158–164. doi:10.1038/s41551-018-0195-0 PMID:31015713

Prosperi, M., Min, J. S., Bian, J., & Modave, F. (2018). Big data hurdles in precision medicine and precision public health. *BMC Medical Informatics and Decision Making, 18*(1), 1–15. doi:10.1186/s12911-018-0719-2 PMID:30594159

Rajkomar, A., Oren, E., Chen, K., Dai, A. M., Hajaj, N., Hardt, M., Liu, P. J., Liu, X., Marcus, J., Sun, M., Sundberg, P., Yee, H., Zhang, K., Zhang, Y., Flores, G., Duggan, G. E., Irvine, J., Le, Q., Litsch, K., & Dean, J. (2018). Scalable and accurate deep learning with electronic health records. *NPJ Digital Medicine*, *1*(1), 18. doi:10.1038/s41746-018-0029-1 PMID:31304302

Rao, M. V. V & Team DHR-ICMR AI Cell. (2023). Ethical guidelines for application of artificial intelligence in biomedical research and healthcare. *Indian Council of Medical Research.* https://main.icmr.nic.in/sites/default/files/upload_documents/Ethical_Guidelines_AI_Healthcare_2023.pdf

Riek, L. D. (2016). Robotics technology in mental health care. In *Artificial intelligence in behavioral and mental health care* (pp. 185–203). Academic Press. doi:10.1016/B978-0-12-420248-1.00008-8

Rolim, C. O., Koch, F. L., Westphall, C. B., Werner, J., Fracalossi, A., & Salvador, G. S. (2010, February). A cloud computing solution for patient's data collection in health care institutions. In *2010 Second International Conference on eHealth, Telemedicine, and Social Medicine* (pp. 95-99). IEEE. 10.1109/eTELEMED.2010.19

Romero-Brufau, S., Whitford, D., Johnson, M. G., Hickman, J., Morlan, B. W., Therneau, T., Naessens, J., & Huddleston, J. M. (2021). Using machine learning to improve the accuracy of patient deterioration predictions: Mayo Clinic Early Warning Score (MC-EWS). *Journal of the American Medical Informatics Association : JAMIA*, *28*(6), 1207–1215. doi:10.1093/jamia/ocaa347 PMID:33638343

Rong, G., Mendez, A., Assi, E. B., Zhao, B., & Sawan, M. (2020). Artificial intelligence in healthcare: Review and prediction case studies. *Engineering (Beijing)*, *6*(3), 291–301. doi:10.1016/j.eng.2019.08.015

Said, A. M., Yahyaoui, A., & Abdellatif, T. (2021). Efficient anomaly detection for smart hospital IoT systems. *Sensors (Basel)*, *21*(4), 1026. doi:10.3390/s21041026 PMID:33546169

Saini, K., & Raj, P. (2022). Edge platforms, frameworks and applications. [). Elsevier.] *Advances in Computers*, *127*, 237–258. doi:10.1016/bs.adcom.2022.02.005

Sarker, I. H. (2021). Deep Learning: A Comprehensive Overview on Techniques, Taxonomy, Applications and Research Directions. *SN Computer Science*, *2*(6), 420. doi:10.1007/s42979-021-00815-1 PMID:34426802

Sarker, I. H. (2021). Machine Learning: Algorithms, Real-World Applications and Research Directions. *SN Computer Science*, *2*(3), 160. doi:10.1007/s42979-021-00592-x PMID:33778771

Sarrouti, M., & Ouatik, E. (2017). A machine learning-based method for question type classification in biomedical question answering. *Methods of Information in Medicine*, *56*(3), 209–216. doi:10.3414/ME16-01-0116 PMID:28361158

Schwalbe, N., & Wahl, B. (2020). Artificial intelligence and the future of global health. *Lancet*, *395*(10236), 1579–1586. doi:10.1016/S0140-6736(20)30226-9 PMID:32416782

Shademan, A., Decker, R. S., Opfermann, J. D., Leonard, S., Krieger, A., & Kim, P. C. (2016). Supervised autonomous robotic soft tissue surgery. *Science Translational Medicine*, *8*(337), 337ra64. doi:10.1126/scitranslmed.aad9398 PMID:27147588

Shafranskaya, D., Chori, A., & Korobeynikov, A. (2021). Graph-based approaches significantly improve the recovery of antibiotic resistance genes from complex metagenomic datasets. *Frontiers in Microbiology, 12*, 714836. doi:10.3389/fmicb.2021.714836 PMID:34690959

Simonite, T. (2018). Google's AI eye doctor gets ready to go to work in India. *Wired Magazine.*

Stahl, B. C., Andreou, A., Brey, P., Hatzakis, T., Kirichenko, A., Macnish, K., Shaelou, S. L., Patel, A., Ryan, M., & Wright, D. (2021). Artificial intelligence for human flourishing–Beyond principles for machine learning. *Journal of Business Research, 124*, 374–388. doi:10.1016/j.jbusres.2020.11.030

Sudlow, C., Gallacher, J., Allen, N., Beral, V., Burton, P., Danesh, J., Downey, P., Elliott, P., Green, J., Landray, M., Liu, B., Matthews, P., Ong, G., Pell, J., Silman, A., Young, A., Sprosen, T., Peakman, T., & Collins, R. (2015). UK biobank: An open access resource for identifying the causes of a wide range of complex diseases of middle and old age. *PLoS Medicine, 12*(3), e1001779. doi:10.1371/journal.pmed.1001779 PMID:25826379

Sun, T. Q. (2021). Adopting artificial intelligence in public healthcare: The effect of social power and learning algorithms. *International Journal of Environmental Research and Public Health, 18*(23), 12682. doi:10.3390/ijerph182312682 PMID:34886404

Talamini, M. A., Chapman, S., Horgan, S., & Melvin, W. S. (2003). A prospective analysis of 211 robotic-assisted surgical procedures. *Surgical Endoscopy, 17*(10), 1521–1524. doi:10.1007/s00464-002-8853-3 PMID:12915974

Tham, S., Koh, F. H., Teo, E. K., Lin, C. L., & Foo, F. J. (2023). Knowledge, perceptions and behaviours of endoscopists towards the use of artificial intelligence-aided colonoscopy. *Surgical Endoscopy, 37*(10), 7395–7400. doi:10.1007/s00464-023-10412-3 PMID:37670191

Thanoon, M. A., Zulkifley, M. A., Mohd Zainuri, M. A. A., & Abdani, S. R. (2023). A Review of Deep Learning Techniques for Lung Cancer Screening and Diagnosis Based on CT Images. *Diagnostics (Basel), 13*(16), 2617. doi:10.3390/diagnostics13162617 PMID:37627876

The Hindu Bureau. (2023). Monitoring app and AI-based model for predictive analysis of dengue to be launched. *The Hindu.* https://www.thehindu.com/news/cities/bangalore/app-and-ai-based-model-for-predictive-analysis-of-dengue-to-be-launched/article67281225.ece

Tomczak, K., Czerwińska, P., & Wiznerowicz, M. (2015). Review The Cancer Genome Atlas (TCGA): an immeasurable source of knowledge. *Contemporary Oncology/Współczesna Onkologia*, 68-77.

Ukil, A., Bandyoapdhyay, S., Puri, C., & Pal, A. (2016, March). IoT healthcare analytics: The importance of anomaly detection. In *2016 IEEE 30th international conference on advanced information networking and applications (AINA)* (pp. 994-997). IEEE.

Vijayan, V., Connolly, J. P., Condell, J., McKelvey, N., & Gardiner, P. (2021). Review of Wearable Devices and Data Collection Considerations for Connected Health. *Sensors (Basel), 21*(16), 5589. doi:10.3390/s21165589 PMID:34451032

Vora, L. K., Gholap, A. D., Jetha, K., Thakur, R. R. S., Solanki, H. K., & Chavda, V. P. (2023). Artificial Intelligence in Pharmaceutical Technology and Drug Delivery Design. *Pharmaceutics*, *15*(7), 1916. doi:10.3390/pharmaceutics15071916 PMID:37514102

Wang, L. (2022). Deep Learning Techniques to Diagnose Lung Cancer. *Cancers (Basel)*, *14*(22), 5569. doi:10.3390/cancers14225569 PMID:36428662

Wang, M., Yang, B., Liu, Y., Yang, Y., Ji, H., & Yang, C. (2023). Emerging infectious disease surveillance using a hierarchical diagnosis model and the Knox algorithm. *Scientific Reports*, *13*(1), 19836. doi:10.1038/s41598-023-47010-1 PMID:37963966

Wang, W. H., & Hsu, W. S. (2023). Integrating artificial intelligence and wearable IoT system in long-term care environments. *Sensors (Basel)*, *23*(13), 5913. doi:10.3390/s23135913 PMID:37447763

Wang, Y., Tetko, I. V., Hall, M. A., Frank, E., Facius, A., Mayer, K. F., & Mewes, H. W. (2005). Gene selection from microarray data for cancer classification—A machine learning approach. *Computational Biology and Chemistry*, *29*(1), 37–46. doi:10.1016/j.compbiolchem.2004.11.001 PMID:15680584

Wei, W., Visweswaran, S., & Cooper, G. F. (2011). The application of naive Bayes model averaging to predict Alzheimer's disease from genome-wide data. *Journal of the American Medical Informatics Association : JAMIA*, *18*(4), 370–375. doi:10.1136/amiajnl-2011-000101 PMID:21672907

Williams, R. (2019). *European perspectives on the anticipatory governance of AI. AI Governance: A Year in Review: Observations of 50 Global Experts*. Institute for Science of Science.

Wilson, E. B., & Sudan, R. (2013). The evolution of robotic bariatric surgery. *World Journal of Surgery*, *37*(12), 2756–2760. doi:10.1007/s00268-013-2125-3 PMID:23778837

Wirtz, J., Hofmeister, J., Chew, P., & Ding, X. (2023). Digital Service Technologies, Service Robots, AI, and the Strategic Pathways to Cost-Effective Service Excellence. *Service Industries Journal*, *43*(15-16), 1173–1196. doi:10.1080/02642069.2023.2226596

Wong, J., Murray Horwitz, M., Zhou, L., & Toh, S. (2018). Using machine learning to identify health outcomes from electronic health record data. *Current Epidemiology Reports*, *5*(4), 331–342. doi:10.1007/s40471-018-0165-9 PMID:30555773

World Health Organization. (2021). *Ethics and governance of artificial intelligence for health: WHO guidance*. WHO.

Wu, Q., Zhang, Y. D., Tao, W., & Amin, M. G. (2015). Radar-based fall detection based on Doppler time–frequency signatures for assisted living. *IET Radar, Sonar & Navigation*, *9*(2), 164–172. doi:10.1049/iet-rsn.2014.0250

Yamashita, R., Nishio, M., Do, R. K. G., & Togashi, K. (2018). Convolutional neural networks: An overview and application in radiology. *Insights Into Imaging*, *9*(4), 611–629. doi:10.1007/s13244-018-0639-9 PMID:29934920

Yao, L., Zhang, H., Zhang, M., Chen, X., Zhang, J., Huang, J., & Zhang, L. (2021). Application of artificial intelligence in renal disease. *Clinical eHealth, 4*, 54-61.

Ye, D., Liu, Z., Sun, M., Shi, B., Zhao, P., Wu, H., Yu, H., Yang, S., Wu, X., Guo, Q., Chen, Q., Yin, Y., Zhang, H., Shi, T., Wang, L., Fu, Q., Yang, W., & Huang, L. (2020, April). Mastering complex control in MOBA games with deep reinforcement learning. *Proceedings of the AAAI Conference on Artificial Intelligence*, *34*(04), 6672–6679. doi:10.1609/aaai.v34i04.6144

Ye, J. (2021). Advancing mental health and psychological support for health care workers using digital technologies and platforms. *JMIR Formative Research*, *5*(6), e22075. doi:10.2196/22075 PMID:34106874

Yu, K. H., Fitzpatrick, M. R., Pappas, L., Chan, W., Kung, J., & Snyder, M. (2018). Omics AnalySIs System for PRecision Oncology (OASISPRO): A web-based omics analysis tool for clinical phenotype prediction. *Bioinformatics (Oxford, England)*, *34*(2), 319–320. doi:10.1093/bioinformatics/btx572 PMID:28968749

Yu, K. H., Levine, D. A., Zhang, H., Chan, D. W., Zhang, Z., & Snyder, M. (2016). Predicting Ovarian Cancer Patients' Clinical Response to Platinum-Based Chemotherapy by Their Tumor Proteomic Signatures. *Journal of Proteome Research*, *15*(8), 2455–2465. doi:10.1021/acs.jproteome.5b01129 PMID:27312948

Yu, K. H., & Snyder, M. (2016). Omics profiling in precision oncology. *Molecular & Cellular Proteomics*, *15*(8), 2525–2536. doi:10.1074/mcp.O116.059253 PMID:27099341

Zhang, J., & Zhang, Z. M. (2023). Ethics and governance of trustworthy medical artificial intelligence. *BMC Medical Informatics and Decision Making*, *23*(1), 7. doi:10.1186/s12911-023-02103-9 PMID:36639799

Zhao, J., & Gómez Fariñas, B. (2023). Artificial intelligence and sustainable decisions. *European Business Organization Law Review*, *24*(1), 1–39. doi:10.1007/s40804-022-00262-2

Chapter 3
Energy Gaps and Bacteriochlorophyll Molecular Graph Representation Based on Machine Learning Algorithm.

Kapil Kumar
Jawaharlal Nehru University, India

Manju Khari
Jawaharlal Nehru University, India

ABSTRACT

Light harvesting applications have proven to be highly valuable in various fields such as photovoltaics, photo catalysis, and photo polymerization. When bacteriochlorophyll is manipulated at the molecular level, it opens up numerous possibilities for improving its physical and chemical properties for harvesting light. . In their research study, the authors suggest using molecular graph representations from the Computational Material Repository (CMR) dataset to forecast the energy gaps of bacteriochlorophyll. The electro-topological-state index is the description for the molecular network that the authors choose expressly. The physical configuration of carbon in methane bonds associated to the attached group, as found by the authors, has a considerable impact on the energy gaps of bacteriochlorophyll. This observation is consistent with the three-dimensional diffusion patterns of the molecule's leading-edge molecular orbitals, as predicted by quantum mechanical calculations.

INTRODUCTION

The authors of the study demonstrated the origin of Bacteriochlorophylls (BChls). These molecules are predominantly found in bacteria capable of synthesizing Bacteriochlorophylls. The researchers(Hiroto et al.2017) found that BChls have two distinct origins: one primary source found in the chloric phylum, including its paralogue BchV, and a secondary source inherited by the enzyme found in other oxygenic

DOI: 10.4018/979-8-3693-1922-2.ch003

phototrophic bacteria. The evolutionary history of BChls predates the radiation event that led to the Chloroflexi, Chloric, Acid bacteria, Proteobacteria, and Gemmatimonadetes phyla, occurring before the divergence of Mode I and Mode II reaction centers. Therefore, the absence of light harvesting or phototropic capabilities in certain groups of bacteria can be considered a derived characteristic.

Bacteriochlorophyllides belong to a family of macrocyclic compounds. Their molecules consist of a bacteriochlorin macrocycle ring with two condensed pyrrole rings (B and D). Bacteriochlorophyllides c, d, e, and f have chlorine atoms, resulting in molecules with a chlorine macrocycle ring and one condensed pyrrole ring (D) (Biesaga et al.2000). Depending on the specific compositions, the physicochemical properties of Bacteriochlorophyllides can be influenced by substituent ligands and chelating metal ions, allowing for various combinations. Importantly, the energy gap is a critical feature for phototropic or light harvesting applications. Bacteriochlorophyllides with well-placed energy levels in their molecular orbitals show significant promise in converting solar energy into electricity, photosensing, photomedicine, photocatalysis, and more (Liao et al.2000).

Finding these resources by trial-and-error experimental approaches has traditionally been a time-consuming, expensive, and labor-intensive procedure. Density functional theory (DFT), for example and other modern advancements in computing infrastructure and electronic structure techniques have made it possible to design materials based on fundamental ideas(Imahori et al.2004). Nevertheless, because to the computational cost of quantum-chemical simulations with exact exchange-correlation properties, the broad configuration space of Bacteriochlorophyllides poses difficulties. Machine learning techniques have recently attracted more attention as a way to research physical chemistry and find new materials. Large datasets are readily available, and methods for classifying or predicting attributes of interest can be used to facilitate this.

Mathematically, in lieu of a resources or its special modelling information in relations of related fleshly limits is significant for creating of machine-learning algorithm. In the concern, numerous entries are delivered (Imahori et al.2011). For instance, the fractional behavior of phototropic applications and the moment's d-band distribution of transition materials produce based on the electronic and physical structure (ornse et al.2013). The effectively put vim blockades of the anterior orbits demonstrate huge confidence in the conversion of solar energy into electrical energy, (Mandel et al.2015) photo sensing, (Berg et al.2005) photomedicine, photo-catalysis etc. Machine learning procedures, which are supervised and rely on labelled data, produce patterns of feature representations based on radial feature distributions, in contrast to quantum-chemical techniques, which entail solving Schrödinger equations for a collection of atoms (Artrith et al.2016).The authors discovered that the Coulomb matrix (Rupp et al.2012) is a straightforward global descriptor that models the electrostatic interaction between atomic nuclei. It provides a kind of fingerprint, represented as a graph of atomic bonds, Morton control curves, mapping of electronic structure, and other techniques used to connect with material properties or output representations. Practical applications can be drawn from these findings for further individual study, rather than solely relying on machine learning experts. In their paper, the authors develop a complex and precise approach, utilizing statistical modeling through graphs to capture variations in the energy of bacteriochlorophyllides across different graphical representations of molecules. This approach helps gather comprehensive information and facilitates informed design through detailed feature analysis and the application of quantum computing methodologies in physical chemistry.

COMPUTATIONAL SECTION

The computational material repository contains a dataset comprising over 9000 molecular diagrams of Bacteriochlorophyllides, along with DFT-computed features such as frontier orbital energy gaps, optical differences, and energy variations. In their research papers, the authors will primarily focus on the assessed energy difference, specifically related to the carboxylate group serving as an anchor point for understanding the semiconducting difference between electron affinity and ionization potential, denoted as Eea – Eip. This parameter is significant due to its implications in assessing the efficiency of capturing solar light and energy conversion. The authors will explore variations in adjacent collections denoted as P1, P2, and P3, as well as the lead group P4 at mesa-steps, external substituents labeled as Z at β-steps, and chelating metal ions referred to as M, as depicted in Figure 1.

Figure 1. Structure of Bacteriochlorophyll ides

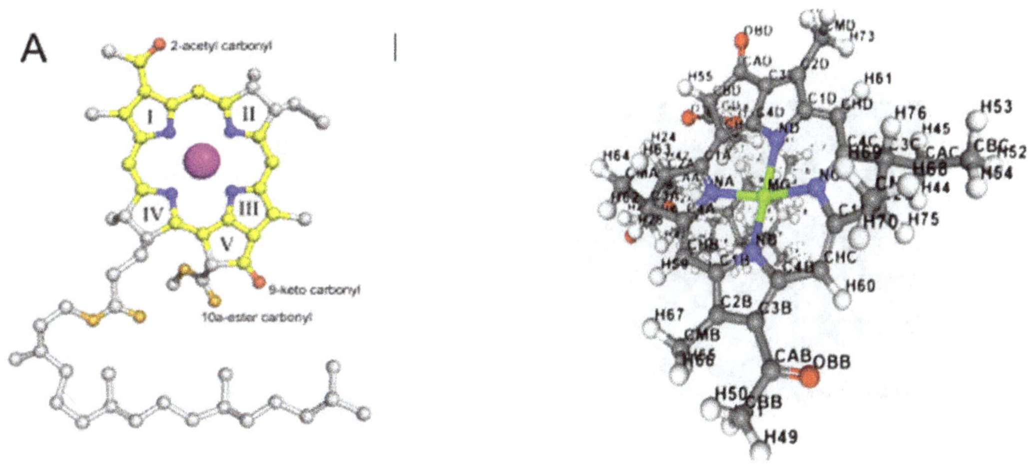

Figure 1 illustrates the structural arrangement of Bacteriochlorophyllides, including their connection to various metal chelating agents and ligands. The carboxylate group serves as the focal point for inducing semiconductor properties, such as those seen in TiO2. In the dataset of Bacteriochlorophyllides stored in the background, P1 and P3 ligands remain the same, while P4 exhibits two different positions, following the criteria of the Bacteriochlorophyllides plane. The complete set of structures used in this dataset can be accessed in the Computational Material Storage (CMS) repository. To understand the analysis of Bacteriochlorophyllides and their characteristics across the dataset, it is essential to represent them using plots, such as the violin plots shown in Figure. These plots demonstrate that the metal chelating agents had a minimal impact on the energy variation of Bacteriochlorophyllides (less than 0.5 electronvolts).

Molecular signatures, Coulomb matrix metrics, chemoinformatics, and electro-topological state indices are four different categories of molecular representations that are used for learning, as shown in Figure 2. In 1024-dimensional binary bit patterns, which each bit represents as the existence either (1) or absence (0) of a certain substructure, molecular signatures are used to specify key properties inside the molecular representation. To gather the chemical fingerprints of Bacteriochlorophyllides, the open-source Babel model is used.

Figure 2. A representation of the strength of the Coulombic interactions between atoms based on their Cartesian coordinates and nuclear characteristics is the Coulomb matrix, which shows a three-dimensional structure of molecules in a two-dimensional array. molecules diagram: (a) molecular diagram, (b) Coulomb electro topological-state index, (c) coulomb matrix (d) cheminformatics.

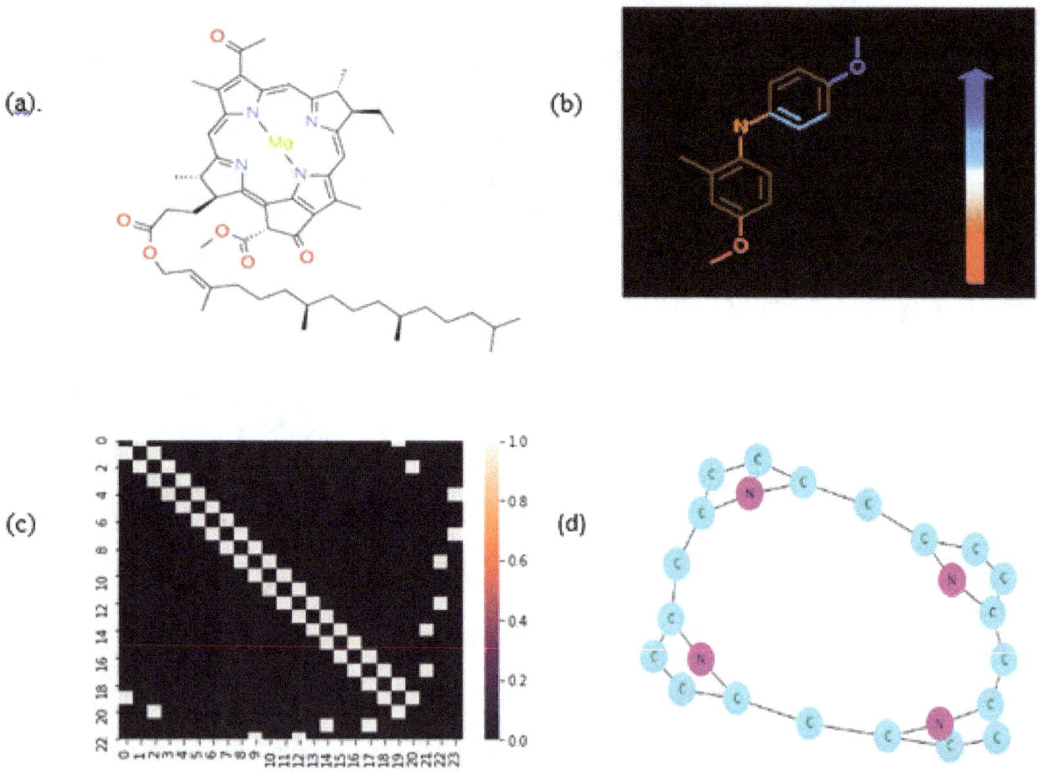

Parity plots, shown in Figure 2, are used to contrast the results of calculations utilizing molecular graph representations and Density Functional Theory (DFT) with predictions of energy differences (measured in electron volts, eV) made by a machine-learning model. The probability distribution of errors produced by bootstrapping is shown in the plot's insets. In the graphic, a shaded area denotes a 95% confidence interval, and a red vertical line represents the mean of the bootstrap errors.

The Coulomb matrix's well-organized eigenvalues play a significant role as important characteristics in a posited molecular description. These qualities, which are influenced by the chemical compound's structural, automated, and quantum-chemical aspects, encompass all of its physical properties. The authors put together a group of 13-dimensional cheminformatics descriptors that relate to various aspects of molecules using the Mordred compendium. Intricate characteristics including molecular connectivity, complexity, spin bond ratio, electrical polarizability, and more are covered by these descriptors. The electro-topological state index (E-state index) stands out among these descriptors and is frequently used in quantifiable structure-property relationship (QSPR) models for forecasting.

boiling temperature, aqueous solubilities, and other kinematics properties of organic molecules.[64–68] It takes into consideration a molecular view as a correlated detailed *field*.[69] Within a given chemical surrounding situation, an atom's internal property electrotopological- state, I_i, is calculated by studying its personal chemical bonding. An intuitive summary of I_i is the valence-state electro- negativity70 defined by eq 2:

$$I_i = [(2/n)^{2\sigma v} + 1]/\sigma \ldots\ldots\ldots\ldots 2.$$

RESULT AND DISCUSSION

Although the Coulomb graph experiences a drop in specificity despite its intrinsic complexity, the eigenvalue spectrum of a Coulomb graph representing molecules has a surprising capacity to maintain a strong model-feature link. Using an artificial neural network (ANN) model, it is possible to investigate how the E-state index can be adjusted to function with a variety of materials. The E-state index is used in this model as a useful tool for forecasting the energy variation of organic synthetic resins. These forecasts are based on information from a large computational repository of materials, where the resin molecules' size and the techniques used to make them show a great degree of flexibility in contrast to the more rigid qualities of Bacteriochlorophyllides.

Figure 3. Energy band gap

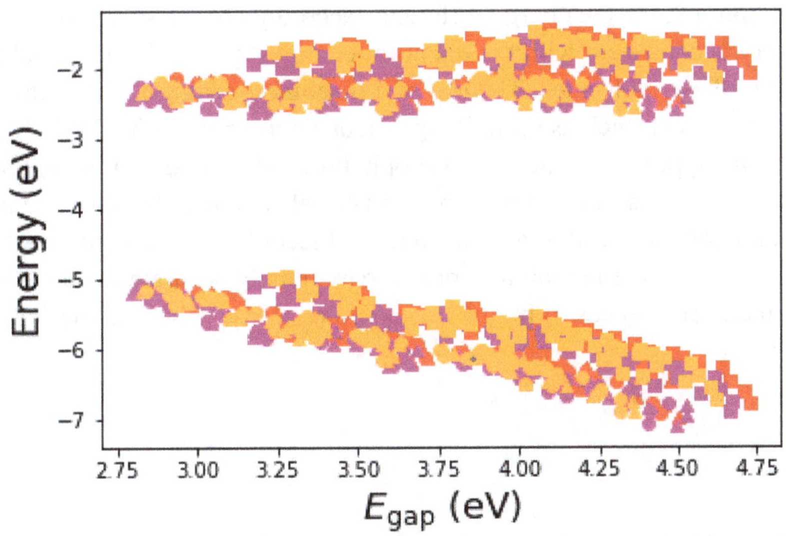

How specific design features affect the energy difference of bacteriochlorophyllides is depicted in the Sabol sensitivity index graph. Single bonds are indicated in this form by the letters "s," "d," "t,"

and aromatic bonds by the letters "a." For determining the molecular frontiers in respect to a typical Bacteriochlorophyllide, the Ethen-theophany-Ciano-acrylic configuration serves as the reference set. The authors used a library that included a principal component analysis (PCA) function for the dataset in order to model the whole range of molecules and comprehend the outstanding performance of the E-state index as a feature. PCA is a dimensional reduction approach that successfully compresses high-dimensional data into a two-dimensional subspace that is easier to read and in which differences in the dataset are highlighted.

One of the additional benefits of the E-state index is its ability to combine atomic-level information with molecular structure, relegating electrotopological properties to the role of supporting a certain fundamental design. The E-state index's noteworthy property enables it to offer useful information on the design-feature linkages of resources. In this situation between the energy difference of bacteriochlorophyllides and characteristic of graph of molecules, the machine learning approach model is equivalent to "black-box" changing of the refined inter-correlations of outcomes with description.

Quantifying the impact of a particular description on a result can be challenging. The authors of this study conducted a statistical investigation of the effects of several feature variables on the energy difference using the widely utilised sensitivity analysis based on variance technique developed by Sobol . This approach has shown to be very beneficial for machine learning engineers who want to understand complex frameworks in applications like biology systems, risk assessment, and pharmacology. Quantifying the contribution of each variable to the total variation in model outcomes is the aim of Sobol sensitivity analysis, which is fundamentally a probability-based methodology. The Sobol sensitivity analysis implementation in the SALib package uses bootstrapping and Monte Carlo to investigate the portions of outcome variation related to each individual.

The arrangement of carbon atoms within the methine bond connected to the primary cluster is identified as the main contributor to minimizing the energy difference according to the sensitivity analysis driven by variance. This discovery is consistent with the quantum-chemically calculated spatial properties of the electron orbits linked to these molecules. The use of a machine learning-based framework has the potential to have significant effects, such as optimizing silicone performance in "DSSCs" (Dye-Sensitized Solar Cells), suggesting appropriate bacteriochlorophyllides with oxidation potentials tailored to a specific I/I - redox pair, and investigating other relevant materials like metal oxides and metal-organic frameworks. Physical chemistry datasets that have been the subject of this thorough investigation using machine-learning approaches have the potential to produce extremely accurate models for elucidating well-established unexplained occurrences. It might also be especially pertinent to the fields of robotics and artificial intelligence.

CONCLUSION

Using machine learning, the authors of the research have produced very precise predictions for the energy difference between bacteriochlorophyllides and essential components for phototropic or light harvesting applications. The E-state index displays a strong, accurate estimate with RMSE 0.06 eV using a variety of ascertaining approaches, including convolution neural networks and multiple chemical representations via graph. The enhanced model based on machine learning can be useful for enhancing the performance of bacteriochlorophyllides when the representation of the molecules is mined from computer-generated bacteriochlorophyllides. Using variance-sensitive analysis, it has been determined that the arrangement of carbon atoms within the methine bond that connects the major cluster appears to be the main cause of the decrease in the energy difference. This result is consistent with the spatial properties of the borderline electron orbits that were calculated using quantum-chemical methods. The use of a machine learning framework has the potential to have a significant impact on a number of applications, such as improving silicone efficiency in "DSSCs" (Dye-Sensitized Solar Cells), suggesting appropriate bacteriochlorophyllides with oxidation potentials pertinent to a particular I/I - redox pair, and investigating various materials like metal oxides and metal-organic frameworks. This thorough investigation may result in the development of highly accurate models for explaining known intriguing phenomena, particularly with implications for artificial intelligence and robotics, by utilizing machine learning techniques on widely available datasets in the field of physical chemistry.

REFERENCES

Artrith, N., & Urban, A. (2016). An implementation of artificial neural-network potentials for atomistic materials simulations: Performance for TiO2. *Computational Materials Science*, *114*, 135–150. doi:10.1016/j.commatsci.2015.11.047

Berg, K., Selbo, P. K., Weyergang, A., Dietze, A., Prasmickaite, L., Bonsted, A., Engesaeter, B. Ø., Angell-Petersen, E., Warloe, T., Frandsen, N., & Høgset, A. (2005). Porphyrin-related photosensitizers for cancer imaging and therapeutic applications. *Journal of Microscopy*, *218*(2), 133–147. doi:10.1111/j.1365-2818.2005.01471.x PMID:15857375

Biesaga, M., Pyrzyńska, K., & Trojanowicz, M. (2000). Porphyrins in analytical chemistry. A review. *Talanta*, *51*(2), 209–224. doi:10.1016/S0039-9140(99)00291-X PMID:18967853

Hiroto, S., Miyake, Y., & Shinokubo, H. (2017). Synthesis and functionalization of porphyrins through organometallic methodologies. *Chemical Reviews*, *117*(4), 2910–3043. doi:10.1021/acs.chemrev.6b00427 PMID:27709907

Imahori, H., Kashiwagi, Y., Endo, Y., Hanada, T., Nishimura, Y., Yamazaki, I., Araki, Y., Ito, O., & Fukuzumi, S. (2004). Structure and photophysical properties of porphyrin-modified metal nanoclusters with different chain lengths. *Langmuir*, *20*(1), 73–81. doi:10.1021/la035435p PMID:15745002

Kumar, K. (2021). Comprehensive Method of Botnet Detection Using Machine Learning. [IJOSSP]. *International Journal of Open Source Software and Processes*, *12*(4), 37–61. doi:10.4018/IJOSSP.287613

Kumar, K., & Bhatnagar, V. (2021). Credit Card Fraud Transaction Detection System Using Neural Network-Based Sequence Classification Technique. [IJOSSP]. *International Journal of Open Source Software and Processes*, *12*(1), 21–40. doi:10.4018/IJOSSP.2021010102

Liao, M.-S., & Scheiner, S. (2002). Electronic structure and bonding in metal porphyrins, metal = Fe, Co, Ni, Cu, Zn. *The Journal of Chemical Physics*, *117*(1), 205–219. doi:10.1063/1.1480872

Mandal, B., Sarkar, S., & Sarkar, P. (2015). Theoretical Studies on Understanding the Feasibility of Porphyrin-Sensitized Graphene Quantum Dot Solar Cell. *The Journal of Physical Chemistry. C, Nanomaterials and Interfaces*, *119*(6), 3400–3407. doi:10.1021/jp511375a

Ornsø, K. B., Garcia-Lastra, J. M., & Thygesen, K. S. (2013). Computational screening of functionalized zinc porphyrins for dye sensitized solar cells. *Physical Chemistry Chemical Physics*, *15*(44), 19478–19486. doi:10.1039/c3cp54050b PMID:24129651

Rupp, M., Tkatchenko, A., Müller, K.-R., & von Lilienfeld, O. A. (2012). Fast and accurate modeling of molecular atomization energies with machine learning. *Physical Review Letters*, *108*(5), 058301. doi:10.1103/PhysRevLett.108.058301 PMID:22400967

Yedukondalu, M., & Ravikanth, M. (2011). Core-modified porphyrin based assemblies. *Coordination Chemistry Reviews*, *255*(5-6), 547–573. doi:10.1016/j.ccr.2010.10.039

Chapter 4
Artificial Intelligence in Protective Gear Design and Maintanence

Rupayan Roy
National Institute of Fashion Technology, Kannur, India

Animesh Laha
Welspun India Limited, India

Apromita Chakraborty
National Institute of Fashion Technology, Kannur, India

ABSTRACT

This chapter investigates the application of artificial intelligence (AI) to the upkeep of safety equipment for employees across several industries. Wear and tear, exposure to dangerous materials, and other factors can reduce the effectiveness of protective equipment, making routine maintenance essential to ensuring that it stays functional and offers sufficient protection. It is explained how AI can be used for preventative maintenance, checks and repairs, and extending the lifespan of safety equipment. AI is capable of identifying tears or other damage in personal protective equipment (PPE), guiding repairs, and extending the lifespan of equipment. Incorporating AI into protective equipment maintenance can improve safety, save time and money, and as technology advances, it has the potential to completely change how protective equipment is maintained and managed. The chapter concludes that in order to preserve worker safety and wellbeing, enterprises should integrate AI into their strategy for maintaining protective equipment.

DOI: 10.4018/979-8-3693-1922-2.ch004

INTRODUCTION

Any clothing or equipment intended to shield people from dangers that could lead to illness, injury, or even death is referred to as protective gear, commonly known as personal protective equipment (PPE). Helmets, gloves, goggles, respirators, body armour, and more sophisticated gear like chemical-resistant suits and firefighting gear are all examples of protective equipment. Protective equipment's main objective is to create a barrier between the user and any potential hazards, lowering the likelihood of disease or injury *(Karim et al., 2020; Damalas & Koutroubas, 2016; Shushila & Talekar 2023)*. It cannot be emphasised how important protective equipment is as a part of occupational safety. Workers in sectors like construction, manufacturing, and mining are exposed to dangers like falls, flying objects, and dangerous chemicals. Medical practitioners in the healthcare sector need protective equipment to safeguard both themselves and their patients from infectious infections *(Perlman et al., 2014; Harrod et al., 2020)*. To carry out their responsibilities safely, emergency responders including firefighters, police officers, and paramedics need specialised protective gear. In the military and in space exploration, where exposure to harsh environments poses serious hazards to human health, protective gear is also required. In addition to safeguarding the wearer, protective equipment lowers the possibility of workplace accidents and injuries, which can increase costs for companies in the form of lost productivity, workers' compensation claims, and legal expenses. The Occupational Safety and Health Administration (OSHA) in the United States and the Directive on Personal Protective Equipment of the European Union are two of the most notable regulatory bodies that recognise the value of protective equipment in preventing occupational injuries and illnesses *(Czarnecki et al., 2024; Kendal, 2020; Hymel et al., 2011)*. AI has the ability to completely transform the development, testing, upkeep, and production of protective equipment. Artificial intelligence (AI) is the emulation of human intellect in computers that are capable of carrying out tasks that traditionally call for human intelligence, such as problem-solving, learning, and judgement *(Li et al., 2017)*. Numerous sectors, including healthcare, transportation, and finance, have already used AI. By providing real-time monitoring, predictive maintenance, and better material design, AI in the field of protective gear can enhance PPE's efficacy *(Dopico et al., 2016; pan & Zhang 2021)*.

The ability of AI to enhance protective gear design is one of the most important advantages of this technology. Large datasets can be analysed using AI to find patterns that would be hard for humans to see *(Hyysalo et al., 2022)*. Artificial intelligence (AI) can pinpoint locations where protective equipment may be insufficient or uncomfortable by studying data on worker posture and movement, for instance. AI may be utilised to create materials for safety gear that are more efficient. AI can assist in determining the optimum materials for particular applications by examining the chemical and physical features of various materials *(García-Herrero et al., 2012; Maleki et al. 2022)*.

AI can be quite useful for evaluating the efficacy of safety equipment. In the past, testing of protective equipment has relied on human observation and opinion. AI can offer a more accurate, quantitative evaluation of PPE effectiveness. For instance, AI can be used to track physiological information like heart rate and body temperature to assess how well cooling systems in safety equipment worn by workers in hot conditions function *(Abduallah et al., 2022)*. AI can be quite useful for evaluating the efficacy of safety equipment. In the past, testing of protective equipment has relied on human observation and opinion. AI can offer a more accurate, quantitative evaluation of PPE effectiveness. For instance, AI can be used to track physiological information like heart rate and body temperature to assess how well cooling systems in safety equipment worn by workers in hot conditions function *(Bernal et al., 2017)*.

Real-time monitoring of protective gear using AI can also provide workers with immediate feedback on the effectiveness of their protective gear. For example, sensors embedded in protective gear can detect damage or wear and tear, alerting workers to the need for replacement. This can improve worker safety by ensuring that protective gear is always in optimal condition *(Nayak et al., 2015)*.

Predictive maintenance is another area where AI can be used to improve the effectiveness of protective gear. Predictive maintenance involves using data and analytics to predict when maintenance is required before equipment fails *(Wang et al., 2017)*. This can help reduce downtime and prevent accidents caused by equipment failure. In the context of protective gear, AI can be used to predict when PPE will need repair or replacement based on data from sensors embedded in the gear. This can include monitoring wear and tear on the gear, as well as monitoring the performance of any electronic components. For example, AI can be used to detect when the battery in a firefighter's breathing apparatus is running low and needs to be replaced. By predicting maintenance needs in advance, AI can help ensure that protective gear is always in optimal condition, reducing the risk of equipment failure and increasing the safety of workers *(Giordano & McClean 1975; Campero-Jurado et al., 2020)*.

In conclusion, protective equipment is crucial to safeguarding the security of individuals and workers across a range of industries. Designing, testing, maintaining, and producing protective equipment with AI has the potential to increase PPE's effectiveness, providing employees with better protection and lowering costs for employers. We can create more sophisticated and useful safety equipment that can shield workers from even the most dangerous situations by utilising the power of AI. As AI technology develops, it's probable that we'll see even more cutting-edge uses of AI in the realm of protective gear, significantly enhancing worker and individual safety in a variety of industries.

AI IN PROTECTIVE GEAR DESIGN

Protective equipment is an essential part of safeguarding worker safety across many industries. Traditional protective gear designs, however, have frequently been constrained by things like price, comfort, and efficacy. The use of artificial intelligence (AI) to improve the design of protective gear has attracted a lot of attention in recent years *(Bello et al., 2015)*. This chapter will look at how AI is being used to create safer, more effective protective gear.

Designing More Effective Protective Gear Using AI

Using AI to design protective equipment has many benefits, one of which is the capacity to produce designs that are more effective and efficient. AI can be used to assist designers in identifying the most important protection-required regions and optimising the design accordingly *(Abiyo et al., 2021)*. AI, for instance, can be used to create equipment that offers superior defence against particular dangers like heat, radiation, or chemical exposure.

AI can also be used to evaluate design efficacy by analysing data gathered from sensors built into the equipment. This includes information on how effectively a piece of safety equipment lowers exposure to a certain hazard. The design can then be improved upon using this information to refine it *(Zhang et al., 2022; Pan & Zhang, 2021)*.

Advancements in Material Science and AI

Recent advances in material science have led to the creation of new materials that can provide better protection while remaining cosy and light. Designers can produce cutting-edge protective gear that can adjust to various environments and hazards by fusing AI with material science.

AI can be used to optimise how materials are used in safety gear. Designers can choose the best materials for certain protective gear applications by examining data on the physical attributes of various materials and their performance under various conditions. This could contribute to the development of more robust and efficient protective gear that can endure adverse weather conditions and offer superior protection from a variety of threats *(Flouris & Cheung, 2006)*.

Improving the Ergonomics of Protective Gear

Enhancing ergonomics is another way AI can be utilised to build better protective equipment. Comfortable, simple-to-wear protective gear can increase worker compliance and lower the risk of accidents. To discover places where equipment might be upgraded to give better comfort and support, AI can be used to evaluate data on worker motions and postures *(Hanson, 2013)*.

AI can be used to build equipment that is better suited to each wearer specifically. AI can design clothing that fits and feels better by studying data about a worker's size and shape. This can increase worker productivity and lower the risk of accidents brought on by improperly fitting equipment.

Role of AI in Testing the Effectiveness of Protective Gear

AI can be used to test the efficacy of pre-existing gear in addition to inventing protective gear. Real-time monitoring of the protective equipment is one of the major benefits of employing AI in testing. This involves keeping track of how the equipment is deteriorating and looking for any damage that can reduce its efficiency. AI can be used to create safety standards for protective gear. By analysing data on the effectiveness of different gear designs and materials, AI can help develop safety standards that can be used to ensure the effectiveness of protective gear in various industries.

AI-based Predictive Maintenance of Protective Gear Utilizing data and analytics, predictive maintenance foresees when repair is necessary before equipment malfunctions. Based on information from sensors incorporated into the protective gear, AI can be used to forecast when personal protective equipment (PPE) will require repair or replacement. By doing this, you can cut downtime and avoid mishaps brought on by broken equipment. Additionally, preventive maintenance may make sure that workers are using safety equipment that is in good condition and offers the necessary level of protection *(Mobley, 2002)*.

Using AI for Inspections and Repairs of Protective Gear

Automation of protective equipment inspections and repairs is another application for AI. AI-assisted inspections and repairs can help minimise human mistake and guarantee that equipment is properly inspected and repaired. Inspections and repairs are frequently done manually using traditional methods, which can be time-consuming and error-prone. These procedures can be automated by AI-based systems, increasing their accuracy and efficiency *(Vrana & Singh, 2021)*.

Data from sensors built into the equipment can be analysed by inspection systems driven by AI to find any flaws or damage. For example, sensors can identify cracks or other structural deterioration to the gear. The extent of the damage may then be assessed using AI algorithms, which can also produce repair recommendations. AI-powered devices can also automate repairs *(Filz et al., 2021)*. For instance, the replacement parts for protective gear can be made using 3D printing technology. The design and composition of the gear can be analysed by AI algorithms to produce precise replacement parts.

AI-driven systems can also monitor the effectiveness of protective gear over time. The system can forecast when maintenance or replacement of the gear will be required by analysing data on the wear and tear of the equipment. This can lessen the chance of worker damage and assist minimise equipment failure.

Improving the Lifespan of Protective Gear with AI

Extending the lifespan of protective gear is another area where AI can be utilised to make improvements. Traditional protective equipment maintenance techniques use a schedule-based maintenance approach, in which equipment is maintained and replaced in accordance with a predetermined schedule. However, this strategy may result in premature gear replacement, wasting resources and raising expenses. By foreseeing when maintenance or replacement is required, AI can be utilised to optimise the maintenance of protective equipment *(Prabhakaran et al., 2022)*. In order to decide when maintenance or replacement is necessary, AI algorithms can assess data on the usage, environmental factors, and wear and tear of the equipment. Protective equipment can be kept in use for a longer period of time by applying AI to optimise maintenance, which lowers costs and saves resources. Additionally, by ensuring that equipment is changed only when necessary and not before, this can increase worker safety *(Lee et al., 2020)*.

In conclusion, the application of AI to the upkeep of protective equipment has enormous potential to increase worker safety and save expenses. AI-powered solutions can increase the precision and effectiveness of inspections and repairs by automating these operations. Additionally, AI can help protective equipment last longer by optimising maintenance schedules, lowering the need for premature replacement and saving resources. An important step toward assuring the safety and wellbeing of workers in diverse industries is the incorporation of AI in protective gear maintenance.

AI IN PROTECTIVE GEAR TESTING

Workers' safety in a variety of industries is mostly dependent on protective equipment. Traditional testing techniques, on the other hand, frequently fall short in their ability to evaluate protective equipment's efficacy with accuracy. Artificial intelligence (AI) has given rise to new opportunities for evaluating the efficacy of protective gear *(Tong et al., 2019)*. This chapter will examine how artificial intelligence (AI) is being used to improve the testing of protective equipment and the potential advantages it could have for a number of different industries.

Role of AI in Testing the Effectiveness of Protective Gear

AI has the potential to completely change how safety equipment is evaluated for performance. Analyzing the information obtained from the sensors built into the equipment is one method AI can accomplish this. The information can reveal how successfully the equipment shields users from certain dangers like heat, radiation, or chemical exposure *(Chung et al., 2023)*. AI can gain important knowledge about the effectiveness of the equipment and identify areas for improvement by evaluating this data.

AI can also evaluate the efficacy of safety equipment by simulating various scenarios. For instance, AI can analyse how successfully the equipment is safeguarding workers in the event of an explosion or chemical leak by simulating the events. By doing so, it will be easier to spot equipment flaws and make the required adjustments to maintain worker safety.

Real-Time Monitoring of Protective Gear Using AI

To make sure that protective equipment is operating properly and offering workers enough protection, real-time monitoring is essential. Real-time performance monitoring of protective equipment is possible with the help of AI, which may also inform staff members and managers about any problems or concerns. AI, for instance, may keep track of the temperature within a piece of safety equipment and notify workers if it becomes too high. By doing this, heat stroke or heat exhaustion may be avoided. AI can also keep an eye on the equipment's integrity, spotting any rips or other flaws that might reduce its effectiveness. Real-time monitoring can assist in spotting possible risks before they develop into major issues, lowering the likelihood of accidents or fatalities.

Additionally, AI can offer proactive maintenance for safety equipment. AI is able to foresee when the equipment will need maintenance or repair by evaluating the data from sensors. By doing so, equipment failure can be avoided and workers' safety can always be guaranteed. By enabling repairs to be planned in advance of a failure, predictive maintenance can help save time and money.

Developing Safety Standards with AI

AI can also be used to develop safety standards for protective gear. By analyzing data on accidents and injuries, AI can identify patterns and trends that can help inform the development of safety standards. This includes identifying the types of hazards that are most common in various industries and the types of protective gear that are most effective at protecting workers against those hazards.

AI can also analyze data on the performance of different types of protective gear to identify areas where improvements can be made. For example, if workers consistently report that a particular type of gear is uncomfortable or too heavy, AI can analyze the data and suggest design improvements to make the gear more comfortable and less burdensome *(Ngarambe et al., 2020)*. This can help to develop safety standards that are based on empirical evidence rather than assumptions.

Moreover, AI can be used to develop customized protective gear based on individual worker needs. By analyzing data on a worker's job tasks, physical attributes, and potential hazards, AI can recommend protective gear that provides optimal protection while also being comfortable and easy to wear. This can help to increase worker satisfaction and reduce injuries.

Challenges of Implementing AI in Testing Protective Gear

While the use of AI in testing protective gear offers significant potential, there are also some challenges that must be overcome. One of the biggest challenges is the lack of data. Many industries may not have enough data on accidents, injuries, and hazards to create effective safety standards or train AI models

(Shneiderman, 2020). To overcome this challenge, industries must collect more data and make it accessible to AI models.

Another challenge is the lack of standardization in protective. There are a wide variety of protective gear options available, each with their own unique design, materials, and intended use *(Shigley et al., 2004).* This can make it difficult to develop standardized testing methods that can be applied across different types of protective gear. Additionally, different industries may have different requirements for protective gear, which can further complicate the development of standardized testing methods *(Kopetz & Steiner, 2022).* However, AI can play a role in addressing this challenge by helping to develop more comprehensive and standardized testing methods. By analyzing data from various types of protective gear, AI can identify commonalities and differences between them. This can help to identify areas where standardization can be improved and can inform the development of more comprehensive testing methods that can be applied across different types of protective gear.

Additionally, AI can be used to analyze data on accidents and injuries in different industries to identify the types of hazards that are most common and the types of protective gear that are most effective at protecting workers against those hazards. This can inform the development of industry-specific standards for protective gear and can help to ensure that workers in different industries are adequately protected *(Van Cott & Knkade, 1972).*

Real-World Examples of AI in Testing Protective Gear

There are already several real-world examples of AI being used to test the effectiveness of protective gear. For example, researchers at the University of Alabama have developed a system that uses AI to assess the effectiveness of firefighting gear. The system uses sensors embedded in the gear to collect data on temperature, humidity, and other environmental factors. This data is then analyzed using AI algorithms to assess the effectiveness of the gear in different firefighting scenarios *(Rodger, 2012).*

Another example comes from the mining industry, where researchers at the University of Arizona have developed an AI-powered system for monitoring the effectiveness of protective gear worn by miners. The system uses sensors to collect data on the movements of miners, as well as environmental factors such as temperature and humidity *(Mardonova & Choi, 2018).* This data is then analyzed using AI algorithms to assess the effectiveness of the protective gear and to identify any potential hazards.

In the construction industry, researchers at the University of Waterloo have developed an AI-powered system for testing the effectiveness of hard hats. The system uses a robotic arm to drop hard hats onto a metal plate from a height of several meters. The impact is measured using sensors, and the data is then analyzed using AI algorithms to assess the effectiveness of the hard hats in protecting workers against head injuries.

Challenges and Limitations of AI in Testing Protective Gear

While there are many potential benefits to using AI in testing protective gear, there are also several challenges and limitations to consider. One challenge is the cost of implementing AI systems for testing protective gear. The development and implementation of AI-powered systems can be expensive, and smaller companies or organizations may not have the resources to invest in these systems *(Panda et al., 2019).* Another challenge is the need for specialized knowledge and expertise to develop and implement AI-powered systems for testing protective gear. This requires a team of experts with knowledge in both

AI and the specific industry in question *(Freire et al., 2022).* This can be a barrier for smaller organizations or those without access to specialized expertise.

Furthermore, there is a risk of overreliance on AI in testing protective gear. While AI can provide valuable insights into the effectiveness of protective gear, it should not be relied on as the sole method of testing. Human oversight and testing are still essential to ensure that protective gear is functioning correctly and providing adequate protection to workers.

The use of AI in testing protective gear has the potential to revolutionize the way we ensure the safety of workers in various industries. By analyzing data from sensors embedded in protective gear and simulating different scenarios, AI can provide valuable insights into the effectiveness of protective gear *(Lim et al., 2020).* Real-time monitoring using AI can help to identify potential hazards before they become significant problems, reducing the risk of injury or death. Additionally, AI can be used to develop

AI IN PROTECTIVE GEAR MAINTENANCE

Personal Protective Equipment (PPE) is essential in ensuring the safety of workers in various industries. However, the effectiveness of protective gear can be compromised by wear and tear, exposure to hazardous materials, and other factors. To ensure that PPE remains functional and provides adequate protection, proper maintenance is critical. In recent years, the use of artificial intelligence (AI) in protective gear maintenance has offered significant potential for enhancing the lifespan and effectiveness of protective gear. In this chapter, we will explore the various ways in which AI is being used for predictive maintenance, inspections and repairs, and improving the lifespan of protective gear.

Predictive Maintenance of Protective Gear using AI

Predictive maintenance involves using data and analytics to predict when maintenance is required before equipment fails. In the context of protective gear, AI can be used to predict when Personal Protective Equipment (PPE) will need repair or replacement based on data from sensors embedded in the gear *(Saidiet al., 2021).* By analyzing data on the temperature, humidity, and exposure to hazardous materials that a piece of protective gear has experienced, AI can predict when the gear will need repairs or replacement, allowing for proactive maintenance rather than reactive maintenance. Predictive maintenance can help identify potential issues with protective gear before they become significant problems, reducing downtime and improving safety *(Fioravanti et al., 2020).*

Using AI for Inspections and Repairs of Protective Gear

AI can also be used for inspections and repairs of protective gear. For example, AI can analyze images of protective gear, detecting any tears or damage that could compromise its effectiveness *(Comiter, 2019).* This allows for quick identification of any issues with the gear and prompt repairs to ensure that it remains functional. Additionally, AI can provide guidance on the types of repairs that are required and the best methods for carrying out those repairs. This can help to ensure that repairs are carried out efficiently and effectively, reducing downtime and improving the lifespan of the gear *(Lai et al., 2019).*

Improving the Lifespan of Protective Gear with AI

The lifespan of protective gear can be improved by using AI to monitor and maintain the gear. AI can analyze data on the usage of protective gear, identifying patterns and trends that can help to optimize the lifespan of the gear *(Baduge et al., 2022)*. This can include factors such as frequency of use, environmental conditions, and exposure to hazardous materials. Based on this data, AI can recommend maintenance schedules for protective gear, ensuring that maintenance is carried out at the right time, reducing downtime and improving the lifespan of the gear. By optimizing the lifespan of protective gear, workers can be assured that their PPE is functional and provides adequate protection.

The Benefits of AI in Protective Gear Maintenance

The integration of AI in protective gear maintenance offers several benefits. First, AI can help to reduce downtime by identifying potential issues with protective gear before they become significant problems. This proactive approach to maintenance can save time and money by reducing the need for reactive maintenance. Second, AI can improve safety by ensuring that PPE remains functional and provides adequate protection. Third, AI can optimize the lifespan of protective gear, ensuring that workers are using PPE that is functional and effective *(Okpalaet al., 2020)*. Finally, AI can improve the efficiency of maintenance by providing guidance on the types of repairs required and the best methods for carrying out those repairs.

Challenges in Implementing AI in Protective Gear Maintenance

While the benefits of AI in protective gear maintenance are significant, there are also challenges to implementing this technology. First, the cost of implementing AI can be high, especially for small businesses or those with limited budgets. Second, there may be resistance from workers who are unfamiliar with the technology or who are concerned about privacy and data security. Third, the technology is constantly evolving, which can make it difficult for businesses to keep up with the latest advancements. Finally, there may be legal and regulatory challenges associated with implementing AI in certain industries or jurisdictions *(Baduge et al., 2022)*.

The use of AI in protective gear maintenance offers significant potential for enhancing the lifespan and effectiveness of protective gear, which is crucial in ensuring the safety of workers in various industries. Protective gear plays a vital role in protecting workers from various occupational hazards such as chemical exposure, electrical shock, heat stress, and physical trauma. However, the effectiveness of protective gear can be compromised by wear and tear, exposure to hazardous materials, and other factors. Proper maintenance of protective gear is, therefore, critical to ensuring that it remains functional and provides adequate protection. The integration of AI in protective gear maintenance is a significant step forward in ensuring the safety and well-being of workers. Predictive maintenance using AI can help to identify potential issues with protective gear before they become significant problems, reducing downtime and improving safety *(Okpalaet al., 2020)*. AI can analyze data on the temperature, humidity, and exposure to hazardous materials that a piece of protective gear has experienced. Based on this data, AI can predict when the gear will need repairs or replacement, allowing for proactive maintenance rather than reactive maintenance.

AI IN EMERGENCY RESPONSE GEAR

Enhancing emergency response gear with AI

The need to lower the number of First Responders (FRs) fatalities and injuries persists despite all efforts, such as the introduction of new training techniques and personal protection equipment. According to reports, technological advancements have not yet made it possible to effectively safeguard FRs from harm, health effects, and odourless hazardous vapours. There are now emerging technologies that can be used to better protect FRs in emergency management contexts. The adoption of AI could be really important, and it is currently one of the most recent trends (*Radianti et al., 2019*). Acute natural disasters, however, whether they be biological, geophysical, coastal, or related to hydrography or airspace, have the power to cause significant damage to both the natural world and even further away. Events of this nature have a disproportionately harmful impact on particular locations and the people who live there. The development of AI has made it feasible to predict natural disasters and estimate their size (*Samuel, 2023*).

Machine learning (ML) has become widely used in the science and engineering of fire in recent years. To address the shortcomings of traditional approaches (such as physics-based or mechanistic) currently used in fire science, a wide range of ML and AI techniques, such as deep learning and decision trees, have been proposed (*Tian et al., 2023; Jain et al., 2020*). These developments have sparked research initiatives to use AI and ML to create and improve contemporary fire assessment tools. For fuel characterization, fire detection, and fire readiness and response, researchers have applied AI & ML techniques (*Jain et al., 2020*). Researchers discovered that implementing and enabling ML models within fire management agencies for practical, operational usage is a significant problem encountered by the community of fire research and management. Artificial neural networks have recently been utilised to develop tools to help in firefighting that forecast environmental temperature changes and alert firefighters to dangerous temperature spikes or abrupt flame spread (*Mozaffari et al., 2023, Garrity & Yusuf, 2021*). In a real-world fire testing setting, these technologies proved successful in warning firefighters of an oncoming flashover by sounding an alert. When used in a physical product or piece of software, ML algorithms will significantly increase the safety of first responders and promote the uptake of similar technologies (*Garrity & Yusuf, 2021*).

Since hazardous materials (HAZMAT) are combustible, explosive, poisonous, corrosive, or radioactive, transportation incidents involving HAZMAT pose serious risks to persons, infrastructure, the community, and the environment. According to the US Department of Transportation's Hazardous Materials Incident Statistic Reports, the frequency of HAZMAT incidents has also grown recently (*US Department of Transportation, Pipeline and Hazardous Materials Safety Administration, 2023*). Studies have been started to develop solutions using cutting-edge digital technologies like ML and AI that aim to improve HAZMAT risk assessment, routing, network design, infrastructure resilience, and emergency response and management in order to reduce the increased distinct hazards of HAZMAT transportation accidents (*Tian et al., 2023*; *Prasenjit, 2021*).

Adopting ML and AI technologies has the potential to be beneficial, particularly in the areas of first responder support and HAZMAT emergency response and management (*Tian et al., 2023*). An AI-enabled smartphone application was created by researchers in this field (*Tian et al. 2023*), capable of voice interaction with users to answer their questions regarding dangerous compounds in a context that makes sense to them. Furthermore, the programme can help with image detection of the HAZMAT placards on freight trucks. Despite the fact that these initiatives resulted in a new capability to support first responders and lessen their stress, the main capability gaps are in raising situational awareness.

AI IN PROTECTIVE GEAR MANUFACTURING

Role of AI in Improving Manufacturing Processes of Protective Gear

Population growth has greatly boosted demand for all goods, particularly protective equipment. It can be difficult to manufacture a product without flaws while using the machine to its greatest capacity. As a result, researchers are developing a variety of manufacturing processes. Intelligent manufacturing is a new manufacturing model and the technical means by which new information and communication technology, intelligent science and technology, large manufacturing technology (including design, production, management, testing, and integration), system engineering technology, and related product technology are integrated with the overall system and lifecycle of product development. In order to integrate and optimise various aspects of a manufacturing enterprise or group, including three elements (people/organizations, operational management, and equipment and technology), five flows, and autonomous sensing, interconnection, collaboration, learning, analysis, cognition, decision-making, control, and execution of human, machine, material, and environmental information are used (information flow, logistics flow, capital flow, knowledge flow, and service flow). This helps with production, offers customers a high-efficiency, high-quality, cost-effective service, and is environmentally friendly. As a result, the manufacturing firm or group is more competitive in the market. In the field of intelligent manufacturing, AI technology makes it easier to build new models, means, and forms as well as system architecture and technological systems *(Li et at., 2014; Pan, 2016; Li et al., 2017).*

Using AI for Quality Control in Protective Gear Manufacturing

Machine learning (ML) and artificial intelligence (AI) can increase the productivity, sustainability, and efficiency of manufacturing. However, implementing AI in manufacturing also comes with a number of difficulties, such as problems with data administration and acquisition, infrastructure, infrastructure concerns, human resources, security risks, and implementation issues. Obtaining the data required to train AI models, for instance, might be costly for big datasets that require labelling or challenging for infrequent events. Security problems might arise when AI models are included into industrial control systems. Additionally, due to a lack of trust or comprehension of how AI functions, certain industry actors can be reluctant to adopt it. Despite these difficulties, applications like predictive maintenance, quality control, and process optimization show great promise for AI in the industrial sector. When determining whether and how to apply AI in manufacturing, it is crucial to take into account the particular requirements and capabilities of each production situation. The objective of this review is to increase understanding of AI/ML technologies available for solving manufacturing problems, provide decision-support for prioritising and selecting appropriate AI/ML technologies, and identify areas where additional research can produce transformational returns for the industry. This review identifies current developments, challenges, and future directions in AI/ML relevant to manufacturing. Early evidence indicates that AI and machine learning can significantly reduce costs and increase productivity in the manufacturing sector, especially when combined with the capacity to collect massive volumes of data from production systems *(Plathottam et al., 2023; Geissbauer et al., 2020; Fahle et al., 2020; Brosset et al., 2019; Abràmoff et al., 2018).*

CHALLENGES AND FUTURE DIRECTIONS

Challenges in Implementing AI in Protective Gear

Despite the advantages of AI implementation that have been previously outlined, there are numerous obstacles to overcome. The great majority of research conducted to date have been retrospective, using historically labelled data to develop and test algorithms. We won't start to comprehend the true value of AI systems until we do prospective research since performance is likely to suffer when dealing with real-world data that is different from the data used to train algorithms. However, there have only been a small number of prospective investigations for various illness categories (*Liu et al., 2019; Lindsey at al., 2018; Long et al., 2017)*. Many works have been published in the machine learning community only on preprint sites and not in peer-reviewed journals. For the wider medical community to accept and trust AI, peer-reviewed research will be crucial. There are very few randomised controlled trials (RCTs) of AI systems to date. These include an algorithm to detect childhood cataracts that performed well in a small prospective study but performed less accurately than senior clinicians in a diagnostic RCT; and a single-blind RCT that demonstrated a significantly lower blind-spot rate in esophagogastroduodenoscopy.

Additionally, AI algorithms may have a number of drawbacks, such as bias, brittleness, and inapplicability outside of the training area (tendency to be easily fooled). Dataset shift, accidentally fitting confounders rather than true signal, spreading unintentional biases in clinical practise, providing algorithms with interpretability, creating reliable measures of model confidence, and the difficulty of generalisation to various populations are important factors to take into account *(Lin et al., 2019; Wu et al., 2019; Marcus, 2018)*. Figure 1 presents typical flow diagram for production process related to protective gear manufacturing and related solutions of different problems.

Figure 1. Flow chart to identify challenges and implementation of AI in protective gear manufacturing

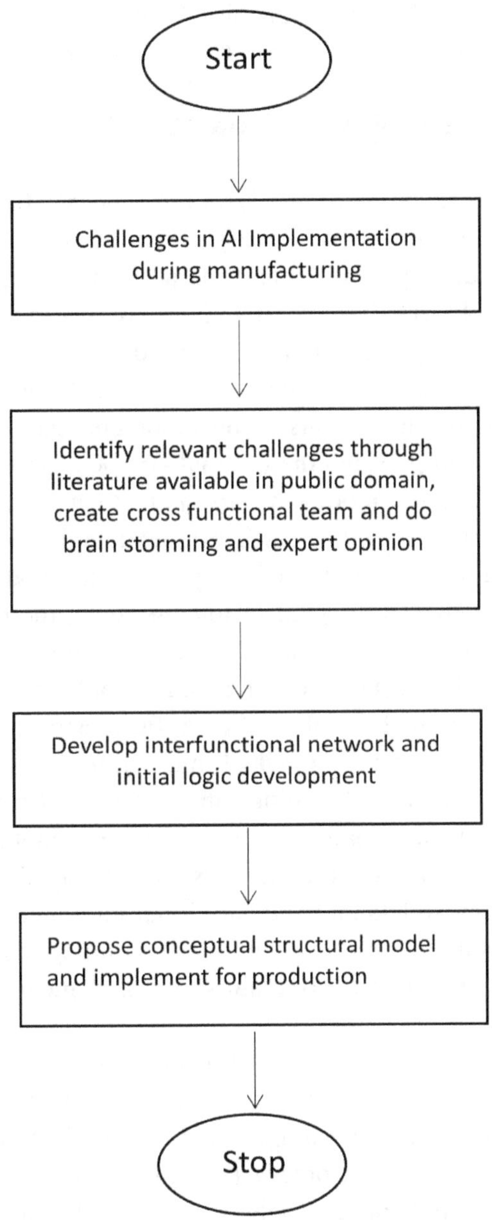

Another problem seen is dataset shift. It is simple to overlook the fact that all input data are generated in a non-stationary environment with moving patient populations, where clinical and operational procedures develop over time (*Lin et al., 2019)*, which is particularly crucial for EHR algorithms. A different distribution from the one used to train the algorithm may emerge as a result of the adoption of a new predictive algorithm in practise. Methods to spot drift and update models as a result of declining performance are crucial. Measures to mitigate this effect include the anticipated need for periodic retraining as well as the thorough assessment of performance over time to proactively identify issues. In order to sustain performance over time, data-driven testing methodologies have been proposed to select the best suitable updating method, from straightforward recalibration to comprehensive model retraining (*Wu et al. 2019*). Other barriers include algorithmic bias, human resistance to AI adoption, and an infant but quickly developing field of algorithmic interpretability (*Kelly et al., 2019*).

Future Directions in the Field of AI in Protective Gear

Above all, AI is vital to our protection, particularly in the field of clinical care, where it can swiftly recognise the patient's CT and X-ray images to identify the patient's kind of pneumonia. To gather clinical information about patients, identify their clinical characteristics, and forecast their present severity level in order to alert the medical staff. The implementation of artificial intelligence algorithms in the field of medicine, according to this study, still faces several difficulties (*Mohamadou et al., 2020*).

For instance, the issue of data imbalance is the primary difficulty in COVID-19 detection. The lack of COVID-19 patient lung imaging data prevents the development model from being tested and assessed on a large number of data sets, and prevents the optimal Artificial Intelligence algorithm from being chosen. For researchers to train and test models, we must create an open and shareable data collection (*Islam & Islam, 2020*). Second, there is still a dearth of label data, necessitating either the extension of current data sets or the use of a limited number of samples during model training at this time. Due of the time and money required for manual tagging of imaging data, the majority of current models are weakly supervised techniques. To handle imaging data in the future, transfer learning techniques and unsupervised deep learning models may be required. It can not only guarantee the algorithm's accuracy but also remove the restriction imposed by labelled data. Furthermore, because artificial intelligence is viewed as a "black box," the diagnosis of medical imaging utilising artificial intelligence requires enough evidence to support its accuracy. Therefore, in the realm of medicine, the interpretability of artificial intelligence models is crucial. Finally, medical imaging cannot accurately depict if COVID-19 is indeed infected or not, hence a multimodal model must be developed. Artificial intelligence can gain knowledge from multimodal clinical data and add additional intelligence to medical systems to better identify illness characteristics and produce accurate COVID-19 diagnosis findings. To create a system that is more versatile and effective in order to attain superior clinical medical goals (*Rahman et al., 2021; Alamo at al., 2020*).

Artificial intelligence also has a significant impact on how medicines, vaccines, treatments, and risk assessment by medical professionals are developed. We need to investigate more uses of artificial intelligence and modelling for COVID-19 in the healthcare industry in the future work in addition to moving in the direction of being more intelligent and precise (*Rahman et al., 2021; Ricci et al., 2021; Bhargava & Bansal 2021*). The COVID-19 epidemic can only be eliminated in this manner. In order to establish the appropriate prevention and control measures, we can also employ artificial intelligence technologies to generate logical forecasts about the COVID-19's future development trend. All models are incorrect, but some models are useful, according to British statistician George E.P. Box. The COVID-19 epidemic's forecasting models are similar to previous ones in that straightforward models of the disease's early phases can still be used as a basis for more intricate transmission models. The prediction model, which can only give us an early warning because it is impossible to determine which model best predicts the spread of COVID-19, should thus be used with caution (*Ulhaq et al., 2020*).

According to this study, the primary trend for future development in terms of infectious disease dynamics models is the merging of mathematical models and machine learning, which may both raise the adaptability of the model and boost the scientific rationality of simulation prediction. Second, in order for the model to more accurately reflect the spread of the epidemic in reality, it is essential to completely understand the key elements influencing the development trend of the epidemic, such as population mobility, seasonal factors, isolation control efforts, etc. *(Wang et al., 2022 and Ahmad et al., 2021)*. Finally, we will discuss multimodal integration. Future studies should include information from additional sources, including mass media and social media.

A number of groups have created guidelines for the appropriate and ethical use of AI in society in an effort to uphold human dignity while reducing potential risk *(Floridi et al., 2018 European Commission, 2020; On artificial intelligence and Azoulay, 2021)*. Such advice can assist OSH experts and practitioners in collaborating with data scientists and computer programmers to create AI systems that are appropriate for the workplace and that are efficient, explicable, accountable, secure, and fair.

Effectiveness: Make sure AI is the appropriate instrument to address the issue. Technology should not be used carelessly; instead, it should be used to increase production or improve working conditions. Even while a specific AI system's incorrect use might not immediately result in damage, it might eventually affect other AI-based systems' ability to be trusted.

Explainable: Stakeholders should be informed clearly and effectively of the reasoning behind and decisions made by AI. This is crucial for risk reduction and evaluating the impact of unanticipated, potentially damaging, consequences.

Accountable: The consequences of the AI systems that organisations and individuals design and use should be held to account. Accountability fosters a connected awareness of the systems built and their possible effects on others in data scientists and computer programmers. Additionally, if unplanned or safety incidents happen, the relevant organisation or people can take note of them and make improvements.

Secure: AI systems need to be secure against outside interference. Although programmers are usually conversant with cybersecurity, a crucial safety factor should be the potential consequences if the system is breached or the data becomes corrupt. The system's code and data access should be known and based on an adequate risk-benefit analysis. These analyses should be taken into account or led by OSH practitioners and professionals.

Fair: AI systems ought to be conscious of potential bias and discrimination and respond to it accordingly. Systems that are trained using a certain population segment may be biassed and generate findings that are different for a different population segment. Systems that protect against this premise can be evaluated and tested to assure safer results and increase worker acceptance.

Numerous potential applications of AI in the workplace are anticipated to shape the nature of employment in the future. The potential risks to worker health are likely to increase along with technological advancements, despite the fact that the advantages are anticipated to be enormous. NIOSH will continue to investigate how AI might benefit the OSH community, advancing knowledge of the causes of how AI may have a negative impact on health, and also enhancing the practical application of worker safety and health risk management in a world that is becoming more and more AI-populated (*The role of artificial intelligence in the future work, 2021)*.

CONCLUSION

The manufacturing sector has an exceptional chance to change thanks to the quick development of AI and machine learning technology. This review detailed the potential of AI/ML to increase the security, effectiveness, productivity, and sustainability of manufacturing protective gear across a wide range of manufacturing applications. It looked at operations, planning, quality assurance, energy consumption forecasts, process optimization, security and safety, product design, automation, and human-machine interaction as applications, possible benefits, and problems of integrating AI/ML in the manufacturing pipeline. In light of these findings, the review highlighted AI/ML technologies available for resolving manufacturing issues and identified areas where additional research can produce transformative results for the industry producing proactive gear. It also identified nascent developments, current challenges, and future directions in AI/ML relevant to manufacturing.

Industrial sensors create a lot of data, and AI/ML can use that data to generate meaningful insights and autonomously execute the best possible actions. As the data, infrastructure, and algorithms are improved over time, AI/ML models can offer the manufacturing sector compounding benefits over the next ten years. The AI/ML solutions also necessitate a deep awareness of the potential trade-offs (such as facility reconfiguration, energy expenses, and expertise) as well as the unique requirements and capacities of the business and stakeholders. According to the trends, AI/ML will continue to be used in conjunction with human expertise while at the same time continuously increasing the level of automation. As long as AI/ML algorithms and techniques continue to show advances in safety, product quality, and operational efficiency, they will continue to be adopted in manufacturing industry applications. The confidence in the effectiveness and productivity potential of this technology will increase across industrial sectors and among practitioners as AI/ML is adopted by industry at a faster rate. The hazards involved will, however, limit the rate of adoption, particularly if the application shifts from analytical support to AI control of industrial operations. Decisions about how to deploy AI/ML must take into account the particular circumstances and requirements of each company. In the future, this field might benefit from longer-term case studies comparing manufacturers who have implemented AI/ML with those who have not.

REFERENCES

Abdullah, M. I., Raya, L., Norazman, M. A. A., & Suprihadi, U. (2022, July). Covid-19 patient health monitoring system using IoT. In *2022 IEEE 13th Control and System Graduate Research Colloquium (ICSGRC)* (pp. 155-158). IEEE. 10.1109/ICSGRC55096.2022.9845162

Abioye, S. O., Oyedele, L. O., Akanbi, L., Ajayi, A., Delgado, J. M. D., Bilal, M., & Ahmed, A. (2021). Artificial intelligence in the construction industry: A review of present status, opportunities and future challenges. *Journal of Building Engineering*, *44*, 103299. doi:10.1016/j.jobe.2021.103299

Abràmoff, M. D., Lavin, P. T., Birch, M., Shah, N., & Folk, J. C. (2018). Pivotal trial of an autonomous AI-based diagnostic system for detection of diabetic retinopathy in primary care offices. *NPJ Digital Medicine*, *1*(1), 39. doi:10.1038/s41746-018-0040-6 PMID:31304320

Ahmad, A., Garhwal, S., Ray, S. K., Kumar, G., Malebary, S. J., & Barukab, O. M. (2021). The number of confirmed cases of COVID-19 by using machine learning: Methods and challenges. *Archives of Computational Methods in Engineering*, *28*(4), 2645–2653. doi:10.1007/s11831-020-09472-8 PMID:32837183

Alamo, T., Reina, D. G., Mammarella, M., & Abella, A. (2020). Covid-19: Open-data resources for monitoring, modeling, and forecasting the epidemic. *Electronics (Basel)*, *9*(5), 827. doi:10.3390/electronics9050827

Azoulay, A. (2021). *Towards an ethics of artificial intelligence*. United Nations. [Report]

Baduge, S. K., Thilakarathna, S., Perera, J. S., Arashpour, M., Sharafi, P., Teodosio, B., Shringi, A., & Mendis, P. (2022). Artificial intelligence and smart vision for building and construction 4.0: Machine and deep learning methods and applications. *Automation in Construction, 141*, 104440. doi:10.1016/j.autcon.2022.104440

Bello, O., Holzmann, J., Yaqoob, T., & Teodoriu, C. (2015). Application of artificial intelligence methods in drilling system design and operations: A review of the state of the art. *Journal of Artificial Intelligence and Soft Computing Research*, *5*(2), 121–139. doi:10.1515/jaiscr-2015-0024

Bernal, G., Colombo, S., Al Ai Baky, M., & Casalegno, F. (2017, June). Safety++ designing IoT and wearable systems for industrial safety through a user centered design approach. In *Proceedings of the 10th international conference on pervasive technologies related to assistive environments* (pp. 163-170). IEEE. 10.1145/3056540.3056557

Bhargava, A., & Bansal, A. (2021). Novel coronavirus (COVID-19) diagnosis using computer vision and artificial intelligence techniques: A review. *Multimedia Tools and Applications*, *80*(13), 19931–19946. doi:10.1007/s11042-021-10714-5 PMID:33686333

Brosset, P., Jain, A., Khemka, Y., Buvat, J., Thieullent, A., Khadikar, A., & Patsko, S. (2019). Scaling AI in manufacturing operations: A practitioners' perspective. Capgemini Research Institute.

Campero-Jurado, I., Márquez-Sánchez, S., Quintanar-Gómez, J., Rodríguez, S., & Corchado, J. M. (2020). Smart helmet 5.0 for industrial internet of things using artificial intelligence. *Sensors (Basel)*, *20*(21), 6241. doi:10.3390/s20216241 PMID:33139608

Centers for Disease Control and Prevention (CDC). (2021). *The Future of Work: Artificial Intelligence.* CDC. https://blogs.cdc.gov/niosh-science-blog/2021/05/24/ai-future-of-work/

Chung, W. W. S., Tariq, S., Mohandes, S. R., & Zayed, T. (2023). IoT-based application for construction site safety monitoring. *International Journal of Construction Management*, *23*(1), 58–74. doi:10.1080/15623599.2020.1847405

Comiter, M. (2019). Attacking artificial intelligence. *Belfer Center Paper*, *8*, 2019–08.

Czarnecki, F., Maguire, B. J., Harrell, M., Samo, D., McKinney, Z. J., Guidotti, T. L., & McLellan, R. K. (2024). Worker Health and Safety in Disaster Response. In *Ciottone's Disaster Medicine* (pp. 206–214). Elsevier. doi:10.1016/B978-0-323-80932-0.00033-1

Damalas, C. A., & Koutroubas, S. D. (2016). Farmers' exposure to pesticides: Toxicity types and ways of prevention. *Toxics*, *4*(1), 1. doi:10.3390/toxics4010001 PMID:29051407

Dopico, M., Gómez, A., De la Fuente, D., García, N., Rosillo, R., & Puche, J. (2016). A vision of industry 4.0 from an artificial intelligence point of view. In *Proceedings on the international conference on artificial intelligence (ICAI)* (p. 407). The Steering Committee of The World Congress in Computer Science, Computer Engineering and Applied Computing (WorldComp).

European Commission. (2020). On artificial intelligence – a European approach to excellence and trust. EC.

Fahle, S., Prinz, C., & Kuhlenkötter, B. (2020). Systematic review on machine learning (ML) methods for manufacturing processes–Identifying artificial intelligence (AI) methods for field application. *Procedia CIRP*, *93*, 413–418. doi:10.1016/j.procir.2020.04.109

Filz, M. A., Langner, J. E. B., Herrmann, C., & Thiede, S. (2021). Data-driven failure mode and effect analysis (FMEA) to enhance maintenance planning. *Computers in Industry*, *129*, 103451. doi:10.1016/j.compind.2021.103451

Fioravanti, R., Kumar, K., Nakata, S., Chalamala, B., & Preger, Y. (2020). Predictive-maintenance practices: For operational safety of battery energy storage systems. *IEEE Power & Energy Magazine*, *18*(6), 86–97. doi:10.1109/MPE.2020.3014542

Floridi, L., Cowls, J., Beltrametti, M., Chatila, R., Chazerand, P., Dignum, V., Luetge, C., Madelin, R., Pagallo, U., Rossi, F., Schafer, B., Valcke, P., & Vayena, E. (2018). Ai4people—an ethical framework for a good ai society: Opportunities, risks, principles, and recommendations. *Minds and Machines*, *28*(4), 689–707. doi:10.1007/s11023-018-9482-5 PMID:30930541

Flouris, A. D., & Cheung, S. S. (2006). Design and control optimization of microclimate liquid cooling systems underneath protective clothing. *Annals of Biomedical Engineering*, *34*(3), 359–372. doi:10.1007/s10439-005-9061-9 PMID:16463083

Freire, S. K., Panicker, S. S., Ruiz-Arenas, S., Rusák, Z., & Niforatos, E. (2022). A Cognitive Assistant for Operators: AI-Powered Knowledge Sharing on Complex Systems. *IEEE Pervasive Computing*, *22*(1), 50–58. doi:10.1109/MPRV.2022.3218600

García-Herrero, S., Mariscal, M. A., García-Rodríguez, J., & Ritzel, D. O. (2012). Working conditions, psychological/physical symptoms and occupational accidents. Bayesian network models. *Safety Science*, *50*(9), 1760–1774. doi:10.1016/j.ssci.2012.04.005

Garrity, D. J., & Yusuf, S. A. (2021). A predictive decision-aid device to warn firefighters of catastrophic temperature increases using an AI-based time-series algorithm. *Safety Science*, *138*, 105237. doi:10.1016/j.ssci.2021.105237

Geissbauer, R., Schrauf, S., Berttram, P., & Cheraghi, F. (2017). Digital Factories 2020: Shaping the future of manufacturing.

Giordano, T. J., & McClean, M. (1975, October). Development of a Voice Projection System for Masked Firefighters. In *Audio Engineering Society Convention 52*. Audio Engineering Society.

Hanson, M. A. (2013). Green ergonomics: Challenges and opportunities. *Ergonomics*, *56*(3), 399–408. doi:10.1080/00140139.2012.751457 PMID:23384154

Harrod, M., Weston, L. E., Gregory, L., Petersen, L., Mayer, J., Drews, F. A., & Krein, S. L. (2020). A qualitative study of factors affecting personal protective equipment use among health care personnel. *American Journal of Infection Control*, *48*(4), 410–415. doi:10.1016/j.ajic.2019.08.031 PMID:31610895

Hymel, P. A., Loeppke, R. R., Baase, C. M., Burton, W. N., Hartenbaum, N. P., Hudson, T. W., McLellan, R. K., Mueller, K. L., Roberts, M. A., Yarborough, C. M., Konicki, D. L., & Larson, P. W. (2011). Workplace health protection and promotion: A new pathway for a healthier—and safer—workforce. *Journal of Occupational and Environmental Medicine*, *53*(6), 695–702. doi:10.1097/JOM.0b013e31822005d0 PMID:21654443

Hyysalo, J., Dasanayake, S., Hannu, J., Schuss, C., Rajanen, M., Leppänen, T., Doermann, D., & Sauvola, J. (2022). Smart mask–Wearable IoT solution for improved protection and personal health. *Internet of Things : Engineering Cyber Physical Human Systems*, *18*, 100511. doi:10.1016/j.iot.2022.100511 PMID:37521492

Islam, M. N., & Islam, A. N. (2020). A systematic review of the digital interventions for fighting COVID-19: The Bangladesh perspective. *IEEE Access : Practical Innovations, Open Solutions*, *8*, 114078–114087. doi:10.1109/ACCESS.2020.3002445 PMID:34192108

Jain, P., Coogan, S. C., Subramanian, S. G., Crowley, M., Taylor, S., & Flannigan, M. D. (2020). A review of machine learning applications in wildfire science and management. *Environmental Reviews*, *28*(4), 478–505. doi:10.1139/er-2020-0019

Karim, N., Afroj, S., Lloyd, K., Oaten, L. C., Andreeva, D. V., Carr, C., Farmery, A. D., Kim, I.-D., & Novoselov, K. S. (2020). Sustainable personal protective clothing for healthcare applications: A review. *ACS Nano*, *14*(10), 12313–12340. doi:10.1021/acsnano.0c05537 PMID:32866368

Kelly, C. J., Karthikesalingam, A., Suleyman, M., Corrado, G., & King, D. (2019). Key challenges for delivering clinical impact with artificial intelligence. *BMC Medicine*, *17*(1), 1–9. doi:10.1186/s12916-019-1426-2 PMID:31665002

Kendal, E. (2020). Biological modification as prophylaxis: How extreme environments challenge the treatment/enhancement divide. *Human enhancements for space missions: Lunar, Martian, and future missions to the outer planets*, 35-46.

Kopetz, H., & Steiner, W. (2022). Real-Time Communication. In *Real-time systems: Design principles for distributed embedded applications* (pp. 177–200). Springer International Publishing. doi:10.1007/978-3-031-11992-7_7

Lai, C. T. A., Jiang, W., & Jackson, P. R. (2019). Internet of Things enabling condition-based maintenance in elevators service. *Journal of Quality in Maintenance Engineering*, *25*(4), 563–588. doi:10.1108/JQME-06-2018-0049

Lee, J., Ni, J., Singh, J., Jiang, B., Azamfar, M., & Feng, J. (2020). Intelligent maintenance systems and predictive manufacturing. *Journal of Manufacturing Science and Engineering*, *142*(11), 110805. doi:10.1115/1.4047856

Li, B. H., Hou, B. C., Yu, W. T., Lu, X. B., & Yang, C. W. (2017). Applications of artificial intelligence in intelligent manufacturing: A review. *Frontiers of Information Technology & Electronic Engineering*, *18*(1), 86–96. doi:10.1631/FITEE.1601885

Li, B. H., Hou, B. C., Yu, W. T., Lu, X. B., & Yang, C. W. (2017). Applications of artificial intelligence in intelligent manufacturing: A review. *Frontiers of Information Technology & Electronic Engineering*, *18*(1), 86–96. doi:10.1631/FITEE.1601885

Li, B. H., Zhang, L., & Chai, X. (2014). Smart Cloud Manufacturing (Cloud Manufacturing 2.0)-A New Paradigm and Approach of Smart Manufacturing. In ISPE CE 2014 Sep 23 (p. 26).

Lim, K. Y. H., Zheng, P., Chen, C. H., & Huang, L. (2020). A digital twin-enhanced system for engineering product family design and optimization. *Journal of Manufacturing Systems*, *57*, 82–93. doi:10.1016/j.jmsy.2020.08.011

Lin, H., Li, R., Liu, Z., Chen, J., Yang, Y., Chen, H., & Wu, X. (2019). Diagnostic efficacy and therapeutic decision-making capacity of an artificial intelligence platform for childhood cataracts in eye clinics: A multicentre randomized controlled trial. *EClinicalMedicine*, *9*, 52–59. doi:10.1016/j.eclinm.2019.03.001 PMID:31143882

Lindsey, R., Daluiski, A., Chopra, S., Lachapelle, A., Mozer, M., Sicular, S., Hanel, D., Gardner, M., Gupta, A., Hotchkiss, R., & Potter, H. (2018). Deep neural network improves fracture detection by clinicians. *Proceedings of the National Academy of Sciences of the United States of America*, *115*(45), 11591–11596. doi:10.1073/pnas.1806905115 PMID:30348771

Liu, Y., Kohlberger, T., Norouzi, M., Dahl, G. E., Smith, J. L., Mohtashamian, A., Olson, N., Peng, L. H., Hipp, J. D., & Stumpe, M. C. (2019). Artificial intelligence–based breast cancer nodal metastasis detection: Insights into the black box for pathologists. *Archives of Pathology & Laboratory Medicine*, *143*(7), 859–868. doi:10.5858/arpa.2018-0147-OA PMID:30295070

Long, E., Lin, H., Liu, Z., Wu, X., Wang, L., Jiang, J., & Li, J. (2017). An artificial intelligence platform for the multihospital collaborative management of congenital cataracts. *Nature Biomedical Engineering*, *1*(2), 0024.

Maleki, R., Asadnia, M., & Razmjou, A. (2022). Artificial Intelligence-Based Material Discovery for Clean Energy Future. *Advanced Intelligent Systems*, *4*(10), 2200073. doi:10.1002/aisy.202200073

MarcusG. (2018). Deep learning: A critical appraisal. arXiv. https://arxiv.org/abs/1801.00631

Mardonova, M., & Choi, Y. (2018). Review of wearable device technology and its applications to the mining industry. *Energies*, *11*(3), 547. doi:10.3390/en11030547

Mobley, R. K. (2002). *An introduction to predictive maintenance*. Elsevier.

Mohamadou, Y., Halidou, A., & Kapen, P. T. (2020). A review of mathematical modeling, artificial intelligence, and datasets used in the study, prediction, and management of COVID-19. *Applied Intelligence*, *50*(11), 3913–3925. doi:10.1007/s10489-020-01770-9 PMID:34764546

Mozaffari, M. H., Li, Y., & Ko, Y. (2023). Real-time detection and forecast of flashovers by the visual room fire features using deep convolutional neural networks. *Journal of Building Engineering*, *64*, 105674. doi:10.1016/j.jobe.2022.105674

Nayak, R., Wang, L., & Padhye, R. (2015). Electronic textiles for military personnel. In *Electronic Textiles* (pp. 239–256). Woodhead Publishing. doi:10.1016/B978-0-08-100201-8.00012-6

Ngarambe, J., Yun, G. Y., & Santamouris, M. (2020). The use of artificial intelligence (AI) methods in the prediction of thermal comfort in buildings: Energy implications of AI-based thermal comfort controls. *Energy and Building*, *211*, 109807. doi:10.1016/j.enbuild.2020.109807

Okpala, I., Nnaji, C., & Karakhan, A. A. (2020). Utilizing emerging technologies for construction safety risk mitigation. *Practice Periodical on Structural Design and Construction*, *25*(2), 04020002. doi:10.1061/(ASCE)SC.1943-5576.0000468

Pan, Y. (2016). Heading toward artificial intelligence 2.0. *Engineering (Beijing)*, *2*(4), 409–413. doi:10.1016/J.ENG.2016.04.018

Pan, Y., & Zhang, L. (2021). Roles of artificial intelligence in construction engineering and management: A critical review and future trends. *Automation in Construction*, *122*, 103517. doi:10.1016/j.autcon.2020.103517

Pan, Y., & Zhang, L. (2021). Roles of artificial intelligence in construction engineering and management: A critical review and future trends. *Automation in Construction*, *122*, 103517. doi:10.1016/j.autcon.2020.103517

Panda, G., Upadhyay, A. K., & Khandelwal, K. (2019). Artificial intelligence: A strategic disruption in public relations. *Journal of Creative Communications*, *14*(3), 196–213. doi:10.1177/0973258619866585

Perlman, A., Sacks, R., & Barak, R. (2014). Hazard recognition and risk perception in construction. *Safety Science*, *64*, 22–31. doi:10.1016/j.ssci.2013.11.019

Plathottam, S. J., Rzonca, A., Lakhnori, R., & Iloeje, C. O. (2023). A review of artificial intelligence applications in manufacturing operations. *Journal of Advanced Manufacturing and Processing*, *5*(3), 10159. doi:10.1002/amp2.10159

Prabhakaran, P., Subbaiyan, A., Gopalakrishnan, D., Ramkumar, S., Veluswamy, S., Murugesan, D. K., ... Bhaskaran, P. (2022). Maintenance Methodologies Embraced for Railroad Systems: A Review. *Advances in Materials Science and Engineering*. Advance online publication. doi:10.1155/2022/7655245

Prasenjit, C. (2021). Model for selecting a route for the transport of hazardous materials using a fuzzy logic system. *Vojnotehnički glasnik, 69*(2), 355-390.

Radianti, J., Dokas, I., Boersma, K., Noori, N. S., Belbachir, N., & Stieglitz, S. (2019). *Enhancing disaster response for hazardous materials using emerging technologies: The role of AI and a research agenda*. In Engineering Applications of Neural Networks: 20th International Conference, EANN 2019, (pp. 368-376). Springer International Publishing.

Rahman, S., Sarker, S., Miraj, M. A., Nihal, R. A., Nadimul Haque, A. K., & Noman, A. A. (2021). Deep learning–driven automated detection of Covid-19 from radiography images: A comparative analysis. *Cognitive Computation*, 1–30. doi:10.1007/s12559-020-09779-5 PMID:33680209

Ricci, L., Maesa, D. D., Favenza, A., & Ferro, E. (2021). Blockchains for covid-19 contact tracing and vaccine support: A systematic review. *IEEE Access : Practical Innovations, Open Solutions*, *9*, 37936–37950. doi:10.1109/ACCESS.2021.3063152 PMID:34812382

Rodger, J. A. (2012). Toward reducing failure risk in an integrated vehicle health maintenance system: A fuzzy multi-sensor data fusion Kalman filter approach for IVHMS. *Expert Systems with Applications, 39*(10), 9821–9836. doi:10.1016/j.eswa.2012.02.171

Saidi, A., Gauvin, C., Ladhari, S., & Nguyen-Tri, P. (2021). Advanced functional materials for intelligent thermoregulation in personal protective equipment. *Polymers, 13*(21), 3711. doi:10.3390/polym13213711 PMID:34771268

Samuel, S. (2023). Artificial intelligence in disaster prediction and risk reduction. *Applied Intelligence in Human-Computer Interaction*, 83-95.

Shigley, J. E., Mischke, C. R., & Brown, T. H. Jr. (2004). *Standard handbook of machine design.* McGraw-Hill Education.

Shneiderman, B. (2020). Bridging the gap between ethics and practice: Guidelines for reliable, safe, and trustworthy human-centered AI systems. [TiiS]. *ACM Transactions on Interactive Intelligent Systems, 10*(4), 1–31. doi:10.1145/3419764

Sushila, S. D., & Talekar, S. C. (2023). Recent Developments of Protective Textiles 6. in India. *Innovation and Recent Advancement in Fashion and Textile*, 63.

Tian, S., Zhang, Y., Feng, Y., Elsagan, N., Ko, Y., Mozaffari, M. H., Xi, D. D., & Lee, C. G. (2023). Time series classification, augmentation and artificial-intelligence-enabled software for emergency response in freight transportation fires. *Expert Systems with Applications, 233*, 120914. doi:10.1016/j.eswa.2023.120914

Tong, W., Hussain, A., Bo, W. X., & Maharjan, S. (2019). Artificial intelligence for vehicle-to-everything: A survey. *IEEE Access : Practical Innovations, Open Solutions, 7*, 10823–10843. doi:10.1109/ACCESS.2019.2891073

Ulhaq, A., Born, J., Khan, A., Gomes, D. P., Chakraborty, S., & Paul, M. (2020). COVID-19 control by computer vision approaches: A survey. *IEEE Access : Practical Innovations, Open Solutions, 8*, 179437–179456. doi:10.1109/ACCESS.2020.3027685 PMID:34812357

Van Cott, H. P., & Kinkade, R. G. (Eds.). (1972). *Human engineering guide to equipment design.* Department of Defense.

Vrana, J., & Singh, R. (2021). NDE 4.0—A design thinking perspective. *Journal of Nondestructive Evaluation, 40*(1), 8. doi:10.1007/s10921-020-00735-9 PMID:33424070

Wang, H., Jia, S., Li, Z., Duan, Y., Tao, G., & Zhao, Z. (2022). A Comprehensive Review of Artificial Intelligence in Prevention and Treatment of COVID-19 Pandemic. *Frontiers in Genetics, 13*, 845305. doi:10.3389/fgene.2022.845305 PMID:35559010

Wang, J., Li, C., Han, S., Sarkar, S., & Zhou, X. (2017). Predictive maintenance based on event-log analysis: A case study. *IBM Journal of Research and Development, 61*(1), 11–121. doi:10.1147/JRD.2017.2648298

Wu, L., Zhang, J., Zhou, W., An, P., Shen, L., Liu, J., Jiang, X., Huang, X., Mu, G., Wan, X., Lv, X., Gao, J., Cui, N., Hu, S., Chen, Y., Hu, X., Li, J., Chen, D., Gong, D., & Yu, H. G. (2019). Randomised controlled trial of WISENSE, a real-time quality improving system for monitoring blind spots during esophagogastroduodenoscopy. *Gut*, *68*(12), 2161–2169. Advance online publication. doi:10.1136/gutjnl-2018-317366 PMID:30858305

Zhang, Q., Zhou, C., Xiong, N., Qin, Y., Li, X., & Huang, S. (2015). Multimodel-based incident prediction and risk assessment in dynamic cybersecurity protection for industrial control systems. *IEEE Transactions on Systems, Man, and Cybernetics. Systems*, *46*(10), 1429–1444. doi:10.1109/TSMC.2015.2503399

Zhang, W., Yang, D., & Wang, H. (2019). Data-driven methods for predictive maintenance of industrial equipment: A survey. *IEEE Systems Journal*, *13*(3), 2213–2227. doi:10.1109/JSYST.2019.2905565

Zhang, Z., Wen, F., Sun, Z., Guo, X., He, T., & Lee, C. (2022). Artificial intelligence-enabled sensing technologies in the 5G/internet of things era: From virtual reality/augmented reality to the digital twin. *Advanced Intelligent Systems*, *4*(7), 2100228. doi:10.1002/aisy.202100228

Chapter 5
COVID-19 Pandemic Personal Safety Is Revolutionized by the Smart Mask

N. Raghu
https://orcid.org/0000-0002-2091-8922
Jain University, India

G. Swetha
https://orcid.org/0000-0001-8620-7149
R.R. Institute of Technology, India

Druva
Dayananda Sagar College of Engineering, India

Shruthishree
https://orcid.org/0000-0002-2906-9090
Jain University, India

Rekha
https://orcid.org/0000-0002-6789-9677
R.R. Institute of Technology, India

Niranjan Kannanugo
https://orcid.org/0009-0002-7710-8565
Powerschool, India

Hemanth Kumar
https://orcid.org/0000-0001-7379-2561
Malnad College of Engineering, India

ABSTRACT

In light of the massive use of face masks during the COVID-19 pandemic, it is clear that these protective gears are crucial in reducing the spread of respiratory viruses. Our research unveils the Smart Mask, a cutting-edge platform built on the Internet of Things (IoT) that aims to proactively regulate and prevent the spread of illnesses such as COVID-19. The Smart Mask creates an all-encompassing environment for monitoring and analysing data in real-time about health by using sophisticated sensor technologies, smart algorithms, wireless connection, and state-of-the-art materials. The Smart Mask gathers and analyses data pertaining to the wearer's and their immediate surroundings' health via the seamless integration of Internet of Things capabilities. Accurate diagnoses, individualised treatment programmes, and health trajectory prediction are all made possible by processing this data using artificial intelligence. By combining individual data with massive public databases.

DOI: 10.4018/979-8-3693-1922-2.ch005

INTRODUCTION

In a world where infectious illnesses have altered everything, finding new ways to keep people safe and improve their quality of life is now more important than ever. Presenting our revolutionary prototype—the most intelligent mask ever made—is something we are really proud of. With this state-of-the-art technology, we can meet the changing needs of our post-pandemic society while simultaneously guaranteeing the greatest degree of user safety. The overarching goal of this project is to help people live healthier lives despite the constant dangers posed by illnesses and germs by changing people's perceptions of and experiences with face masks (Hyysalo, Dasanayake, Hannu, Schuss, Rajanen, Leppänen, Doermann, & Sauvola, 2021; Lazaro, Lázaro, Villarino et al, 2021). The importance of masks in preventing the spread of the COVID-19 pandemic has been highlighted by the worldwide outbreak. Reducing the transmission of harmful germs and safeguarding vulnerable people, mask-wearing has had a tremendous influence on public health via its broad acceptance as a preventative strategy as shown in fig 1. Although mask use has spiked due to the epidemic, their importance will surely remain long after the crisis is passed. People are taking more precautions to avoid spreading infectious illnesses in public places as our knowledge of these diseases and how they spread grows (Panarello et al., 2018; Pandya et al., 2021).

Figure 1. The illustration in which epidemics force everyone to wear masks.

A lot of hard work in the fields of engineering, research, and creativity went into making our intelligent mask. It combines cutting-edge technology to provide an unrivalled degree of protection and convenience, going above and beyond the typical functions of a face mask. The mask actively seeks for dangerous microbes and removes them from the air using cutting-edge sensors and materials, guaranteeing that the user is breathing in pure air. The mask also has the ability to track the wearer's vitals in real time, so they can know how the air is, how hot it is, and anything else that might affect their health (Lee et al., 2022; Ye, Ling, Yang, Xu, Zhu, Yan, & Chen, 2022; Zhao et al., 2022). Also, our smart mask is designed with the user in mind, making sure it's easy to wear and comfortable to use. Its painstaking design ensures a snug fit for a wide range of face sizes and shapes, reducing irritation and increasing comfort while wearing it. The modern, comfortable form of the mask makes it easy to breathe in and out, and it fits snugly enough that wearers may go about their day without any discomfort.

The Study Aims

The aim of this research chapter is to develop a technologically advanced smart mask that provides comprehensive protection against diseases and pathogens while incorporating features that enhance user safety, comfort, and long-term usability. By leveraging cutting-edge materials, filters, and integrated sensors, this smart mask aims to revolutionize the field of personal protective equipment (PPE) by offering real-time data on the mask's condition, durability, and effectiveness. Furthermore, the project aims to create an affordable solution that can be widely accessible, particularly in resource-constrained settings, without compromising on performance or quality.

The Study Objectives

1. Develop a robust and durable smart mask: The project will focus on designing a smart mask that can withstand harsh conditions during decontamination processes and maintain its functionality even after prolonged storage. Emphasis will be placed on selecting materials and components that ensure durability, flexibility, and resistance to wear and tear.
2. Enhance user safety and comfort: The smart mask will prioritize user comfort by integrating ergonomic design principles, adjustable fit options, and breathable materials to facilitate extended wear without compromising protection. The aim is to create a mask that users can wear for extended periods without discomfort or impairment.
3. Implement advanced filtration and protection technologies: The project will explore innovative filtration techniques, such as high-efficiency particulate air (HEPA) filters and antimicrobial coatings, to maximize the mask's ability to filter out pathogens and provide optimal protection against airborne diseases.
4. Integrate sensor technology for real-time monitoring: By incorporating sensors into the smart mask, the project aims to provide users with real-time information on the mask's condition, including filter efficiency, breathability, and any potential wear or damage. This data will enable users to make informed decisions about the mask's usage and replacement.
5. Ensure affordability and accessibility: A key objective of the project is to develop a smart mask that is cost-effective and accessible to a wide range of users, including individuals in rural or economically disadvantaged areas. This will involve optimizing production processes, exploring cost-efficient materials, and considering scalability for mass production.
6. Conduct rigorous testing and validation: The smart mask will undergo rigorous testing to ensure its compliance with relevant safety and performance standards. Through extensive trials and evaluations, the project aims to validate the mask's effectiveness, durability, and user satisfaction.

By achieving these objectives, the project endeavors to make significant strides in the field of personal protective equipment, providing individuals with an innovative and reliable tool to protect against diseases and pathogens, while promoting public health and safety.

LITERATURE REVIEW

Lazaro et al. (2021) have stated that many problems have been brought to light by the COVID-19 epidemic. Wearing face masks is one of the precautions used to stop the spread of the illness. The effectiveness and accuracy of temperature controls installed at public building entrances to deter the admission of virus carriers have been shown to be questionable. An intelligent mask that monitors vitals like respiration rate and core temperature is introduced in this research. An insulating substance separates the four sensors that make up a non-invasive dual-heat-flux device, which measures body temperature. A thermistor placed close to the nose measures the temperature changes inside the mask, which are used to calculate the breathing rate. Reduced average power consumption is the result of the system's communication via long-range (LoRa) backscattering. Its purpose is to determine the smart mask's relative position by comparing signals received by two LoRa receivers placed inside and outside an entrance door. To gather data and send it to a server, the prototype uses inexpensive LoRa transceivers that can connect to WiFi. A thermistor placed under the armpit consistently yields accurate readings of core body temperature. By analysing the recorded temperatures and breathing rate readings, the system verifies that the mask is properly placed. In addition, if the thermistor temperature suddenly drops during a cough episode, it will be able to identify it.

Li et al. (2023) have illustrated that wearing a mask is a basic way to shield yourself from harmful airborne particles and other environmental hazards. Masks may detect, reflect, or react to environmental factors in addition to filtering infections and dust particles; they are an essential piece of personal protective equipment (PPE) for protecting the respiratory system. The potential of this intelligence in areas like as illness detection, health monitoring, and care has piqued the attention of both academics and industry. From basic structural designs to advanced functional modules that enable the mask to receive and transmit physiological or environmental data, we cover it all in this overview of present-day air filtration technologies utilised in masks. We spent some time talking about how masks can now detect macroscopic physiological cues from the user and how they may be used to identify diseases like COVID-19. In addition, we set out the groundwork for smart masks of the future and the criteria for material choice and functional design that will allow masks to communicate and perform vital functions in health-care devices.

Ivanoska-Dacikj and Stachewicz (2020) have demonstrated that a new global epidemic of COVID-19 has broken out recently, threatening not only our freedom but also our healthcare system, economy, social life, and culture. So far, this pandemic has shown us that, despite the amazing breakthroughs in pharmaceuticals and medicine, the role of non-pharmaceutical measures is critical in the early stages of a pandemic, when containment is most important. These steps are vital for the amount of lives that may be saved since they bridge the gap between when a pandemic occurs and when medications or vaccinations are developed. Important components of healthcare systems include telemedicine and smart fabrics and innovative materials used in PPE. Here, we provide a synopsis of the textiles employed in the battle against pandemics, including the COVID-19 pandemics of the past and present, and we examine the morphology of the most popular face masks, which are often made of cotton or polypropylene (PP). Additionally, we introduce the viewpoint that innovative materials, smart fabrics, and wearable technology are providing in the battle against future pandemics, primarily via PPE and telemedicine.

Ye et al. (2022) have conveyed that we present a zero-power smart face mask that is both lightweight and able to detect whether people are wearing the correct masks in public during a pandemic by wirelessly tracking coughs in real-time. The smart face mask's ability to identify when it is detached from the wearer's face is dependent on a small, battery-free radio frequency (RF) harmonic transponder that is connected to the mask's inner layer. The miniature antennas and passive frequency multiplier of the RF transponder are constructed from spray-printed silver nanowires (AgNWs) covered with a PEDOT:PSS passivation layer and the newly-discovered multiscale porous polystyrene-block-poly(ethylene-ran-butylene)-block-polystyrene (SEBS) substrate. The SEBS-AgNWs/PEDOT:PSS RF transponder is more advantageous than traditional on-chip or on-board wireless sensors due to its portability, flexibility, breathability, and comfort. Not only that, but this wireless gadget is very durable and long-lasting, even after several uses (such as putting and taking off the soft transponder from the mask). By locating the potentially infectious individual and determining whether or not they are wearing a mask, this wireless smart face mask has the ability to monitor both coughing and the wearer's condition, which might reduce the occurrence of virus-transmissive events. Also, certain disorders, such chronic obstructive pulmonary disease (COPD), may be better diagnosed with the use of wirelessly assessed cough frequencies.

Baluprithviraj et al. (2021) have illustrated that being healthy is crucial for everyone in this pandemic scenario. There are ways to protect yourself and your environment against this epidemic, but most people don't know about them. To protect ourselves and others, a face mask is a must-have. Thus, it is necessary for individuals to consistently use face masks. Because they aren't aware of the potential effects, visitors won't wear masks when they come home. When people aren't home, it could be hard for them to tell whether someone has been there. In this project, we propose an AI-based smart device—a Raspberry Pi with an AI model and a camera—that can detect whether someone is wearing a face mask and send us an alarm message—through a mobile app. This gadget works in tandem with an app on your phone. When users aren't physically present in their homes, the app may detect when someone enters. Only when individuals put on face masks will this smart gadget unlock the door. Day or night, this gadget will do its job. A variety of settings, including retail stores, hospitals, and temples, may make use of it.

Ghatak et al. (2020) have showcased that due to the established effectiveness of face masks in preventing the spread of the SARS-CoV-2 pandemic, their usage has been made obligatory in several nations since the virus's emergence. This article details a face mask that uses a basic textile triboelectric nanogenerator (TENG) to filter out SARS-CoV-2, one of several recent developments in face mask design. The suggested mask has a layered construction with protective sheets that function as a triboelectric (TE) filter on the inner and a smart filter on the outer. The suggested smart mask's conjugated action of touch electrification and electrostatic induction effectively inactivates the range of aerosols laden with viruses in a two-way fashion. Based on their effective tribo-electric charge densities of 83.13, 211.48, 38.62, 69, and 74.25 nC/m2, respectively, five pairs of triboseries fabrics—nylon - polyester, cotton - polyester, poly(methyl methacrylate) - PVDF, lylon - PVDF, and polypropylene - polyester—were optimised in this study. Because of its user-friendly design, self-powered operation (which draws power from the user's breathing, talking, or other facial movements), and high filtration efficiency, this smart mask has the potential to help mitigate the catastrophic effects of the COVID-19 pandemic.

Pandya et al. (2021) have stated that due to the rapid emergence and spread of novel virus strains in today's interconnected world, the possibility of another influenza pandemic is a real possibility. Hygiene, social distance, and medical masks are non-pharmacological approaches that should be prioritised in the absence of vaccinations and effective therapies to prevent or slow down the transmission. When it comes to public health, many professionals consider masks a last resort measure. A multi-pronged approach to reducing the spread of the disease and saving lives should include the usage of masks. Thus, in the present case, we have a smart mask that protects us from the coronavirus. This mask not only reveals the other person's temperature (in Celsius), but it also protects us from viruses, germs, and pollutants.

Li et al. (2022) have demonstrated that while dealing with a pandemic, masks are a must-have piece of PPE. Traditional face masks, which serve as physical barriers against size-dependence filtration and electrostatic absorption, may rapidly become ineffective when their surfaces deteriorate, leading to a large number of masks that are no longer relevant and environmental pollution. A self-cleaning smart mask is created here that actively protects against airborne particles and dangerous gases while simultaneously monitoring in real-time. To avoid extra electret procedures, electrospun ZIF-8/polyacrylonitrile membranes may be utilised in lieu of melt-blown non-woven materials. The captured pathogens and total volatile organic compounds (TVOC) may combine with the active components of ZIF-8, leading to the breakdown of TVOC and the death of the intercepted pathogens. After 1 hour of electrospinning, the membranes had a filtering efficiency of more than 99.58% for particulate materials and a 92.57% efficiency for TVOC absorption. It has a 99.07% antibacterial rate and may keep disinfecting for 72 hours or more. Further, by incorporating flexible circuits into masks, it is possible to measure the efficacy of filtering and keep tabs on environmental variables. One way to prolong the life of everyday masks and cut down on single-use masks is to have a mask that can regenerate its surface. They are versatile enough to meet the needs of biohazard defence, occupational disease prevention, and public health crises.

Masna et al. (2021) have showcased that due to the fast spread of COVID-19, several people took measures to limit the spread of the virus. Because the virus may be transmitted by respiratory droplets, the most important of them is to protect one's face, such as with a face mask. Smart mask technology is improving, but thus far it only offers passive protection against viruses by filtering out harmful particles and decontaminating the air inside. This research outlines a novel smart mask solution that actively mitigates the wearer's exposure to airborne infections by monitoring them using a PM sensor. This method makes use of a mist spray to lower the pathogen's airborne persistence by making the particles smaller. A smartphone app that shows the monitoring data may also be used to remotely operate the device. The smart mask may be used in many settings, such as schools, hospitals, and workplaces. Our experiments show that pathogen settling time is decreased when droplet sizes are increased. Our next projects include enhancing pathogen identification and mitigation deployment via the use of AI and fine-grained aerosol analysis.

Hyysalo et al. (2021) have stated that One effective strategy in the battle against the COVID-19 epidemic is the widespread usage of face masks. This study presents the Smart Mask, an ecosystem and platform backed by the Internet of Things (IoT) that aims to manage and prevent the transmission of respiratory viruses like COVID-19. Health data and health-related event detection in real time from the person and their surroundings may be facilitated by integrating sensors, materials, AI, wireless, the internet of things (IoT), and software. On a grander scale, health data analytics powered by AI may help cut down on medical expenses by providing more precise diagnoses and treatment plans; for instance, by comparing individual data sets to public databases, one can create a personalised health trajectory. Not only are future research topics and important research challenges for smart respiratory protection equipment recognised, but they are also prioritised.

METHODOLOGY

The purpose of our mask is to enable users to identify the body temperature of others around, serving as a vital instrument for ensuring safety in these difficult circumstances as shown in fig 2. The mask is fitted with a warning mechanism that notifies the user if somebody in close proximity has a temperature beyond the usual range. This prompts the wearer to maintain a safe distance to prevent possible contact (Ye et al., 2022). The mask has a removable silicone cushion, which can be easily cleaned, ensuring both cleanliness and usability. The ergonomic design of the product guarantees a safe and pleasant fit on the face, especially for those wearing spectacles, due to the use of skin-friendly silicone that forms a tight seal. This electric mask ensures secure respiration throughout everyday activities and sports, with a weight of just 105g, which is similar to the weight of two eggs. In addition, the one-way breathing air duct effectively removes exhaust gases, reducing the growth of germs caused by humid heat, so improving overall protection and providing reassurance as shown in figure 2.

Figure 2. Vital instrument for ensuring safety.

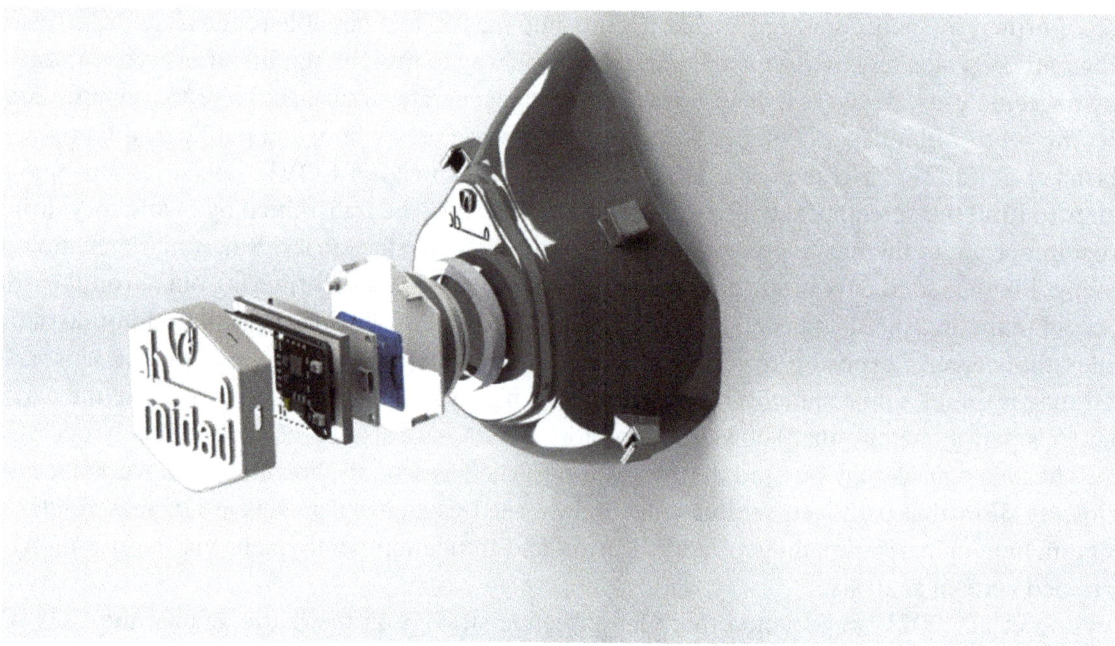

Figure 3. Analysis of overall protection.

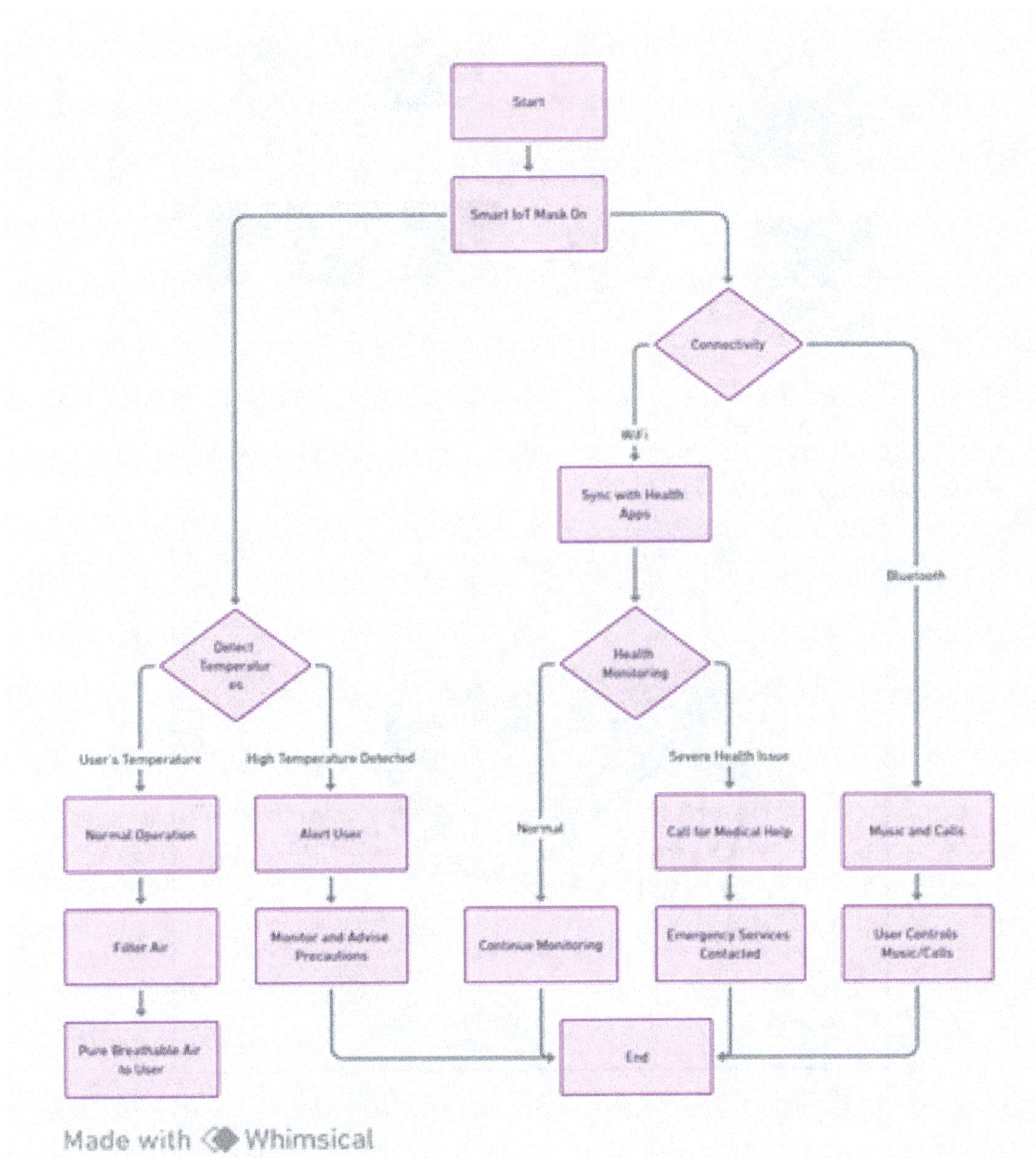

STRUCTURAL DESIGN

4.1 Compatibility

The compatibility of compact design extends to a diverse array of 3M goods as shown in fig 4. We have specifically selected 3M as our foundation due to their global reach and the fact that their products are mostly standardised rather than limited to certain regions. The assembling of smart mask components is shown in figure 4 and the input and output parameters are shown in figure 5.

Figure 4. The compact design of 3M products

Figure 5. Prototype design of safety mask

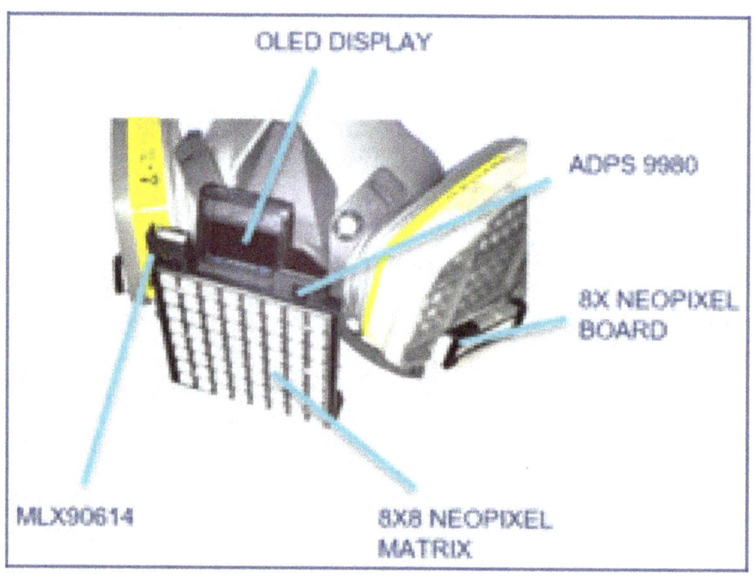

Figure 6. Assembly of safety mask components

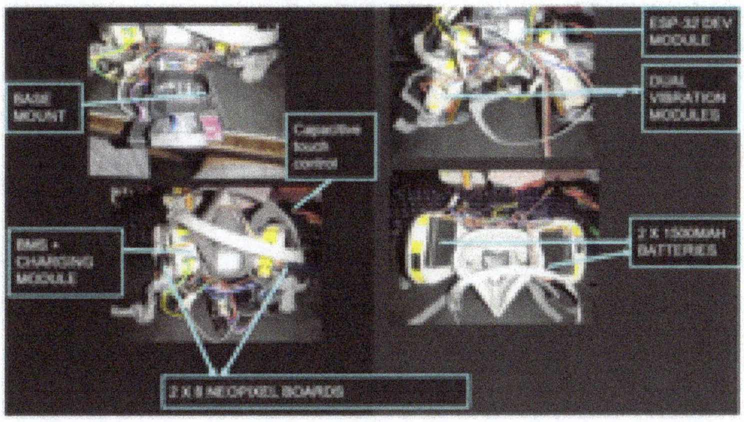

4.2 Hardware and Software Requirements

1. Arduino Nano Microcontroller

Function: The Arduino Nano acts as the main processing unit of the mask, managing the input and output data from several sensors and modules as shown in fig 7.

Figure 7. Core processor of smart mask (Selvadass et al., 2022)

Approach: Comprehensive programming is needed for the device's versatility. The Arduino IDE will configure the Arduino to handle sensor data, operate the OLED display, manage Bluetooth connection, and analyse GPS module data. This precise programming procedure may need specialised libraries, especially for GPS and Bluetooth components, to facilitate their maintenance. The Arduino IDE controls these different duties, providing component synergy and a unified and responsive user experience. These programming helps the gadgets works smoothly and perform its many purposes.

2. The HC-05 Bluetooth Module

Function: It facilitates wireless communication, enabling the mask to establish a connection with a smartphone or other devices that use Bluetooth technology as shown in fig 8.

Figure 8. Connecting module (Selvadass et al., 2022)

Approach: The HC-05 module will connect to the Arduino Nano in a very important way, making wireless contact easier. This module can send important information to a connected device, such as measures of air quality and temperature. At the same time, it can get orders through a smartphone app, which makes it more useful in both directions. This function for wireless connection makes the system much more flexible by letting users view real-time info or handle the Arduino Nano from their smartphones. The addition of the HC-05 module not only speeds up data transfer, but it also makes it possible for engaging and dynamic control, which makes the Arduino-based system even easier to use and more convenient.

3. OLED Display

Function: The OLED display is responsible for presenting data such as air quality, temperature measurements, and GPS locations are shown in fig 9.

Figure 9. Display device (Kanaga Suba Raja et al., 2023)

Approach: The Arduino Nano will be closely connected to the display system through a key connection. When you have to turn data from different sensors into meaningful written and maybe even visual representations, you have to do a lot of careful programming. This code layer connects the raw sensor data to information on the screen that is easy for the user to understand. Through this complicated process, the Arduino Nano turns complicated data streams into insights that are easy to understand, making it simple for users to grasp and make sense of data. The Arduino Nano and the display work well together, which not only makes collaboration easier but also shows how important programming is for turning different types of data into a user experience that is unified and easy to use.

4. The AIT1000 Non-Contact Infrared Temperature Sensor

Function: This sensor is designed to measure the ambient temperature of people in proximity to the user as shown in fig 10.

Figure 10. Temperature sensor (Lazaro et al., 2021)

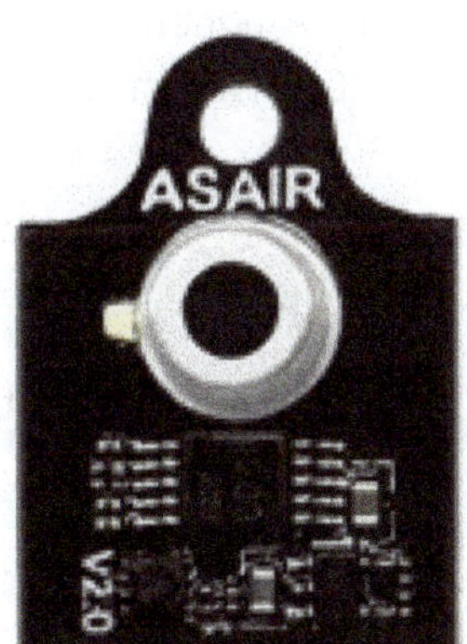

Approach: The sensor is carefully put on the front of the mask so that it works best. It faces outward. This setting makes it easier for the monitor to gather useful information about the surroundings. The data is sent smoothly to the Arduino so that it can be analysed in detail, taking advantage of the device's processing power. After this study, the results are quickly shown on the OLED screen, giving people real-time information about the situations they are travelling. The OLED screen's easy-to-use interface makes it simple for people to understand and act on the sensor's results, which improves their environmental awareness and helps them make better decisions in a variety of situations.

5. Rapid Charging Batteries

Function: These batteries provide the required energy to operate all the components in the mask is shown in fig 11.

Figure 11. Power backup device (Kanaga Suba Raja et al., 2023)

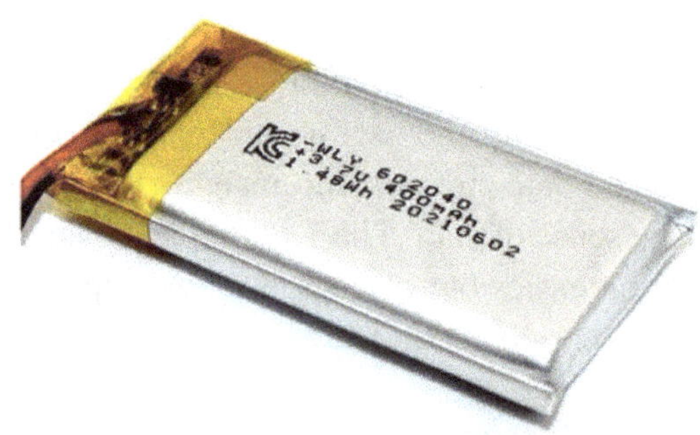

Approach: For the mask to work properly, the batteries must be placed firmly so that it stays stable, and the user's comfort must be taken into account when distributing weight. This location is very important for a smooth wearing experience. A strong charging connection should also be built in so that the batteries can be charged quickly and effectively. The charging system's effectiveness is very important because it keeps the mask ready to use while reducing downtime. This well-thought-out mix of safe battery placement and fast charging makes the mask easier to use and more convenient, which supports its usefulness in a range of situations.

6. Air Quality sensor.

The ZP07-MP503 Air Quality Detection Module has the function of detecting the air quality in the user's surroundings is shown in fig 12.

Figure 12. Air quality measuring device (Lazaro et al., 2021)

Approach: Adding this module to the Arduino system makes it easy to get data, just like adding the temperature sensor did. The collected information, especially about the quality of the air, can be shown clearly on the OLED screen, which provides an easy-to-use interface for quick reference. In addition to the local display, the device is more useful because it can send information about the air quality index to a smartphone via Bluetooth. This function makes real-time info easier to access, so users can check on and study air quality conditions from afar. There are many possible uses, such as situations where getting fast alerts or in-depth analyses of air quality factors are very important. This combination of sensor data with a range of output choices gives users a full picture of air quality, which helps them make smart decisions and take action. This is useful for personal health, environmental knowledge, and workplace safety.

7. GPS Module

The NEO-6M GPS Module serves the purpose of delivering precise position information for the user is shown in fig 13.

Figure 13. Position mapping module (Lazaro et al., 2021)

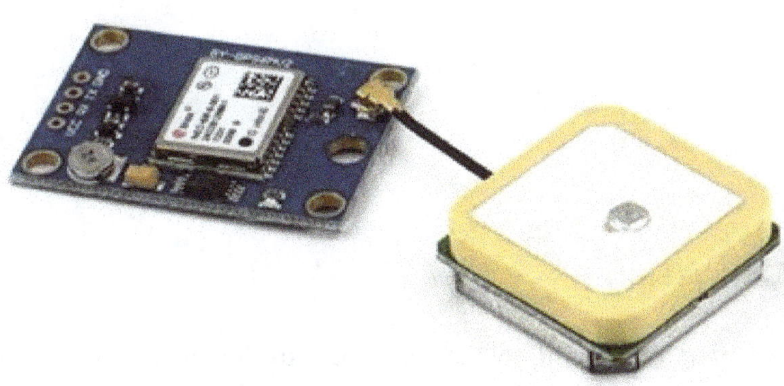

Approach: When the GPS module is connected to the Arduino, it makes the link smooth, which lets you get accurate location data. This information is useful for both location-based warnings and actions that keep an eye on things. This setup's flexibility makes it especially useful in a wide range of situations, from monitoring for pandemics to places where there are natural risks. In the context of pandemic monitoring, the method can make it easier to track and analyse regional trends, which can help find breakouts and keep them under control. In high-risk areas, the ability to watch in real time is very useful because it lets you respond quickly to possible threats. So, when Arduino and GPS work together, they make a powerful tool that can help people be more aware of their surroundings and take proactive steps in many situations where location-based data is important for making smart decisions and managing risk.

8. Embedded C for Programming

Embedded C is a variant of the C programming language that is tailored for programming microcontrollers and embedded devices, such as the Smart Mask is shown in fig 14. It is very suitable for this project because of its efficiency, ability to access low-level hardware, and relatively short runtime. This makes it great for devices with limited processing resources and power. When programming the Arduino Nano microcontroller in your mask, using Embedded C will facilitate direct manipulation of hardware components, enabling meticulous control over the operation of each module, including the HC-05 Bluetooth module, the OLED display, temperature sensors, and the GPS module. This programming language excels in managing real-time activities and communicating with specific hardware via its input/output functions, making it an ideal option for developing a responsive and efficient Smart Mask.

Figure 14. Programming to connect modules

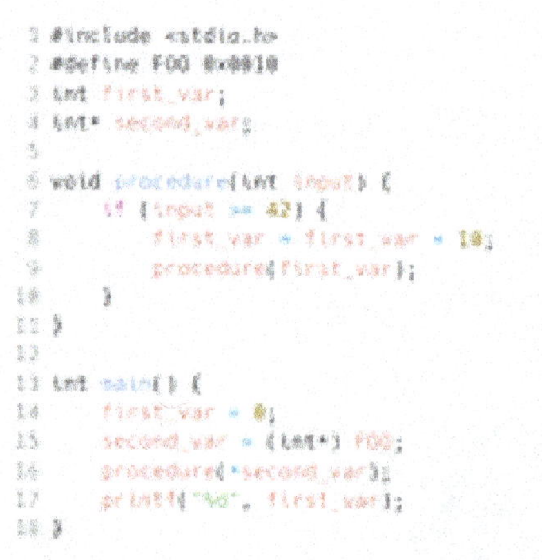

4.3 Results and Discussion

The Smart Mask, empowered by an Arduino Nano microcontroller and programmed with Embedded C, is a breakthrough in wearable technology, offering an array of advanced features are shown in fig 15. It includes real-time health monitoring, enabled by a non-contact infrared temperature sensor to track the temperature of people nearby, aiding in early detection of health issues like fever is shown in fig 16 and fig 17. The air quality is constantly assessed using a specialized module, alerting the wearer to pollutants or hazardous conditions in their immediate environment. Additionally, the integrated GPS module provides precise location tracking, which could be vital for safety in emergencies or for navigation in unfamiliar areas. All this information, along with battery status, temperature and other essential data, is conveniently displayed on an OLED screen is shown in table 1, while the HC-05 Bluetooth module ensures seamless connectivity with other devices for data sharing and remote monitoring. This Smart Mask is a remarkable blend of health, safety, and connectivity, designed for the modern, health-conscious individual.

Figure 15. Smart mask

Figure 16. Temperature results given by the temperature sensor

Table 1. Temperature data collected with percentage calculation

Distance (cm)	Reference Temperature (°C)	Measured Temperature (°C)	Error (°C)	Result: Reference Temperature	Result: Measured Temperature
3.61	39.30	39.16	-0.14	High	High
4.57	38.80	37.49	-1.31	High	High
4.37	38.60	38.45	-0.15	High	High
4.89	38.30	36.89	-1.41	High	Normal
4.52	38.10	36.83	-1.27	High	Normal
4.59	37.60	36.36	-1.24	High	Normal
4.63	37.90	36.53	-1.37	High	Normal
3.83	37.70	37.51	-0.19	High	High
4.82	37.30	36.88	-0.42	High	Normal
3.41	37.00	37.04	0.04	High	High
3.42	36.90	36.76	-0.14	Normal	Normal
3.91	36.40	36.26	-0.14	Normal	Normal
4.55	36.30	36.26	-0.04	Normal	Normal
3.62	36.30	36.28	-0.02	Normal	Normal

Figure 17. The infrared temperature sensor can detect temperatures with range of onemeter

4.4 Market Survey

Many people wear masks frequently and are concerned about their wellbeing, while simultaneously feeling uncomfortable. Majorities of the people are also fascinated by the idea of a smart mask and are willing to invest in it. By making the quality product available, everyone can benefit from the untapped market potential are shown in fig 18. Based on the survey we conducted, we obtained the following results,

Figure 18. Analysis of our market survey of our mask

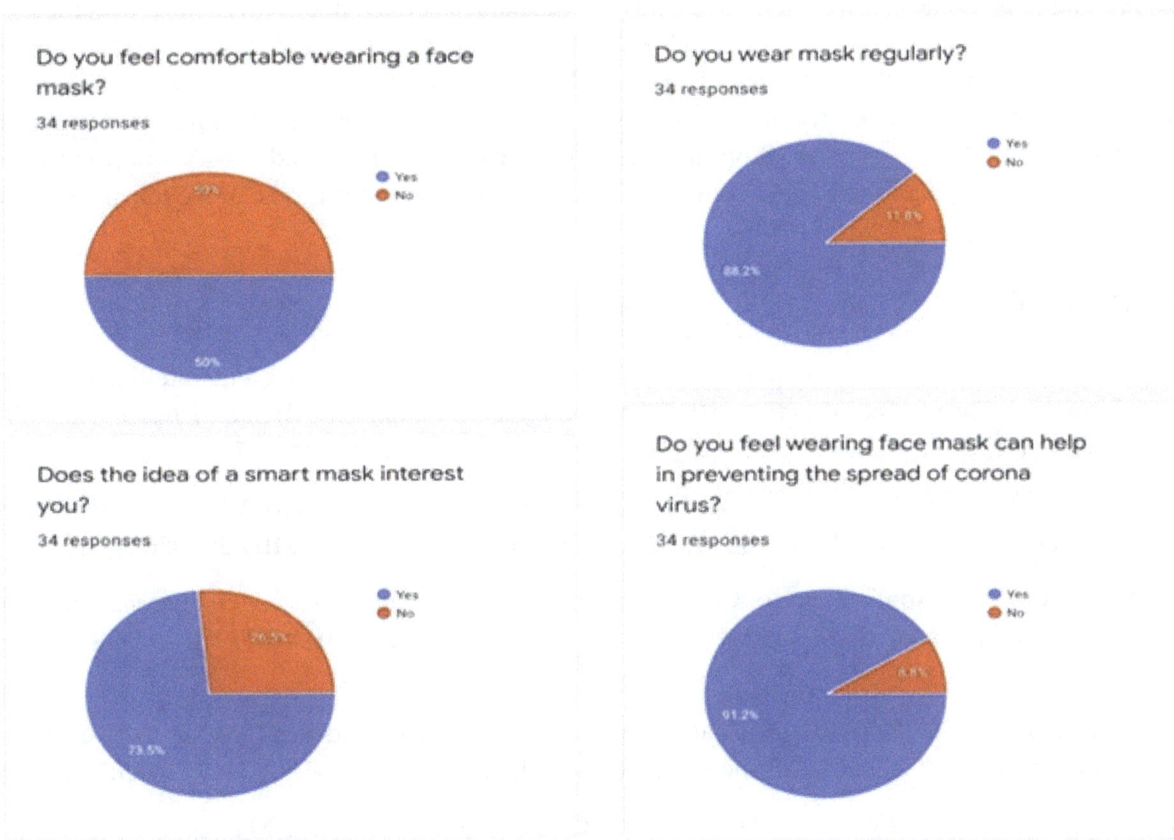

CONCLUSION AND FUTURE WORK

A notable reaction to the continuing COVID-19 epidemic and the growing need for efficient personal protection has been the creation of a smart mask equipped with enhanced features and the ability to identify pathogens. With this initiative, we were able to meet the need for a cutting-edge answer that improves public health and safety while simultaneously offering more robust protection. We have taken the initiative to combat the global spread of illness and provide a better atmosphere for people everywhere by equipping the smart mask with sensors and real-time pathogen detection systems. In low and middle-income nations, where smart masks are not yet widely used, our initiative has shown their business potential. Improving public health outcomes and opening up new possibilities for manufacturers and entrepreneurs may be achieved by making smart masks more accessible and available in these places. The fast industrialization in these nations also bodes well for the potential acceptance and growth of the smart mask market.

There are a number of promising future directions for smart mask research and development. The ongoing enhancement of pathogen detection skills is a key area of concentration. We may improve the smart mask's capacity to identify a wide range of infections, not just COVID-19, by enhancing its sensors and algorithms. Because of this, people will be able to take precautions in a timely manner, which will help in the early diagnosis and control of epidemics. Improving smart mask functionality is another area that might use some research into the use of cutting-edge tech like AI and ML. Customers may get significant insights into their health state and possible exposure hazards via the use of these technologies, which can allow real-time data analysis, personalised health monitoring, and predictive modelling. In order to advance smart mask technology, it is essential to work together with researchers, healthcare providers, and regulatory agencies. Working together, we can standardise and certify smart masks to make sure they are safe, effective, and comply with regulations in various public and clinical contexts.

REFERENCES

Baluprithviraj, K., Bharathi, K., Chendhuran, S., & Lokeshwaran, P. (2021). Artificial Intelligence based Smart Door with Face Mask Detection. *2021 International Conference on Artificial Intelligence and Smart Systems (ICAIS)* (pp. 543-548). IEEE. 10.1109/ICAIS50930.2021.9395807

Cai, S., Xu, X., Yang, W., Chen, J., & Fang, X. (2019). Materials and Designs for Wearable Photodetectors. *Advanced Materials, 31*(18), 1808138. doi:10.1002/adma.201808138 PMID:30785644

Daoud, M. (2022). IoT Applications for Coronavirus Industry Protection. *Advances in Systems Analysis, Software Engineering, and High Performance Computing*, 225–237. doi:10.4018/978-1-7998-8367-8.ch012

Dias, D., & Cunha, J. (2018). Wearable Health Devices—Vital Sign Monitoring, Systems and Technologies. *Sensors (Basel), 18*(8), 2414. Advance online publication. doi:10.3390/s18082414 PMID:30044415

Fischer, H., Wittmann, D., Costa, A., Zhou, B., Joost, G., & Lukowicz, P. (2021). Masquare: A Functional Smart Mask Design for Health Monitoring. *Proceedings of the 2021 ACM International Symposium on Wearable Computers*. ACM. 10.1145/3460421.3478831

Ghatak, B., Banerjee, S., Ali, S., Bandyopadhyay, R., Das, N., Mandal, D., & Tudu, B. (2020). Design of a Self-powered Smart Mask for COVID-19. arXiv: Medical Physics.

Goar, V., Sharma, A., Yadav, N. S., Chowdhury, S., & Hu, Y. C. (2023). IOT-based smart mask protection against the waves of covid-19. *Journal of Ambient Intelligence and Humanized Computing, 14*(8), 11153–11164. doi:10.1007/s12652-022-04395-7 PMID:36117515

Gorbalenya, A., Baker, S., Baric, R., Groot, R., Drosten, C., Gulyaeva, A., Haagmans, B., Lauber, C., Leontovich, A., Neuman, B., Penzar, D., Perlman, S., Poon, L., Samborskiy, D., Sidorov, I., Sola, I., & Ziebuhr, J. (2020). The species Severe acute respiratory syndrome-related coronavirus: classifying 2019-nCoV and naming it SARS-CoV-2. *Nature Microbiology, 5,* 536 - 544. . doi:10.1038/s41564-020-0695-z

Hu, B., & Koo, S. (2022). Development of wearable device designs for respiratory infection prevention. *International Journal of Clothing Science and Technology, 34*(5), 764–781. doi:10.1108/IJCST-07-2021-0094

Hyysalo, J., Dasanayake, S., Hannu, J., Schuss, C., Rajanen, M., Leppänen, T., Doermann, D., & Sauvola, J. (2021). Smart mask – Wearable IoT solution for improved protection and personal health. *Internet of Things : Engineering Cyber Physical Human Systems, 18*, 100511–100511. doi:10.1016/j.iot.2022.100511 PMID:37521492

Hyysalo, J., Dasanayake, S., Hannu, J., Schuss, C., Rajanen, M., Leppänen, T., Doermann, D., & Sauvola, J. (2021). Smart mask – Wearable IoT solution for improved protection and personal health. *Internet of Things : Engineering Cyber Physical Human Systems*, *18*, 100511–100511. doi:10.1016/j.iot.2022.100511 PMID:37521492

Ivanoska-Dacikj, A., & Stachewicz, U. (2020). Smart textiles and wearable technologies – opportunities offered in the fight against pandemics in relation to current COVID-19 state. *Reviews on Advanced Materials Science*, *59*(1), 487–505. doi:10.1515/rams-2020-0048

Kanaga Suba Raja, S., Sandhiya, B., Kausalya, K., Ramya, B., Monish, S., & Sai Raja Gopal, R. (2023). IoT based Smart Door System for Monitoring Human Body Temperature and Face Mask Detection. *International Conference on Advances in Computing, Communication and Applied Informatics (ACCAI)*, Chennai, India. 10.1109/ACCAI58221.2023.10199273

Lazaro, M., Lázaro, A., Villarino, R., & Girbau, D. (2021a). Smart Face Mask with an Integrated Heat Flux Sensor for Fast and Remote People's Healthcare Monitoring. *Sensors (Basel)*, *21*(22), 7472. doi:10.3390/s21227472 PMID:34833547

Lazaro, M., Lazaro, A., Villarino, R., & Girbau, D. (2021b). *Smart mask for temperature monitoring with LoRa backscattering communication*. 2021 6th International Conference on Smart and Sustainable Technologies (SpliTech), Bol and Split, Croatia. 10.23919/SpliTech52315.2021.9566327

Lee, P., Kim, H., Kim, Y., Choi, W., Zitouni, M., Khandoker, A., Jelinek, H., Hadjileontiadis, L., Lee, U., & Jeong, Y. (2022). Beyond Pathogen Filtration: Possibility of Smart Masks as Wearable Devices for Personal and Group Health and Safety Management. *JMIR mHealth and uHealth*, *10*(6), e38614. doi:10.2196/38614 PMID:35679029

Li, J., Yin, J., Ramakrishna, S., & Ji, D. (2023, January 30). Smart Mask as Wearable for Post-Pandemic Personal Healthcare. *Biosensors*, *13*(2), 205. doi:10.3390/bios13020205 PMID:36831971

Li, Ya., Yang, Q., Liu, S., Liu, X., Wu, Z., Wang, J., Zhou, M., Ren, M., Wan, C., Huo, W., Gao, Y., & Huang, X. (2022). A Self-Purifying Smart Mask Integrated With Metal–Organic Framework Membranes and Flexible Circuits. *Advanced Materials Interfaces*, *10*(4), 2201895. doi:10.1002/admi.202201895

Lok, R., Karacali, H., Varol, A., Camli, U., & Yilmaz, E. (2022). Fabrication and characterization of resistance temperature detector by smart mask design. *International Journal of Advanced Manufacturing Technology*, *122*(1), 147–158. doi:10.1007/s00170-022-09041-2

Masna, N. V. R., Kalavakonda, R. R., Dizon-Paradis, R. N., & Bhunia, S. (2021). Smart and Connected Mask for Protection beyond the Pandemic: (Invited Paper). *IEEE International Midwest Symposium on Circuits and Systems (MWSCAS)*. 10.1109/MWSCAS47672.2021.9531802

Matsui, S., Terada, T., & Tsukamoto, M. (2017). Smart eye mask: eye-mask shaped sleep monitoring device. *Proceedings of the 2017 ACM International Joint Conference on Pervasive and Ubiquitous Computing and Proceedings of the 2017 ACM International Symposium on Wearable Computers*. ACM. 10.1145/3123024.3129264

Mekid, S., & Chenaoua, K. (2023). IoT-Enabled Smart Mask for Monitoring Body Parameters and Location through Cloud. *Internet of Things : Engineering Cyber Physical Human Systems*, *22*, 100794. doi:10.1016/j.iot.2023.100794 PMID:37266184

Panarello, A., Tapas, N., Merlino, G., Longo, F., & Puliafito, A. (2018). Blockchain and IoT Integration: A Systematic Survey. *Sensors (Basel)*, *18*(8), 2575. doi:10.3390/s18082575 PMID:30082633

Pandya, S., Sur, A., & Solke, N. (2021, November 30). COVIDSAVIOR: A Novel Sensor-Fusion and Deep Learning Based Framework for Virus Outbreaks. *Frontiers in Public Health*, *9*, 797808. doi:10.3389/fpubh.2021.797808 PMID:34917585

Perera, C., Liu, C., Jayawardena, S., & Chen, M. (2015). A Survey on Internet of Things From Industrial Market Perspective. *IEEE Access : Practical Innovations, Open Solutions*, *2*, 1660–1679. doi:10.1109/ACCESS.2015.2389854

Prabha, K. R., Nataraj, B., Swathi, P., Tharani, V. R., & Vinola, A. (2022, October). IoT Based Smart Mask for Health Monitoring. In *2022 3rd International Conference on Smart Electronics and Communication (ICOSEC)* (pp. 568-572). IEEE. 10.1109/ICOSEC54921.2022.9952121

Rahman, M. M., Manik, M. M. H., Islam, M. M., Mahmud, S., & Kim, J. H. (2020, September). An automated system to limit COVID-19 using facial mask detection in smart city network. In *2020 IEEE International IOT, Electronics and Mechatronics Conference (IEMTRONICS)* (pp. 1-5). IEEE.

Rothan, H., & Byrareddy, S. (2020). The epidemiology and pathogenesis of coronavirus disease (COVID-19) outbreak. *Journal of Autoimmunity*, *109*, 102433–102433. doi:10.1016/j.jaut.2020.102433 PMID:32113704

Selvadass, S., Paul, J. J., & Bella Mary, I. (2022). IoT-Enabled smart mask to detect COVID19 outbreak. *Health and Technology*, *12*(5), 1025–1036. doi:10.1007/s12553-022-00695-2 PMID:36000088

Sisinni, E., Saifullah, A., Han, S., Jennehag, U., & Gidlund, M. (2018). Industrial Internet of Things: Challenges, Opportunities, and Directions. *IEEE Transactions on Industrial Informatics*, *14*(11), 4724–4734. doi:10.1109/TII.2018.2852491

Suo, J., Liu, Y., Wu, C., Chen, M., Huang, Q., Liu, Y., Yao, K., Chen, Y., Pan, Q., Chang, X., Leung, A., Chan, H., Zhang, G., Yang, Z., Daoud, W., Li, X., Roy, V., Shen, J., Yu, X., & Li, W. (2022). Wide-Bandwidth Nanocomposite-Sensor Integrated Smart Mask for Tracking Multiphase Respiratory Activities (Adv. Sci. 31/2022). *Advanced Science (Weinheim, Baden-Wurttemberg, Germany)*, *9*(31), 2270200. Advance online publication. doi:10.1002/advs.202270200

Wölfel, R., Corman, V., Guggemos, W., Seilmaier, M., Zange, S., Müller, M., Niemeyer, D., Jones, T., Vollmar, P., Rothe, C., Hoelscher, M., Bleicker, T., Brünink, S., Schneider, J., Ehmann, R., Zwirglmaier, K., Drosten, C., & Wendtner, C. (2020). Virological assessment of hospitalized patients with COVID-2019. *Nature*, *581*(7809), 465–469. doi:10.1038/s41586-020-2196-x PMID:32235945

Ye, Z., Ling, Y., Yang, M., Xu, Y., Zhu, L., Yan, Z., & Chen, P. (2022). A Breathable, Reusable, and Zero-Power Smart Face Mask for Wireless Cough and Mask-Wearing Monitoring. *ACS Nano*, *16*(4), 5874–5884. doi:10.1021/acsnano.1c11041 PMID:35298138

Zhang, B., Zheng, Y., Qiu, Y., Yang, L., Zhang, R., Liao, Y., & Xu, J. (2022). Overview of smart masks and research on new technology. *Wearable Technologies*, *2*(2), 46. doi:10.54517/wt.v2i2.1648

Zhao, L., Ferraro, P., & Shorten, R. (2022). A DLT enabled smart mask system to enable social compliance. ArXiv, abs/2205.13256. https://doi.org//arXiv.2205.13256. doi:10.48550

Chapter 6
An Analysis of AI in Healthcare: A Comprehensive Study on India

R. Dilip
Dayananda Sagar Academy of Technology and Management, India

V. Deepti
Dayananda Sagar Academy of Technology and Management, India

S. Mahadev
Dayananda Sagar Academy of Technology and Management, India

N. Tejashwini
Sai Vidya Institute of Technology, India

G. Muthugurunathan
Madanapalle Institute of Technology and Science, India

Manasa
Sapthagiri College of Engineering, India

Sabyasachi Pramanik
https://orcid.org/0000-0002-9431-8751
Haldia Institute of Technology, India

ABSTRACT

Building computer devices and systems that can demonstrate learning, reasoning, and decision-making skills similar to those often associated with human intelligence is the focus of the field of artificial intelligence. It also includes working with data sets larger than what the human mind can process. A variety of academic disciplines, including computer science, data analytics, statistics, hardware and software engineering, languages, neuroscience, philosophy, and psychology, are all included in the interdisciplinary area of artificial intelligence. The development of artificial intelligence is increasingly taking centre stage in many countries' governmental agendas. Policymakers are primarily concerned with the application of artificial intelligence to social welfare, legal framework creation, economic consequences, implications for global security, and equity issues, among other things. Artificial intelligence has been the main driver of recent large and rapid breakthroughs in the technological area.

DOI: 10.4018/979-8-3693-1922-2.ch006

APPLICATION

The term "artificial intelligence" describes computer programs that replicate human psychology. It is particularly made to mimic cognitive abilities and think like a person. Any machine that exhibits traits often connected to human cognition, such learning and problem-solving, may also be referred to as "artificial intelligence." AI, a technique meant to simulating human thought processes to solve problems or speed up learning has a big influence on a lot of aspects of life. When robots are created to display unique intelligence that is different from human intelligence, this specific aspect of intelligence is often referred to as machine intelligence. Over the course of its long history—more than 70 years—computer scientists like Alan Turing, Marvin Minsky, and John McCarthy have made major contributions to the theoretical and technological underpinnings of artificial intelligence. Alan Turing, a British mathematician, led the way in developing the foundational ideas of computer science. His main goal when he started this project was to increase the effectiveness of decoding encrypted German messages sent during World War II (Turing, 2015).

After the war ended, he began thinking about artificial intelligence. One important academic book that explores artificial intelligence is Alan Turing's "Computing Machinery and Intelligence" (Turing, 2009). Artificial Intelligence is always evolving to provide benefits in several domains and industries. A cross-disciplinary approach that incorporates elements of mathematics, computer science, linguistics, psychology, and other fields links machines. AI is becoming more prevalent across a wide range of sectors and businesses, including but not limited to the banking and financial industries, smart cities, mobility and transportation, healthcare, education, and agriculture, among many other fields that are adopting this technology at an accelerating rate. These days, AI is widely used to solve socioeconomic problems. It is acknowledged as a viable means of addressing the key initiatives, targets, and objectives delineated in the UN Framework for Sustainable Development Goals (SDGs). In many countries, the rise of AI has become a major policy worry. When dealing with AI, policymakers take into account a number of factors, including the use of AI for societal improvement, the creation of regulatory frameworks, economic ramifications, concerns about global security, and challenges related to justice. There are a few things that must be taken into account while making AI plans. These include forming a nationwide task group and establishing challenging objectives. In order to use a pragmatic technique, it is essential to ascertain the variables that enable or bolster the intended result. Identifying the people or organizations having a stake in the issue is also very important. It is crucial to educate these stakeholders of the pertinent facts. In addition, interacting with different stakeholders is required to guarantee a thorough comprehension of the problem. It is advised that an extensive evaluation of the existing situation be conducted. For a realistic analysis, the information must be consolidated. Furthermore, building credibility and taking ethical values into account are essential steps in this process. The next phase of the process is to put the developed plan into practice.

In India, the Use of AI in the Healthcare Industry Is Growing Quickly

This is because there is a dearth of qualified medical professionals in the healthcare industry as a whole (Saha, 2018). Artificial Intelligence (AI) has several applications, including predictive modeling in Electronic Health Records (EHRs), and may help solve issues over the integrity of data in EHRs.

Efforts for preventative healthcare and predictive treatment (Alnsour et al., 2019; Eubanks, 2017; Khumalo & Mnjama, 2019). It is noteworthy that the healthcare system in India has been shown to benefit greatly from the use of AI technology. But it's essential to take a rational and scientific approach to integrating AI technology into India's healthcare system. At this point, India's plans for using AI technology in the healthcare industry are still in the planning stages (Chatterjee & Kar, 2018). A thorough and workable healthcare strategy that tackles the many problems related to AI technology is required.

But this work still has to be done. This research intends to close the current knowledge gap and build AI-healthcare strategy for India. The author has worked hard to provide policymakers valuable insights, addressing the ethical dilemma surrounding AI and offering workable solutions that could be helpful in developing AI-assisted health policy in India.

THE WORLDWIDE VIEWPOINT ON ARTIFICIAL INTELLIGENCE IN MEDICINE

In recent years, artificial intelligence has gained prominence in the global healthcare arena. Healthcare delivery might undergo a radical change because to new technology like blockchain and artificial intelligence. AI is being used extensively in the healthcare industry in a variety of areas, including as robotic surgery, pharmaceuticals, medical technology, drug research, diagnostics, hospital administration, health insurance operations, and clinical decision-making. Chatbots for customer service, population health monitoring, care routing, data-driven and image-based diagnosis, clinical decision support, compliance monitoring, medical record maintenance, fraud prevention, capacity planning, personnel management, quality assurance, and training are a few real-world uses of AI in healthcare. GE Healthcare, Remedy Health, Oncora Medical, Google Health/DeepMind, and Augmedix are some of the top healthcare organizations in the world that integrate AI in their operations.

In 2023, Globe Newswire issued a research estimating that the healthcare industry's worth of artificial intelligence (AI) in 2021 was around $11 billion. By 2030, it is predicted to have grown significantly to $188 billion. Between 2022 and 2030, this increment is anticipated to expand at a compound annual growth rate of 37%.

Some of the key elements of AI in healthcare are as follows:

1. AI Shaping the Future of Medicine: With the introduction of Artificial Intelligence, the healthcare industry is undergoing a significant transformation. AI offers a bright future for improving health care, streamlining processes, and advancing R&D. In the healthcare industry, its ability to analyze enormous volumes of data, automate processes, and make predictions is nothing short of revolutionary (Subbiah, 2023). Among its many benefits, AI may improve the accuracy of medical diagnosis, therapeutic processes, administrative tasks, and patient involvement.
2. Medical Imaging and Diagnostics: Among the most common applications of AI in healthcare are imaging and diagnostic processes. Algorithms based on artificial intelligence provide the capacity to evaluate medical images with unmatched accuracy and speed. By using artificial intelligence technologies, doctors and radiologists may now diagnose patients more accurately and efficiently. Artificial intelligence (AI) systems have shown the capacity to recognize and highlight abnormalities in X-rays, MRIs, and CT scans. This method has shown to be quite valuable in the prompt detection of health problems including cancer and fractures. Furthermore, in pathology and histology, artificial intelligence is essential. Artificial intelligence-enabled tools may be used by pathologists to analyze tissue samples, which can lead to improved and quicker detection of many illnesses,

including cancer (Rana & Bhushan, 2023). Systems using artificial intelligence are able to identify complex patterns and anomalies that would escape human observers.

3. Drug Discovery and Development: By using artificial intelligence, pharmaceutical firms are now able to effectively analyze large datasets, simulate molecular interactions, and anticipate possible medicine candidates, therefore revolutionizing the process of drug research and development. By giving patients quicker access to cutting-edge treatments, this creative strategy may considerably cut down on the time and money needed to bring new medications to market. Better patient outcomes and more efficient drug development processes have resulted from the use of machine learning models, which have shown to be especially successful in the analysis of biological data, the identification of prospective therapeutic targets, and the optimization of clinical trial design (Sadybekov & Katritch, 2023).
4. Personalized Medicine: The growing popularity of personalized medicine may be attributed to artificial intelligence. AI systems can tailor treatment for each patient by looking at their genetic makeup, past medical histories, lifestyle decisions, and other pertinent data. Because medications and treatments are customized to the unique characteristics of each patient, this approach improves treatment efficacy and reduces the likelihood of adverse events. Predictive analytics, which uses artificial intelligence, makes it easier to identify people who are more likely to have specific diseases. Artificial intelligence has the capability to detect individuals who may have a genetic susceptibility to diseases like diabetes by analyzing large datasets (Serrano et al., 2023). This makes it possible to put preventive and timely actions into action.
5. Electronic health record: In today's healthcare system, maintaining Electronic Health Records (EHRs) has become crucial. Artificial intelligence is essential to this process because it gives medical practitioners a faster and more effective method to arrange and use Electronic Health Records. AI is able to via Natural Language Processing (NLP) get insightful knowledge from unstructured clinical notes, facilitating better access to vital patient data. Furthermore, AI technology aids in the standardization of EHRs, guaranteeing their interoperability with different healthcare systems and promoting the smooth transfer of patient data. By ensuring that medical professionals have access to accurate and up-to-date patient data, they can make better decisions and, in turn, provide patients with higher-quality treatment.
6. Remote Monitoring and Telemedicine: The field of telemedicine has had a notable upsurge, mostly attributed to artificial intelligence. Artificial intelligence (AI)-driven chatbots and virtual assistants play a crucial role in helping patients by providing them with information, setting up appointments, and quickly and effectively handling medical problems. These state-of-the-art tools greatly improve accessibility to healthcare and patient contact. Additionally, people's health state may be continually monitored with the use of AI algorithms and remote monitoring devices, with a focus on chronic illnesses like diabetes and hypertension (Ahmed Kamal et al., 2023). The ability to get warnings and take prompt action thanks to this real-time data lowers the risk of readmissions to hospitals.
7. Genomics and DNA Sequencing: The analysis of genetic data is a very complex process, and artificial intelligence plays a critical role in this process. Artificial intelligence (AI) systems have the capacity to identify genetic markers linked to various ailments and may provide customized treatment choices based on an individual's unique genetic makeup. Pharmacogenomics is a technique that guarantees that medications are tailored based on an individual's genetic makeup, resulting in better therapeutic results. Furthermore, by analyzing genetic data, AI may help identify uncommon illnesses. AI may expedite the process of identifying medical conditions by comparing the genetic

profiles of individuals with large genetic information databases (Warburton & Sebra, 2023). With this technology, diseases that could have gone undetected otherwise can be found.

AI, MEDICINE, AND THE INDIAN SCENE

The Indian economy benefits greatly from the healthcare sector, which also provides a wealth of job and revenue prospects. In 2022, the Indian healthcare sector was valued at $372 billion, and NASSCOM research indicates that it is expected to grow at a compound annual growth rate (CAGR) of nearly 20% from 2023 to 2030. The COVID-19 pandemic has resulted in significant changes to the healthcare sector, making it more sophisticated and resilient. In an effort to provide more effective and efficient patient care, the epidemic has accelerated technology development and led Indian healthcare start-ups to invest more in artificial intelligence. The Indian government has set out INR 89,155 crore in the Union Budget for 2023–2024 to improve the country's healthcare system. The funds will be used to support a range of studies and innovative healthcare initiatives on a massive scale, with INR 2,980 crore going to the Department of Health Research and INR 86,175 crore going to the Department of Health and Family Welfare. The government has aggressively encouraged different corporate players to engage in vital industries like healthcare.

High-end diagnostic services from India are well-known and draw significant financial investment. Indians are becoming more conscious of their health and the value of taking preventive actions, which has increased demand for healthcare services across the nation. The rising trend of healthcare spending has also been attributed to the growing usage of insurance in the healthcare industry, with predictions for future expansion. Artificial intelligence in healthcare might result in employment losses but also open up new career prospects, according to a 2017 TCS poll. Accenture estimated that integrating AI into US healthcare will save $150 billion in costs. India's healthcare sector is having trouble since there aren't enough qualified workers. AI may benefit the industry by empowering less experienced workers to make more accurate decisions. In addition, AI can help diagnose common medical disorders, and skilled medical professionals may work with AI to diagnose difficult instances. This strategy could lower treatment costs in India. For diagnosis, hospitals like Fortis Healthcare, Apollo Hospital, LV Prasad Eye Care, and Aravind Eye Care System employ AI technologies created by IBM, Google, and Microsoft, which lowers costs. Throughout the healthcare industry, AI technology is being used in hospitals, pharmacies, and diagnostic facilities.

Robots and artificial intelligence have transcended science fiction and scientific investigation to become an essential element of our everyday lives. AI has a lot to offer the healthcare sector, in particular. Artificial intelligence has the potential to save costs while improving the effectiveness of diagnostic and treatment processes. Additionally, it may reduce the burden for medical professionals, including doctors, surgeons, and nurses, freeing them up to concentrate on important patient care facets.

Furthermore, AI is essential to the efficient operation of hospitals. Medical record administration, revenue management, invoicing, and documentation processes at hospitals may all be streamlined by AI-based technology. Numerous AI-based initiatives and apps have shown to significantly enhance the Indian healthcare system. These include the following: mining medical information; creating treatment plans; anticipating early life-threatening illness detection; screening for diabetes and cancer; and deploying AI-based instruments for ophthalmic and associated therapies. Artificial intelligence (AI) has also been used to the detection and evaluation of COVID-19 as well as the diagnosis and treatment of chronic obstructive pulmonary disease (COPD). AI may also help with the process of designing new medications and provide accurate and timely results for health diagnostics. Lastly, compared to conventional approaches, results may be improved by using sophisticated statistical tools to predict results for expectant mothers and their unborn children in a timely manner. For newborns who are identified as having low birth weight, routine monitoring may also be put into place.

In the Indian healthcare sector, artificial intelligence is being used more and more to speed up diagnosis times and provide reliable findings using a variety of technologies, including image analysis, clinical diagnosis, and laboratory tests. Among the AI technologies most often used in the Indian healthcare sector are machine learning (ML), deep learning (DL), and natural language processing (NLP) (Shastry & Shastry, 2023).

AI AND INDIAGOVERNMENT INITIATIVES

The Indian government has taken strong steps to promote technological innovation in a variety of areas, including the development of artificial intelligence. The creation of an AI-integrated healthcare system in India is one of the main areas of attention for the government and has been identified as a crucial effort. To aid in the accomplishment of this objective, the government has formed a number of organizations, including task forces, policy groups, and authorities (Das et al., 2023). We'll provide a quick rundown of some of the most important subjects being discussed in this context in this section.

National Authority for e-Health

The Government of India's Ministry of Health and Family Welfare founded the National e-Health Authority (NeHA) in 2015. Creating an integrated healthcare information system for India is its main duty. NeHA will provide guidance to all pertinent parties about the usage of artificial intelligence applications in the healthcare industry, as well as other novel technology. It is anticipated that a large number of the process's stakeholders would support this governing body. The NeHA is in charge of making sure that laws and standards pertaining to the privacy and security of patient health data are followed. Its job is to support and encourage customs and traditions that have been developed. As India moves towards e-health, effective use of information and communication technologies (ICTs) offers potential for improving the country's healthcare system. This body's goal is to create a thorough National e-Health Policy and Strategy that would facilitate the incorporation of e-health, including artificial intelligence technology, into every facet of India's healthcare system. The National Law School of India, University of Bengaluru and the Government of India (GOI) have worked together to draught the Health Data Privacy and Security Act of 2016. This legislation will provide guidance on how to maintain and protect ownership, confidentiality, and privacy of different types of health-related data. According to the Government of India, the draught bill was submitted to the Ministry of Health and Family Welfare in July 2016. Before the collected data is used for Big Data analytics, it will be anonymized.

It is important to recognize the potential for re-identification of the anonymised data, especially in situations when the anonymization method is weak and unreliable. This is a worry that has to be addressed by the NeHA, especially when it comes to artificial intelligence-powered data analysis.

Task Force on AI

In order to investigate the possible uses of AI in several sectors to support economic development, the Indian government formed a task group in 2017 in collaboration with the Ministry of Commerce and Industry. The task group is made up of experts from the industry, academic institutions, and government officials from the Ministry of Electronics and Information Technology and NITI Aayog. The administration has reportedly received suggestions from the task group on the use of AI in various industries and research facilities (Navaratna & Saxena, 2023). The task group was assisted by NITI Aayog, a government think tank housed under the same ministry, and the government is likely to incorporate its recommendations. India has established a dedicated task force to apply emerging technology, including artificial intelligence, to the agriculture industry. This task force's primary goals are to advance crop production, optimize agricultural practices, and solve agricultural problems using AI-driven solutions.

A specialized committee has been established to investigate the possible uses of artificial intelligence and machine learning methods in the medical field. Improving patient care, medical diagnosis, and healthcare delivery are the goals. The AI for Education Task Force was formed in the field of education to investigate the possibilities of AI technology in raising educational standards and offering individualized learning opportunities. Artificial Intelligence (AI) is being examined in the financial sector, which includes banking, insurance, and fintech, to increase fraud detection capabilities, optimize operational efficiency, and improve service quality. A task group has been formed to look at the possible uses of artificial intelligence in public administration and e-governance. The main emphasis is on how AI can boost public participation, raise the caliber of government services, and make government procedures more transparent and efficient. India's manufacturing and industrial sectors are also investigating the use of AI technology to boost productivity and competitiveness. A task team has been formed to assess the possible benefits and difficulties in this area. A task force to address the ethical and regulatory aspects of AI is being discussed, in addition to various sector-specific task groups. The task group that has been suggested seeks to uphold ethical standards and encourage the proper use of AI. Task groups are formed by knowledgeable persons in senior roles in industry, academia, and government to work together on developing strategies and proposals for integrating artificial intelligence in their respective domains. The objective is to use AI's potential to improve services, encourage economic development, and address a range of issues affecting India's many sectors.

Formation of an AI Policy Group

The Ministry of Electronics and Information Technology has formed a policy committee including representatives from NASSCOM and members from academic institutions. This group's main goal is to provide a comprehensive and workable framework for integrating artificial intelligence across several businesses. The group is primarily focused on tackling issues with talent development, security, and privacy protection in relation to the use of AI in many fields. A Centre of Excellence for Data Science and AI has been established in India via a partnership between NASSCOM and the Karnataka government. According to NASSCOM's 2017 report, the program has attracted a total investment of INR 40 crore and runs within the framework of a public-private partnership (PPP). This organization acts as a focal point for knowledge, creativity, and teamwork in order to advance these fields and incorporate them into many facets of an organization's operations. Fostering competence and skill in data science and AI approaches is the primary goal of a CoE-DS&AI. A wide range of specialists, including data professionals, AI scientists, researchers, domain experts, and data analysts, are encouraged to collaborate via this effort.

By sharing and combining ideas, this interdisciplinary cohort can handle difficult data and AI-related concerns and capitalize on accompanying possibilities, fostering an environment that encourages creativity and innovation. The Centre of Excellence teams that specialize in Computer Engineering, Data Science, and Artificial Intelligence rely heavily on research and development. To stay ahead of the competition, they aggressively pioneer new developments and are always investigating cutting-edge algorithms, techniques, and technologies. Their objective is to increase the capabilities of artificial intelligence and data-driven decision-making, which will help the company and add to the body of knowledge. A CoE-DS&AI, which formalizes established processes and ensures adherence to tight standards in data collecting, administration, model development, and AI application, cannot exist without best practices.

Maintaining data integrity, guaranteeing project repeatability, and promoting teamwork all depend on consistency. In a CoE-DS&AI, education and training are essential because they provide employees with the skills they need to actively participate in data science and artificial intelligence projects. The democratization of data and AI literacy is aided through workshops, courses, and mentoring program, which help to create a workforce with a solid grasp of data.

A CoE-DS&AI is in charge of setting up and overseeing the essential infrastructure, such as high-performance computer clusters, data storage facilities, and a wide range of software tools, required to support data science and AI projects. These resources are essential for growing AI projects and producing noteworthy results. In addition, the CoE supports projects by serving as an advisory committee, helping various business units define project limits, developing data-driven solutions, and implementing AI plans that align with company objectives. Collaborative approaches guarantee that data and AI projects successfully address real-world problems and provide tangible results. A CoE-DS&AI must have effective governance in order to limit the inherent risks of artificial intelligence while fostering trust among the stakeholders. It does this by establishing frameworks and guidelines for the moral use of artificial intelligence and data, assuring compliance with legal requirements and ethical principles. Additionally, a CoE-DS&AI is essential for evaluating the impact of data in science and AI projects by evaluating key performance indicators (KPIs) and giving organizational leaders insightful data on the ROI and business value generated. These indicators are essential for enhancing data-driven decision-making and strategy refinement.

AI and IPR in India

A campaign to increase public awareness of the value of intellectual property rights (IPR) as important financial assets in the market has been launched by the Department of Industrial Policy and Promotion (DIPP) of the Government of India. The idea of the intellectual property system is anticipated to change as artificial intelligence applications become more prevalent, especially considering how many applications is cross-border and raise jurisdictional issues. The changing environment of IP protection is further enhanced by AI applications' ability to identify and resolve IP rights infringements. Patent protection is granted by the legal system to ideas that meet three criteria: novelty, non-obviousness, and industrial application. Innovations related to artificial intelligence, such as novel algorithms, hardware developments, or AI-driven processes, might be eligible for patent protection if they meet certain criteria. However, figuring out whether software and algorithms qualify for patent protection may be difficult and dependent on the specific details of the invention. Original works—including writing, painting, and music—are safeguarded under India's copyright laws. Copyright issues in artificial intelligence arise when work produced by AI systems is involved, notably in the fields of music, literature, and art. The identification of authorship and ownership of works produced by artificial intelligence may provide substantial issues due to the legal complications involved. Protecting trade secrets in artificial intelligence requires storing datasets, proprietary algorithms, and other critical data securely. India has put in place a thorough legal framework of laws and guidelines to protect private data and trade secrets. The Personal Data Protection Bill (PDPB), which was introduced in India, governs the collection, storage, and use of personal data, including that which is used in artificial intelligence applications, even if it does not specifically address intellectual property rights (IPR) (Chakrabarti & Ray, 2023). Ensuring the protection of individuals' privacy is crucial when artificial intelligence systems that use personal data are developed. The Agreement on Trade-Related Aspects of Intellectual Property Rights is one of the several international agreements on intellectual property that India has ratified. The aforementioned agreements are important for managing intellectual property rights (IPR) in the field of artificial intelligence, including patents, copyrights, and trademarks. The intellectual property rights (IPR) policy of India creates an environment that is favorable to the advancement of innovation and research, particularly AI-related research.

American-Indian Joint Cooperation

India and the United States have collaborated on the US-India Science and Technology Endowment Fund, or Indo-US Science and Technology Fund, or IUSSTF. It was founded in 2009 as a component of the two countries' strategic relationship. The fund's objective is to promote and assist collaborative research and development initiatives in the fields of science, technology, engineering, and mathematics (STEM) among scientists, researchers, and institutions from both nations. The fund provides funding for initiatives in a number of fields, such as biotechnology, cybersecurity, healthcare, renewable energy, and agriculture. Through the facilitation of technology transfer and the commercialization of research products, the main goal is to encourage innovation and entrepreneurship. The foundation offers financial assistance through fellowships, grants, and other program. It acts as a platform to connect investors, partners, entrepreneurs, and innovators. A cooperative board including representatives from the US and India government is in charge of the fund. The practical functioning of the fund is heavily influenced by the ministries or departments of science and technology of both nations. Numerous research initiatives with substantial scientific and technical accomplishments have benefited financially from the fund's support (Bandyopadhyay, 2023). The bilateral ties in science and technology between India and the US have improved because to the Indo-US Science and Technology Fund. The fund's funded efforts have made a significant contribution to the body of knowledge in a variety of scientific domains and will tangibly benefit society.

LEGAL CONCERNS IN AI AND INDIA

There is a chance to solve a wide variety of issues as artificial intelligence and India's legal system come together. We may contribute to the development of a future in which artificial intelligence (AI) is securely and successfully incorporated into our legal system by carefully negotiating the complex and dynamic environment of privacy, bias, liability, intellectual property, regulation, cybersecurity, employment, antitrust, and international collaboration.

1. Data privacy and protection: In India, data privacy and protection are the main legal issues relating to AI. In response, the government introduced the Personal Data Protection Bill 2019, which aims to control how personal data—including that used by AI systems—is handled. Safeguards for data privacy, consent, and security must be established since AI depends on massive datasets. Finding a balance between using AI's potential and guaranteeing the privacy of personal information is difficult, however.
2. Justice and Fairness: It is crucial to uphold justice and fairness in AI systems within the confines of Indian law. AI algorithms may reinforce biases found in training data and provide biased results if they are not carefully designed and assessed. Consequently, it is essential to identify and fix any prejudices and to put in place legal frameworks that ensure AI systems handle all protected groups equally and without bias.
3. Liability and accountability: The legal system in India is debating how to handle situations involving artificial intelligence and liability and responsibility. Because AI systems are complex, it may be difficult to decide who should be held responsible when they make mistakes or do damage. Legal frameworks need to specify the roles and obligations of developers and users in order to provide clear liability standards for AI errors.
4. Copyright concerns: With the increasing use of AI by Indian enterprises, it is becoming more and more important to handle ownership and copyright concerns pertaining to AI-generated inventions and material. It could be necessary to make changes to India's present legal system, which addresses trade secrets, copyrights, patents, and trademarks, in order to effectively address the particular difficulties related to AI-generated innovations and breakthroughs. Intellectual property issues are thus becoming more and more important in the Indian AI scene.
5. Frameworks for regulation and certification: Creating frameworks for regulation and certification is essential to handling legal issues arising from the use of AI. As AI systems proliferate across many sectors, it is imperative to keep a close eye on their development, deployment, and use. Adopting certification and compliance standards might be crucial to ensuring AI systems adhere to strict ethical and safety requirements, especially in industries like banking, healthcare, and autonomous cars.
6. Cybersecurity: One of the main issues with artificial intelligence is making sure cybersecurity. Protecting AI systems against cyberattacks and data breaches is crucial as AI is being incorporated into vital infrastructure and industries. Legal frameworks must be changed in order to adequately address the protection of AI systems and the data they handle.
7. Employment and labor laws: The development of artificial intelligence has resulted in a substantial modification of employment and labor laws in India. AI's growing use in the workplace has the potential to have a big influence on things like worker rights, job classifications, and job displacement. Consequently, it is essential to modify legislative frameworks to properly address evolving circumstances and safeguard the well-being of workers.
8. Competition and antitrust laws have become more complicated as a result of the use of artificial intelligence in business operations. Antitrust and competition laws are under serious scrutiny in light of the use of AI to rig markets or get an unfair advantage. India's legal system has to change fundamentally in order to address these expanding issues in a comprehensive manner. Addressing

the possibility of antitrust and competition breaches is crucial given the exponential rise of AI technology and its growing usage in business practices. The use of AI to restrict market entrance or establish market monopolies may be the cause of these infractions. For the legal system to successfully handle these problems, it must keep up with the fast advancements in artificial intelligence technology. India's legal system must change and adapt to take into consideration the difficulties that AI technology in business practices presents. The goal of this development should be to provide the required instruments for effectively detecting and resolving any antitrust and competition breaches. By doing this, India can make sure that its business practices continue to be fair, competitive, and compliant with international norms.

9. International cooperation: Overcoming cross-border legal barriers pertaining to artificial intelligence requires effective international cooperation. This is particularly important for fostering data interchange, harmonizing AI rules, and addressing legal concerns related to AI across several areas. Since AI knows no territorial bounds, it is imperative to build efficient collaboration with international players. To provide a coherent and harmonized approach to the global governance of AI, such cooperation is required.

RULES AND REGULATORY LANDSCAPING

The Indian government has made significant progress in developing policies for using AI into healthcare. The existing regulatory framework has to be improved in order to develop successful policies (Paul et al., 2018). We provide a brief synopsis of the laws, policy issues, and standards that are now being examined in this area.

The IT Act of 2000 and the 2011 Rules

In order to create AI-integrated healthcare systems, physicians, patients, and service providers must communicate effectively and openly. This is especially crucial when it comes to contacts between doctors and patients, since private information about patients, such as

Medical history, mental health state, and personal information are disclosed. The Information Technology (Reasonable Security Practises and Procedures and Sensitive Personal Data or Information) Rules, 2011 set out guidelines for the gathering, storing, and transmission of SPDI in order to guarantee the secure management of sensitive personal data. Any organization that gathers, maintains, or transmits SPDI must have the patient's approval in writing before using their information. Notifying the patient of the purposes for which their data is being collected is also crucial (Shah & Srivastava, 2014). Security processes and methods must be in place to ensure that SPDI is protected, and the ISO 27001 standard is generally acknowledged as the established methodology for doing so.

Nevertheless, it has been noted that weak governance practices and improper application of existing laws sometimes jeopardize SPDI security. The Supreme Court of India's recognition of the necessity for an all-encompassing, unbiased data protection framework that takes various technologies into account is positive. In response to the Puttaswamy decision in 2012, which upheld privacy's constitutional validity as a basic right, the Indian Parliament tabled a draught Data Protection Bill in 2018.

The 1956 Medical Council of India Act and the 2002 MCI Code

The MCI Code, 2002 is the general term for the Indian Medical Council (Professional Conduct, Etiquette and Ethics) Regulations, 2002. The ethical and professional requirements for the interaction between patients and healthcare practitioners are outlined in this collection of rules. It provides instructions on how to communicate prognoses, protect patient privacy, and digitize medical data so they may be easily retrieved. These rules are monitored for conformity by the Medical Council of India (MCI) (Mukhopadhyay, 2009). However, the Medical Council of India (MCI) has been superseded by the National Medical Commission (NMC) as the main regulatory agency in charge of managing medical education and the medical profession in the nation. The NMC is now in charge of overseeing, coordinating, and upholding medical education and practice standards throughout the country of India.

Approach to Open Data

The Indian government has put in place the National Data Sharing and Accessibility Policy (NDSAP) to improve the openness, availability, and usability of data generated by various departments and agencies for use by different stakeholders, such as businesses, researchers, and the general public (Shah et al., 2016). Although it includes concepts that support open data practices, the NDSAP is not a "open data policy." Guidelines for government data sharing are provided by the policy, which prioritizes privacy and security concerns while highlighting proactive sharing of non-sensitive or secret data. The guideline also emphasizes how crucial it is to provide consumers access to metadata and cataloguing details so they can comprehend the context of the material. In order to improve accessibility, the NDSAP encourages the adoption of data standards and formats. It also suggests updating datasets on a regular basis to guarantee their relevance and quality. The guideline promotes open formats and APIs for data access and the use of digital platforms and technology to improve data accessible. In terms of copyright, ownership, and licensing, the policy suggests using open licenses to make it easier to reuse and redistribute data.

The NDSAP encourages data exchange but also acknowledges the need to protect private and sensitive information. It does this by putting policies in place to preserve data privacy and improve security. It is important to remember that restrictions may apply to certain government datasets because of security, privacy, or other relevant factors. In general, the NDSAP creates the conditions for improved data availability, accessibility, and transparency for different stakeholders.

Digital Health Records

An important milestone for India's healthcare sector was the adoption of the Electronic Health Records (EHR) Standards of 2016. By creating a thorough framework for the adoption and use of electronic health records, these recommendations sought to bring about a revolutionary transformation in the administration of healthcare information. The main aim of the EHR Standards 2016 was to provide a structure that would enable the seamless transfer of health-related data across diverse healthcare settings. The objectives of this framework were to advance interoperability, strengthen data security, and improve patient care. According to the standards, an Electronic Health Record (EHR) is a digitalized version of a patient's medical record that contains a wide range of health-related data that has been collected over time. The foundation for creating and preserving standardized Electronic Health Records (EHRs) in the healthcare sector was set by this definition (Pai et al., 2021). The recommendations also outlined an organized process for data elements inside electronic health records (EHRs), which include critical information such patient demographics, medical history, allergies, prescriptions, test results, and clinical notes. The purpose of the recommendations was to encourage consistency and thoroughness in the information included in electronic health records (EHRs), hence improving their dependability and functioning as a complete source of patient data.

The inclusion of interoperability was a fundamental component of the 2016 EHR Standards. In order to achieve this goal, the guidelines encouraged the use of technology and open standards that allowed health data to be easily shared across various healthcare systems and providers. Prioritizing interoperability was primarily done to break down siloed information repositories so that medical professionals could get all patient data, regardless of where it came from. The recommendations recognized the need of maintaining patient privacy and data security. In order to safeguard sensitive patient data, the author outlined guidelines for guaranteeing the safety of health information, with special emphasis on the need of enacting stringent access controls, encryption, and other security measures. In order to protect patient privacy and promote trust in electronic health record (EHR) systems, the recommendations attempted to address three essential components.

The adoption of Electronic Health Record (EHR) Standards has been made possible in large part by Health Information Exchanges (HIEs). The development of Health Information Exchanges (HIEs), which served as safe channels for the interchange of health information across different healthcare organizations, was made easier by the use of standards. Health Information Exchanges (HIEs) were envisioned as essential components of the healthcare system, promoting collaboration and smooth information sharing. The recommendations also included the important topic of managing patient consent. It was emphasized how important it is to provide patients the power to decide who may access their electronic health records (EHRs) and under what circumstances. The purpose of this policy is to protect patients' rights and freedom of choice when it comes to sharing their health information.

The installation of audit trails was necessary to keep an eye on how patient data was being used and altered inside electronic health record (EHR) systems. By taking this action, healthcare procedures will become more accountable and transparent (Sahney & Sharma, 2018). All things considered, India's EHR Standards of 2016 demonstrated a comprehensive and forward-thinking approach to improving healthcare information management. The goals of these standards were to guarantee patient data security and confidentiality, promote seamless communication and exchange of information across various systems, standardize the content of electronic health records (EHRs), and give individuals more control and authority over their personal health information.

While these recommendations were a significant step forward, further improvements and developments could have been spurred by the constantly changing healthcare environment and technology breakthroughs. Therefore, in order to guarantee compliance and alignment with modern healthcare practices in India, it is advised to consult the most current regulations and legislation.

RISKIER HEALTH LINKS IN INDIA

A lot of attention is being paid to the use of artificial intelligence in the field of healthcare and medical systems. These days, a growing number of businesses are actively working to disrupt the healthcare sector by using artificial intelligence. Governmental organizations, tech companies, and conventional healthcare service providers now have many important opportunities to collaborate as a result of the adoption and use of AI-based solutions in the healthcare industry.

By using the newest technology and collaborating with top organizations, India is making major advancements in the healthcare industry by offering easily accessible and reasonably priced healthcare services. In order to save diabetic patients from becoming blind, NITI Aayog is working with Microsoft and Forus Health to develop automated methods for the early identification of diabetic retinopathy. Furthermore, to enhance healthcare services in rural regions, the International Centre for Transformational Artificial Intelligence (ICTAI) is being built. By working with the Central Tuberculosis Division to provide interventions for TB patient care and support India's National TB Elimination Program in incorporating AI technology, Wadhwani AI is having an effect.

By introducing a COVID-19 live monitoring app for the state and establishing the Microsoft Intelligent Network for Eyecare, the Telangana government is leading by example. The introduction of Healthcare Pods by Vevra by the Karnataka government is a creative way to manage and stop the spread of infectious illnesses like TB and Covid-19. The Kerala government announced the AI Machine Learning Developer Program to solve the skills gap in the industry. The program will provide people with the necessary AI and data science capabilities to fulfill the rising demand in the healthcare sector. Finally, AI-driven startups offering cutting-edge healthcare services and solutions with an emphasis on care management and patient interaction include Netmeds, Innovaccer, Tata Health, PharmEasy, HealthifyMe, Practo, NIRMAI, HealthPlix, Lybrate, and HealthKart.

PRESENTATIONS AND GUIDELINES

India's Artificial Intelligence Challenges

India's healthcare system might undergo a change thanks to artificial intelligence, which could also improve the system's effectiveness and efficiency. However, there are some challenges that come with the adoption and widespread usage of this technology that need to be successfully addressed (Sharma et al., 2022).

Here are a few major obstacles to overcome:

1. Data privacy: The healthcare sector places a great deal of weight on this problem, which is equally relevant in the Indian setting. With regard to AI-driven healthcare solutions, protecting patient data from misuse, breaches, and unauthorized access is crucial. Maintaining patient trust requires strict respect to data privacy laws, such as the Personal Data privacy Bill.
2. Data Quality: There is a great deal of variation in the quality of healthcare data in India, which is typified by inaccuracies and inconsistencies in medical records. Healthcare organizations must improve the way they gather, store, and manage data if they want to maximize the use of AI in healthcare. For data exchange to be seamless and effective, standardized data formats must be established, and interoperability between different systems must be encouraged.
3. Facilities and Connection: More healthcare facilities and dependable internet connections are needed in many Indian areas, especially the rural ones. Deploying AI applications often requires access to advanced equipment and high-speed internet, which makes it challenging to extend the reach of these solutions to underserved areas. To guarantee equitable access to AI-powered healthcare, the digital divide must be closed.

4. Skilled workers: There are currently too few people in the workforce who are proficient in both artificial intelligence and healthcare. The personnel must be trained and developed to be competent in developing, executing, and maintaining artificial intelligence programs in the field of medicine. Initiatives to support interdisciplinary education and encourage collaboration across various academic fields must be put into action immediately.
5. Regulatory Framework: A thorough and well-defined framework is required to address the existing condition of artificial intelligence regulations in India's healthcare industry. To maintain patient safety, protect data privacy, and encourage the moral use of artificial intelligence technology, clear rules and regulations must be established. Such regulations are able to keep up with the rapid advances in technology and healthcare practices.
6. Ethical Considerations: The necessity to protect patient autonomy and ethical issues about AI systems that may display bias must be addressed. It is imperative that justice, accountability, and transparency be given top priority when creating and deploying AI systems in order to avoid discrimination and preserve public confidence.
7. Cost-effectiveness: Using AI solutions in the healthcare industry may be prohibitively expensive, which would restrict their accessibility, especially for smaller healthcare providers and underserved populations. It is necessary to make an effort to create plans for lowering costs and improving the accessibility of AI technology.
8. Degree of reluctance: When implementing artificial intelligence technology, the healthcare sector often experiences resistance to change. Using artificial intelligence for advice or changing standard operating practices may cause some hesitation among healthcare professionals. Investing in initiatives that educate, train, and build the confidence of healthcare professionals in artificial intelligence is imperative.
9. Data Localization: Indian data localization regulations might complicate data sharing and impede cross-border cooperation, which are essential for developing and investigating artificial intelligence. Finding a balance between localizing data and guaranteeing access to it for research and innovation is the issue.
10. Patient Knowledge and Trust: A greater understanding of the capabilities and limitations of artificial intelligence in the healthcare industry may be necessary for a large number of patients in India. For artificial intelligence-powered healthcare solutions to be implemented and accepted, patients must be informed about them and given trust in their abilities.

It will need a team effort to resolve these problems, including the government, medical facilities, IT companies, and other pertinent organizations. So as to completely fully use artificial intelligence's benefits for India's healthcare industry, cooperative initiatives, funding for infrastructure and education, and the establishment of a comprehensive regulatory framework are essential. Despite these challenges, artificial intelligence is a revolutionary force that India's healthcare sector should actively pursue due to its potential to improve healthcare outcomes, reduce costs, and increase access to high-quality treatment. Suggestions for Artificial Intelligence in India

In order to overcome the challenges and fully realize the potential benefits of this technological revolution, it is imperative that suggestions be developed for the efficient integration of artificial intelligence into the healthcare sector in India (Chatterjee et al., 2022).

Here are a few of noteworthy recommendations:

1. Data Governance and Security: It is crucial to create thorough frameworks for data governance those priorities protecting data integrity, security, and privacy. To protect patient data, encryption, access controls, and recurring audits are crucial. Establishing and adhering to data protection laws, like the Personal Data Protection Bill, is crucial to promoting patient trust and protecting sensitive data.
2. Data Quality Improvement: To raise the precision and thoroughness of healthcare data, funding initiatives to enhance data quality are advised. Therefore, encouraging the use of standardized data formats and facilitating interoperability across various healthcare institutions are crucial.
3. Infrastructure Development: Prioritizing the expansion of healthcare facilities in impoverished areas is crucial, with a particular emphasis on rural areas. It is recommended that funds be set aside for the acquisition of a stable internet connection and technological infrastructure in order to enable the successful deployment of AI applications.
4. Workforce Development: Putting in place incentives and training programs is one way to start addressing the skills gap in AI and healthcare. These programs provide people with the information and abilities needed to successfully explore and contribute to both disciplines. By creating these kinds of programs, people would be able to get the knowledge and skills required to close the gap between AI and healthcare. Furthermore, offering incentives might be a driving force behind people's active participation in these training programs, boosting their skill development and eventually closing the skills gap that now exists. Promoting interdisciplinary collaboration between data scientists and medical professionals is crucial.
5. Regulatory Framework: To achieve a balance between encouraging innovation and guaranteeing patient safety, a thorough regulatory framework specifically designed for artificial intelligence in healthcare has to be established. Ensuring that regulations are flexible enough to keep up with the rapid advancements in technology and the always changing field of healthcare practices is imperative.
6. Ethical Considerations: It's critical to address prejudice, responsibility, and openness in order to encourage the ethical growth of AI. Committees for ethical reviews need to assess how AI affects patient privacy and care.
7. Patient Awareness and Involvement: It's critical to enlighten patients about the relevance of data protection, the role of artificial intelligence in the healthcare sector, and its possible advantages. It is critical to include patients in the decision-making process when it comes to AI-powered healthcare solutions.
8. Collaborative Initiatives: To enable the broad use of artificial intelligence, cooperation between eminent international organizations, governmental bodies, healthcare facilities, technology companies, and research institutions is necessary. The author suggests setting up research centers and innovation hubs for artificial intelligence as a means of fostering collaboration and idea sharing. These hubs will facilitate cooperative relationships and help integrate AI into a wider range of industries.
9. Telemedicine Expansion: Promoting and improving the availability of telemedicine services is essential to improving healthcare accessible in underserved and rural areas. Ensuring the seamless integration of artificial intelligence technology with telemedicine systems is crucial for remote monitoring and diagnosis.
10. Data Sharing for Research: Encourage healthcare organizations to share de-identified patient data for research purposes while maintaining the requirement to protect patient privacy. Promoting the

use of data-sharing programs and platforms is crucial if we are to increase the effectiveness of collaborative research projects.

THE STUDY'S CONCLUSION AND IMPLICATIONS

The research explores the possible uses of AI in India's healthcare sector, providing important information that may guide policy decisions in this area. The paper offers a series of suggestions for creating AI- relevant Indian health policy. Policymakers may find these recommendations useful in helping to formulate relevant policies. Nevertheless, there can be difficulties in putting some of the theoretical recommendations into practice. Because of this, this study may be a crucial resource for researchers in the future, providing insightful information to help them improve their proposals. Moreover, the results might help other developing countries formulate rules about the use of AI in healthcare.

India's healthcare sector is already reaping the benefits of artificial intelligence's many uses. While artificial intelligence (AI) has not yet fully replaced human labor, it has greatly improved human capabilities. The healthcare system still has a number of difficulties, however. The integration of AI applications into the healthcare system must be done strategically and methodically in order to realize the full potential of this technology. This calls for the creation of a suitable legislative framework for managing privacy and security issues in an efficient manner. As both Indian and foreign firms are aggressively adopting AI to better the healthcare landscape in India, the present state of things surrounding the usage of AI in the healthcare industry is promising. In order to effectively incorporate AI into India's healthcare system, appropriate legislative frameworks that place a high priority on accountability and transparency must be developed.

But it's also critical to make sure that these rules don't impede the development of innovative AI. For India to see the growth and success of a healthcare ecosystem using AI technology, regulatory measures to avoid possible exploitation must be balanced with the creative potential of AI. Policymakers may use this research's insightful findings to create a thorough and workable plan for integrating AI technologies in the healthcare sector. This study effectively makes important recommendations that might help the government create a plan for integrating AI in India's healthcare system, even if it acknowledges certain expected limits.

REFERENCES

Ahmed Kamal, M., Ismail, Z., Shehata, I. M., Djirar, S., Talbot, N. C., Ahmadzadeh, S., & Kaye, A. D. (2023). Telemedicine, e-health, and multi-agent systems for chronic pain management. *Clinics and Practice*, *13*(2), 470–482. doi:10.3390/clinpract13020042 PMID:36961067

Alnsour, Y., Hadidi, R., & Singh, N. (2019). Using data analytics to predict hospital mortality in sepsis patients. *International Journal of Healthcare Information Systems and Informatics*, *14*(3), 40–57. doi:10.4018/IJHISI.2019070104

Bandyopadhyay, M. (2023). Development Paradigm of India as an Enabler for Practicing Science, Technology and Innovation Diplomacy. In *Science, Technology and Innovation Diplomacy in Developing Countries: Perceptions and Practice* (pp. 115–139). Springer Nature Singapore. doi:10.1007/978-981-19-6802-0_8

Chakrabarti, M. S., & Ray, M. R. K. (2023). Artificial Intelligence And The Law. *Journal of Pharmaceutical Negative Results*, 87–95.

Chatterjee, S., Chaudhuri, R., Thrassou, A., & Vrontis, D. (2022). Technology disruption in healthcare: Artificial intelligence application, challenges, and policy recommendations in India. *International Journal of Internet Marketing and Advertising*, *17*(3-4), 394–414. doi:10.1504/IJIMA.2022.126716

Chatterjee, S., & Kar, A. K. (2018). Regulation and governance of the Internet of Things in India. *Digital Policy. Regulation & Governance*, *20*(5), 399–412. doi:10.1108/DPRG-04-2018-0017

Das, S., Garg, V., & Joseph, J. J. (2023, March). Advancements in e-Governance Initiatives: Digitalizing Healthcare in India. In *International Conference on Advances in IoT and Security with AI* (pp. 287-296). Singapore: Springer Nature Singapore. 10.1007/978-981-99-5085-0_28

Eubanks, R. (2017). *Artificial Intelligence and the Healthcare Ecosystem–Part One.*

Khumalo, N. B., & Mnjama, N. (2019). The effect of eHealth information systems on health information management in hospitals in Bulawayo, Zimbabwe. *International Journal of Healthcare Information Systems and Informatics*, *14*(2), 17–27. doi:10.4018/IJHISI.2019040102

Mukhopadhyay, B. (2009). Undergraduate medical education in India and scope of pediatric surgery teaching. *Journal of Indian Association of Pediatric Surgeons*, *14*(3), 83. doi:10.4103/0971-9261.57697 PMID:20376246

Navaratna, A. R., & Saxena, D. (2023). An Indian Approach to AI Policy: A Comparative Study Between Three Sectors. Handbook of Evidence Based Management Practices in Business, 440.

Pai, M. M., Ganiga, R., Pai, R. M., & Sinha, R. K. (2021). Standard electronic health record (EHR) framework for Indian healthcare system. *Health Services and Outcomes Research Methodology*, *21*(3), 339–362. doi:10.1007/s10742-020-00238-0

Paul, Y., Hickok, E., Sinha, A., Tiwari, U., Mohandas, S., Ray, S., & Bidare, P. M. (2018). *Artificial intelligence in the healthcare industry in India.* The Centre for Internet and Society.

Rana, M., & Bhushan, M. (2023). Machine learning and deep learning approach for medical image analysis: Diagnosis to detection. *Multimedia Tools and Applications*, *82*(17), 26731–26769. doi:10.1007/s11042-022-14305-w PMID:36588765

Sadybekov, A. V., & Katritch, V. (2023). Computational approaches streamlining drug discovery. *Nature*, *616*(7958), 673–685. doi:10.1038/s41586-023-05905-z PMID:37100941

Saha, T. (2018). *Artificial intelligence based healthcare startups in India.*

Sahney, R., & Sharma, M. (2018). Electronic health records: A general overview. *Current Medicine Research and Practice*, *8*(2), 67–70. doi:10.1016/j.cmrp.2018.03.004

Serrano, D. R., Kara, A., Yuste, I., Luciano, F. C., Ongoren, B., Anaya, B. J., & Lalatsa, A. (2023). 3D printing technologies in personalized medicine, nanomedicines, and biopharmaceuticals. *Pharmaceutics, 15*(2), 313. doi:10.3390/pharmaceutics15020313 PMID:36839636

Shah, H., & Srivastava, A. (2014). Signature provisions in the amended Indian Information Technology Act 2000: Legislative chaos. *Common Law World Review, 43*(3), 208–230. doi:10.1350/clwr.2014.43.3.0271

Shah, L., Dandawate, V., & Gutam, S. (2016). *The changing landscape and future of Open Access in India*. Not Available.

Sharma, M., Luthra, S., Joshi, S., & Kumar, A. (2022). Implementing challenges of artificial intelligence: Evidence from public manufacturing sector of an emerging economy. *Government Information Quarterly, 39*(4), 101624. doi:10.1016/j.giq.2021.101624

Shastry, K. A., & Shastry, A. (2023). An integrated deep learning and natural language processing approach for continuous remote monitoring in digital health. *Decision Analytics Journal, 8*, 100301. doi:10.1016/j.dajour.2023.100301

Subbiah, V. (2023). The next generation of evidence-based medicine. *Nature Medicine, 29*(1), 49–58. doi:10.1038/s41591-022-02160-z PMID:36646803

Turing, A. M. (2009). *Computing machinery and intelligence*. Springer Netherlands. doi:10.1007/978-1-4020-6710-5_3

Turing, D. (2015). *Prof: Alan Turing Decoded*. Pitkin.

Warburton, P. E., & Sebra, R. P. (2023). Long-Read DNA Sequencing: Recent Advances and Remaining Challenges. *Annual Review of Genomics and Human Genetics, 24*(1), 24. doi:10.1146/annurev-genom-101722-103045 PMID:37075062

Chapter 7
Harnessing the Potential of Metabolomic Biomarkers for Metabolic Health

Shifali G. Gupta
Chitkara College of Pharmacy, Chitkara University, India

Paranjeet Kaur
Chitkara College of pharmacy, Chitkara University, India

Rajwinder G. Kaur
Chitkara College of Pharmacy, Chitkara University, India

Thakur G. Singh
Chitkara College of Pharmacy, Chitkara University, India

ABSTRACT

Metabolomics is a fascinating area that specializes in analyzing the small molecules in our frame, like hormones and enzymes, that play an important role in our metabolism. With the aid of reading these molecules, scientists can gain insights into our metabolic fitness and the way our body processes nutrients. It entails different factors such as energy production, nutrient usage, and waste elimination. Through metabolomics, researchers can become aware of precise metabolic biomarkers that imply the state of our fitness. Through studying these markers, they are able to hit upon early symptoms of metabolic disorders like diabetes, obesity, and cardiovascular ailment. This knowledge can then be used to broaden customized interventions and treatments to improve metabolic health. It is able to assist in the development of centered treatment plans, personalized nutrition plans, and preventive strategies. In this chapter, we will attempt towards achieving and maintaining optimal health with the study of metabolomics biomarkers.

DOI: 10.4018/979-8-3693-1922-2.ch007

INTRODUCTION

Metabolomics is a rapidly growing 'omics' subject that follows genomics, transcriptomics, and proteomics, and it's far an important issue of systems biology. In metabolomics, the foremost common investigation things are natural liquids (e.g., serum, plasma, pee, spit, cerebrospinal liquid, bile, amniotic liquid, tears, pancreatic juice, intestinal liquid, and breast drain), human tissues, and cells. Metabolites are the cease products of complex cell regulatory networks, and they are able to affect or maybe alternate rules via comment loops.

The terms metabolomics and metabolomic are frequently utilized interchangeably and both are categorized as 'technology'. Sequence genomics recognizes qualities, transcriptomics shows which qualities are being changed over into RNA, proteomics proposes whether the RNA is deciphered into protein and what submit-translation alterations are made to the protein, and metabolomics/metabolomics recommends whether protein expression results in metabolic alterations. Metabonomic is described as the "quantitative size of time-related multiparametric metabolic responses of multicellular systems to pathophysiological stimuli or genetic amendment". Fiehn defines metabolomics as "a comprehensive and quantitative evaluation of all metabolites that would help researchers apprehend such (organic) systems." Because such an evaluation famous the metabolome of the biological system under investigation, this technique must be called metabolomics (Tomita & Kami, 2012).

A Historical Perspective

Despite the truth that metabolic profiling or metabolomics could be a noticeably unused subject in frameworks of science, the primary surveys of metabolic ponders may be followed lower back to noteworthy China, wherein ants have been utilized to identify diabetes through evaluating glucose degrees in pee tests. "Urine charts", which correlated the scent, taste, and coloration of pee, have been utilized within the Center Ages to analyze a repercussion of metabolic-associated therapeutic circumstances. Roger Williams and his colleagues proposed and tried the thought that people may additionally have a superb "metabolic pattern" that can be 'fingerprinted' through analyzing their natural liquids within the late 1940s. They utilized paper chromatography to discover that metabolic styles contrasted significantly among subjects but remained very standard inside an individual. Their research on a ramification of topics, counting heavy drinkers and schizophrenics, has found that each of those offices encompasses a wonderful metabolic test. In the 1960s and1970's, mechanical progress in gas chromatography (GC), liquid chromatography (LC), and mass spectrometry (MS) empowered quantitative metabolic profiling. Horning and partners utilized GC-MS to grade metabolites in human pee and tissue extricates effectively in 1971 (Zhang, Sun, & Wang, 2012). Through the 1970s and early 1980s, Morning, Pauling, Robinson, and their investigate bunches drove the improvement of GC-MS-based techniques for metabolic estimations in organic liquids. Afterward, advances in excessive-resolution/sensitivity MS and NMR instrumented, mixed with multivariate measurable assessment, empowered metabolomics to develop to be an out of the blue creating subject in framework science over the past decade.

Metabolomics Method

The concept of metabolomics was first put forth by educator Jeremy Nicholson in 1999. Metabolism may provide a means of quantitatively examining every molecule in living organisms and determining the correlation between metabolites and metabolic and physiological changes, making use of the advances in genomes and proteomics. Particles of a relative atomic weight of fewer than 1000 make up the majority of the objects studied in metabolomics, which is an essential component of framework science. Progressed analytical strategies blended with sample notoriety and proficient frameworks are the elemental strategies of metabolomics to investigate. The investigation method comprises design instruction, design assessment, measurements assessment and metabolic pathway assessment. Organic tests analysed primarily incorporate blood, pee, faeces and the like. Sample guidance for the most part consisted of test extraction, processing of compounds, and partitioning (Tsutsui et al., 2011). To win the exam, the design must be of a high caliber. Before the material is analyzed, extensive pretreatment is often necessary. Pretreatment is an essential component of sample preparation. In pattern training, sample analysis includes sample extraction, sample supply management, and sample capacity to ensure that sample processing does not change the metabolic records. Remarkable examinations have unmistakable necessities for sample handling. The records of the test may moreover exchange with the environment in which it is utilized. Mass spectrometry is used to monitor metabolism, but there are positive requirements for the pH and ionic quality of the test, which may be elite. Within the test case, macromolecules need to be eliminated and small particles need to be recovered as thoroughly as possible. Gas chromatography-mass spectrometry (GC-MS), liquid chromatography-mass spectrometry (LC-MS), and nuclear magnetic resonance spectroscopy (NMR) are examples of the instruments used in metabolomics research. Among them, the NMR approach frequently uses a phosphorus range (31P NMR), a carbon range (13C NMR), or a hydrogen range (1H NMR), the most widely used of which is 1H NMR. Metabolomics study is still a whole rationalization of the realities. Due to the truth the one-of-a-kind outline of metabolomics investigation is complex and the information amount is expansive, it can't be analysed by way of ordinary information preparing strategies, so metabolomics examination incorporates an extraordinary investigation with 3 steps: insights extraction, information preprocessing and design ubiquity. Noise filtering, peak resolution, peak arrangement, top coordinating, normalization, and many other preprocessing steps are included in the preprocessing processes (Zhang et al., 2014). Employing either an administered or an unmonitored mode can be achieved by pattern recognition. Unsupervised models include principal component analysis (PCA), cluster analysis (HCA), and nonlinear mapping (NLM). Supervisory models include artificial neural networks (ANN), positive and negative least two-square discriminant analysis (OILS-DA), partial least squares discriminant analysis (PLS-DA), and many more. Metabolomics has to handle a gigantic sum of crude truths and extricate the capacity records via statistical processing and provide an explanation for the biological merit included inside the records, so as to be the important part of metabolomics research. Utilizing a measurable approach to profoundly investigate the data in metabolic tests will emphatically exceed expectations in the improvement of metabolomics generation (Haviland et al., 2012).

Metabolomics investigates endogenous little particle compounds, comprising sugars, characteristic lipids, acids, amino acids, and so forth. The goal of metabolomics is to identify the affected microbes and then target them to certain linked diseases. Metabolomics era has been widely applied in clinically connected fields as a utility-driven growing mechanical information, with significant effects on lifestyle, nutrition and plant sciences, drug development, toxicity, environmental science, and pharmaceuticals (Tenori et al., 2012).

Mass Spectrometry Technology

British physicists effectively considered the sector to begin in 1919 with mass spectrometers. Mass spectrometers became popular explanatory instruments used by chemists by the mid-1900s, and they are increasingly being used for natural science research. Compared with in vogue expository strategies, one explanatory technique for determining the mass-to-charge ratio of particles is mass spectrometry. It provides important supplementary information in a single evaluation and is crucial for the identification of natural atoms. It can quickly and precisely determine the atomic weight of biomacromolecules, enabling proteomics considerations ranging from protein identification evidence to advanced fundamental research and analyses into the interrelationships between different proteins. It is constantly evolving with the use of mass spectrometry innovation, and the integration of mass spectrometry and partition innovation could lead to a breakthrough in clinical partition processes. In chemistry, the environment, power, pharmaceutical, sports exercises, criminal mechanical period, life sciences, and other fields, mass spectrometers are widely used due to their intemperate, moo test admissions, quick assessment, and advantages of division and character (Yoshida et al., 2012).

The advancement of science and innovation in the twenty-first century presents new obstacles for exploratory testing methods. In contrast to classical chemical investigation procedures and ordinary instrumental assessment strategies, in situ, real-time, on-line, non-destructive, high-throughput, excessive-sensitivity, excessive-selectivity, and occasional-loss are ceaselessly the objectives sought after by means of investigators in cutting-edge explanatory innovation. Among the several explanatory test procedures, the MS strategy is one of the most widely used and is regarded as an extremely specific procedure. The current demands of clinical and technological development have been satisfied by electrospray desorption ionisation technology, corona discharge real-time direct analysis ionisation technology, and electrospray ionisation technology. These technologies have also created an opportunity for quick MS evaluation of complex samples.

Metabolic Mechanism

One of the most important tasks in the treatment of clutter is to make the metabolic mechanism of the clutter crystal clear. Due to the truth the event of illnesses is laid moo with numerous components, counting natural and hereditary components, analysts utilize metabolomics techniques to clarify the comes about of different variables on sickness components. Experts have shown that breviscapine is beneficial in mice with Alzheimer's disease caused by amyloid β1-42, and they have discovered that breviscapine specifically accelerates the metabolism of phospholipids via controlling serotonin levels and other factors. Zhao, G. et al. investigated the hypoglycemic influence of Crassostrea gigas polysaccharides using UPLC-Q-TOF-MS for the case of analysing acing and memory impedance. Nineteen metabolites, specifically related to the metabolism of amino acids, carbohydrates, and purines, have been employed as capacity-restoring indicators. Ji, H. et al. discovered that aloe emodin has a fantastic healing effect on HFD-induced hyperlipidemia by directing metabolic problems using the UPLC-ESI-QTOF-MS timeframe. The metabolomics procedure is moreover utilized to require a see at ailments comprehensive of joint pain. Pang, Z. et al. found that methotrexate can restrain the advancement of joint pain by restraining the enactment of NF-kappa B and NLRP3/caspase-1 pathways and directing inflammatory-associated metabolic systems. An infection is a result of the sickness of the activity mechanism inside the living outline. But distinctive ailments may be motivated through different mechanisms of activity. Thus, analysts talk about the mechanism of activity of the malady to be able to discover a modern approach for affliction treatment. According to Rodrigues et al., a GC-MS metabolomics-based approach for evaluating unstable natural chemicals may be a financially advantageous process for identifying underutilized and unique biomarkers that can advance the detection of bladder cancer (Zhang, Sun, Wu et al, 2012). To measure the fecal metabolism of Clostridium difficile, high quality and bad diarrhea, and wholesome manipulation of stool samples, Zhou, P. et al. used ultra-overall execution liquid chromatography-mass spectrometry. They saw that Clostridium difficile possesses definite fecal metabolization traits. In order to examine the role of paeoniflorin in the management of adversely susceptible reactions, Shou, Q. et al. linked metabolomics techniques. By regulating fatty acid metabolism, aggravation, and system-level adhesion, they discovered that it can treat asthma that is predisposed to negative outcomes. In its typical state, the body's natural mechanism is largely in a balanced state. A disease of corresponding severity would disrupt this state of relative harmony. Evidence has shown that angiotensin-(1–7) can exacerbate metabolic irritations associated with cardiac fibroblast activation that are triggered by angiotensin II. Arachidonic acid and pyroglutamate may be useful indicators for assessing the anti-fibrotic effects of angiotensin-1–7. A single dosage of S-1-propenyl cysteine (S1PC) had an antihypertensive effect, according to research by Matsumoto, T. et al. Using a metabolomic technique, they examined the possible component of S1PC's antihypertensive effect in rats that had unexpectedly developed hypertension. They concluded that S1PC alters histidine metabolism and then exerts its antihypertensive effect through the crucial histamine H3 receptor (Arakaki et al., 2008). The metabolic mechanism's equilibrium is dependent no longer best upon the steadiness of the inside surroundings, but moreover on man or lady variations. A way to see at the mechanism of activity will be a pivotal figure to advance propels in restorative medicate and science.

Metabolic Markers

In order to further the treatment of ailments, additional remedial demonstration and evaluation techniques are needed as soon as possible. To this end, metabolomics provides current potential results for biomarker identification in complex diseases and may provide neurotic data of diseases prior to conventional innovation. It is the systematic study of low-molecular-weight metabolites in natural experiments and conclusion as a crucial tool in therapeutic thinking and the study of human disease, and it has been realized in the disclosure and unique evidence of the irritation pathways (Zhang, Wu, Yin et al, 2012). It provides a comprehensive method with the assurance of improving diagnostics clinically and oversees an engaging process for finding biomarkers in organic frameworks.

Figure 1. Potential biomarker sources

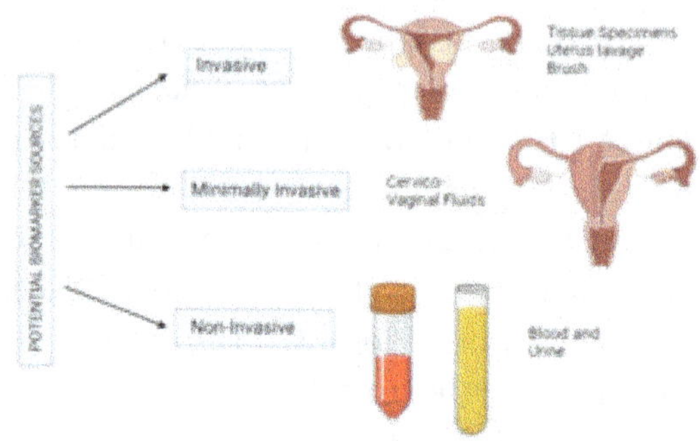

Potential Biomarker Sources

Metabolomics for Biomarker Discovery

Preclinical disease biomarkers may play a critical role in improving treatments that modify or perhaps prevent disease. Early disease detection is essential for efficient understanding of administration and therapy. The publication of biomarkers has garnered unusual attention in light of recent advancements in technology. Metabolite alterations are indicators of physiologic or pathological forms. A well-established and constantly evolving field of study, metabolomics is concerned with the assessment of metabolites within organic structures, both quantitatively and subjectively (Zhang, Sun, Han et al, 2012). The metabolize refers to the comics cascade's endpoint and is also the point in the cascade that comes closest to the phenotype. Therefore, metabolize assessment could be a useful method for identifying effective demonstrative markers and investigating less common neurotic illnesses. The field of metabolomics focuses on the relationship between phenotype and a metabolic signature, which are important aspects of natural characteristics. This approach has the potential to explain biochemical processes that lead to disease progression and treatment outcomes. It may provide insight into the pathophysiology of infection states and identify symptom markers for recovery goals. Benefits from this approach, such as the predictive potential of metabolomics, include increased accuracy and specificity, which could pave the way for the identification of biomarkers. Moreover, metabolic profiling must be shown to be substantially beneficial in metabolomic consideration applications. It is especially straightforward, accurate, and modular (Yuan et al., 2012).

Figure 2. Metabolomics biomarkers discovery

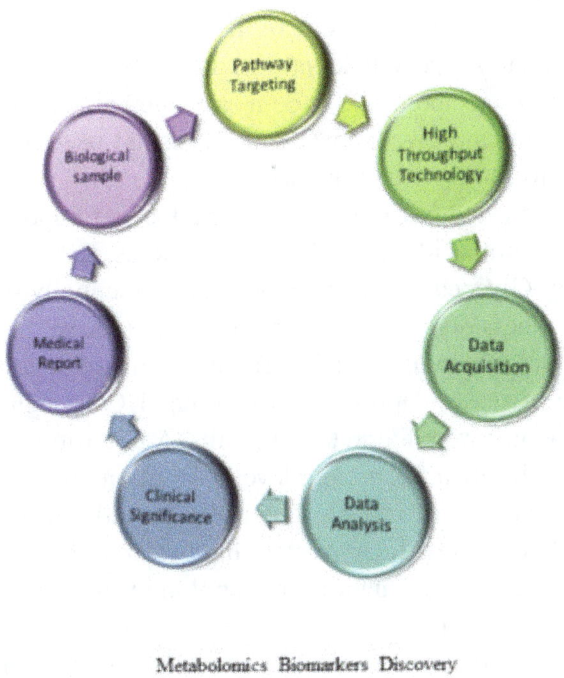

Application in Ailment Biomarker Discovery

Alzheimer's disease (AD) may be a major contributor to morbidity and mortality as well as a major global health concern. Unfortunately, existing biomarkers for early clutter are limited because they are expensive, time-consuming, or invasive. An attempt was made to investigate AD by combining chemometrics and high-resolution mass spectrometry. The metabolomics approach's predictive control for identifying AD cases was confirmed, resulting in complete demonstration exactness. In order to examine the metabolic variations of individuals with unique cognitive status associated with AD movement, a metabolomic technique was developed. As plausible movement biomarkers, choline, dimethyl arginine, arginine, valine, and so forth had been examined. They had far superior predictive values than those obtained using traditional AD biomarkers.

Although blood-based indicators may be an attractive option as well, none can currently detect preclinical AD clutter with the required precision. Crucially, a metabolomic approach was described by Capstone et al. to detect preclinical AD clutter in a group of cognitively normal but more experienced adults. They chose and tested a collection of ten lipids from peripheral blood that predicted, with over 90% accuracy, phenoconversion to either AD or amnesic mild cognitive impairment in a span of two to three years. In the long run, these biomarkers will more closely mimic the pathophysiology of AD, increasing the likelihood that preventive treatments will be effective (Nicholson & Wilson, 2003).

Early detection of hepatocarcinoma (HCC) is one of the most important factors in determining a patient's poor prognosis. Therefore, it is crucial to identify sensitive biomarkers for an accurate and timely diagnosis of HCC. Xiao et al. used metabolomics to compare metabolite ranges in the period of patients with cirrhosis and HCC. Glycolic and glycodeoxycholic acids are used along with selected putative distinguishing pieces of proof to illustrate their respective characteristics. In a Wang et al. review, 13 capacity indicators were identified, and full-size unpleasant effects of important metabolic pathways—which include phospholipids, natural acids, greasy acids, bile acids, and intestinal fluid metabolism—were observed in HCC patients. Glycochenodeoxycholic acid is essentially suggested to be an essential marker for HCC research and illness diagnosis. Metabolite markers have the potential to be useful and beneficial for early detection and screening of HCC. Tan et al. investigated the usefulness of the possible biomarkers from the HCC disease adaption in early human HCC prognosis using a metabolomics approach. Three marker metabolites were found to have an 80.5% sensitivity and an 80.1% specificity, which can be significantly superior than fetoprotein levels. Additionally, serum metabolites from both healthy controls and HCC patients were investigated using the advanced metabolomics technique. 1-Methyladenosine was identified as the primary metabolite for HCC that showed significant advancement. In addition to multivariate factual evaluations, metabolomics was used to profile the metabolites in human HCC. In HCC patients, 43 serum metabolites and 31 urine metabolites have been examined in relation to several important metabolic pathways, including methionine metabolism, the urea cycle, and bile acids. The various factors confirmed that the metabolomics process may be a promising tool for HCC patient screening. Shaw et al. developed a urine technique based mostly on LC-MS that was used for HCC and cirrhosis (CIR) testing (Chen, Xue, Zhou et al, 2011). Urinary nucleosides, bile acids, citric corrosive, and different amino acids were seen to be significantly altered in clutter organisations when compared to the controls, suggesting that liver diseases are caused by dysregulation of purine metabolism, quality metabolism, and amino metabolism. Furthermore, hydantoin-five-propionic acid and butyryl carnitine have been identified as indicators to distinguish HCC from CIR. In order to investigate the steroid hormone difficulties and identify possible urinary steroid hormone indicators of early HCC, a unique steroid hormone metabolomic technique based entirely on liquid chromatography-mass spectrometry combined with computed relapse investigation revealed too connected. A combination of tetrahydrocortisol and epitestosterone appears to have demonstrated extraordinarily high symptomatic utility for identifying early HCC, according to calculated relapse evaluation. Ensuring the metabolic changes associated with the proximity of HCC can improve our understanding of the aetiology of the majority of malignancies (Woo et al., 2009).

Because they are not sensitive enough, conventional biological indicators of renal failure were elevated to a whole new level following the occurrence of significant, ongoing kidney illnesses (CKD). The use of metabolomics was linked to the establishment of a human CKD serum metabolic profile and the large endogenous metabolites that helped identify CKD in unique ranges. The form for diagnosing patients with CKD completed the sensitivity and specificity of 100% based on those metabolites and may provide useful information for the early diagnosis of CKD. It described how the progression of CKD and renal congestion altered the serum metabolic profile. Several indicators surfaced during multiple studies to identify CKD before conventional chemical therapy and histological approaches. Researchers looking for the most current demonstrative biomarkers for ovarian endometriosis have the opportunity to benefit from underutilised experiences into the pathophysiological components thanks to the utility of metabolomics in CKD studies (Patterson et al., 2011). A hypothesis-generating centred metabolomics approach was employed by Souk et al. It has been determined that eight lipid metabolites are biomarkers associated with endometriosis. In a demonstration, the ratio of 16:1 to the ether-phospholipid 34:2 resulted in a sensitivity of 90.0% and a specificity of 84.3%. The findings imply that endometriosis is associated with longer sphingomyelin and phosphatidylcholine tiers, which may impact signalling pathways connected to lipids and help mask apoptosis. The metabolites included in the non-invasive study of hepatic stagnation and spleen deficiency syndrome-like illness were selected using metabolite profiling. In addition to Penrose, cysteine, methionine, tyrosine, tryptophan, and nucleotide sugar metabolism, twelve urine differential metabolites that contributed to the entire partition of LSS patients from matched healthy controls had been identified. Imperatively, four metabolite markers with a 93.0% success rate for the analysis of human LSS were successful. Patients with metastatic colorectal cancer (MCC) may need to have their treatment and survival rates advanced before they are identified. In their investigation, Bernini et al. used NMR to profile serum metabolise in patients with MCC and determine whether a distinct signature that is reliable enough to predict overall survival may exist. Within the preparation set, 100% accuracy NMR metabolomic profiling may be required to distinguish patients with MCC from healthy ones. However, 96.7% of the subjects in the approved set had proper labels. Multivariate analysis has been used with an LC-MS-based technique to separate the global serum profiles of individuals with renal cell cancer. As a result, renal versatile carcinoma has 30 identified capacity biomarkers (Holmes et al., 2008).

Using an NMR metabolomics technique, Jung et al. analysed the altered metabolic architecture in patients with sensitivity and searched for potential indicators. Patients with a history of hypersensitivity have been linked to higher levels of histidine, glutamine, and methionine as well as lower levels of arginine, choline, methanol, and acetic acid derivation. Potential biomarkers appeared to have long-term predictive power, and the presence of bronchial asthma in external approval models became accurately predicted. Promising and healing biomarkers beneficial for oesophageal squamous cell carcinoma (ESC) can increase cancer patients' chances of long-term survival. Metabolomics was used in a pilot study to identify global alterations in the metabolic profiles and identify biomarkers that are unquestionably relevant for identification. Naturally, the 18 most significantly altered plasma metabolites in patients with ESC were most likely identified as fatty acids, l-carnitine, acyl carnitine, natural acids, and a sterol metabolite. Additionally, variants of these three capacity indicators evolved to study during the healing process and could be useful in diagnosing, verifying beneficial reactions, and projecting effects. Lam et al. misapplied metabolomic profiling in their study to identify biomarkers for the majority of malignancies and assess the risk of damaging pleural radiations. Insignificant indicators have been eliminated through the application of a metabolize-wide significance threshold. Investigating hazardous pleural radiations is supported by the symptomatic use of the FFA 18:1-to-ceramide ratio. Gas chromatography coupled to mass spectrometry was used to carry out metabolomic profiling, and the serum tests identified 48 metabolites (Wang et al., 2011). Itaconic acid has the potential to function as a unique biomarker in the early stages of pregnancy to predict the development of gestational diabetes mellitus. It was discovered to be noticeably more abundant in women who eventually developed the disease. These findings demonstrate the potential of tough metabolomics as a non-invasive approach and a viable screening tool to assess the ability of those metabolites in the early evaluation of patients and provide previously untapped information about pathophysiology components.

METABOLOMICS PROSPECT OF METABOLISM HEALTH

The spectrum of metabolomics investigations based on MS has expanded. MS-based metabolomics techniques were applied to observe the effects of drugs, toxins, and diseases on metabolites. Numerous infections, including breast and kidney cancers, cardiovascular diseases, bladder, esophageal, and gastric cancers, kidney clutter, herbal metabolic errors, toxicological findings, and supplements have been examined using MS-based metabolomics techniques (Graham et al., 2013).

Disease Diagnosis

Metabolomics revolves around the similarities among particular constituents and ultimately encompasses examining the similarities, characteristics, and regulations of every metabolic additive. Physiology and metabolism are more closely related than genetics and proteomics. The illness modifies the outline's pathophysiological approach, which in turn leads to corresponding changes in metabolites. The infection's biomarkers are found by looking through certain metabolites and contrasting them with normal human metabolites. Strategies for metabolomics will function as an underutilised tool for sickness prediction. Johno, H. et al. established a demonstrative method for computerised atherosclerotic affliction discovery using electrostatic ionisation mass spectrometry. Phospholipids and cholesterol sulphate are considered to be underutilised indicators of arteriosclerosis. Tavares, G. et al. performed metabolomics on 56 DKD patients' plasma. They seem to imply that decreased concentrations of valine, polyaspartic acid, and 1,5-anhydroglucitol are connected to the progression of DKD. As a cutting-edge tool in DKD, reverse biomarkers of hyperglycaemia may be studied (Ibáñez et al., 2012). Hao, M. et al. used community pharmacology and plasma metabolomics to elucidate Yishu's mode of action in treating blood stasis caused by a metabolic component of sea. It was discovered that Curcuma wenyujin rhizomes improved blood stasis, and that this improvement was particularly associated with lipid and amino acid metabolism. The current human population has a limited knowledge base regarding infections, and many people are unaware that many diseases can lurk in a patient's body for years following a contamination. Certain illnesses don't immediately affect the outline, and most patients tend to have an unnoticed mindset. This could result in the timely and silent loss of the best treatment, making it extreme. Understanding the aetiology of diseases and the factors that influence them using metabolomics approaches based on mass spectrometry is crucial for understanding the active components of diseases. Abdulwahab, R.A. et al. found widespread immunoglobulin up-regulation and unutilized connections between 37 proteins and T2DM by the use of LC-MS/MS assessment. Yang, G. et al. discovered 33 amazing metabolites in the urine of both healthy people and patients with femoral head necrosis using UPLC-QTOF/MS and multivariate quantifiable analysis. When Lin, S. et al. used gas chromatography-MS to identify plasma metabolites, they discovered that, across all patient groups with breath problems classified as direct to serious, the ranges of alanine, aspartic acid, and carbonate were remarkably infrequent (Mapstone et al., 2014).

Sun, H. et al. used UPLC/MS to perform a metabolomic analysis of ischemic stroke and found significant correlation in 12 metabolites related to heme and fatty acid metabolism. Lu, H. et. al. used

HPLC-MS to investigate the effect of hydrogen on serum and urine metabolites on renal disease outcome. The abnormal metabolites found were mainly related to the metabolism of amino acids, fatty acids and phospholipids (Xiao et al., 2012).

Biomarkers have many different uses at some point in the research process. It can be used to diagnose disease, determine disease levels, and evaluate new drugs. For informative biomarkers, it is important to understand the impact of metabolomics strategies on biomarkers. The following examples show the benefit of metabolomics strategies in the process of detecting disease biomarkers in unique guidelines and lay the foundation for near-integration of metabolomics and medical medicinal drugs. Zhao, H. et al. using UPLC-Q-TOF/MS, detected specific metabolites of functional biomarkers in the urine of cancer patients. The metabolic pathways involved are amino acid metabolism, fatty acid metabolism, steroid hormones, biosynthesis and oxidative strain (Wang et al., 2012). Dittarrot, K. et al. investigated the urinary metabolize profile of patients with chronic hepatitis B infection using GC/MS. 12 different metabolites compared to normal people in particular, it has been found to be related to fatty acids, amino acids, bile acids and intestinal microbial metabolic pathways. Zhao, H. et al. using LC-MS-based metabolomics, investigated potential biomarkers for Parkinson's disease and identified 17 important metabolites, especially related to tyrosine biosynthesis, metabolism, carnitine metabolism, and folic acid metabolism. Li, X. Z. et al. used UPLC-QTOF-MS to monitor the pathogenesis of oral lichen planus and identified 12 established metabolites as putative functional biomarkers. UPLC/MS was used to study urine samples from patients with acute coronary syndrome and healthy controls. Fatty acids, fatty acid oxidation metabolism, amino acids, and the TCA cycle have been found to play important roles in coronary syndrome (Tan et al., 2012). On the basis of the metabolomic method of Gas Chromatography-MS, Wei, J. et al. studied the disease characteristics of systemic lupus erythematosus and found it to be strongly related to energy and amino acid pathways. The era of metabolomics is very important for diagnosis, but this diagnosis remains one-sided. It must be fully integrated in today's clinical era to better understand diseases, understand and convey the pathogenesis of diseases to protect the patient's health.

Cancer Research

Tens of millions of lives are lost to cancer each year, making it a severe issue that still exists today. Effective communication about cancer requires an understanding of its aetiology and contributing factors. Because most cancers are caused by a combination of factors, such as genetics, nutrition, and environment, researchers utilise metabolomics to explain how different factors interact to cause cancer. In order to examine metabolic alterations in ovarian cancer, for instance, Yang, W. et al. employed MS and looked into 18 metabolites that were connected to the disease. Using LC/MS, An, Y. et al. examined metabolic alterations in pancreatic cancer tissue and discovered seven compounds that served as dosage biomarkers. Using LC-MS/MS, Jason, P. et al. discovered metabolic indicators in breast cancer and discovered a close relationship between the immune system, metastasis, and growth loss pathways impacted by tumors. Fan, X. et al. examined the impact of Macaranthoidin B on all metabolites in colorectal cancer cells using a combination of LC-MS and GC-MS. It was found to stop colorectal cancer from growing by causing reactive oxygen species-induced cell death. People are impacted by the issues that most malignancies cause, and the annual percentage of deaths from most cancers rises. It will take a long time for technological advancements to fulfil the urgent demands of cancer patients, who need to understand most cancer mechanisms and effective treatments. Metabolomics is used to determine mechanisms of action in most cancers, most cancers are a unique sickness that are difficult to treat, and medical samples of most cancers' sufferers are hard to reap. Therefore, in addition improvement has been hindered (Chen, Xie, Wang et al, 2011).

Nutrition

Nutritional metabolomics, a term used to describe the application of metabolomics to the supplement industry, refers to the methodical investigation of the relationship between a weight-loss plan and biological metabolism using metabolomics in specific health and illness states of living things. All of the trace elements needed by the human outline are found in vegetables and natural goods, which also appear to complement all of the trace elements needed by the human outline. The trace component composition of fruits is influenced by a multitude of factors. Numerous scenarios can be used to assist characterise the features of metabolomics in vitamins. Among them, Zhao, L. et al. examined the metabolite alterations in cucumber fruit under nano-Cu stress using GC-MS and 1H NMR methods. They concluded that the metabolic profile of metabolites from natural products was affected by exposure to nano-copper (Shao et al., 2014). A UHPLC-MS/MS-based metabolomics technique was employed by Wang, Y.Q. et al. to examine the supplement profiles of Chinese kale. It was discovered that the additions in certain cabbage assortments were unique in nature. Rahman, M.M. et al. investigated dinitrofuran and its three metabolites in plums using LC-MS/MS, which provided a foundation for further research on plum additives. The differentiation between overpowering steel substance material and auxiliary metabolites in dried ginger from uncommon habitats was found by Alolga, R. N. et al. using Q/TOF-MS innovation. Variations in metabolite wealth may be caused by abiotic stretch caused by metals, according to supposition (Dai et al., 2014). Using GC-MS innovation, Ramírez-Acosta, S. et al. examined the metabolic changes in strawberry natural product due to treatment in a controlled environment and found that some of the metabolites connected with it were present in particular biologically active components. Kim, G. D. et al. identified danshensu and caffeine acid subordinates as the primary anti-inflammatory compounds in the leaves of kiwi natural product plants using LC-MS/MS. Nevertheless, the body receives fundamental proteins, fats, and carbohydrates from a variety of additional sources in addition to veggies and culmination. According to a study, essential royal jelly proteins improve adult rats' spatial memory by affecting the metabolism of niacin and nicotine, cysteine and taurine, and strength. In order to differentiate between the various metabolites in black and white sesame, Wang, D. et al. used the greenhorn metabolomics approach. They found differences in the metabolism of tyrosine, riboflavin, and phenyl propanol.

ADVANCEMENTS OF METABOLOMICS

Drug Efficacy and Screening

Chinese medicine is a complex hypothetical device because traditional Chinese medical guidelines and Western clinical guidelines are essentially specific in nature. Additionally, Chinese medicine uses metabolomics innovation to identify the improvement handle of a clutter, and because of the concept of the all-encompassing sea, framework view, and argumentative view of Chinese pharmaceuticals conflict with one another. The inquiry offer's contemporary ideas and tactics. The use of metabolomics innovation in traditional Chinese medicine is expanding, and it has steadily permeated a number of traditional Chinese pharmaceutical research domains, including medicine, compatibility, and components of action.

The considerations of standard Chinese medicinal medication are divided into two categories: unit sedate research and Chinese restorative sedate compound considerations. Examining single-flavoured tablets focuses on the specific pharmacological outcomes, adverse effects, and metabolic mechanism of the drug; the compound's compatibility guidelines and metabolic pathways are of special interest to researchers. Li, C. R. et al. discovered that the utilisation of metabolic ways to medicate with various pharmaceuticals will result in an uncommon strategy. Botulinum acid is the purported anticancer component in Forsythia suspense, according to Bao, J. et al.'s rapid discovery of anticancer chemicals in Forsythia using UPLC/MS-based metabolomics (Qi et al., 2012). By examining the anti-inflammatory mechanism of forsythia oil and extracts using UPC-Q-TOF-MS/MS, Yuan, A. et al. discovered that forsythia exhibits anti-inflammatory properties through unique metabolic pathways. Zhang, Y. et al. used HPLC/MS to investigate the possible role of Xanthii fructose in allergic reactions that are not well tolerated. Within certain challenged metabolic pathways that include the metabolism of branched-chain amino acids and glycerophospholipids, 27 capacity indicators have been identified. According to Wang, Y. et al., the net saponins in Tribulus terrestris fruit have the ability to reroute many metabolic pathways in order to counteract the effects of cerebral artery occlusion-induced serum metabolic deviations. Zhou, J. et al. discovered that taurine in the diet, phenylalanine, the metabolism of arachidonic acid, and metabolism B6 were all essentially linked to gastrointestinal injury. Fructus gardenia affects the typical highlights of the gastrointestinal system by diverting the metabolism to a strange and unique place. It has evolved over time to extend and preserve the traditional Chinese medicinal medicine subculture, to safeguard the influence and element of the effectiveness of traditional Chinese medicine, and to safeguard the element of routine Chinese medicine with greater assiduity in order to gain recognition both domestically and internationally. Chinese drugs and western medicine are expanding together these days. They also take into consideration one another's treatment approaches within the same period of time. Chinese pharmaceuticals and western medicine differ in that the former uses an atomic perspective while the latter uses complicated ingredients and incorporates the framework into the treatment of illness. Although it is well known that western medicine has a significant impact on the treatment of illnesses, long-term usage of western medicine alone will undoubtedly harm the affected person's body. In this way, examining the metabolic component of western medications is especially fundamental. Considerations about sedates have been made possible by metabolomics research based on MS. Liu, Z. et al. identified nine common sulfonates in tablets using GC-MS/MS. These alkyl sulfonates can be practically attached to various capsules and were successfully used for the assurance of capacitive and imaging desolate (Zhao, 2013).

Applications for metabolomics in animals are numerous and include approval, sedate screening, assessing the viability and harmfulness of medications, and therapeutic evaluation. The remarkable and persistent hepatotoxicity of gourd extract on mice was investigated, utilising the metabolomics technique of ultra-performed liquid chromatography combined with quadrupole time-of-flight mass spectrometry. Kim, J. H. et al. investigated the metabolic properties of ketoconazole in mice and humans and identified the fundamental pathways of its hepatotoxicity using liquid chromatography-mass spectrometry-primarily based metabolomics.

Yan, H. et al. investigated the toxicity of brodifacoum using HPLC-TOF-MS and concluded that it specifically caused harm by agitating metabolic pathways such as glycolysis and the carboxylic acid cycle. Wang, Z. et al. used UPLC-MS to monitor the hepatotoxicity of the sheath Genkwa Flos and discovered that problems with oily acid metabolism and aminoglycan-1-phosphate metabolism may be the main factors causing liver damage. After examining the hepatotoxicity of Dioscorea bulbifera rhizomes using gas chromatography-mass spectrometry, Zhao, D. S. et al. concluded that the changes in amino acids, bile acids, and strength metabolism had an effect on the physiological and natural features of the liver. On the other hand, taking medication to treat the illness will have a variety of intriguing side effects on the body, especially when it comes to symptoms of kidney and liver damage, among many others (Bertini et al., 2012; Vouk et al., 2012). Determining the ability biomarkers associated with toxic quality requires a method to apply metabolomics to select the tissues, organs, and their components that survive from the toxic quality, speculative capsules, and the activity component. Liu, S. Y. et al. discovered that the production of phenylalanine, tyrosine, and tryptophan may be connected to the protective factor against ischemia brain reperfusion injury caused by Kudieze injection. After studying the mechanism of schisandra polysaccharide treatment for Alzheimer's disease, Zhu, B. et al. concluded that the treatment reduces tau protein phosphorylation, a sign of aβ and oxidative damage.

Agronomy and Plant Biology

Metabolomics fundamentally based on MS is progressively more utilized in unmistakable therapeutic areas, in specific agronomy and plant science, to secure the conduct of vegetation underneath interesting strain conditions. Interior the method of plant development, a colossal assortment of components is habitually required, and it's a distant imperative to investigate the impacts of variables at the development laws of vegetation. With UPLC-MS/MS, Xu, S. et al. examined the metabolic properties of soybean seedlings' leaves and roots under normal and extreme situation conditions. It was hypothesised that natural substances, such as 2-oxo-arginine, l-nicotine, gluconic acid, d-gluconic acid, and citric acid, had a major role in chelating molybdenum and reducing its toxicity. The results of sulphur deficiency on metabolites were analysed by Thomson, H. et al. using metabolomics based on MS, which established the foundation for further research in agriculture. That is in favour of controlling fertilisation in gardening. Yang, N. et al. investigated how a magnesium deficit affected the metabolism of soybean roots and leaves. Underneath magnesium deprivation, it was discovered that there were infinite differences in the carbon and nitrogen metabolic processes among leaves and roots. A fundamental role for nitrogen is played in plant development. The main element in protein is nitrogen. The best material that gives rise to plant life is protein. Another essential component of chlorophyll is nitrogen. Grain quality abdication and the reaction of grain metabolites to increased nitrogen application were examined by Allwood, J. W. et al., who observed a significant impact on grain abdication. Zhang, Y.et al. done metabolomic assessment on wheat assortments with uncommon brought on nitrogen stages, and found that flavonoids and their related subsidiaries have been taken into thought as potential biomarkers underneath moo nitrogen pressure. Herbicide buildups are one of the critical issues in present day horticulture. A way to control herbicide buildup interior to a reasonable extent isn't most compelling a crucial way to extend generation, but moreover a security ensures for customers. In order to determine the content of sulfonyl urea and its metabolites in rice and discuss the problem of herbicide buildups, Du, P. et al. used UPLC and triple tandem quadrupole mass spectrometry (Jung et al., 2013; Lin et al., 2011).

Similarly, to their colossal utility in agribusiness, the plant and natural components in any case have a totally broad assortment of bundles. As a result, the metabolomics technique based mostly on LC-MS was optimised for the cyanobacterium Synechocystis P. PCC 6803, and Pei, G. et al. identified 24 fundamental carbohydrates and fundamental metabolites associated with the metabolism of vitality. Fuel chromatography-mass metabolomics techniques were employed by Mibei, E. Ok., and other researchers to infer the effects of drought pressure on metabolites, specifically in relation to aubergine boom levels. Significant effects of drought pressure have been noted for metabolites, which include natural acids, sugars, and amino acids. Qian Y. et al. investigated the effects of in vitro sucrose on the flavourful components of tea plant life using UPLC/MS (Corb Aron et al., 2021; Singh et al., 2014; Thapa et al., 2021). They discovered that sucrose controls the expression of translation perspective qualities and pathway qualities, which in turn regulates the manufacture of polyphenols in tea flowers. Not only that, but the application trend in the fields of plant science and agribusiness has also gained increasing recognition. The answer to teaching farming and plant science may lie in finding a method to integrate horticulture and plant science with the innovative field of metabolomics. Metabolomics time is broadly utilized in crop facts size, affliction analysis, and numerous others., and contributes to expanding agrarian profit. The common integration of the two has spread out a modern course for the improvement of the zone of metabolomics (Agrawal et al., 2021; Kareem et al., 2021; Kaur et al., 2016; Mujwar et al., 2022; Pradhan et al., 2015).

Future Perspectives

To provide a guarantee for the advancement of metabolomics, discoveries in the field may be combined with more well-known emerging innovations. Further optimization of metabolomics based on MS is possible, and various combinations of techniques, including LC/MS and GC/MS, will provide a rational foundation for resolving the annoyance of metabolite analysis and elucidating metabolic pathways. After all, mass spectrometry continues to be a key explanatory tool in the field of metabolomics. Advances in this subject have facilitated the identification of multiple potential disease indicators and equipped understanding of the etiology of various illnesses. Numerous inquiries have moreover been tried, which incorporate early malady discovery, cure forecast and determination, following treatment and repeat attempting out. Given the strong complexity of natural combos, most MS expository procedures contain LC, GC or CE. But hurriedly developing programs utilizing advancing division methods and conventions offer openings and challenges. Features of chromatographic techniques have unquestionably proven successful in isolating metabolites and producing the diverse range of metabolites identified. But, one basic venture is the need of capacity to look at and connect the results of these things about, which might be performed on the rise to or comparable tests procured through fair-minded inquiry about organizations. That's the essential bottleneck inside the improvement of this discipline. Diverse variables such as test arrangement, test network and leftover results can too result in data changeability (Kareem et al., 2021). To overcome those obstacles, regardless of the explanation stage, technique, or convention used, it is necessary to switch from the commonly used relative metabolite consideration estimations to a more prominent and reliable supreme mindfulness determination. This approach isn't continuously true for MS, but it is exceptionally critical.

CONCLUSION

Ever since the field of metabolomics was founded, scientists have focused on improving rate. It too tackles the challenges of methodology and applications as an evolving approach inquiry. In terms of methodology, advance development is needed for the explanatory devices, expository processes, actualities securing, and actualities assessment. From the application point of sea, in spite of the fact that an expansive wide assortment of basic point of interest metabolites related with hereditary variation or physiological and obsessive alterations were procured, it's miles a colossal wander to set up a scientific predictive demonstrative master framework to realize demonstrative clarification. An assignment is taken after through possibilities, so the challenges right now experienced in metabolomics are openings for future improvement. It's distantly accepted that with the continuous optimization and advancement of metabolomics ponders strategies, sedate security assessment might be more noteworthy, exact and productive, and the knowledge of infection strategies can be enormously comprehensive, and it can indeed rise as a compelling implies to manual human wellbeing and screen the human environment.

REFERENCES

Agrawal, N., Mujwar, S., Goyal, A., & Gupta, J. K. (2021). Phytoestrogens as Potential Antiandrogenic Agents Against Prostate Cancer: An In Silico Analysis. *Letters in Drug Design & Discovery*, *19*(1), 69–78. doi:10.2174/1570180818666210813121431

Arakaki, A. K., Skolnick, J., & McDonald, J. F. (2008). Marker metabolites can be therapeutic targets as well. *Nature*, *456*(7221), 443. doi:10.1038/456443c PMID:19037294

Bertini, I., Cacciatore, S., Jensen, B. V., Schou, J. V., Johansen, J. S., Kruhøffer, M., Luchinat, C., Nielsen, D. L., & Turano, P. (2012). Metabolomic NMR fingerprinting to identify and predict survival of patients with metastatic colorectal cancer. *Cancer Research*, *72*(1), 356–364. doi:10.1158/0008-5472.CAN-11-1543 PMID:22080567

Chen, F., Xue, J., Zhou, L., Wu, S., & Chen, Z. (2011). Identification of serum biomarkers of hepatocarcinoma through liquid chromatography/mass spectrometry-based metabonomic method. *Analytical and Bioanalytical Chemistry*, *401*(6), 1899–1904. doi:10.1007/s00216-011-5245-3 PMID:21833635

Chen, T., Xie, G., Wang, X., Fan, J., Qiu, Y., Zheng, X., Qi, X., Cao, Y., Su, M., Wang, X., Xu, L. X., Yen, Y., Liu, P., & Jia, W. (2011). Serum and urine metabolite profiling reveals potential biomarkers of human hepatocellular carcinoma. *Molecular & Cellular Proteomics*, *10*(7), M110.004945. Advance online publication. doi:10.1074/mcp.M110.004945 PMID:21518826

Corb Aron, R. A., Abid, A., Vesa, C. M., Nechifor, A. C., Behl, T., Ghitea, T. C., Munteanu, M. A., Fratila, O., Andronie-Cioara, F. L., Toma, M. M., & Bungau, S. (2021, March 17). Recognizing the benefits of pre-/probiotics in metabolic syndrome and type 2 diabetes mellitus considering the influence of akkermansia muciniphila as a key gut bacterium. *Microorganisms*, *9*(3), 618. doi:10.3390/microorganisms9030618 PMID:33802777

Dai, W., Yin, P., Chen, P., Kong, H., Luo, P., Xu, Z., Lu, X., & Xu, G. (2014). Study of urinary steroid hormone disorders: Difference between hepatocellular carcinoma in early stage and cirrhosis. *Analytical and Bioanalytical Chemistry*, *406*(18), 4325–4335. doi:10.1007/s00216-014-7843-3 PMID:24817358

Graham, S. F., Chevallier, O. P., Roberts, D., Hölscher, C., Elliott, C. T., & Green, B. D. (2013). Investigation of the human brain metabolome to identify potential markers for early diagnosis and therapeutic targets of Alzheimer's disease. *Analytical Chemistry*, *85*(3), 1803–1811. doi:10.1021/ac303163f PMID:23252551

Haviland, J. A., Tonelli, M., Haughey, D. T., Porter, W. P., & Assadi-Porter, F. M. (2012). Novel diagnostics of metabolic dysfunction detected in breath and plasma by selective isotope-assisted labeling. *Metabolism: Clinical and Experimental*, *61*(8), 1162–1170. doi:10.1016/j.metabol.2011.12.010 PMID:22304834

Holmes, E., Loo, R. L., Stamler, J., Bictash, M., Yap, I. K. S., Chan, Q., Ebbels, T., De Iorio, M., Brown, I. J., Veselkov, K. A., Daviglus, M. L., Kesteloot, H., Ueshima, H., Zhao, L., Nicholson, J. K., & Elliott, P. (2008). Human metabolic phenotype diversity and its association with diet and blood pressure. *Nature*, *453*(7193), 396–400. doi:10.1038/nature06882 PMID:18425110

Ibáñez, C., Simó, C., Martín-Álvarez, P. J., Kivipelto, M., Winblad, B., Cedazo-Mínguez, A., & Cifuentes, A. (2012). Toward a predictive model of Alzheimer's disease progression using capillary electrophoresis-mass spectrometry metabolomics. *Analytical Chemistry*, *84*(20), 8532–8540. doi:10.1021/ac301243k PMID:22967182

Jung, J., Kim, S.-H., Lee, H.-S., Choi, G. S., Jung, Y.-S., Ryu, D. H., Park, H.-S., & Hwang, G.-S. (2013). Serum metabolomics reveals pathways and biomarkers associated with asthma pathogenesis. *Clinical and Experimental Allergy*, *43*(4), 425–433. doi:10.1111/cea.12089 PMID:23517038

Kareem, S. M., Al-Kadmy, I. M., Kazaal, S. S., Mohammed Ali, A. N., Aziz, S. N., Makharita, R. R., Algammal, A. M., Al-Rejaie, S., Behl, T., Batiha, G. E., El-Mokhtar, M. A., & Hetta, H. F. (2021, February 12). Detection of gyrA and parC mutations and prevalence of plasmid-mediated quinolone resistance genes in Klebsiella pneumoniae. *Infection and Drug Resistance*, *14*, 555–563. doi:10.2147/IDR.S275852 PMID:33603418

Kaur, A., Mujwar, S., & Adlakha, N. (2016). In-silico analysis of riboswitch of Nocardia farcinica for design of its inhibitors and pharmacophores. *International Journal of Computational Biology and Drug Design*, *9*(3), 9. doi:10.1504/IJCBDD.2016.078278

Lin, L., Huang, Z., Gao, Y., Yan, X., Xing, J., & Hang, W. (2011). LC-MS based serum metabonomic analysis for renal cell carcinoma diagnosis, staging, and biomarker discovery. *Journal of Proteome Research*, *10*(3), 1396–1405. doi:10.1021/pr101161u PMID:21186845

Mapstone, M., Cheema, A. K., Fiandaca, M. S., Zhong, X., Mhyre, T. R., MacArthur, L. H., Hall, W. J., Fisher, S. G., Peterson, D. R., Haley, J. M., Nazar, M. D., Rich, S. A., Berlau, D. J., Peltz, C. B., Tan, M. T., Kawas, C. H., & Federoff, H. J. (2014). Plasma phospholipids identify antecedent memory impairment in older adults. *Nature Medicine*, *20*(4), 415–418. doi:10.1038/nm.3466 PMID:24608097

Mujwar, S., Shah, K., Gupta, J. K. & Gour, A. (2022). *Docking Based Screening of Curcumin Derivatives: A Novel Approach in the Inhibition of Tubercular DHFR "Docking Based Screening of Curcumin.* Research Gate.

Nicholson, J. K., & Wilson, I. D. (2003). Understanding 'global' systems biology: Metabonomics and the continuum of metabolism. *Nature Reviews. Drug Discovery*, *2*(8), 668–676. doi:10.1038/nrd1157 PMID:12904817

Patterson, A. D., Maurhofer, O., Beyoglu, D., Lanz, C., Krausz, K. W., Pabst, T., Gonzalez, F. J., Dufour, J.-F., & Idle, J. R. (2011). Aberrant lipid metabolism in hepatocellular carcinoma revealed by plasma metabolomics and lipid profiling. *Cancer Research, 71*(21), 6590–6600. doi:10.1158/0008-5472.CAN-11-0885 PMID:21900402

Pradhan, P., Soni, N. K., Chaudhary, L., Mujwar, S., & Pardasani, K. R. (2015). In-silico prediction of riboswitches and design of their potent inhibitors for H1N1, H2N2 and H3N2 strains of influenza virus. *Biosciences Biotechnology Research Asia, 12*(3), 2173–2186. doi:10.13005/bbra/1889

Qi, S., Ouyang, X., Wang, L., Peng, W., Wen, J., & Dai, Y. (2012). A pilot metabolic profiling study in serum of patients with chronic kidney disease based on 1H-NMR-spectroscopy. *Clinical and Translational Science, 5*(5), 379–385. doi:10.1111/j.1752-8062.2012.00437.x PMID:23067349

Shao, Y., Zhu, B., & Zheng, R. (2014). Development of urinary pseudotargeted LC-MS-based metabolomics method and its application in hepatocellular carcinoma biomarker discovery. *Journal of Proteome Research.* doi:10.1021/pr500973d PMID:25483141

Singh, A. K., Kumar, S., & Singh, V. P. (2014, March 4). Optimization of parameters using conductive powder in dielectric for EDM of super Co 605 with multiple quality characteristics. *Materials and Manufacturing Processes, 29*(3), 267–273. doi:10.1080/10426914.2013.864397

Tan, Y., Yin, P., Tang, L., Xing, W., Huang, Q., Cao, D., Zhao, X., Wang, W., Lu, X., Xu, Z., Wang, H., & Xu, G. (2012). Metabolomics study of stepwise hepatocarcinogenesis from the model rats to patients: Potential biomarkers effective for small hepatocellular carcinoma diagnosis. *Molecular & Cellular Proteomics, 11*(2), M111.010694. doi:10.1074/mcp.M111.010694 PMID:22084000

Tenori, L., Oakman, C., Claudino, W. M., Bernini, P., Cappadona, S., Nepi, S., Biganzoli, L., Arbushites, M. C., Luchinat, C., Bertini, I., & Di Leo, A. (2012). Exploration of serum metabolomic profiles and outcomes in women with metastatic breast cancer: A Pilot Study. *Molecular Oncology, 6*(4), 437–444. doi:10.1016/j.molonc.2012.05.003 PMID:22687601

Thapa, K., Khan, H., Singh, T. G., & Kaur, A. (2021, September). Traumatic brain injury: Mechanistic insight on pathophysiology and potential therapeutic targets. *Journal of Molecular Neuroscience, 71*(9), 1725–1742. doi:10.1007/s12031-021-01841-7 PMID:33956297

Tomita, M., & Kami, K. (2012). Systems biology, metabolomics, and cancer metabolism. *Science, 336*(6084), 990–991. doi:10.1126/science.1223066 PMID:22628644

Tsutsui, H., Maeda, T., Min, J. Z., Inagaki, S., Higashi, T., Kagawa, Y., & Toyo'oka, T. (2011). Biomarker discovery in biological specimens (plasma, hair, liver and kidney) of diabetic mice based upon metabolite profiling using ultra-performance liquid chromatography with electrospray ionization time-of-flight mass spectrometry. *Clinica Chimica Acta, 412*(11-12), 861–872. doi:10.1016/j.cca.2010.12.023 PMID:21185819

Vouk, K., Hevir, N., Ribič-Pucelj, M., Haarpaintner, G., Scherb, H., Osredkar, J., Möller, G., Prehn, C., Rižner, T. L., & Adamski, J. (2012). Discovery of phosphatidylcholines and sphingomyelins as biomarkers for ovarian endometriosis. *Human Reproduction (Oxford, England), 27*(10), 2955–2965. doi:10.1093/humrep/des152 PMID:22859507

Wang, B., Chen, D., Chen, Y., Hu, Z., Cao, M., Xie, Q., Chen, Y., Xu, J., Zheng, S., & Li, L. (2012). Metabonomic profiles discriminate hepatocellular carcinoma from liver cirrhosis by ultraperformance liquid chromatography-mass spectrometry. *Journal of Proteome Research*, *11*(2), 1217–1227. doi:10.1021/pr2009252 PMID:22200553

Wang, T. J., Larson, M. G., Vasan, R. S., Cheng, S., Rhee, E. P., McCabe, E., Lewis, G. D., Fox, C. S., Jacques, P. F., Fernandez, C., O'Donnell, C. J., Carr, S. A., Mootha, V. K., Florez, J. C., Souza, A., Melander, O., Clish, C. B., & Gerszten, R. E. (2011). Metabolite profiles and the risk of developing diabetes. *Nature Medicine*, *17*(4), 448–453. doi:10.1038/nm.2307 PMID:21423183

Woo, H. M., Kim, K. M., Choi, M. H., Jung, B. H., Lee, J., Kong, G., Nam, S. J., Kim, S., Bai, S. W., & Chung, B. C. (2009). Mass spectrometry based metabolomic approaches in urinary biomarker study of women's cancers. *Clinica Chimica Acta*, *400*(1-2), 63–69. doi:10.1016/j.cca.2008.10.014 PMID:19010317

Xiao, J. F., Varghese, R. S., Zhou, B., Nezami Ranjbar, M. R., Zhao, Y., Tsai, T.-H., Di Poto, C., Wang, J., Goerlitz, D., Luo, Y., Cheema, A. K., Sarhan, N., Soliman, H., Tadesse, M. G., Ziada, D. H., & Ressom, H. W. (2012). LC-MS based serum metabolomics for identification of hepatocellular carcinoma biomarkers in Egyptian cohort. *Journal of Proteome Research*, *11*(12), 5914–5923. doi:10.1021/pr300673x PMID:23078175

Yoshida, M., Hatano, N., Nishiumi, S., Irino, Y., Izumi, Y., Takenawa, T., & Azuma, T. (2012). Diagnosis of gastroenterological diseases by metabolome analysis using gas chromatography-mass spectrometry. *Journal of Gastroenterology*, *47*(1), 9–20. doi:10.1007/s00535-011-0493-8 PMID:22041921

Yuan, M., Breitkopf, S. B., Yang, X., & Asara, J. M. (2012). A positive/negative ion-switching, targeted mass spectrometry-based metabolomics platform for bodily fluids, cells, and fresh and fixed tissue. *Nature Protocols*, *7*(5), 872–881. doi:10.1038/nprot.2012.024 PMID:22498707

Zhang, A., Sun, H., Han, Y., Yuan, Y., Wang, P., Song, G., Yuan, X., Zhang, M., Xie, N., & Wang, X. (2012). Exploratory urinary metabolic biomarkers and pathways using UPLC-Q-TOF-HDMS coupled with pattern recognition approach. *Analyst*, *137*(18), 4200–4208. doi:10.1039/c2an35780a PMID:22852134

Zhang, A., Sun, H., & Wang, X. (2012). Saliva metabolomics opens door to biomarker discovery, disease diagnosis, and treatment. *Applied Biochemistry and Biotechnology*, *168*(6), 1718–1727. doi:10.1007/s12010-012-9891-5 PMID:22971835

Zhang, A., Sun, H., Wu, X., & Wang, X. (2012). Urine metabolomics. *Clinica Chimica Acta*, *414*, 65–69. doi:10.1016/j.cca.2012.08.016 PMID:22971357

Zhang, A., Sun, H., Yan, G., Wang, P., Han, Y., & Wang, X. (2014). Metabolomics in diagnosis and biomarker discovery of colorectal cancer. *Cancer Letters*, *345*(1), 17–20. doi:10.1016/j.canlet.2013.11.011 PMID:24333717

Zhang, T., Wu, X., Yin, M., Fan, L., Zhang, H., Zhao, F., Zhang, W., Ke, C., Zhang, G., Hou, Y., Zhou, X., Lou, G., & Li, K. (2012). Discrimination between malignant and benign ovarian tumors by plasma metabolomic profiling using ultra performance liquid chromatography/mass spectrometry. *Clinica Chimica Acta*, *413*(9-10), 861–868. doi:10.1016/j.cca.2012.01.026 PMID:22309680

Zhao, Y. Y. (2013). Metabolomics in chronic kidney disease. *Clinica Chimica Acta*, *422*, 59–69. doi:10.1016/j.cca.2013.03.033 PMID:23570820

Chapter 8
Biomarker-Based Personalized Management for Cardiovascular Disorders

Taruna
Chitkara College of Pharmacy, Chitkara University, India

Paranjeet Kaur
Chitkara University, India

Thakur G. Singh
Chitkara College of Pharmacy, Chitkara University, India

ABSTRACT

People who are at higher risk for cardiovascular disease can be found using biomarkers. Acute coronary syndrome must be diagnosed and treated promptly and accurately, and cardiac biomarkers are crucial for both. For therapy to start as soon as possible, and possibly, a diagnosis within the golden period, is crucial. Among the numerous biomarkers, the myocardial enzymes, a number of myocardial proteins, peptides, and many other compounds were the first ones to be studied. The stable molecules known as microRNAs are the newest addition to the repertoire and can be found in circulation. Many cardiological conditions are still incurable, and traditional treatments frequently have unfavourable side effects or only slightly improve patient outcomes. Hence, personalised therapy of cardiovascular diseases (CVD), also known as personalised or precision cardiology, is the process of choosing the most appropriate course of action for a given patient.

INTRODUCTION

The current definition of acute myocardial infarction (AMI), as established by the American College of Cardiology and the European Society of Cardiology, places a major emphasis on cardiac biomarkers. The definition of a biomarker is "a characteristic that is objectively measured and quantified as an indicator of normal biological processes, pathogenic processes, or pharmacological response to a therapeutic

DOI: 10.4018/979-8-3693-1922-2.ch008

intervention." Our knowledge base has grown as a result of this research, which has also increased our understanding of the underlying pathologic mechanisms that affect patients. Personalized medicine is essentially the practice of prescribing the precise medications and medicines that are most appropriate for a certain patient, taking into environment in which they live environment in which they live.

Figure 1. Biomarkers

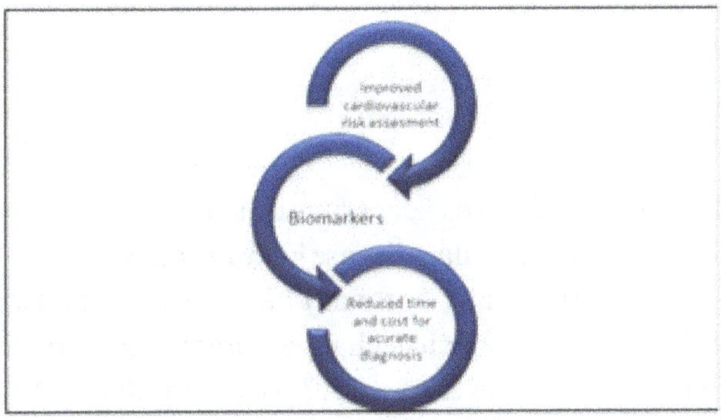

Figure 1 : Biomarkers

Cardiovascular illness has been diagnosed and treated using a number of cardiac indicators. However, there is still a need to search for newer, more specialized molecules due to a lack of sensitivity and specificity to cardiac muscle necrosis (Steyerberg et al., 2010). Cardiac biomarkers have a crucial role in the prompt, correct diagnosis, treatment, and prognosis of ACS. Making a diagnosis within this "golden period" is critical to initiating therapy as soon as feasible and perhaps repairing heart damage. An additional useful triaging technique is the use of cardiac biomarkers.

CARDIOVASCULAR DISORDER

The term "cardiovascular disease" is used to refer generally to conditions that affect the heart or blood vessels. Reduced blood flow to the body, brain, or heart can be brought on by inside an artery that causes it to harden and narrow thrombosis and atherosclerosis The accumulation of fatty deposits. A condition affecting the brain's blood arteries known as **cerebrovascular disease**; Peripheral arterial disease affects the blood vessels that supply the arms and legs. Because rheumatic fever is caused by streptococcal bacteria, rheumatic heart disease destroys the heart muscle and heart valves (Balagopal et al., 2011).

Birth defects resulting from congenital heart disease impact the normal development and function of the heart. Blood clots from deep vein thrombosis and pulmonary embolism can come loose from the veins in the legs.

Epedimiology

It has lately been clear that chronic diseases account for the majority of deaths overall, and this has been covered in-depth elsewhere. Cardiovascular disease (CVD) caused 17.5 million deaths worldwide in 2005, mostly from rheumatic heart disease, coronary heart disease, and stroke (Goff et al., 2014).

In India, the epidemiological landscape has changed dramatically over the past 20 years. Currently, diseases connected to CVD account for over two thirds of the burden of NCD mortality in India. Despite notable geographical variations in the frequency of risk factors, cardiovascular disease (CVD) remains the leading cause of death throughout India, particularly the poorest states and rural areas. In India, the epidemiological transition to the age of delayed chronic illnesses has occurred rapidly over the previous two decades, resembling the accelerated epidemiological transition model (Thygesen et al., 2007).

Approximately 17.5 million people worldwide pass away from different cardiac illnesses. Furthermore, over 75% of deaths globally are expected to be caused by non-communicable illnesses. (Vasan, 2006). According to recent research on CVDs in Europe in 2014, epidemiological elements have been revised. There has been prior evidence of CVDs, particularly CHD and stroke, in European nations. Compared to other nations, Europeans die from CVDs more frequently and more than twice as frequently as they die from cancer. In several Eastern European nations, such as Russia and the Ukraine, the death rate for CHD in those aged 55 to 60 is higher than it is in France, according to reports (Hlatky et al., 2009). In 2014, the UK's CVD epidemiology was discussed in terms of mortality, prevalence, and treatment costs. 2012 saw CVD as the leading cause of death for women in the UK, accounting for 28% of all deaths (Wang, 2011).

Cardiovascular Biomarkers and Their Role

Biomarkers

A biomarker is described as a "characteristic that is objectively measured and evaluated as an indicator of normal biological processes, pathogenic processes, or pharmacologic responses to a therapeutic intervention" by the Consortium in 2001. The American Heart Association then detailed the stringent requirements for how more recent biomarkers should be assessed uniformly before their clinical use may be advised in 2009. When the heart is injured or under stress, chemicals known as cardiac biomarkers are released into the circulation. Acute coronary syndrome (ACS) and cardiac ischemia are two disorders linked to inadequate blood supply to the heart that are identified by measurements of these biomarkers. Cardiac biomarker testing is also useful for monitoring and treating individuals with suspected acute coronary syndrome (ACS) and myocardial ischemia, as well as for estimating an individual's risk of developing these diseases. Biomarkers play a major role in both illness diagnosis and drug development for medical diseases (White House, 2016). In the latter phases of a drug's development, biomarkers could even be helpful in determining the exact doses of a particular medication. A nationwide movement toward precision medicine development has occurred more recently, with a particular focus on the creation of novel cancer treatments.

Types of Biomarkers

Figure 2. Types of biomarkers

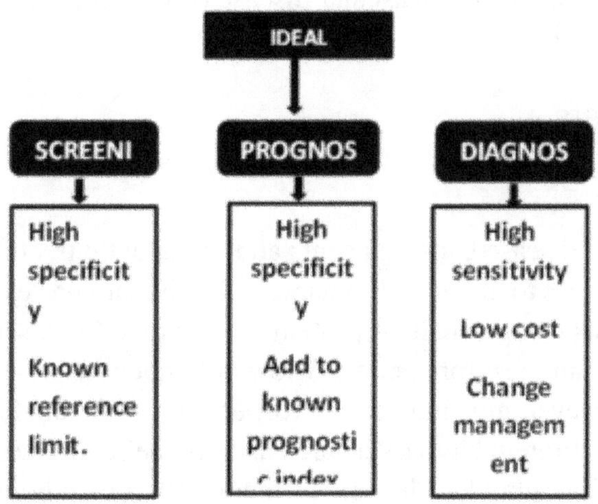

Figure 2 : Types of Biomarkers

A predictive biomarker is one that predicts how a medical state will likely develop in a person who has not received any treatment or in a person who has received conventional therapy. A predictive biomarker, on the other hand, can be used to choose patients who are most likely to react to a certain medication or to identify candidates who can be thought of for particular targeted therapies . Predictive biomarkers aid in modifying therapy to meet the demands of the patient. Until now, only the area of cancer has used clinical trial designs based on analyzing a biomarker for prognostic or predictive usefulness; however, other medical specialties, such as cardiovascular medicine Finally, pharmacodynamics biomarkers gauge how a medicine affects the disease state directly. Blood natriuretic peptide levels are currently being recommended as a surrogate endpoint to measure the effectiveness of pharmacological treatment, for instance, since changes in circulating natriuretic peptide levels are indicative of the severity of heart failure . Another example of how the concentration of a biomarker (low density lipoprotein [LDL] cholesterol) might fluctuate and be used to direct therapy to lower the risk of CVD in the future is the use of statins to lower blood cholesterol levels (Collins & Varmus, 2015). An ideal biomarker should have excellent sensitivity, specificity, accuracy, and precision. According to the illness state under study, discrimination limits enable separation of aberrant levels from normal levels of a biomarker. More crucially, before a biomarker is deemed "fit" to be employed in clinical practice for a "given purpose," it is essential to distinguish between "unwanted" or "abnormal" levels and "levels that require treatment." Last but not least, calibration is a test to evaluate a biomarker's capacity to predict risk in a particular sample compared to the actual observed risk in the same sample of people or in a different population as a whole. For instance, the risk of cardiovascular disease (CVD) depending on cutoff values for waist size or body mass index .

Protein Based Biomarkers

Protein Biomarkers

Due to their frequent roles as disease-causing agents and therapeutic treatment targets, proteins are especially advantageous molecules to utilize as biomarkers. Healthcare professionals may conduct precise non-invasive illness diagnosis using panels of protein biomarkers. Such screening allows for the early detection of illness in donor samples from people who would not otherwise exhibit any odd symptoms.

Protein biomarkers, however, give important prospects for individualized therapeutics in addition to early illness identification. Protein biomarkers are currently being utilized to direct therapy decisions in the treatment of cancer, for example. The sort of therapy that may be most successful can already be predicted by looking for proteins linked to tumor drug resistance or susceptibility to chemotherapy, hormone therapy, or immunotherapy (La Thangue & Kerr, 2011). Targeted proteomics can aid medical professionals in providing more effective care that is suited to a patient's condition when used in conjunction with genome and transcriptome sequencing.

Protein whose presence, absence, or alteration can be used to signify normal or pathological biological processes. Those that have normal biomarker proteins are in good health. individual with aberrant biomarker proteins are diseased.

Monitoring the levels of many biomarkers in clinical trials is now feasible because to recent developments in multiplexing. According to Houser, biomarkers are used by Bio-Rad's pharmaceutical clients at every stage of the medication development process, ranging from safety indicators for preclinical and prognostic markers for clinical trial choices to early cell-based effectiveness markers. They hunt for a specific group of informative biomarkers after doing an initial multiplex screening with the greatest number of probable biomarkers. "Finding a strong, smaller set of biomarkers early in the development [of a drug] is important, and pharmaceutical companies have shared this with me," adds Houser. In preclinical and clinical studies, "they prefer smaller plex levels that allow them to scale up for high-throughput monitoring." The Bio-Plex System from Bio-Rad can identify up to 100.

Pharmaceutical firms depend on the diagnostic, prognostic, and predictive capabilities of protein biomarkers (Malottki et al., 2014). Predictive biomarkers can identify individuals who may benefit from certain medicines; both can result in earlier treatment for a better patient outcome. Biomarkers are helping scientists uncover diagnostic indicators for early-stage illness. For instance, the early detection of cancer using biomarkers to determine its stage and potential for treatment with a favorable prognosis is becoming more and more important. The Alzheimer's Drug Discovery Foundation (ADDF) funds studies to create trustworthy biomarkers for the diagnosis of Alzheimer's. The ADDF and pharmaceutical firms collaborated to develop a bio bank sharing scheme last year. Samples of blood and cerebral spinal fluid from Alzheimer's patients collected at various stages of the disease and utilized in clinical studies are stored in the bio bank.

As seen by the recent introduction of the first blood test for Alzheimer's disease, this has great potential for use in the identification of biomarkers for the condition. The Precivity ADTM blood test, developed by C2N Diagnostics, was introduced in late 2020. It uses mass spectrometry to identify a polio protein E isoforms and assess levels of amyloid beta 40 and 42 proteins (Tajik et al., 2013).

Monitoring the levels of many biomarkers in clinical trials is now feasible because to recent developments in multiplexing. According to Houser, biomarkers are used by Bio-Rad's pharmaceutical clients at every stage of the medication development process, ranging from safety indicators for preclinical and prognostic markers for clinical trial choices to early cell-based effectiveness markers. They hunt for a specific group of informative biomarkers after doing an initial multiplex screening with the greatest number of probable biomarkers. "Finding a strong, smaller set of biomarkers early in the development [of a drug] is important, and pharmaceutical companies have shared this with me," adds Houser. In preclinical and clinical studies, "they prefer smaller plex levels that allow them to scale up for high-throughput monitoring." The Bio-Plex System from Bio-Rad can identify up to 100.

Biomarkers are essential instruments in the prevention and treatment of disease, whether they are used to track elevated blood pressure, control renal function, or identify cancer that was previously undetected. However, not every sufferer is made equal. Protein biomarkers will be crucial to enabling more accurate and tailored medical treatment, which will enable precision medicine to enhance health on a much larger scale.

Precision medicine has made the biggest strides in the field of cancer thus far. However, protein biomarkers hold the potential to revolutionize other medical fields such as infectious illness, cardiovascular disease, and autoimmune. Research that use these new techniques will be fruitful as long as precision medicine is a top medical concern (Corb Aron et al., 2021). The conclusion about protein biomarkers is that their identification and validation might be difficult.

Genome Based Biomarkers

The discovery of biomarkers that forecast a person's risk for a disease, allow for early illness diagnosis, and enhance diagnostic categorization to better guide tailored therapy are all made possible by advances in genetics and biotechnology. I talk about these goals and offer my thoughts on the achievements made thus far as well as potential roadblocks. Examples from cancer are mostly used in the debate, where the nature of the disease has accelerated the adoption of genetic techniques for biomarker development. On the other hand, many of the lessons being learnt in cancer ought to apply to other chronic illnesses.

The term's vagueness and the misconception that "biomarker validity" refers to suitability for intended use are two of the biggest issues facing the creation and validation of biomarkers. Simply said, a great deal of resources are squandered when researchers fail to concentrate intently (Kumar et al., 2021). This may be observed, for instance, in the disparity between the vast body of research on prognostic biomarkers and their little use outside of study. Ignorance also leads to false claims about biomarkers for early detection that are based on research using unsuitable controls. I will now address various broad categories of intended usage in isolation, as shown in Figure 3.

Figure 3. Broad categories of intended use of biomarkers

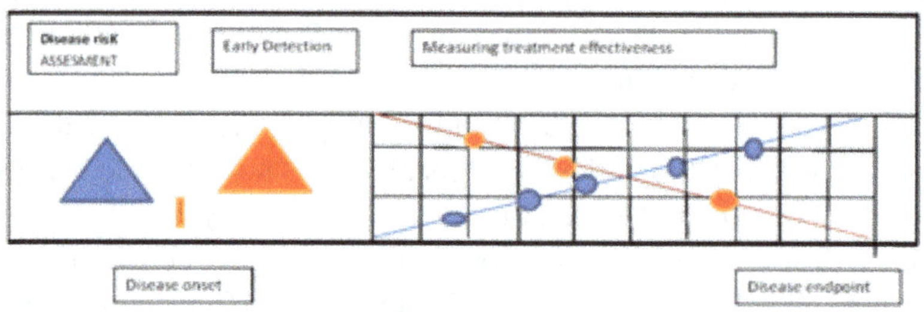

Figure 3 : Broad categories of intended use of Biomarkers

The human genome sequencing project has produced a substantial amount of data that will be used in the creation of biomarkers for each of the aforementioned uses. I offer a brief and subjective evaluation of the advancements made in these domains of the creation of genetic biomarkers in this study.

The following are mentioned: (i) early detection and tailored risk prediction have advanced slowly; (ii) genome-wide association studies are more likely to provide leads for comprehending the etiology of illnesses than helpful data on customized risk assessment.

More progress has been made in the creation of biomarkers to personalize therapy choices, especially in cancer. Important roadblocks that impede the advancement of genetic findings into biomarkers and therapies that lower chronic disease-related death and morbidity are also covered.

Early intervention is often more successful in treating many ailments. Since most solid tumors have a protracted pre-clinical phase before being diagnosed, there should be plenty of chance for early identification. However, the creation and validation of early detection biomarkers with practical applications in medicine has met with limited success thus far. Poor study methodology has been a major hindrance to early detection studies. Comparing the amounts of candidate proteins in tumor tissue obtained during diagnosis to those in normal tissue leads to the "discovery" of several cancer biomarkers. Based on such evidence, a number of "discoveries" have been published and made public (Solberg, 1987). However, the fact that such a difference was found provides very little support for the marker's potential utility in early identification.

Predictive biomarkers identify patients who are most or least likely to benefit from a certain treatment. In non-small-cell lung cancer, EGFR mutation is used to select patients for small-molecule EGFR inhibitors, while in advanced colorectal cancer, KRAS mutation is utilized to remove patients from anti-EGFR antibody treatment. Imatinib was developed as a result of the discovery of the BCR-ABL fusion protein in CML, and second-line therapy are determined by analyzing mutations in that. A kinase inhibitor that targets the EML4-ALK fusion gene was used with very promising results when it was discovered in a small subset of patients with non-small-cell lung cancer (Kwak et al, 2010). Similar to this, an inhibitor with enhanced selectivity for the mutant protein was developed in response to the finding that 60% of patients with metastatic melanoma had a single point mutation in the BRAF gene (Flaherty et al, 2010). The findings of this development were highly encouraging.

It has been less common to employ gene expression profiles as predictive biomarkers for novel medications. It is more fulfilling from a scientific standpoint to use a genetic mutation or gene amplification linked to the drug's mechanism of action as a predictive biomarker. While the development of prognostic signatures has made extensive use of gene expression profiling, the development of prediction indicators of treatment efficacy has been far less common (Lott et al., 1992). Until recently, FFPP tissues were unsuitable for microarray gene analysis due to RNA degradation.

Lipid Based Biomarkers

Organic substances based on carbon-carbon chains that are insoluble in water are known as lipid molecules, and they are an essential component of biological cell membranes. Lipids are therefore thought to be good biomarkers for life detection in terrestrial environments since they are widespread in Earthly life. Numerous novel lipid biomarkers are being discovered; some, like lipoproteins and a polipo proteins, have been studied to some extent before. Research on other lipid compounds is also growing; they include sphingo lipids and their oxidized versions resulting from oxidative inflammation in atherosclerosis. Many, even the less recent ones, have unclear atherogenic risk, and there are no established clinical guidelines to help clinicians use them in routine clinical practice. Moreover, it is still unclear how lipids contribute to childhood atherogenesis. The current understanding of lipids as indicators of cardiovascular risk in the pediatric population is compiled in this study. Hydrophobic substances known as lipids are essential for the function of cell membranes, cell communication, and cell metabolism (Morrow & Antman, 2009). Fatty acids, glycerol lipids, glycerophospholipids, poly ketides, prenol lipids, saccharo lipids, sphingo lipids, and sterol lipids are the eight lipid groups into which they may be divided. The human genome's sequencing has made it possible to do more study, and recently, focus has switched to postgenomic technologies like metabolomics. The complete examination of metabolites in a biological material is known as metabolomics, and modern metabolomic technologies allow for the accurate identification of hundreds to thousands of metabolites. However, because of their complicated composition, the large number of distinct lipid specimens, and their variations, lipids pose a difficult task for thorough examination. The importance of lipids in determining a child's cardiovascular risk is highlighted in this review. The methodical search strategy is first explained, along with a comparison of earlier, comparable review studies. As far as is known, each physiological function has been discussed together with the most recent data about cardiovascular risk in the pediatric population. We stress the importance of more lipid biomarker research in the evaluation of cardiovascular risk in children as we wrap up.

Lipid Determination

Tandem mass spectrometry and nuclear magnetic resonance are the two most used techniques for determining lipid content. The first offers a technique for identifying and measuring a larger variety of lipoprotein subtypes. Different lipoprotein subclasses release a different signal in direct proportion to their concentration when exposed to a magnetic field. When compared to mass spectrometry, it offers a valuable tool for lipid structure investigation in both solution and solid forms, although it is less sensitive and needs more samples. For the very sensitive and selective identification and measurement of lipid species, the latter approach is preferred.Liquid chromatography is the most used method for separating lipid species in order to prevent the ionization of coeluting low-abundant molecules prior to detection. Both top-down and bottom-up methods are applied in lipidomics. While intact lipid species are recognized and quantified using lipid discrimination by their precise mass in the top-down technique, lipids are studied in the bottom-up approach based on their unique fragmentation pattern as determined by tandem mass spectrometry. Relative or absolute quantification are both possible (Sunderman, 1975).

Total cholesterol is the most well-known lipid that increases the risk of cardiovascular disease. The molecular structure of cholesterol is lipoprotein, as seen schematically in figure 4. It consists of a hydrophobic core mostly made up of triglycerides and cholesterol esters, and a hydrophilic membrane made up of phospholipids, free cholesterol, and apolipoproteins.

Figure 4. Lipid based biomarkers and Atherosclerosis

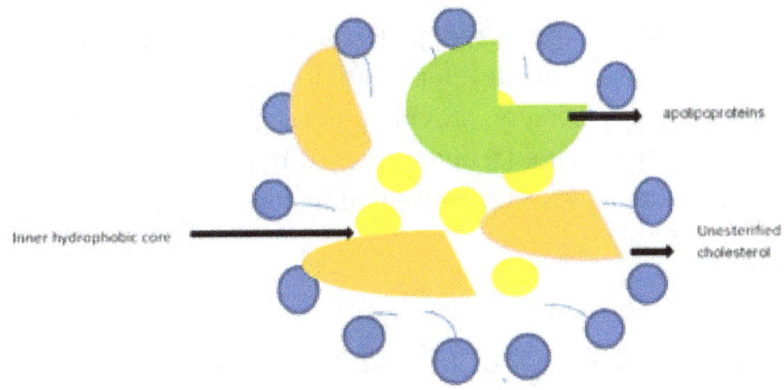

Figure 4 : Lipid based biomarkers and Atherosclerosis

Triglycerides, which are fatty acids, are not included in the category of lipoproteins; instead, they are sterol lipid . when total cholesterol is measured. Major lipids found in chylomicrons (and their remains), VLDL, and IDL are triglycerides; cholesterols, on the other hand, are found mostly in LDL and HDL, but are also found in IDL and chylomicron remnants.

Because existing markers lack adequate specificity, there is an unmet demand for novel or better biomarkers for enhanced disease risk assessment and monitoring for numerous illnesses. A lot of attention is being paid to lipids as possible solutions to these deficiencies. The discovery of specific ceramides for cardiovascular mortality prediction has recently shown this, allowing for better risk assessment of cardiovascular illness in comparison to traditional clinical indicators. In this review, we address the latest research on lipid biomarkers and the potential link between aging, chronic obstructive pulmonary disease, and cardiovascular illness. Furthermore, we talk about how to go above the obstacles that lipid biomarker research is now experiencing.

Sedimentary rocks that contain fossilized lipids provide unique insights into the palaeobiology of the Earth. Information on the oxygenation of the atmosphere and seas, changes in ocean plankton, the greening of continents, major extinctions, and climatic change is encoded in these "biomarkers." In the past, the interpretation of biomarkers was dependent on lists of lipids found in both natural settings and living microbes. But because only a small portion of Earth's microbes can be grown, many historically important bacteria for the environment are extinct, and there are vast gaps in our understanding of lipid production, progress has been hindered. We now have more resources to learn more about lipid biomarkers, their biosynthesis processes, and natural distributions because to the genomics and bioinformatics revolution (Gupta et al., 2006).

We examine in this how preserved organic molecules offer a distinctive viewpoint on the evolution of microbial life on Earth. We go over how developments in molecular biology have made it possible to better understand the origins of biomarkers, provided more reliable interpretations of ancient lipids, and how the rock record is essential for molecular clock calibration. With more sequenced microbial genomes available in searchable databases, these research may be further utilized.

Carbohydrate Based Biomarkers

One of the most significant families of biomarkers for cell types, illness states, protein activities, and developmental stages is thought to be carbohydrates. The development of innovative site-specific delivery vehicles and imaging agents may benefit from the application of carbohydrate "binders" that possess the ability to selectively identify a carbohydrate biomarker. In this study, we highlight significant carbohydrate indicators and discuss how to target them to create new therapeutic and diagnostic products. Though certain problems span therapeutic borders, examples are organized according to illness categories, such as (1) infectious diseases, (2) cancer, (3) inflammation and immunological responses, (4) signal transmission, (5) stem cell transformation, (6) embryo development, and (7) cardiovascular disorders.

In the US, cardiovascular diseases (CVD) constitute the main cause of morbidity and mortality

Public health priorities include primary and secondary prevention of CVD

A substantial body of research suggests that CVD is a disease of the life cycle that starts with the emergence of risk factors, which subsequently aid in the development of subclinical atherosclerosis.3

Manifest CVD is the result of subclinical illness.5

A poor prognosis with an increased chance of recurrent episodes, morbidity, and death is foretold at the very beginning of CVD.

Furthermore, it is becoming more obvious that clinical assessment has its limits even if it serves as the cornerstone of patient treatment.

According to a study, clinicians have employed extra instruments to support clinical evaluation and improve their capacity to recognize the "vulnerable" patient at risk for CVD. As recommended by a recent panel of the National Institutes of Health (NIH). One such technology is biomarkers, which may be used to more reliably and quickly identify high-risk individuals, diagnose diseases, and prognosticate and treat patients (Zeng et al., 2014).

Our knowledge of glycan-associated abnormalities in sick states has greatly benefited from recent developments in glycomics and glycoproteomics. For example, aberrant glycosylations unique to cancer are present in the majority of cancer types, indicating aberrant expression of enzymes such glycosyltransferases and glycosidase that are involved in the manufacture of glycans.

.Glycated hemoglobin (GlcHb) and fibrinogen are two important glycoprotein biomarkers for cardiovascular diseases (CVDs) that are well known. For example, gly-cated valine HbA1c (GlcHb) is significantly greater in diabetes patients than in non-glycemic people. As the erythrocyte lifetime approaches its conclusion, approximately 24% of the total Hb in these individuals is GlcHb (Kaplan et al., 2014). Increased levels of GlcHb have been linked to a considerable rise in CVDs, indicating the significance of GlcHb levels as biomarkers for the prognosis and development of illness. Fibrinogen is a 340 kDa soluble plasma glycoprotein that functions as a precursor to fibrin, the main protein involved in vertebrate blood clotting. Fibrinogen's carbohydrate content is around 3% and consists of NeuAc, Gal, Man, and GlcNAc connected to the γ-chain on Asn52 and the β-chain on Asn364.The regulation of fibrinogen's functionality is influenced by its glycosylation. Deglycosylated fibrinogen, for example, causes fibrin fibers to aggregate more quickly and accelerates fibrin polymerization, while increased negative charge on sialic acids associated with fibrinogen negatively affects fibrin polymerization. Removing the sialyl moiety from fibrinogen also causes plasma clearance through the hepatic galactose/galactosamine binding lectin. Numerous pathologic diseases, including hepatoma, pancreatic and other malignancies, tumor meta-stasis, human hemopoietic cell proliferation, and embryogenesis and reproduction, are also associated with anomalies connected to fibrinogen (D'Agostino, Grundy, Sullivan, & Wilson, 2001).

The traditional definition of a biomarker is "a biological molecule that represents health and disease states." The three main risk factors for cardiovascular disease (CVD), the world's largest cause of morbidity and death, are type 2 diabetes, dyslipidemia, and hypertension. As a result, biomarkers that are often employed in clinical practice, such as blood pressure, blood glucose, traditional lipid profiles, and inflammation (such as C-reactive protein, or CRP), are useful in predicting cardiovascular disease. In most cases, functional foods have positive impacts on traditional lipids and have been found in certain studies to reduce blood glucose and enhance insulin sensitivity, especially when it comes to berries, chocolate, and tea. Adults who consume soy as a functional diet have shown reduced levels of both total and LDL cholesterol. Though we did not see similar results after supplementing persons with "prediabetes" with berries and tea, there is emerging evidence that these foods can lower CRP. These foods include fruits and vegetables, chocolate, and tea. In persons with "prediabetes" or severe cardiovascular disease, berries, cocoa, and tea have been shown to have antihypertensive properties. It has been demonstrated that dietary bioactive substances, particularly polyphenols, mediate biological processes that result in the modification of clinical indicators. Therefore, a few functional foods that are typically included in a regular diet show promise for treating CVD and lowering the levels of biomarkers linked to the advancement of the illness. Cancers of the ovaries, lungs, and gastrointestinal tract have been linked to elevated blood CA 125 levels (4), CA 125 is now a commonly utilized biomarker. Furthermore, there is growing evidence connecting blood CA 125 levels to non-cancerous illnesses such coronary artery disease, pericardial disorders, and cardiovascular disease (8). More precisely, functional class (9), echocardiography (10), and hemodynamics (11) in heart failure were found to be correlated with serum CA 125 levels. Since the cause of cardiac aggressiveness is unrelated to the production of CA 125, it is best seen as a last indicator of organ damage (8). Consequently, pulmonary hypertension (PH) may be affected by it as well. Regretfully, not much is still known about this subject. According to Rahimi-Rad et al. (16), individuals with chronic obstructive pulmonary disease who had PH had greater blood CA 125 levels than those who did not (16). Congenital cardiac disease also showed a similar pattern (17). It is unknown if Blood CA 125 levels are linked to both the prognosis and severity of PH. In this study, we aimed to examine the associations between CA125 and hemodynamics, functional status, echocardiography, and pH-related prognosis in a retrospective cohort.

Nucleic Acid Derived Biomarkers

Important biomarkers for illness identification, tracking, and therapy are nucleic acids. Nucleic acid biomarker development and clinical application have been made possible by technological advancements in nucleic acid analysis. Nonetheless, there are still issues with nucleic acid analysis technology, which restricts the application of nucleic acid biomarkers in specific situations. Here, we examine single-molecule nucleic acid analysis tools that potentially help address these issues. First, we go over the several kinds of nucleic acid biomarkers that are significant for clinical use as well as traditional nucleic acid analysis tools. Next, we go into technology for in situ and single-molecule nucleic acid biomarker analysis. Lastly, we go over other incredibly sensitive methods for finding nucleic acid biomarkers.

Figure 5. Nucleic acid biomarker detection

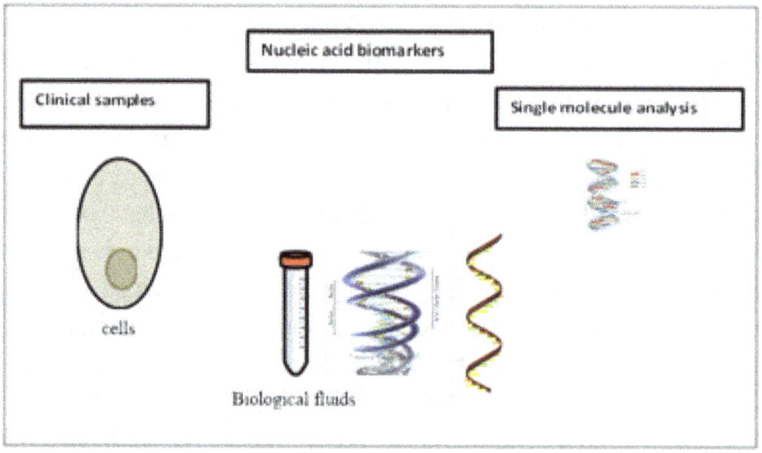

Figure 5: Nucleic acid biomarker Detection

It is widespread practice to employ nucleic acids as biomarkers for a variety of illnesses, including infectious, neurodegenerative, and cancerous disorders . Numerous clinical uses for nucleic acid biomarkers exist, including illness diagnosis, treatment selection, and severity assessment. The development of biomarkers for research and clinical application has been facilitated by conventional nucleic acid analysis technology.

DNA

Since these nucleic acids may be found in tissues or fluids (liquid biopsies, such as saliva, urine, serum, or blood), they are frequently used as biomarkers for cancer diagnosis, tumor progression tracking, and treatment response prediction, including molecular resistance or responsiveness. Encryption of genes is found in DNA, which is the matrix of all an organism's cells. These genes are expressed as RNA, with around 2% of the human genome coding for proteins and the remaining 20% coding for non-coding RNA. These genes facilitate the formation of the fundamental components of life, encourage specific enzyme reactions, and enable very intricate control of each of these processes

RNA

There are two families that make up the enormous RNA family:

The RNAs that code for proteins (mRNA or coding RNA). The messenger RNA (mRNA), which makes up less than 3–5% of all RNA molecules in cells, is the most researched kind. They contribute to the synthesis of non-coding RNAs, or RNAs without the ability to code for proteins, as well as proteins, hormones, enzymes, and other molecules. The ability of non-coding RNAs to physically interact with other RNAs, DNA, or proteins to alter their functions is used to categorize them into distinct groups based on their location or mode of action. For example, tiny nuclear RNA is engaged in RNA modification, long noncoding RNA is involved in translation processes, and long transfer or ribosomal RNA (which make up over 90% of all RNA molecules.

The process known as "alternative splicing" confers variation by enabling a single gene to generate distinct RNA molecules in response to various stimuli (such as changes in the environment or biology, alterations in cell behavior, pathologies, pharmacological reactions, etc.). Therefore, any RNA can be translated—that is, used to create a protein from an RNA molecule—into a variety of proteins, which in turn can provide a range of biological activities, both beneficial and detrimental;

RNA editing, also known as post-transcriptional alterations of RNAs, is a method that allows for the introduction of structural RNA changes (polymorphism, methylation, etc.) at the time of RNA maturation and, subsequently, the modification of the resulting protein;RNA degradation, often known as the "degradome," is a mechanism that regulates an RNA molecule's lifetime and activity, which can range from a few seconds to many days (Lemeshow & Hosmer, 1982).

Epigenetic Changes

The natural history of human illness may be influenced by epigenetic changes and regulators, which are putative molecular components that regulate pertinent physiological and pathological characteristics. These epigenetic modulators can be exploited as illness biomarkers, as they demonstrate various benefits and give information about gene function, thereby explaining discrepancies across patient Endophenotypes . Moreover, epigenetic biomarkers can track the effectiveness of applied medicines and combine data on how lifestyle and environment affect health and illness. The natural history of human illness may be influenced by epigenetic changes and regulators, which are putative molecular components that regulate pertinent physiological and pathological characteristics. These epigenetic modulators can be exploited as illness biomarkers, as they demonstrate various benefits and give information about gene function, thereby explaining discrepancies across patient Endophenotypes . Moreover, epigenetic biomarkers can track the effectiveness of applied medicines and combine data on how lifestyle and environment affect health and illness. The tools used to examine these epigenetic biomarkers are getting easier to use and better all the time. Professionals in laboratories may readily get expertise, and procedures are getting cheaper. In addition, we will go over a few lab techniques and tools to hasten the integration of epigenetic biomarkers into standard clinical practice (Wade, 2000). Better markers are probably going to be found throughout time and added to multi-target assays in the future, which might assist to maximize their use in clinical laboratories. This will increase cost-effectiveness, which will in turn promote the use of precision medicine and the advancement of theragnosis.

Numerous illnesses have been linked to oxidatively produced DNA damage as part of their pathophysiology. The consequences of harm to RNA and the (2′-deoxy-)ribo nucleotide pools—the other nucleic acids—are also receiving more attention, and there is mounting evidence that these nucleic acid pools may potentially play a significant role in illness, may be quantified exactly using LC-MS/MS. Nevertheless, large amounts of tissue are required, which restricts its application in research on human bio monitoring. Conversely, the comet test uses a lot less material—as little as 5 µL of blood may be used—and provides a less intrusive way to measure oxidative stress in vivo, but it has limitations. Significant progress has been made by the European Standards Committee on Urinary (DNA) Lesion Analysis in significantly validating these assays.Urine is an appropriate matrix in which to research nucleic acid-derived biomarkers of oxidative stress non-invasively. The gold standard method for urine analysis is LC-MS/MS; while the ELISA methodology has improved, it is still primarily confined to 8-oxodG. New techniques in DNA adductomics, such as mapping DNA damage throughout the nuclear and mitochondrial genomes or comprehensively evaluating the entirety of adducts in DNA, have the potential to significantly improve our comprehension of the mechanistic involvement of oxidatively damaged nucleic acids in disease.

Nucleotides, circulating cell-free DNA, circulating tumor DNA, mRNA, microRNAs, circular RNA, exosomal RNAs, and long non-coding RNA in tumor cells and serum are all covered by nucleotide and nucleic acid-associated biomarkers. They can be utilized for cancer risk assessment, diagnosis, prognosis, treatment effectiveness, and recurrence prediction. They are playing crucial roles as therapeutic or diagnostic agents for cancer the ranostics. Furthermore, the sensitive biomarkers can be targeted therapeutically for a variety of tumor types. Thus, in order to significantly enhance clinical targeted treatments and immunotherapies for cancer, it is imperative to discover novel biomarkers and investigate the biological roles of these biomarkers. Important illness biomarkers are nucleic acids. Numerous nucleic acid biomarkers have been discovered and put into therapeutic use because to technological advancements. However, there are a number of difficulties with nucleic acid analysis. One difficulty is that in many biological samples, nucleic acids are present in extremely low amounts. As a result, extremely delicate techniques could be required. Another problem is that the sample can include a large number of additional off-target nucleic acids (Bossuyt et al., 2003).

Nucleic acid biomarkers' range of applications may increase with developments in nucleic acid technology. For nucleic acid analysis, single-molecule methods are very promising and offer a few noteworthy benefits . Ultra-sensitive detection is made possible by counting individual molecules. Furthermore, the examination of heterogeneous samples is particularly pertinent to single-molecule technologies. Because bulk averages might obscure the distribution of the molecules in the sample, ensemble measurements are not ideal for these kinds of applications.

Pre- Disease Biological Markers

The need for innovative treatment approaches to treat metabolic syndrome is critical since this condition, which includes a number of problems including hypertension, raises the risk of diseases linked to a certain lifestyle. One strategy is to concentrate on the pre-disease state, which is thought to be the ideal time for preventative therapy and is characterized as a condition of high vulnerability prior to the beginning of the disease. Recently, we have introduced a mathematical theory to identify the pre-disease state: the dynamical network biomarker (DNB) theory. This theory is based on the critical transition paradigm. Here, we used 64 whole-genome microarrays based on the DNB theory to examine the time-course gene expression patterns of a mouse model of metabolic syndrome. The usual behavior of 147 DNB genes was used to describe a pre-disease state that existed before the onset of metabolic syndrome (McShane et al., 2005). Only when we completely If one knew the normal physiology of a biological process, the pathophysiology of that process in the disease state, and the effects of an intervention—pharmacological, device, or otherwise—biomarkers may be used as precise stand-ins for clinically significant endpoints. In the absence of ongoing assessments of the correspondence between clinical endpoints and surrogate endpoints .

Figure 6. Stages during disease

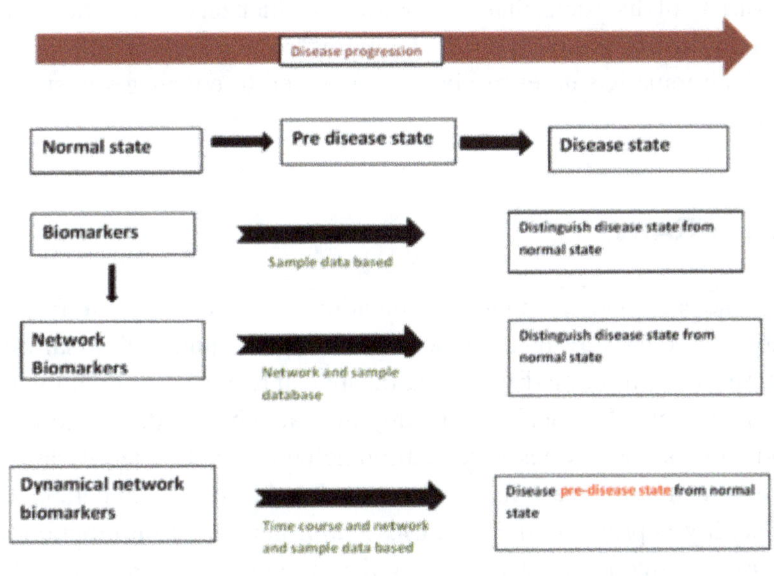

Figure6 : Stages during Disease

Disease state and biomarkers, There are three stage during disease....

One strategy is to concentrate on the pre-disease state, or the state that is most vulnerable to the advancement of the illness, prior to the onset of metabolic syndrome and to clarify the intricate processes that underlie the change from a relatively healthy state or moderate obesity to metabolic syndrome. It is crucial to understand that the lack of any discernible alterations in any portion of the body does not equate to the pre-disease condition. Rather, the condition in question is

(1) believed to have a high sensitivity or vulnerability to illnesses, but is not classified as a disease state according to the current standards.

Within the framework of the Disease Notification Body (DNB) theory, a state transition from a healthy to a diseased state is considered a critical transition if:

(2) rapid recovery to the healthy state becomes difficult or even impossible once the transition occurs. The literature has generally accepted the notion that a disease's development marks a key transition.

The DNB theory offers statistical techniques for choosing pertinent variables in order to identify the pre-disease state. A number of methods for identifying the pre-disease state have also been suggested by the DNB theory and its extensions. Concurrent changes in these metrics are seen to be a trustworthy early warning indicator, indicating the impending crucial shift into a disease state (Ware, 2006).

Real data from a variety of disorders, diabetes mellitus, influenza, cancer, and Alzheimer's disease16, as well as experimental data from cell biology, including cell differentiation18, have been subjected to the DNB theory. Additionally, studies are being conducted to enhance statistical techniques and to improve theory.

PERSONALIZE THERAPY

Clinicians can determine each patient's unique foundation for vulnerability in response to an infection by using tailored and focused therapy. It is also possible to explain inter individual variability, diagnosis, and treatment response . Therefore, modifying healthcare can result in the creation of therapeutic treatments and prophylactic actions. Personalized therapy may also be included into the healthcare system to effect change. Individuals' meticulous vpreparation helped both the identification of epidemics and the local definition of transmission patterns. There is much to be covered in personalized medicine. The most promising possibility is probably the identification of genetic or proteomic biomarkers that may be used to target a disease directly and expose its underlying etiology. These targeted therapies are, of course, well-established for treating infectious illnesses, and they have gained popularity recently in cancer, where they are gradually replacing widely used cytotoxic drugs. epidermal growth factor receptor tyrosine kinase inhibitors for lung cancer. Recently, drugs that target additional genetically defined patient subsets have been approved. These medications include particular genetically different hepatitis C viruses and ivacaftor, which is used to treat people with cystic fibrosis who have certain genetic abnormalities in the cystic fibrosis transmembrane conductance regulator (CFTR) gene that impair the channel's gating action. In each of these cases, laboratory testing enables the responsive patient subgroups to be distinguished from the nonresponders prior to beginning medication.

Precursors of Personalized Medicines

Precision medicine, that classifies individuals into distinct groups and then tailors medical decisions, practices, therapies, and/or products to each patient according to their anticipated response or risk of disease (Kaptoge et al., 2012). This notion is referred to by several names, including precision medicine, stratified medicine, and P4 medicine.

The development of customized medicine is foreshadowed in many ways by the history of western medicine. To keep things brief, we won't cover every one of these events; instead, we'll concentrate on a select number that we believe best capture the fundamental ideas of customized medicine. The English physician Archibald Garrod started researching illnesses that would eventually be referred to as "inborn errors of metabolism" . Garrod researched several uncommon disorders including alkaptonuria, albinism, cystinuria, and pentosuria that have overt, obvious phenotypic symptoms (31).

Traditional pharmacotherapies, or medications, are used by the human body to treat illness in two main ways. First, a medication must cause the body to react. This reaction happens in phases, starting with the body absorbing the medication. The medication must then be dispersed throughout the body (perhaps "biotransformed" or broken down into usable components during this phase) before starting to have an effect. Lastly, any leftover medication or medication components are eliminated. These procedures are sometimes grouped together under the term "pharmacokinetics" and are known as a drug's "ADME" (Absorption, Distribution, Metabolism, and Excretion). A distinct group of genes, such as drug-metabolizing enzymes, frequently regulate pharmacokinetic activity. These genes may contain naturally occurring genetic variations, or "polymorphisms," that affect how these genes operate and, ultimately, how the body responds to a certain medication. The pharmacokinetic properties of pharmaceuticals that are mediated by genetics were linked to several early instances of customized medicine. The knowledge of drug metabolizing enzymes and their function in the body's reaction to medications by the biomedical scientific community was partly to blame for this. Weber's book is a great resource for learning about the pharmacogenetic characteristics of medications and how genetic variations in genes affect therapeutic efficacy and adverse effects, particularly in relation to genetic variants in drug-metabolizing enzymes. Twenty) An improper dosage of the commonly used blood thinner warfarin might result in a possibly fatal adverse drug response. Warfarin is partially metabolized by the gene CYP2C9 and targets a specific gene called VKORC1. Variations in the pharmacodynamics and pharmacokinetic characteristics of warfarin across individuals are caused by naturally existing genetic diversity in both the VKORC1 and CYP2C9 genes, which in turn causes variations in the reactions of individuals to warfarin (Kamath et al., 2001). Therefore, the US Food and Drug Administration has advised that when prescribing warfarin, one's genotype be taken into account (that is, the dose has to be customized for each individual based on the unique genetic variations that each person possesses in the VKORC1 and CYP2C9 genes).

Role of Diagnostic in Personalized Management of cardiovascular Disease

The current medical trend focuses on increasing the accuracy of biomarkers and molecular diagnostics to direct the administration of medicines and biological treatments to specific illness sites. There is a growing recognition of the variation in how each patient reacts to typical therapy methods. The use of personalized medicine concepts will expedite the translation of biotechnologies and advancements in "omics," such as proteomics and genomes, to improve the safety and efficacy of different therapies and disease prevention. Examples of tactics and technological integrations used to customize the treatment of common cardiovascular conditions including hypertension and ischemic heart disease are given in this overview. The amount of information on the pathophysiology of cardiovascular diseases (CVD) that is becoming accessible on a regular basis will necessitate a systematic interpretation of changes in proteins, mRNA, and DNA at the individual and community levels. Biotechnologies advancements have been used to enhance CVD diagnosis and therapy (Cox, 1972). Personalized medicine puts medicine on a scientific basis even if it is an art .

Figure 7. Personalized cardiology

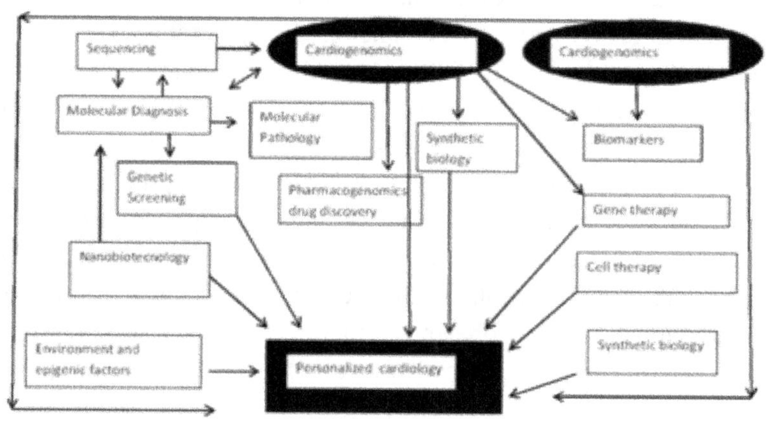

Figure 7 : Personalized Cardiology

Cardiovascular disease (CVD) can be diagnosed using cardiac imaging, which involves the use of CT, MRI, and PET scans as well as regular coronary angiography and cardiac scintigraphy. Understanding the pathomechanisms of different illnesses, especially via the use of molecular imaging, and diagnosing diseases with a genetic component both depend on molecular diagnostics.

In a large morbidity and mortality study, the cholesteryl ester transfer protein inhibitor dacletrapib increased HDL cholesterol levels without influencing the risk of cardiovascular events. A genetic analysis was conducted in the dal-OUTCOMES trial and the dal-PLAQUE-2 imaging study to see if it might still have clinical benefits. The results showed that people homozygous for the ADCY9 gene variant responded to dalcetrapib treatment with a significant reduction in cardiovascular risk. In a phase III study of dalcetrapib, people with this genetic profile are being identified using a companion diagnostic.

Numerous overlapping disorders fall within the difficult treatment field of cardiovascular disease. Low-cost indicators like blood pressure (BP) and cholesterol readings were employed in the past. Imaging investigations for plaque stability and composition in vivo as well as intravascular ultrasonography have identified clinical biomarkers for cardiovascular disorders. In recent years, several novel biomarkers have been identified for use in medication research and development, as well as providing a foundation for diagnostic testing. Improved diagnostic procedures are required, particularly for the metabolic syndrome, a group of conditions that includes obesity, diabetes, and cardiovascular disease. Biomarkers will be crucial in the creation of customized treatments for cardiovascular disease.

Systems based on nanobiotechnology are crucial to the advancement of personalized medicine because they allow for both pathological diagnosis and tailored therapeutic administration for therapy (Baker, 2003). Biomarker-targeted nanoparticles have the potential to be used to both pharmacologic treatment of susceptible atherosclerotic lesions causing MI and molecular imaging. Therapeutics can be made more individualized with the potential combined use of nanoparticles for imaging and site-targeted delivery of therapeutic medicines to cardiovascular disease. Verification of the drug's targeting and molecular action should be possible with image-based therapies that use site-selective agents. By identifying plaques at high risk for acute consequences, magnetic nanoparticles (MNP) used as a contrast agent for magnetic resonance imaging (MRI) can be used to target vascular inflammation or thrombosis and enhance the evaluation of atherosclerosis risk. Annexin-labeled magnetic nanoparticles (MNPs), specifically AnxCLIO-Cy5.5, can be used to visualize cardiac cell death in vivo. Empirical research has demonstrated the viability of utilizing MNP for both diagnosis and treatment (Hanley & McNeil, 1982).

Intensive Personalised Cardiovascular Health Monitoring

Many people in India are unable to pay for general therapy for cardiac illnesses due to the high expense of treating these conditions. Yes, there are people who are unable to find a qualified specialist since they reside in rural areas. The web and innovation have brought about an upgrade to all the customary developments. Things appear more favorable to use and less demanding when they are innovative. In keeping with this, we want to use machine learning algorithms and unique classifiers to predict the likelihood of cardiac problems, study them comprehensively, and provide workouts and drugs to prevent them. Around the world, heart disease is regarded as one of the leading causes of death. The majority of situations include insufficient funds for periodic checkups and inadequate transportation, while most issues may be resolved with medicine and exercise .

Each of these algorithms' performances were evaluated and contrasted in terms of accuracy, confusion matrix, ROC curve, and AUC value. The user-provided factors are utilized to determine the likelihood of heart disease using the most effective algorithm, which then produces a report that includes exercise regimens and prescription drugs to avoid heart attacks.

Keeping an eye on cardiac health to prevent:

Applications for patient monitoring to avoid future disease or harm are numerous. In order to stop more incidents, monitoring of any person with cardiovascular disease is part of the traditional secondary prevention monitoring protocol. Usually, cardiac maintenance and rehabilitation (CR) regimens include these initiative, real-time monitoring of patients receiving continuous left ventricular assist device (LVAD) therapy, monitoring specific populations, such as those who might not be able to travel, and integrating monitors in population-scale m Health preventive health programs .Keeping an eye on cardiac health to prevent

Applications for patient monitoring to avoid future disease or harm are numerous. In order to stop more incidents, monitoring of any person with cardiovascular disease is part of the traditional secondary prevention monitoring protocol (Pepe, 2000). Usually, cardiac maintenance and rehabilitation (CR) regimens include these initiatives, monitoring specific populations, such as those who might not be able to travel, and integrating monitors in population-scale mHealth preventive health programs. Transferability of health information for research and evidence-based health policy is facilitated by the implementation of standard methodology, which yields dependable and comparable data. Even while registries are very helpful, they need a lot of resources to set up and keep up, as well as a lot of work and money to provide solid and trustworthy indications. Brown et al. conducted a comprehensive analysis of 12 studies, 10 of which were RCTs, that used pharmacist interventions in smoking cessation trials.43 While controls got standard treatment, interventions included photoaging, nicotine replacement therapy, and behavioral support. Out of the twelve trials, five shown a noteworthy improvement in abstinence rates with the intervention, and the four that included a cost-effectiveness analysis concluded that the intervention was economical. Brown et al. also examined five trials on weight reduction programs in community pharmacy as part of their evaluation. The weight reduction intervention offered by the community pharmacy (which consists of meal replacement, low-calorie diet, exercise, and orlistat) was compared in all the trials to other weight loss therapies offered by the community pharmacy, commercial weight loss programs, and similar interventions in other primary care settings. i.e., "active" interventions were given to individuals in the intervention and control arms. Weight reductions from pharmacy treatments were comparable to those from active interventions in other healthcare settings. However, pharmacy-based therapies were more costly than commercial weight reduction programs and cost about the same as interventions in other healthcare settings when it came to cost-effectiveness. The length of the intervention seemed to be a significant determinant of the success of weight loss and smoking cessation programs provided by community pharmacies44.

CONCLUSION

Globally, represent a significant healthcare burden on the populace. In order to reduce morbidity and death, early diagnosis of this condition is crucial for prevention and treatment. Biomarkers are an essential tool for pathological illnesses, either for prognostic purposes or for diagnosis or screening. The applicability and limits of the historical cardiac biomarkers used to identify these diseases are covered in this review. These have now been supplanted by the discovery of new biomarkers, which are currently used in standard clinical practice but still do not identify every illness. Early research on future cardiac biomarkers is encouraging, but more work is needed to demonstrate their potential to enhance CVD diagnosis beyond what is no (Cook, 2007).

Furthermore, the analytical tools that would enablew known about biomarkersIn conclusion, with a growing number of existing and new CVD biomarkers, assessing CVD biomarkers has become more and more important for clinical practice (Table 1). With improved sensitivity and specificity, these biomarkers are becoming more and more tailored to the pathophysiology of the cardiovascular system. For instance, in order to guide the treatment approach for individuals with suspected HF, When the NT-proBNP level is more than 2000 ng/L, it is advised to refer the patient immediately for transthoracic echocardiography and a professional examination, the function of recognized CVD biomarkers has evolved from being limited to the retrospective validation of a disease that has previously been identified to taking center stage in clinical algorithms that are used for screening, diagnosis, and prognosis. Many innovative CVD biomarkers haven't yet been used in clinical practice, nevertheless. This is partially caused by the fact that we still don't fully understand their precise clinical use and possible benefits over currently recognized biomarkers. In addition, there are technological problems (such as the dearth of clinical tests that are easily commercialized) and difficulties in accurately determining any novel biomarker as an analyte (Pencina et al., 2008). In fact, convincing evidence is required to demonstrate the potential advantages of these new biomarkers for personalized/biomarker-guided therapy as well as for CVD screening, diagnosis, and prognostics. In this context, multi-biomarker techniques that use a variety of patho-biologic biomarkers may be very influential in terms of CVD diagnosis and/or treatment.

REFERENCES

Atkinson, A. J., Colburn, W. A., DeGruttola, V. G., DeMets, D. L., Downing, G. J., & Hoth, D. F. (2001). Biomarkers and surrogate endpoints: Preferred definitions and conceptual framework. *Clinical Pharmacology and Therapeutics*, *69*(3), 89–95. doi:10.1067/mcp.2001.113989 PMID:11240971

Baker, S. G. (2003). The Central Role of Receiver Operating Characteristic (ROC) Curves in Evaluating Tests for the Early Detection of Cancer. *Journal of the National Cancer Institute*, *95*(7), 511–515. doi:10.1093/jnci/95.7.511 PMID:12671018

Balagopal, P. B., de Ferranti, S. D., Cook, S., Daniels, S. R., Gidding, S. S., Hayman, L. L., McCrindle, B. W., Mietus-Snyder, M. L., & Steinberger, J. (2011). Nontraditional risk factors and biomarkers for cardiovascular disease: mechanistic, research, and clinical considerations for youth: a scientific statement from the American Heart Association. *Circulation*, *123*(23), 2749–2769. doi:10.1161/CIR.0b013e31821c7c64 PMID:21555711

Bossuyt, P. M., Reitsma, J. B., Bruns, D. E., Gatsonis, C. A., Glasziou, P. P., Irwig, L. M., Lijmer, J. G., Moher, D., Rennie, D., & de Vet, H. C. W. (2003). Towards Complete and Accurate Reporting of Studies of Diagnostic Accuracy: The STARD Initiative. *Clinical Chemistry*, *49*(1), 1–6. doi:10.1373/49.1.1 PMID:12507953

Collins, F. S., & Varmus, H. (2015). A New Initiative on Precision Medicine. *The New England Journal of Medicine*, *372*(9), 793–795. doi:10.1056/NEJMp1500523 PMID:25635347

Cook, N. R. (2007). Use and Misuse of the Receiver Operating Characteristic Curve in Risk Prediction. *Circulation*, *115*(7), 928–935. doi:10.1161/CIRCULATIONAHA.106.672402 PMID:17309939

Corb Aron, R. A., Abid, A., Vesa, C. M., Nechifor, A. C., Behl, T., Ghitea, T. C., Munteanu, M. A., Fratila, O., Andronie-Cioara, F. L., Toma, M. M., & Bungau, S. (2021, March 17). Recognizing the benefits of pre-/probiotics in metabolic syndrome and type 2 diabetes mellitus considering the influence of akkermansia muciniphila as a key gut bacterium. *Microorganisms*, *9*(3), 618. doi:10.3390/microorganisms9030618 PMID:33802777

Cox, D. R. (1972). Regression models and life tables. *Journal of the Royal Statistical Society. Series A (General)*, *34*, 187–220.

D'Agostino, R. B. Sr, Grundy, S., Sullivan, L. M., & Wilson, P. (2001). Validation of the Framingham coronary heart disease prediction scores: Results of a multiple ethnic groups investigation. *Journal of the American Medical Association*, *286*(2), 180–187. doi:10.1001/jama.286.2.180 PMID:11448281

Goff, D. C. Jr, Lloyd-Jones, D. M., Bennett, G., Coady, S., D'Agostino, R. B., Gibbons, R., Greenland, P., Lackland, D. T., Levy, D., O'Donnell, C. J., Robinson, J. G., Schwartz, J. S., Shero, S. T., Smith, S. C. Jr, Sorlie, P., Stone, N. J., & Wilson, P. W. F. (2014). 2013 ACC/AHA Guideline on the Assessment of Cardiovascular Risk: A Report of the American College of Cardiology/American Heart Association Task Force on Practice Guidelines. *Circulation*, *129*(25_suppl_2), S49–S73. doi:10.1161/01.cir.0000437741.48606.98 PMID:24222018

Gupta, M., Singh, N., & Verma, S. (2006). South Asians and Cardiovascular Risk: What Clinicians Should Know. *Circulation*, *113*(25), e924–e929. doi:10.1161/CIRCULATIONAHA.105.583815 PMID:16801466

Hanley, J. A., & McNeil, B. J. (1982). The meaning and use of the area under a receiver operating characteristic (ROC) curve. *Radiology*, *143*(1), 29–36. doi:10.1148/radiology.143.1.7063747 PMID:7063747

Hlatky, M. A., Greenland, P., Arnett, D. K., Ballantyne, C. M., Criqui, M. H., Elkind, M. S., Go, A. S., Harrell, F. E. Jr, Hong, Y., Howard, B. V., Howard, V. J., Hsue, P. Y., Kramer, C. M., McConnell, J. P., Normand, S.-L. T., O'Donnell, C. J., Smith, S. C. Jr, & Wilson, P. W. F. (2009). Criteria for evaluation of novel markers of cardiovascular risk: A scientific statement from the American Heart Association. *Circulation*, *119*(17), 2408–2416. doi:10.1161/CIRCULATIONAHA.109.192278 PMID:19364974

Kamath, P. S., Wiesner, R. H., Malinchoc, M., Kremers, W., Therneau, T. M., & Kosberg, C. L. (2001). A model to predict survival in patients with end-stage liver disease. *Hepatology (Baltimore, Md.)*, *33*(2), 464–470. doi:10.1053/jhep.2001.22172 PMID:11172350

Kaplan, R. C., Aviles-Santa, M. L., Parrinello, C. M., Hanna, D. B., Jung, M., Castaneda, S. F., Hankinson, A. L., Isasi, C. R., Birnbaum-Weitzman, O., Kim, R. S., Daviglus, M. L., Talavera, G. A., Schneiderman, N., & Cai, J. (2014). Body Mass Index, Sex, and Cardiovascular Disease Risk Factors Among Hispanic/Latino Adults: Hispanic Community Health Study/Study of Latinos. *Journal of the American Heart Association*, *3*(4), e000923. doi:10.1161/JAHA.114.000923 PMID:25008353

Kaptoge, S., Di Angelantonio, E., Pennells, L., Wood, A. M., White, I. R., & Gao, P. (2012). C-Reactive Protein, Fibrinogen, and Cardiovascular Disease Prediction. *The New England Journal of Medicine*, *367*(14), 1310–1320. doi:10.1056/NEJMoa1107477 PMID:23034020

Kumar, S., Behl, T., Sachdeva, M., Sehgal, A., Kumari, S., Kumar, A., Kaur, G., Yadav, H. N., & Bungau, S. (2021, January 1). Implicating the effect of ketogenic diet as a preventive measure to obesity and diabetes mellitus. *Life Sciences*, *264*, 118661. doi:10.1016/j.lfs.2020.118661 PMID:33121986

La Thangue, N. B., & Kerr, D. J. (2011). Predictive biomarkers: A paradigm shift towards personalized cancer medicine. *Nature Reviews. Clinical Oncology*, *8*(10), 587–596. doi:10.1038/nrclinonc.2011.121 PMID:21862978

Lemeshow, S. T. A. N., & Hosmer, D. W. Jr. (1982). A Review of Goodness of fit Statistics for use in the Development of Logistic Regression Models. *American Journal of Epidemiology*, *115*(1), 92–106. doi:10.1093/oxfordjournals.aje.a113284 PMID:7055134

Lott, J. A., Mitchell, L. C., Moeschberger, M. L., & Sutherland, D. E. (1992). Estimation of reference ranges: How many subjects are needed? *Clinical Chemistry*, *38*(5), 648–650. doi:10.1093/clinchem/38.5.648 PMID:1582014

Malottki, K., Biswas, M., Deeks, J. J., Riley, R. D., Craddock, C., Johnson, P., & Billingham, L. (2014). Stratified medicine in European Medicines Agency licensing: A systematic review of predictive biomarkers. *BMJ Open*, *4*(1), 4. doi:10.1136/bmjopen-2013-004188 PMID:24468721

McShane, L. M., Altman, D. G., Sauerbrei, W., Taube, S. E., Gion, M., & Clark, G. M. (2005). Reporting Recommendations for Tumor Marker Prognostic Studies. *Journal of Clinical Oncology*, *23*(36), 9067–9072. doi:10.1200/JCO.2004.01.0454 PMID:16172462

Morrow, D. A., & Antman, E. M. (2009). Evaluation of High-Sensitivity Assays for Cardiac Troponin. *Clinical Chemistry*, *55*(1), 5–8. doi:10.1373/clinchem.2008.117218 PMID:19028812

Pencina, M. J., D'Agostino, R. B. Sr, D'Agostino, R. B. Jr, & Vasan, R. S. (2008). Evaluating the added predictive ability of a new marker: From area under the ROC curve to reclassification and beyond. *Statistics in Medicine*, *27*(2), 157–172. doi:10.1002/sim.2929 PMID:17569110

Pepe, M. S. (2000). An Interpretation for the ROC Curve and Inference Using GLM Procedures. *Biometrics*, *56*(2), 352–359. doi:10.1111/j.0006-341X.2000.00352.x PMID:10877289

Solberg, H. E. (1987). The concept of reference values. *Journal of Clinical Chemistry and Clinical Biochemistry. Zeitschrift fur Klinische Chemie und Klinische Biochemie*, *25*, 337–342. PMID:3612033

Steyerberg, E. W., Vickers, A. J., Cook, N. R., Gerds, T., Gonen, M., Obuchowski, N., Pencina, M. J., & Kattan, M. W. (2010). Assessing the performance of prediction models: A framework for traditional and novel measures. *Epidemiology (Cambridge, Mass.)*, *21*(1), 128–138. doi:10.1097/EDE.0b013e3181c30fb2 PMID:20010215

Sunderman, F. W. Jr. (1975). Current Concepts of "Normal Values," "Reference Values," and "Discrimination Values" in Clinical Chemistry. *Clinical Chemistry*, *21*(13), 1873–1877. doi:10.1093/clinchem/21.13.1873 PMID:1192578

Tajik, P., Zwinderman, A. H., Mol, B. W., & Bossuyt, P. M. (2013). Trial Designs for Personalizing Cancer Care: A Systematic Review and Classification. [PubMed] [Google Scholar]. *Clinical Cancer Research*, *19*(17), 4578–4588. doi:10.1158/1078-0432.CCR-12-3722 PMID:23788580

Thygesen, K., Alpert, J. S., & White, H. D. (2007). TASK FORCE MEMBERS: Chairpersons: Kristian Thygesen (Denmark), Biomarker Group: Allan S. Jaffe CU. Universal Definition of Myocardial Infarction. *Circulation*, *116*(22), 2634–2653. doi:10.1161/CIRCULATIONAHA.107.187397 PMID:17951284

Vasan, R. S. (2006). Biomarkers of Cardiovascular Disease: Molecular Basis and Practical Considerations. *Circulation*, *113*(19), 2335–2362. doi:10.1161/CIRCULATIONAHA.104.482570 PMID:16702488

Wade, A. N. G. I. (2000). Derivation versus validation. *Archives of Disease in Childhood*, *83*(6), 459–460. doi:10.1136/adc.83.6.459 PMID:11087274

Wang, T. J. (2011). Assessing the Role of Circulating, Genetic, and Imaging Biomarkers in Cardiovascular Risk Prediction. *Circulation*, *123*(5), 551–565. doi:10.1161/CIRCULATIONAHA.109.912568 PMID:21300963

Wang, T. J., Gona, P., Larson, M. G., Tofler, G. H., Levy, D., Newton-Cheh, C., Jacques, P. F., Rifai, N., Selhub, J., Robins, S. J., Benjamin, E. J., D'Agostino, R. B., & Vasan, R. S. (2006). Multiple Biomarkers for the Prediction of First Major Cardiovascular Events and Death. *The New England Journal of Medicine*, *355*(25), 2631–2639. doi:10.1056/NEJMoa055373 PMID:17182988

Ware, J. H. (2006). The Limitations of Risk Factors as Prognostic Tools. *The New England Journal of Medicine*, *355*(25), 2615–2617. doi:10.1056/NEJMp068249 PMID:17182986

White House. (2016). *Precision Medicine*. White House. https://www.whitehouse.gov/precision-medicine

Zeng, Q., He, Y., Dong, S., Zhao, X., Chen, Z., Song, Z., Chang, G., Yang, F., & Wang, Y. (2014). Optimal cut-off values of BMI, waist circumference and waist:height ratio for defining obesity in Chinese adults. *British Journal of Nutrition*, *112*(10), 1735–1744. doi:10.1017/S0007114514002657 PMID:25300318

Chapter 9
Application of CFD in Healthcare or the Biomedical Field

Tejas Chalke
University Institute of Chemical Technology, Kavayitri Bahinabai Chaudhari North Maharashtra University, India

Madhura Bodkhe
https://orcid.org/0009-0002-0066-3405
University Institute of Chemical Technology, Kavayitri Bahinabai Chaudhari North Maharashtra University, India

Sunil Jayant Kulkarni
https://orcid.org/0000-0002-5988-3448
Gharda Institute of Technology, India

Ajaygiri K. Goswami
University Institute of Chemical Technology, Kavayitri Bahinabai Chaudhari North Maharashtra University, India

ABSTRACT

CFD is not merely a computational tool; it is an indispensable asset with versatile applications, particularly in the biomedical field. Its role in understanding complex physiological systems and guiding decision-making in medical design and interventions through advanced simulations cannot be overstated. Whether in drug delivery systems, physiological flow analysis, surgical planning, or medical device development, CFD emerges as a rapidly developing and essential tool in biomedical research. Despite biomedical applications engage with the nuances of human physiology and internal fluid actions, the accession of high-performance hardware and software together with discoveries in computer sciences have rendered recent computational fluid dynamics (CFD) applications in the biomedical field more accessible and viable. The review paints a comprehensive picture of CFD's journey from a specialized area of mathematics to a transformative force in healthcare.

DOI: 10.4018/979-8-3693-1922-2.ch009

INTRODUCTION

Recent Development in Biomedical or Healthcare

The human body and its functioning have always intrigued humankind. The legendary Father of Medicine, Hippocrates, initially identified physical ailments as the outcome of biological processes as opposed to mystical or mystic forces in 400 BC. Through the numerous works of the eminent Italian engineer Leonardo da Vinci, the very first attempt to integrate engineering and physics theories to describe the human body could be dated back to the fifteenth century. Leonardo da Vinci has been credited as the first bioengineer (Armentano and Kun 2019) and the first healthcare physicist (Kron and Krishnan 2019) thanks to his innumerable sketches and drawings which helped improve the comprehension of human physiology and anatomy. The multidisciplinary field of biomedical sciences incorporates knowledge from diverse engineering fields to look into biology, medicine, and healthcare. Biomaterials, bioinstrumentation, bioinformatics, biomechanics, medical imaging, and medical devices are merely a few instances among the multitude of subjects that can be classified within the wide umbrella of research in the biomedical field. The use of modern technology in healthcare serves as one of the primary explanations for why biomedical engineering continues to be a major field of study today and in the years to come. For instance, engineering is employed in multiple contemporary surgical procedures, including laparoscopic surgery (Ooi et al.,2023). Significant improvements in medical treatment, particularly regarding remote monitoring of health, have been brought about by the growth of information technology (IT) (Li et al., 2019). Preventing illness and identifying high-risk individuals early on are among the primary objectives of utilising physical sensor networks (Mohammadzadeh et al., 2020). To be able to track and cure patients' conditions quickly through immediate access and continual monitoring of their health indicators, intelligent technologies and smart gadgets (As an example wearable and Astute wireless sensors) have become becoming increasingly common (Gries et al., 2018). The primary objectives of wearable sensors are for recording and tracking patients' medical problems and conditions while also performing a range of additional functions concerning well-being (Li et al., 2019; Mohammadzadeh et al., 2020).

As a way to foster more promptly and continual assessment of patients' well-being and the associated health sub-systems, an array of smart devices, Internet of Things (IoT), telemedicine, smart pills, and AI-based technologies have been envisioned and created (Junaid et al., 2022). Over the last ten years, a significant rise in interest in smart devices, specifically wearable sensors, mostly in healthcare settings. From physical (body) signs like the rate of your heart (HR), arterial blood pressure (BP), temperature of the body, oxygen consumption, and internally mobility were numerous variables to think about, these instruments search to acquire clinically meaningful health-related data (Majumder et al., 2017). In a nutshell, relevant sensible sensors and networks of wearable sensors are used to extract and share fundamental health information. There's a greater likelihood that the Internet of Things (IoT) could improve healthcare because of its increasing popularity (Muratyan et al., 2022). The expression "Internet of Medical Things" was popularized by many to describe the latest developments in micro processing, which has ramifications for the medical field (Dwivedi et al.,2022). There were 11.3 billion IoT devices in operation (Markets and Markets,2021). The medical device IoT market is projected to rising to $94.2 billion in 2026 from $26.5 billion in 2021(Sinha et al.,2023). With the aid of these advances, the healthcare sector has grown increasingly interconnected, making it hard to ignore IoT. In the years as of late the epidemic commenced in 2020, A lot has changed in telehealth. Patients and treatment teams regularly participate in online video conferences where problems are discussed, and advice is given. To make this possible, the foundational structure is in far better shape. The healthcare study connects 2021 projects that the worldwide value of telehealth solutions would reach $185.6 billion by 2026 (Fortune Business Insights,2023). One of the most innovative applications of internet of things (IoT) the concept of a "smart pill" is used in healthcare technology. (Minaam et al., 2018), which essentially shifts to the Internet of Bodies from the Internet of Things. With its many potential uses, artificial intelligence (AI) represents a fast-emerging field of research and study that is especially relevant to healthcare. Artificial intelligence's application in the battle over COVID-19, diagnosis and development of medications, psychological wellness, chatbots, and various other critical fields has shown its value in recent years (Patel et al.,2023). In 2023, the technologies that has gained attention of Audiences are given in Fig.1.

Figure 1. Healthcare technology trend in 2023

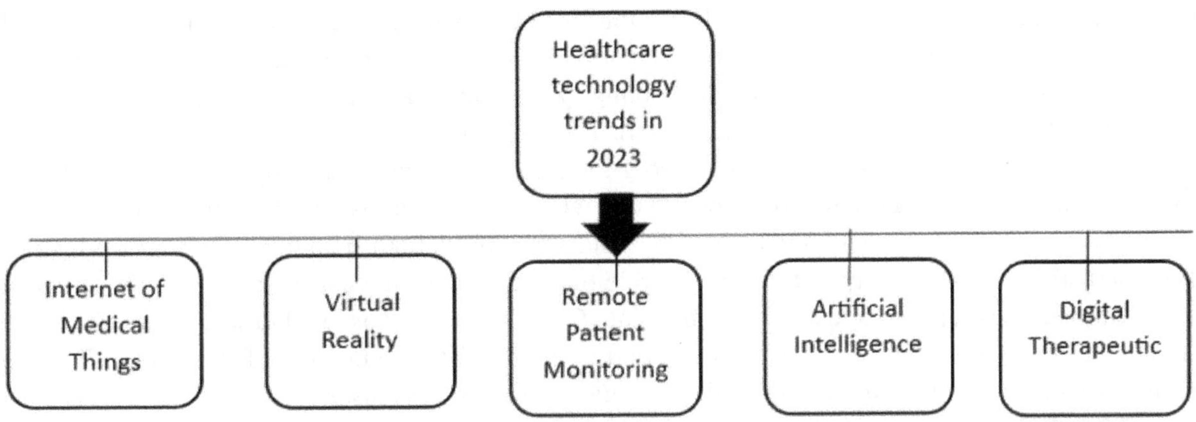

The medical sector has developed in a unique approach to preserve the same high level of assistance owing to major improvements in technology, methods essential to cater to the rising need for access to healthcare, and the accelerating the digitization of private medical data (Robert et al., 2021). Contemporary hospitals depend on its remaining infrastructure and technology, but it's important to consider how it might be updated or completely replaced by newer options. Setting priorities for efficacy, assurances, profitability, and efficiency while preserving dependability and accessibility is crucial (Patel et al., 2023). This study aims to investigate the current trends in technical advancements, such as CFD, that are propelling the the healthcare industry towards digital change-up in the near future.

Introduction to Computational Fluid Dynamic

The theories of fluid dynamics, fluid kinematics, and fluid statics all describe the flow of fluid. Fluid dynamics is the study of the influence of forces on fluid flow. Computational fluid dynamics (CFD) is the use of a computer to simulate the impact of forces on liquid flow in order to approximate the real situation (Bhattacharyya et al., 2021). In fluid dynamics, CFD is regarded as the third dimension. Usually, it's shortened to CFD. The application of CFD was demonstrated during World War II at Los Alamos National Laboratory during the creation of the atomic bomb. Kopal made the first attempt at CFD in 1947. The integration of Math, computer applications, physics, flow technology, and mechanics are together referred as computational fluid dynamics, or CFD. Computational Fluid Dynamics (CFD) is a scientific field that encompasses physicists, mathematicians, and engineers. It creates a virtual environment for any fluid, regardless of its size. It might involve molecules shifting through microchannels, the flames advancing through a chamber of combustion, aircraft motion, the motion of Earth's water bodies or restless atmosphere, biomedicine, inhalation therapy, and more (Bhattacharyya et al., 2021). CFD initially limited to technologically advanced engineering applications, yet it has since become widely utilized for resolving difficult design issues in modern devices (Ferziger et al., 2002). The objective is to figure out the Navier-Stokes equations, which are frequently referred to as the Reynolds-Averaged Navier-Stokes equations. These will fulfil the requirements for energy, momentum, and mass conservation permitting the motion of fluidic systems to be anticipated. By eliminating a few of the parameters taken from the Navier-Stokes equations, one can obtain the Euler equations, which describe viscosity. Removing the parameters that characterise vorticity leads to all of the potential equations, and can be further simplified. Whenever these equations have been simplified, they is linearizable to produce the transformed possible formulas. CFD is a helpful tool for examining domain area for system physical design and operation variables, and also for debugging or troubleshooting system performance when used as computer-aided engineering (CAE) software (Raman et al., 2018). The CFD Principles are displayed in Table 1.

A number of researchers have contributed contributions to the subject of computational fluid dynamics (CFD) in earlier. Sharda et al., 2021 noticed that whenever two fluids with varying the temperatures mix together, there's is a variation in temperature that takes place both spatially and chronologically. If this fluctuation is high, it may result in high cycle thermal fatigue and cause damage to the structure; the phenomenon is referred to as thermal stripping. CFD is also used as an alternative since it reduces the number of experiments, costs, and time required to complete the process of designing. The three primary elements of CFD are pre-processing, solution, and post processors.

CFD analysis in medical studies has become easier thanks to readily accessible high-performance hardware and software. Numerous biomedical researchers are currently demonstrating how effectively CFD can be used to study intricate physiological flow dynamics. The application of CFD demonstrating in cardiovascular and neurovascular medicine has become more prevalent. It may enhance diagnostic assessment, device design through the inclusion of new features, and clinical trials. In order to comprehend and track disease and health, CFD-based technologies are further developed to create intricate computer representations of human beings, or "in silico models" (Sikkandar et al., 2019). For the purpose of to design graft to facilitate the repair of aneurysms, for illustration, the geometric characteristics of the main arteries, such the aorta in the abdomen, are investigated and obtained with the support of CFD methods of modelling (Khalafvand et al., 2017; Hung et al., 2015; Rispoli et al., 2015; Wendell et al., 2016). CFD have been found to have a major influence on the application of biomedical engineering research. Fig. 2 Shows Model construction for application of CFD in biomedical Application.

Figure 2. Construction model for application of CFD in biomedical application

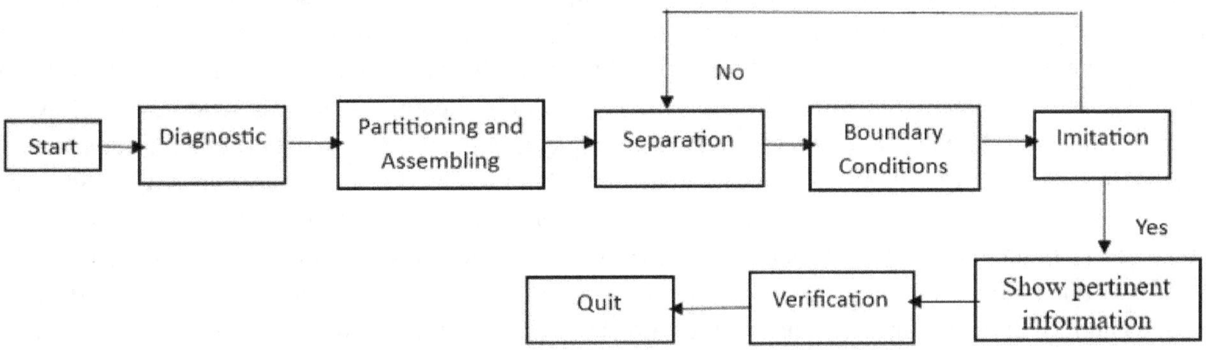

The second section of this paper examines the numerous biomedical applications of computational fluid dynamics (CFD). These include numerical analyses of heart conditions, ophthalmology, cerebral flow of cerebrospinal fluid in the brain, lung airflow and aerosol deposition in the lungs, aneurysm and stenosis in the abdominal and renal arteries, and artificial organ design analysis. These examples show how widely and successfully CFD has been used in biomedical engineering core areas. In Section 3, the limitations and future research scope are finally presented.

Table 1. Principles of CFD

Principle	Definition	Software's
Pre-processing	Preprocessing is the process of discovering a field of interest, pinpointing its unique properties, and constructing under a correct assumption through the use of various techniques for modelling. This leads to in the detection of an understood issue area, while interpretation aids facilitate achieving the most beneficial possible solution.	Hypermesh Femap, ANSYS Meshing, CFD-GEOM, ANSYS ICEM CFD, TGrid. Gridgen.
Solution	Following the recognition of the laws of physics, the next phase is to develop boundary conditions based on the features of the fluid medium, the motion of the fluid physics concept (compressible vs. incompressible, viscous vs. inviscid, laminar vs. turbulent, steady vs. transient, and, most importantly, whether blood flow ought to be viewed as Newtonian or Non-Newtonian), as well as resolve this issue using a computer.	ANSYS CFX, OpenFOAM ANSYS FLUENT, Star CCM, CFD++.
Postprocessing	The process of reviewing the data in order to generate appropriate illustrations and visualized reports is commonly referred to as post-processing. Plenty information regarding the pressure, flow and likely wall dynamics of the vessel during its geometric limitations must be provided throughout this process.	FieldView, ParaView, ANSYS CFD-Post, EnSight, Tecplot 360.

APPLICATION OF CFD IN BIOMEDICAL FIELD

In the ever-evolving environment of healthcare and biomedical engineering, technological breakthroughs are crucial in generating improvements that improve patient outcomes and enhance medical operations. Among these technological wonders, Computational Fluid Dynamics (CFD) stands out as a strong and adaptable instrument, affording unparalleled insights into the complexity of fluid flow inside biological systems. This introduction analyses the expanding subject of CFD in healthcare and biological applications, shining light on its transformational potential and the enormous influence it may have on medical research, diagnostics, and treatment methods. The human body is a complex network of dynamic fluid systems, from the regular pulsations of blood vessels to the complicated paths of air inside the respiratory system. Understanding the fluid dynamics inherent in physiological processes is crucial for understanding the mysteries of many illnesses, improving medical therapies, and inventing cutting-edge healthcare technology. Enter computer Fluid Dynamics - a computer technique that enables researchers and practitioners to simulate, predict, and evaluate fluid flow in biological systems with astonishing accuracy.

One of the major areas where CFD has enormous potential is cardiovascular research. The cardiovascular system, with its complicated network of arteries, veins, and the heart, is a dynamic environment where fluid dynamics plays a vital role in maintaining optimum physiological performance. CFD simulations allow researchers to dive into the hemodynamic of blood flow, studying aspects such as shear stress, pressure gradients, and turbulence. This in-depth knowledge adds considerably to the detection and comprehension of cardiovascular illnesses, opening the path for tailored treatment methods and revolutionary medical device designs.

Respiratory health is another arena where CFD has emerged as a valued ally. By simulating airflow inside the human respiratory system, CFD assists in uncovering the intricacies of respiratory illnesses such as asthma and COPD. The modelling of airflow patterns, pressure changes, and particle deposition in the lungs gives vital insights for diagnosing and treating these disorders. Furthermore, CFD helps to the refining of inhaler designs, providing optimum drug distribution to the targeted parts of the respiratory tract, thereby boosting the effectiveness of therapeutic treatments.

In the domain of medication delivery, CFD is a game-changer. The optimization of drug transport via biological barriers is a complicated undertaking, and CFD models provide a sophisticated way to forecast and increase drug delivery efficiency. From the design of drug delivery systems to the study of personalized medicine methods, CFD plays a crucial role in changing pharmaceutical research and development. This offers the potential to usher in a new age of customized therapeutics, eliminating side effects and increasing the therapeutic effectiveness of drugs.

Surgical procedures, too, are benefitting from the incorporation of CFD into medical practices. Prior to operations, CFD models aid in planning procedures by offering insights into fluid dynamics inside organs or tissues. Surgeons can foresee future obstacles, improve their procedures, and increase the overall accuracy of surgical treatments. Virtual surgical simulations, driven by CFD, serve as vital training tools, enabling medical practitioners to perfect their abilities in a virtual environment before entering the operating room.

The integration of CFD and medical imaging technologies is altering the landscape of diagnosis. Simulating blood flow in vessels, CFD helps the interpretation of medical imaging data, leading to more accurate diagnosis and informed treatment choices. This interaction between CFD and medical imaging offers enormous promise for increasing diagnostic skills and developing a better knowledge of diverse medical diseases.

As we stand on the verge of the future, the integration of CFD with emerging technologies like artificial intelligence (AI) offers new vistas in healthcare and biological research. Machine learning algorithms, trained on CFD simulation data, have the capacity to uncover complicated patterns, anticipate outcomes, and enhance decision-making processes. This merger of CFD and AI promises to usher in an age of tailored, data-driven healthcare solutions, transforming the way we approach medical research, diagnosis, and treatment.

In conclusion, the use of CFD in healthcare and biomedical engineering marks a paradigm changes in our approach to understanding and solving complicated biological processes. From deciphering the secrets of cardiovascular dynamics to improving medication delivery systems and boosting surgical accuracy, CFD is a transformational force that holds the key to unlocking new possibilities in the world of healthcare. As multidisciplinary partnerships develop and technology continues to improve, the marriage of CFD with biological sciences offers a future where precision medicine and individualized healthcare are not just ambitions but realistic realities.

Use of CFD in Blood Flow

Computational fluid dynamics, which utilizes mathematical methods linked to continuity, momentum, and energy equations, may be used to forecast fluid flow fields (Augier et al., 2003). Using the CFD technique to analyze physiological flows, especially blood flow, is a subject that many academics find intriguing. For tiny vessels, the CFD method's premise of unconventional blood flow behavior has been validated, but massive arteries with high coefficients of shear may withstand the Hydrodynamic flow approximation (Bessonov et al., 2016). For medium-sized vessels, the Newtonian assumption's validity is still up for question, notably in the stenotic situation. There isn't currently a consistent depiction of the viscosity of blood. unanimity on the proper model (Shibeshi et al., 2005). As a consequence, blood non-Newtonian behavior is considerably studied, and a range of unconventional simulations are employed to explore how these models impact blood flow characteristics such as flow field, wall shear stresses, and secondary flow patterns (Guerciotti et al., 2016). The circulation of blood via the arteries, which exhibit a variety of characteristics including velocity, static viscosity, enthalpy, and pressure are shown using CFD in Fig 3.

CFD Study of Blood Flow in Arterial Stenosis

The movement of fluid condition in constriction simulations has previously been explored for different Reynolds numbers and cross-sectional area constrictions. The constant state precondition appropriately represents the behaviour of stream and pressure around blood artery limitations within arterial pipes for numerical solutions (Back et al.,1984). Numerous researchers have undertaken computational and experimental investigations of flow of blood within arteries in relation to lesion formation and arterial constriction (Ballyk et al.,1994). The artery's carotid confluence was the focus of the bulk of them, notably the numerical modelling (Berthier et al.,2002). This concentration was adjusted to the right coronary vessel for the purpose of modelling (Feldman et al., 2002). In Manimaran's work, non-Newtonian blood flow assumption was selected when performing CFD simulations in artery stenoses with 48% areal occlusion. Three cases were taken into account for different Reynolds numbers in his analysis.

Figure 3. The circulation of blood via the arteries wrt velocity, pressure, eddy viscosity, static enthalpy

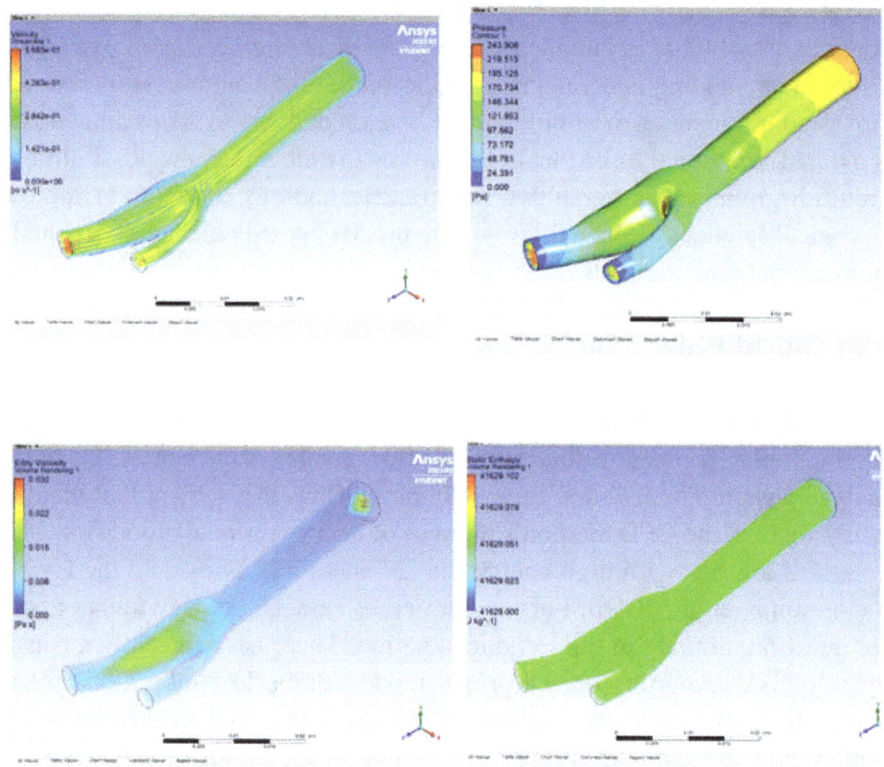

Ophthalmology

The application of CFD in ocular research has risen in the past few years. In order to minimize mistakes produced by researchers during the creation retinal mathematical models are always built with the objective of generating the most realistic depiction of the actual system (Fitt and Gonzalez, 2006). A basic summary of the use of CFD in ophthalmological disorders may be found in an article on the American Academy of Ophthalmology website (Tamm et al.,2007). Glaucoma is the name for the process that affects the optic nerve by rising intraocular pressure and restricting the outflow. Many nuanced problems surround the delicate balance between the influx and outflow of aqueous humor in a normal human eye that are still not completely understood (Kwon et al.,2009). Sultan et al.,2009 stated that despite while intraocular pressure (IOP) is not included in the definition of glaucoma, it is nonetheless a possible risk factor (Sultan et al,2009). IOP is connected with increased resistance to aqueous humor outflow and is not different from normal persons and these findings have led to a better grasp of aqueous humor properties and support for CFD. Many scholars have tried to explain the elements of aqueous fluid outflow using distinct properties of the human eye in the anterior chamber (Schachar et al., 2006). A Newtonian fluid was utilized to represent the aqueous humor, and a linear elastic solid was used to calculate the iris contour in the eyes. Ooi and Ng, 2011 built a 2-D model of the human eye to explore the flow effects within the anterior chamber and to understand the existence of natural aqueous humor flow (Ooi and Ng, 2011). Ooi et al., 2011did research employing the aforementioned model to learn about the influence of natural convection in the anterior chamber on ocular heat transfer. CFD is used to examine the architecture of crystalline lenses and ciliary bodies. According to Sukuma et al., 1996 a lens is a transparent biconvex shape in the eyes that, in combination with the cornea, serves to refract light that has to be focused on the retina. A lens is also used to adjust the focal distance of the eye so that it can accurately focus on things at varying focal length distances. Lens epithelial samples were gathered for investigation of catalase activities, and analysts tried to employ CFD to induce heat, which caused damage to the lens.

Fluid and Air Flow in Lungs

The Navier-Stokes formula for compressible flow well predicts velocity of air in the respiratory tract at a slower rate (Nowak et al.,2003). CFD is a powerful tool for estimating gas and particle flow and accumulation within the pulmonary system. Because of the intricate structure of the airways in the pulmonary system, it is confined to very tiny places. Numerous studies were done to imitate the lungs using the CFD approach. The approach involves a precise CAD model built from CT and magnetic resonance imaging data to construct a network framework to represent the contours (Tian and Longest, 2010). This assists physicians in creating medical equipment and treatment techniques. A commercial CFD program was utilized in research on mathematical simulation to show circulation properties in a vacant Rochester-type breathing apparatus pursuant to a constant state and abrupt amounts of pollutants (Longest et al.,2012). A unique CFD approach was developed to study the sporadic effects, polydisperse aerosol variations in size, and multiple pulmonary region accumulation in the mouth-throat (MT) and full tracheobronchial (TB) passageways (Longest et al.,2012). To anticipate the travel and accumulation of full lung dust particles, the scientists constructed CFD modeling of the sequential individual route.

Using distinctive patterns acquired from Computed Tomography (CT) scan images, this CFD approach simulates the most prominent passageways via the longitudinal bronchi. These models are also promptly prototyped in order to obtain relevant in-vitro implantation records. The SIP technique also predicts transit and accumulation in the remaining tuberculosis (TB) the airway passageways employing combinations to create and compute precise paths (Tian et al.,2011; Malve et al., 2013). Numerous recent investigations from this group present detailed under controlled conditions accumulation findings from genuine respirators using MT, breathing airway, and upper TB region characteristic models (Walenga et al, 2013). In accordance with computational system adjustments made via user defined functions (UDFs), extremely strong concordance was obtained between under controlled conditions accumulation results and CFD estimates for medicinal nanoparticles. When compared to modeling all tracheobronchial airways, the SIP methodology was proved to be an effective technique for modelling pulmonary accumulation of medicinal nanoparticles with a cognitive numerous scale of amount. An examination of tailored to patient's pulmonary avenues under physiological limit conditions was conducted using a computational fluid dynamics model of a sound, stenotic, and post-operatively stented human trachea to estimate the exit stress signals, allowing calculation of supplementary obstruction of diminished pulmonary development and illustrating the respiratory system as fragmented coalitions (Paz et al.,2017). A CFD intermittent modeling of the cough-clearing process was recently studied using the Eulerian wall film model (Clarke et al.,2013). In this work, a system for forecasting splutter waste was developed, which effectively facilitated the modeling and measurement of cough efficiency. Investigators demonstrated that mathematical modeling may be used to simulate and anticipate fluid and air circulation in respiratory system wholly physiologically regarded to be extremely complicated (Zachow et al.,2007).

Cerebrospinal Fluid Flow

Within intracranial space, the cerebral cortex, hemoglobin, and cerebrospinal fluid (CSF) coexist and maintain an equilibrium proportion. The clearance of waste substances from the brain, transport of nutrients, and protection against harm are all critical roles of CSF. CSF kinematics in the posterior spinal subarachnoid cavity (SSS) may be useful in the identification and assessment of diseases affecting the skull and spine, such as Chiari malformation. (Sforza et al.,2012). The pulsatile flow of CSF within the craniospinal cavities, generated by the periodic relaxation and dilation of brain arteries, is a difficult phenomenon in fluid dynamics. The extensive multifactorial processes underlying the genesis, progression, and rupture of cerebral aneurysms remain unclear from a clinical viewpoint. Research on brain aneurysms has been the focus of various scientists recently. The mechanics of mathematical simulations customized for each patient assessment of brain aortic breach hazard and individual treatment were emphasized, and advancements were made employing patient-specific CFD models for brain aneurysms. (Sforza et al.,2012). Additionally, a key role of hemodynamics in brain aneurysm is clinically demonstrated. Researchers and medical practitioners, however, cannot agree on the relevance of any hemodynamic component in these processes. Vascular geometry has a major impact on hemodynamics, and additional study is essential to completely appreciate how hemodynamic stress interacts with mechano-biological barrier adaptations. These govern the biological evolution of cortical stenosis and a complete knowledge of the aneurysmal hemodynamic environment in vivo is essential (Sweetman and Linninger,2011). Additional anatomical data and 2D phase contrast (PC) MRI images serve as the basis for both CFD simulations and in-vitro flow models. In both healthy and ill cases, these computations and models provide an unobtrusive evaluation of the circulatory microenvironment. Nevertheless, measuring 3D complexity using 2D phase contrast MRI CSF flow data with unidirectional encoding within the CSF stream is problematic. CFD research of the circulatory system in the neck area has been carried out using subject-specific 3D models that are geometrically simplified and lack comprehensive anatomical components. These models also contain idealized denticulate ligaments and spinal cord nerve rootlets (Sweetman and Linninger,2011; Sforza et al.,2012). Time-resolved three-dimensional velocity-encoded phase-contrast Magnetic resonance imaging (4D PC MRI) was applied by Yiallourou et al.,2012 in 3 volunteers who were healthy and 4 CM sufferers. The 4D PC outcomes of MRI were compared to specific to the topic of a three-dimensional CFD simulations (Yiallourou et al.,2012). They did not take into consideration microscopic anatomical characteristics as denticulate tendons, nerve endings, and the cerebellum vessels and instead concentrated on rigid-walled geometry. Then, utilizing T2-weighted anatomical MRI pictures and 4D PC MRI flow data at the cervical-medullary confluence of robust individuals, they generated CFD simulations by assuming that the media was Newtonian and impermeable.

Artificial Organ Design

Synthetic organs are constructed gadgets that are integrated or implanted into the body of an individual to substitute for an authentic organ that is no longer functioning, allowing the patient to continue their usual life. In order to make devices clinically practical, multiple numerical simulations on hemodynamic with perfect shape are necessary throughout the prototype development stage of artificial organs. The location and amplitude of stationary regions, and therefore the regional rate of shear, may be assessed via the use of simulated flow. The degree of thrombus formation and hemolysis, which are crucial markers of a blood pump's effectiveness, may be associated with these properties (Sukuma et al.,1996).

Mechanical pumps termed ventricular assist devices (VADs) complement or substitute the operations of any number of failing heart chambers. Despite major efforts, suitable blood damage models are missing in computational evaluation of VADs, which limits the full CFD's potential. Implantable VADs are thought to be a potentially valuable tool in the therapeutic treatment for people with major cardiac issues. Chua et al.,2006 presented a three-dimensional (3D) model of the Kyoto-NTN strongly hanging rotating fluid pump and offered a computer-assisted fluid (CFD) approach for the internal circulation area of the pump, which included the blood's shear stress distributions, static pressure distributions, and velocity profiles (Fraser et al.,2011). In addition to giving a practical critical analysis of the study, Fraser et al.,2011 analyzed the use of CFD in the design of VADs and identified the most current CFD studies of blood pumps.

Additionally, the paper described blood damage models, explored implementation issues in CFD, and defined knowledge gaps that will be filled by future research (Fraser et al., 2011). Review research by Farag et al., 2014 utilized CFD models to provide data on patients with end-stage heart failure who were obtaining mechanical support devices (Farag et al.,2014). CFD is used to examine artificial organ function, including forecasting the bodily functions of a mechanical cardiac system. Using CFD study, Capelli et al., 2017 studied the differences in the cardiovascular capabilities while applying different anchorage strategies (Capelli et .al,2017). For bioprosthetic aortic valves, they combined the use of commercially available computational and experimental approaches. The areas of substantial shear in the stream that injure cell membranes and need adequate border circumstances may be determined by numerical simulations. Numerous more biological applications, such as auditory cord assessment, sinus and nasal surges, grease for joints, neural fluid circulation, etc., are constantly growing the utilization of CFD technology. Apart from these therapeutic applications, computer-aided fluid dynamics (CFD) is also applied in the manufacture of surgical equipment.

Cardiovascular Systems

With the potential to forecast bodily reactions to treatment, determine previously unmeasured hemodynamic criteria, and improve diagnostic assessment and device design, the application of CFD is theoretically highly beneficial in the fields of cardiovascular medicine, clinical trials, and device design (Morris et al., 2015).Studies on the application of computational numerical modeling in the cardiovascular system explored the methodological challenges, analytical assessments, and results connected to the three basic heart physiologies—the ventricle, arteries, and valves. Cardiac illness, which is the largest cause of death globally, is referred to as cardiovascular (Jamuna et al., 2011). Aortic valve constriction or blood flow leakage on the valve leaflet are the basic causes of heart valve disease. Basri et al.'s., 2014 latest study evaluated the physiological properties of the influence of variable valve opening. To determine the blood behavior of severed aortic stenosis, the authors studied the hemodynamic characteristics with respect to of tension, speed, and membrane stresses caused by shear. The conclusion illustrates the large reduction in blood pressure at the small channel debuting, leading to the narrowing of the valve and the blocking of blood ejection. As a consequence, the analysis revealed that the smaller leaflet aperture raised leaflet stress and had a detrimental influence on blood flow. In addition, Basri et al.,2014 studied hemodynamic features to compare the totally open normal aortic valve with the stenosed aortic valve (62.5ˎ opening). The scientists imported the three-dimensional cardiovascular valve from an MRI data scan and did CFD modeling on it. In compared to a normal aortic valve, the research indicated that the quantity of blood accessing the cardiac lobes of a stenosed valve in the aorta reduced by 2.9% and climbed by 13.7%. The research consequently revealed a considerable drop in the blood flow required to supply blood to the human head, neck, and arm. Tan et al., 2012 studied the aorta flow pattern before to and throughout the placement of a transcatheter arterial closure while concentrating on a tailored to patient evaluation of stenosed cardiac valves. The authors investigated the circulation dynamics of the pulmonary arteries as a function of trajectory and wall stress caused by shear using CFD simulation that incorporated MRI data. Following implantation, the flow patterns reveal a 20% drop in jet flow at an immediately apparent streamline and a lowered time-averaged surface stress due to shear. Therefore, a psychological evaluation of the disturbed circulatory rhythms and shear stresses placed on the heart prior to and following executing the incision surgery was made feasible by the combination of imaging and modeling approach employed in this study.

Under different situations, Jamuna and Abnurajan, 2011 measured the blood flow velocity and pressure via a patient's unique aorta. Aortas that are normal, have plaque on the side of the valve sinus, and necessitate a bi-leaflet valve implant are those instances. The scientists employed CFD simulation to assess a computed tomography (CT) image of a specific patient. Following valve implantation, the blood pattern is shown to be equivalent to that of a normal aorta, with raised blood pressure and velocity percentages of 81.8% and 58.5%, respectively. Sirois et al.,2011 also employed CT scans and CFD simulations to examine aortic valve implantation on a patient-specific basis. Prior to and after the implantation procedure, the investigators conducted a statistical evaluation of the physicochemical characteristics of blood circulation dynamics. A striking consequence of the valve implantation was a reduction in decrease in pressure to 25.27 mmHg and an increase in effective orifice area from 0.53 to 1.595 cm^2.Conversely, Sun and Xu, 2014 explored the application of CFD simulations with regard to cardiovascular conditions that are centered on three-dimensional mucosal models. The ultimate objective of the research is to determine how differences in coronary artery geometry impact regional circulation fields and circulation targeting, which in turn alter the dynamics of systemic flow through the body. The average surface shear stress gradient, rhythmic peak measure, and structure peak tension parameters are all factored into account when analyzing the hemodynamic science features. Gao et al.,2014 also looked at the placing of stents as a percutaneous treatment for coronary artery disease. The researchers analyzed the features with regard of blood speed and wall stress due to shear and compared the blood flow before and after stent installation. According to the research, the area of stenosis exhibited increased blood velocity and wall shear stress before the stent was inserted. As a consequence, the optimum flow rate in the coronary artery was lowered, and the wall shear stress rose after the implantation procedure. The physiological impacts of generated debris in left coronary artery simulations of specific to the patient arterial occlusion were examined by Chaichana et al.,2012. Using the method of computational fluid dynamics (CFD) analysis, three parameters—wall stress from shear, pressure curve, and stream velocities—are analyzed and compared between the existence and the lack of amyloid in left artery models throughout the cardiovascular cycle. Wall stress due to shear is comparable in the stenotic regions despite the highest-pressure gradient imposed on by the lesions and lower rate of flow areas reported at post-plaque locations. Table 2 Summarizes the key Important CFD application in Biomedical.

Table 2. summary of CFD application in biomedical: Compiled by author

S/No.	Biomedical Applications	Summary
1.	Blood flow in arterial stenosis	This study's primary emphasis was the bifurcation of the carotid artery. This concentration was adjusted to the right coronary artery in terms of simulation. In the course of this study, three distinct instances were considered based on distinct Reynolds numbers.
2.	Ophthalmology	The exact equilibrium between the inflow and outflow of aqueous humour in a normal human eye continues to be insufficiently understood and was the focus of many intricate worries. The aqueous humour was described by a Newtonian fluid, and the iris shape in the retina was determined using a linear elastic solid. 2-D was developed to comprehend the existence of an inherent movement of aqueous humour.
3.	Fluid and Air Flow in Lungs	The respiratory tract system's air and particle movement and deposition can be estimated with outstanding precision via CFD. In order to generate a networked model for the geometry, the method requires a precise CAD model made from MRI and CT images. In an investigation on computational modelling, flow variables in an empty Rochester-style inhalation chamber beneath a steady-state and transient levels of pollutants have been demonstrated employing a commercial CFD program. To foresee the transit and deposition of complete lung aerosols, the scientists built CFD modelling of the probabilistic individual path (SIP).
4.	Cerebrospinal Fluid Flow	A variety of specialists are currently concentrating their studies upon brain tumours. In addition, a clinically verified critical function of hemodynamic in skull aneurysm Yet, gauging 3D complexity within the CSF flow field using 2D contrast phase MRI CSF flow information with unidirectional encoding is a challenging endeavour. In addition, the denticulate ligaments and cervical nerve rootlets in these models are exaggerated.
5.	Artificial Organ Design	Artificial organs are synthetic implants that have been injected or integrated within the body of an individual to operate in the spot of a dysfunctional real organ, letting the patient to get back to leading a life as usual. A summary of research illustrates how data on patients undergoing support from mechanical devices for end-stage coronary artery disease has been presented using CFD models. The variation in hemodynamic performances when employing various anchoring approaches were examined using CFD investigation.
6.	Cardiovascular systems	Investigation on the utilization of computational fluid dynamics (CFD) to the cardiovascular system evaluated methodological worries, analytical inspections, as well as outcomes pertaining to the three primary physiologies of the heart: the pulmonary artery, artery walls, and connectors. Following integrating a three-dimensional heart valve using an MRI image, the researchers used CFD models to imitate it. The quantity of blood reaching the aorta limbs of a stenosed valve in the aorta was shown to drop by 2.9% and rise by 13.7% in contrasted to a healthy valve in the aorta. As a consequence, the study showed a substantial decrease in the blood flow sufficient for delivering blood to the human neck, torso, and forearm.

FUTURE OF CFD IN BIOMEDICAL ENGINEERING AND HEALTHCARE

Computational Fluid Dynamics (CFD) has grown as a formidable tool in various engineering fields, and its future potential in biomedical engineering and healthcare is particularly intriguing. As technology develops and our understanding of physiological processes grows, CFD plays a crucial role in simulating and analyzing fluid flow in biological systems, giving major insights for medical research, diagnosis, and treatment.

One crucial area where CFD provides significant potential is in understanding blood flow and cardiovascular dynamics. The circulatory system is a sophisticated network of arteries, veins, and capillaries that deliver blood throughout the body. CFD models enable researchers to examine the hemodynamic of blood flow, including components such as shear stress, turbulence, and pressure gradients. This information is vital for predicting and diagnosing cardiovascular problems, boosting the design of medical devices like stents and heart valves, and assisting in surgical planning.

In the subject of respiratory health, CFD may be applied to model airflow within the respiratory system of humans. This is particularly crucial for understanding respiratory disorders such as asthma and chronic obstructive pulmonary disease (COPD). By predicting airflow patterns and pressure changes in the lungs, CFD helps researchers find likely areas of airway obstruction and better the design of inhalers and respiratory devices. Moreover, tailored medicine may benefit from CFD simulations by customizing medicines to individual patients based on their unique respiratory properties.

In the field of medication delivery, CFD plays a significant role in enhancing the effectiveness of drug delivery systems. By imitating the transportation of medicines over varied biological barriers, researchers may maximize the efficacy of drug transport to specific target tissues or cells. This has the potential to change medical research, making treatments more effective while reducing undesirable effects. Additionally, CFD may aid in the design of novel medication delivery systems, delivering precise and controlled release of pharmaceuticals.

Surgical procedures also stand to benefit from CFD applications. Prior to surgeries, CFD models may assist in planning procedures by delivering insights into fluid dynamics within organs or tissues. This permits doctors to identify possible complications and modify their technique, ultimately enhancing surgical outcomes. Virtual surgical simulations applying CFD may also work as training aids for medical staff, allowing them to practice and refine their expertise in a risk-free context.

CFD's relevance extends to the realm of medical imaging, where it may increase the accuracy of diagnostic equipment. For instance, CFD may be used to imitate blood flow in arteries, aiding in the interpretation of medical imaging data such as magnetic resonance angiography (MRA) or computed tomography angiography (CTA). This may lead to more precise diagnosis and better-informed treatment choices.

As technology continues to advance, the coupling of CFD with other emerging technologies like artificial intelligence (AI) offers even greater potential. Machine learning algorithms may be trained on CFD simulation data to detect patterns, forecast outcomes, and help in decision-making processes. This fusion of CFD and AI could lead to more sophisticated prediction models, allowing for individualized and data-driven approaches to healthcare.

In conclusion, the future scope of CFD in biomedical engineering and healthcare is large and carries the prospect of revolutionary advances. From extending our understanding of physiological processes to enhancing medical equipment and boosting diagnostic and therapeutic procedures, CFD is positioned as serving an indispensable part in defining the future of healthcare. As interdisciplinary relationships continue to grow, the synergy between CFD, medical research, and technological innovation is predicted to produce breakthrough solutions that will benefit patients and healthcare practitioners alike.

MERITS AND LIMITATIONS OF BIOMEDICAL APPLICATIONS IN CFD

In the last few decades, computational fluid dynamics (CFD) has drawn plenty of interest from mathematic aficionados as an excellent instrument to analyse nuanced physiological flows and revealing their potential, especially in the cardiovascular and respiratory systems. Medical scientists have so far utilized CFD to assist with predict the traits of the body's cardiovascular flow of blood and the flow of air while nasal breathing. As a consequence, it has benefits such diminishing the likelihood of complications during surgery, participating in the growth of better surgery, maintaining high efficiency with fewer adverse effects medical equipment, and delivering a solid understanding of biological phenomena.

From the viewpoint of theory, CFD offers strengths by concentrating upon the definition and resolve of governing equations and additionally analysing of various assumptions to these equations. In between, the numerical and experimental approaches revealed the worth of CFD as a distinct and less costly way to simulate real-life fluid flow, particularly whilst it pertains to the systems in the human body. As a consequence, it supplies comprehensive visual information that's detailed while assessing the dynamics of fluids statistical and experimental techniques. Despite its benefits, CFD, or computational fluid dynamics, has certain drawbacks as well. As an instance, it is unable to accurately forecast physical representations or the accuracy of input data about real-world phenomena such as flexibility, multiphase flow, and turbulence. It is therefore vital to meticulously analyse and review numerical results in order to make advised critical judgements with regard to the computed results. Additionally, while solving mathematical problems on a computer, computation mistakes involving truncation & round-off errors might occur. The justification for this is since the feasibility of CFD hinges on many factors, namely specific materials and steps, exact algorithms for the equations that govern, sturdy CFD software, thereby rapid computation speeds, and sizable servers (Candreva et al., 2022).

CONCLUSION

In the persistent quest of improvements in healthcare and biomedical engineering, the integration of Computational Fluid Dynamics (CFD) emerges as a vital force, transforming the landscape of medical research, diagnostics, and therapeutic interventions. As we span the worlds of cardiovascular dynamics, respiratory health, medication delivery, surgical planning, and medical imaging, the significant influence of CFD becomes obvious, offering a future where precision and individualized healthcare are not only ambitions but concrete realities.

The voyage through the intricacy of cardiovascular research exposes CFD as a guiding light, lighting the route to a greater knowledge of the dynamic fluid systems that support life. The modelling of blood flow, encompassing elements such as shear stress, pressure gradients, and turbulence, helps researchers to grasp the complexity of cardiovascular disorders. This acquired information acts as a compass for creating tailored treatment methods, boosting the effectiveness of medical therapies, and refining the design of cardiovascular devices. As we stand on the threshold of an age when cardiac therapy is personalized to individual variations, CFD is at the vanguard, leading the march toward precision medicine in the area of heart health.

Respiratory health, too, undergoes a radical transformation under the impact of CFD. From the modelling of airflow inside the complicated tubes of the respiratory system to the explanation of airflow patterns in the lungs, CFD offers a lens through which we may appreciate the intricacies of respiratory illnesses. Asthma and COPD, formerly cloaked in mystery, are now exposed to the scrutiny of CFD models, uncovering the subtleties of airway obstruction and influencing the creation of more effective therapies. The improvement of inhaler designs, driven by CFD insights, heralds a new age when respiratory medicines are not only effective but precisely targeted, reducing side effects and increasing patient outcomes.

In the field of medication delivery, CFD emerges as a catalyst for innovation. The delicate dance of drugs via biological barriers, formerly a tough issue, is now orchestrated with precision by CFD models. From the conception of drug delivery systems to the investigation of customized treatment, CFD reshapes the landscape of pharmaceutical research. The promise of personalized therapies, led by unique patient characteristics, becomes a concrete possibility, signalling a paradigm change in the way we approach and perceive medical treatments.

Surgical procedures, typically a delicate dance between expertise and uncertainty, find a trusted companion in CFD. The simulations that precede procedures give surgeons with a virtual blueprint, delivering insights into the fluid dynamics inside organs or tissues. This virtual rehearsal, driven by CFD, boosts surgical accuracy, allowing for better approaches and improved results. In the hands of medical practitioners, CFD becomes a tool for mastery, a bridge between theoretical understanding and practical application, eventually helping patients via safer and more successful procedures.

The coupling of CFD with medical imaging technology redefines the boundaries of diagnosis. Simulating blood flow in arteries, CFD augments the interpretation of medical imaging data, increasing diagnostic accuracy to new levels. This synergy between CFD and medical imaging reflects a harmonic combination of computational capability and clinical intuition, suggesting a future where illnesses are diagnosed sooner, diagnoses are more accurate, and treatment choices are better informed.

As we peek into the future, the combination of CFD with artificial intelligence (AI) accentuates the revolutionary potential of both technologies. Machine learning algorithms, trained on enormous datasets created by CFD simulations, grow into strong tools for forecasting outcomes, finding patterns, and improving decision-making processes. The union of CFD and AI takes us into a future when healthcare becomes not just precise but also predictive, predicting individual reactions to treatments and ushering in a new frontier of customized medicine.

In conclusion, the use of CFD in healthcare and biomedical engineering signals a fundamental change in our approach to understanding, diagnosing, and treating complicated biological events. From the microscopic dance of blood cells in the circulatory system to the macroscopic interaction of airflow in the lungs, CFD acts as a bridge between theoretical understanding and practical application. As multidisciplinary partnerships bloom and technology continues its unrelenting march ahead, the legacy of CFD in healthcare is not simply one of technical capability but a monument to our capacity to harness the power of computational insights for the good of human health. The future beckons, offering a healthcare environment where the precision of CFD converges with the customization of medicine, providing a path toward optimum health and well-being for people throughout the world.

REFERENCES

Armentano, R. L., & Kun, L. (2019). Leonardo da Vinci—the first bioengineer: Educational innovation to meet his desire for knowledge and promote his concept of interdisciplinarity. *Creative Education*, *10*(6), 1180–1191. doi:10.4236/ce.2019.106089

Augier, F., Masbernat, O., & Guiraud, P. (2003). Slip velocity and drag law in a liquid-liquid homogeneous dispersed flow. *AIChE Journal. American Institute of Chemical Engineers*, *49*(9), 2300–2316. doi:10.1002/aic.690490907

Back, L. H., Cho, Y. I., Crawford, D. W., & Cuffel, R. F. (1984). *Effect of mild atherosclerosis on flow resistance in a coronary artery casting of man.*

Ballyk, P. D., Steinman, D. A., & Ethier, C. R. (1994). Simulation of non-Newtonian blood flow in an end-to-side anastomosis. *Biorheology*, *31*(5), 565–586. doi:10.3233/BIR-1994-31505 PMID:7833458

Basri, A. A., Zubair, M., Aziz, A. F., Ali, R. M., Tamagawa, M., & Ahmad, K. A. (2014, December). Computational fluid dynamics study of the aortic valve opening on hemodynamics characteristics. In *2014 IEEE Conference on Biomedical Engineering and Sciences (IECBES)* (pp. 99-102). IEEE. 10.1109/IECBES.2014.7047660

Berthier, B., Bouzerar, R., & Legallais, C.CBerthier. (2002). Blood flow patterns in an anatomically realistic coronary vessel: Influence of three different reconstruction methods. *Journal of Biomechanics*, *35*(10), 1347–1356. doi:10.1016/S0021-9290(02)00179-3 PMID:12231280

Bessonov, N., Sequeira, A., Simakov, S., Vassilevskii, Y., & Volpert, V. (2016). Methods of blood flow modelling. *Mathematical Modelling of Natural Phenomena*, *11*(1), 1–25. doi:10.1051/mmnp/201611101

Bhattacharyya, S. (2021). Computational Fluid Dynamics-The Futuristic Innovation in Pharmaceutical Industry. *Indian Journal of Pharmaceutical Education & Research*, *55*(4).

Candreva, A., De Nisco, G., Rizzini, M. L., D'Ascenzo, F., De Ferrari, G. M., Gallo, D., & Chiastra, C. (2022). Current and future applications of computational fluid dynamics in coronary artery disease. *Reviews in Cardiovascular Medicine*, *23*(11), 377. doi:10.31083/j.rcm2311377

Capelli, C., Corsini, C., Biscarini, D., Ruffini, F., Migliavacca, F., Kocher, A., Laufer, G., Taylor, A. M., Schievano, S., Andreas, M., Burriesci, G., & Rath, C. (2017). Pledget-armed sutures affect the haemodynamic performance of biologic aortic valve substitutes: A preliminary experimental and computational study. *Cardiovascular Engineering and Technology*, *8*(1), 17–29. doi:10.1007/s13239-016-0284-8 PMID:27873180

Chaichana, T., Sun, Z., & Jewkes, J. (2012). Computational fluid dynamics analysis of the effect of plaques in the left coronary artery. *Computational and Mathematical Methods in Medicine*, *2012*, 2012. doi:10.1155/2012/504367 PMID:22400051

Chua, L. P., Song, G., Lim, T. M., & Zhou, T. (2006). Numerical analysis of the inner flow field of a biocentrifugal blood pump. *Artificial Organs*, *30*(6), 467–477. doi:10.1111/j.1525-1594.2006.00243.x PMID:16734599

Clarke, E. C., Fletcher, D. F., Stoodley, M. A., & Bilston, L. E. (2013). Computational fluid dynamics modelling of cerebrospinal fluid pressure in Chiari malformation and syringomyelia. *Journal of Biomechanics*, *46*(11), 1801–1809. doi:10.1016/j.jbiomech.2013.05.013 PMID:23769174

Dwivedi, R., Mehrotra, D., & Chandra, S. (2022). Potential of Internet of Medical Things (IoMT) applications in building a smart healthcare system: A systematic review. *Journal of Oral Biology and Craniofacial Research*, *12*(2), 302–318. doi:10.1016/j.jobcr.2021.11.010 PMID:34926140

Farag, M. B., Karmonik, C., Rengier, F., Loebe, M., Karck, M., von Tengg-Kobligk, H., Ruhparwar, A., & Partovi, S. (2014). Review of recent results using computational fluid dynamics simulations in patients receiving mechanical assist devices for end-stage heart failure. *Methodist DeBakey Cardiovascular Journal*, *10*(3), 185. doi:10.14797/mdcj-10-3-185 PMID:25574347

Feldman, C. L., Ilegbusi, O. J., Hu, Z., Nesto, R., Waxman, S., & Stone, P. H. (2002). Determination of in vivo velocity and endothelial shear stress patterns with phasic flow in human coronary arteries: A methodology to predict progression of coronary atherosclerosis. *American Heart Journal, 143*(6), 931–939. doi:10.1067/mhj.2002.123118 PMID:12075241

Ferziger, J. H., & Peri, C. M. (2002). *Computational methods for fluid dynamics.*

Fitt, A. D., & Gonzalez, G. (2006). Fluid mechanics of the human eye: Aqueous humour flow in the anterior chamber. *Bulletin of Mathematical Biology, 68*(1), 53–71. doi:10.1007/s11538-005-9015-2 PMID:16794921

Fortune Business Insights. (2023). The global telemedicine market size was valued at $87.41 billion in 2022 & is projected to grow from $94.44 billion in 2023 to $286.22 billion by 2030. *Market research Report.* https://www.fortunebusinessinsights.com/industry-reports/telemedicine-market- 101067

Fraser, K. H., Taskin, M. E., Griffith, B. P., & Wu, Z. J. (2011). The use of computational fluid dynamics in the development of ventricular assist devices. *Medical Engineering & Physics, 33*(3), 263–280. doi:10.1016/j.medengphy.2010.10.014 PMID:21075669

Gao, F., Li, G., Hu, R., & Okada, H. (2014). Computational Fluid Dynamic Analysis of Coronary Artery Stenting. *International Journal of Bioscience, Biochemistry, Bioinformatics, 4*(3), 155–159. doi:10.7763/IJBBB.2014.V4.330

Giddens, D. P., Zarins, C. K., & Glagov, S. (1990). *Response of arteries to near-wall fluid dynamic behavior.*

Gries, A., Seekamp, A., Wrede, C., & Dodt, C. (2018). Training in clinical acute and emergency medicine–Supraspeciality in Germany: A concept for nationwide implementation! *Der Anaesthesist, 67*(12), 895–900. doi:10.1007/s00101-018-0515-5 PMID:30511109

Guerciotti, B., & Vergara, C. (2018). Computational comparison between Newtonian and non-Newtonian blood rheologies in stenotic vessels. *Biomedical Technology: Modeling, Experiments and Simulation,* 169-183.

Hung, T. K., Khalafvand, S. S., & Ng, E. Y. K. (2015). Fluid dynamic characteristics of systolic blood flow of the left ventricle. *Journal of Mechanics in Medicine and Biology, 15*(03), 1550047. doi:10.1142/S0219519415500475

Jamuna, C., & Anburajan, M. (2011, April). Design of patient specific prosthetic aortic valve and to study its computational fluid dynamics. In *2011 3rd International Conference on Electronics Computer Technology* (Vol. 3, pp. 355-360). IEEE. 10.1109/ICECTECH.2011.5941771

Junaid, S. B., Imam, A. A., Balogun, A. O., De Silva, L. C., Surakat, Y. A., Kumar, G., & Mahamad, S. (2022, October). Recent advancements in emerging technologies for healthcare management systems: A survey. [). MDPI.]. *Health Care, 10*(10), 1940. PMID:36292387

Khalafvand, S. S., Ng, E. Y. K., Zhong, L., & Hung, T. K. (2017). Three-dimensional diastolic blood flow in the left ventricle. *Journal of Biomechanics*, *50*, 71–76. doi:10.1016/j.jbiomech.2016.11.032 PMID:27939353

Kron, T., & Krishnan, P. (2019). Leonardo DaVinci's contributions to medical physics and biomedical engineering: Celebrating the life of a 'Polymath'. *Australasian Physical & Engineering Sciences in Medicine*, *42*(2), 403–405. doi:10.1007/s13246-019-00757-2 PMID:31049840

Kwon, Y. H., Fingert, J. H., Kuehn, M. H., & Alward, W. L. (2009). Primary open-angle glaucoma. *The New England Journal of Medicine*, *360*(11), 1113–1124. doi:10.1056/NEJMra0804630 PMID:19279343

Li, J., Ma, Q., Chan, A. H., & Man, S. (2019). Health monitoring through wearable technologies for older adults: Smart wearables acceptance model. *Applied Ergonomics*, *75*, 162–169. doi:10.1016/j.apergo.2018.10.006 PMID:30509522

Longest, P. W., Tian, G., Walenga, R. L., & Hindle, M. (2012). Comparing MDI and DPI aerosol deposition using in vitro experiments and a new stochastic individual path (SIP) model of the conducting airways. *Pharmaceutical Research*, *29*(6), 1670–1688. doi:10.1007/s11095-012-0691-y PMID:22290350

Majumder, S., Mondal, T., & Deen, M. J. (2017). Wearable sensors for remote health monitoring. *Sensors (Basel)*, *17*(1), 130. doi:10.3390/s17010130 PMID:28085085

Malve, M., Chandra, S., Lopez-Villalobos, J. L., Finol, E. A., Ginel, A., & Doblare, M. (2013). CFD analysis of the human airways under impedance-based boundary conditions: Application to healthy, diseased and stented trachea. *Computer Methods in Biomechanics and Biomedical Engineering*, *16*(2), 198–216. doi:10.1080/10255842.2011.615743 PMID:22220946

Markets and Markets. (2021). The global peripheral vascular devices market in terms of revenue was estimated to be worth $10.3 billion in 2021 and is poised to reach $13.4 billion by 2026, growing at a CAGR of 5.3% from 2021 to 2026. *Market research Report*. https://www.marketsandmarkets.com/Market-Reports/interventional-cardiology-peripheral-vascular-devices-market-790.html

Minaam, D. S. A., & Abd-ELfattah, M. (2018). Smart drugs: Improving healthcare using smart pill box for medicine reminder and monitoring system. *Future Computing and Informatics Journal*, *3*(2), 443–456. doi:10.1016/j.fcij.2018.11.008

Mohammadzadeh, N., Gholamzadeh, M., Saeedi, S., & Rezayi, S. (2020). The application of wearable smart sensors for monitoring the vital signs of patients in epidemics: A systematic literature review. *Journal of Ambient Intelligence and Humanized Computing*, 1–15. PMID:33224305

Morris, P. D., Narracott, A., von Tengg-Kobligk, H., Soto, D. A. S., Hsiao, S., Lungu, A., & Gunn, J. P. (2015). Computational fluid dynamics modelling in cardiovascular medicine. *Heart (British Cardiac Society)*. PMID:26512019

Muratyan, A., Cheung, W., Dibbo, S. V., & Vhaduri, S. (2022, January). Opportunistic multi-modal user authentication for health-tracking IoT wearables. In *The Fifth International Conference on Safety and Security with IoT: SaSeIoT 2021* (pp. 1-18). Cham: Springer International Publishing.

Nowak, N., Kakade, P. P., & Annapragada, A. V. (2003). Computational fluid dynamics simulation of airflow and aerosol deposition in human lungs. *Annals of Biomedical Engineering, 31*(4), 374–390. doi:10.1114/1.1560632 PMID:12723679

Ooi, E. H., & Chiew, Y. S. (2023). *Model-Based Approaches in Biomedical Engineering*. IOP Publishing. doi:10.1088/978-0-7503-4016-8

Ooi, E. H., & Ng, E. Y. K. (2011). Effects of natural convection within the anterior chamber on the ocular heat transfer. *International Journal for Numerical Methods in Biomedical Engineering*, *27*(3), 408–423. doi:10.1002/cnm.1411

Patel, S. R., Pundge, A., Zebanaaz, S., & Akther, N. (2023, May). An Analysis of Recent Developments in Healthcare Technology from the Standpoints of Artificial Intelligence, the Internet of Things, and Remote Health Monitoring. In *International Conference on Applications of Machine Intelligence and Data Analytics (ICAMIDA 2022)* (pp. 520-526). Atlantis Press. 10.2991/978-94-6463-136-4_45

Paz, C., Suárez, E., & Vence, J. (2017). CFD transient simulation of the cough clearance process using an Eulerian wall film model. *Computer Methods in Biomechanics and Biomedical Engineering*, *20*(2), 142–152. doi:10.1080/10255842.2016.1206532 PMID:27380322

Raman, R. K., Dewang, Y., & Raghuwanshi, J. (2018). A review on applications of computational fluid dynamics. *International Journal of LNCT*, *2*(6), 137–143.

Rispoli, V. C., Nielsen, J. F., Nayak, K. S., & Carvalho, J. L. (2015). Computational fluid dynamics simulations of blood flow regularized by 3D phase contrast MRI. *Biomedical Engineering Online*, *14*(1), 1–23. doi:10.1186/s12938-015-0104-7 PMID:26611470

Robert, G., Donetto, S., & Williams, O. (2021). Co-designing healthcare services with patients. The Palgrave Handbook of Co-Production of Public Services and Outcomes, 313-333. Palgrave.

Schachar, R. A. (2006). The mechanism of accommodation and presbyopia. *International Ophthalmology Clinics*, *46*(3), 39–61. doi:10.1097/00004397-200604630-00006 PMID:16929224

Sforza, D. M., Putman, C. M., & Cebral, J. R. (2012). Computational fluid dynamics in brain aneurysms. *International Journal for Numerical Methods in Biomedical Engineering*, *28*(6-7), 801–808. doi:10.1002/cnm.1481 PMID:25364852

Sharada, K., Kiran, P. U., Katipelli, R., Maddi, Y., & Rajendar, P. (2021). Investigation Of Compressible Fluid Flow Between Spinning Rotor Cascades In Elevated Aerodynamics Using Numerical Techniques. *Turkish Online Journal of Qualitative Inquiry, 12*(10)

Shibeshi, S. S., & Collins, W. E. (2005). The rheology of blood flow in a branched arterial system. *Applied Rheology (Lappersdorf, Germany)*, *15*(6), 398–405. doi:10.1515/arh-2005-0020 PMID:16932804

Sikkandar, M. Y., Sudharsan, N. M., Begum, S. S., & Ng, E. Y. K. (2019). Computational Fluid Dynamics: A Technique to Solve Complex Biomedical Engineering Problems-A Review. *WSEAS Transactions on Biology and Biomedicine*, *16*, 121–137.

Sinha, S. (2023). State of IoT 2023: Number of connected IoT devices growing 16% to 16.7 billion globally. *Iot Analytics*. https://iot-analytics.com/number-connected-iot-devices/

Sirois, E., Wang, Q., & Sun, W. (2011). Fluid simulation of a transcatheter aortic valve deployment into a patient-specific aortic root. *Cardiovascular Engineering and Technology*, *2*(3), 186–195. doi:10.1007/s13239-011-0037-7

Sukuma, R., Athavale, M. M., Makhijani, V. B., & Przekwas, A. J. (1996). Application of computational fluid dynamics techniques to blood pumps. *Artificial Organs*, *20*(5), 529–533. doi:10.1111/j.1525-1594.1996.tb04475.x PMID:28868695

Sultan, M. B., Mansberger, S. L., & Lee, P. P. (2009). Understanding the importance of IOP variables in glaucoma: A systematic review. *Survey of Ophthalmology*, *54*(6), 643–662. doi:10.1016/j.survophthal.2009.05.001 PMID:19665744

Sun, Z., & Xu, L. (2014). Computational fluid dynamics in coronary artery disease. *Computerized Medical Imaging and Graphics*, *38*(8), 651–663. doi:10.1016/j.compmedimag.2014.09.002 PMID:25262321

Sweetman, B., & Linninger, A. A. (2011). Cerebrospinal fluid flow dynamics in the central nervous system. *Annals of Biomedical Engineering*, *39*(1), 484–496. doi:10.1007/s10439-010-0141-0 PMID:20737291

Tamm, E. R., & Fuchshofer, R. (2007). What increases outflow resistance in primary open-angle glaucoma? *Survey of Ophthalmology*, *52*(6), S101–S104. doi:10.1016/j.survophthal.2007.08.002 PMID:17998032

Tan, F. P. P., Xu, X. Y., Torii, R., Wood, N. B., Delahunty, N., Mullen, M., Moat, N., & Mohiaddin, R. (2012). Comparison of aortic flow patterns before and after transcatheter aortic valve implantation. *Cardiovascular Engineering and Technology*, *3*(1), 123–135. doi:10.1007/s13239-011-0073-3

Tian, G., & Longest, P. W. (2010). *Development of a CFD boundary condition to model transient vapor absorption in the respiratory airways.*

Tian, G., Longest, P. W., Su, G., Walenga, R. L., & Hindle, M. (2011). Development of a stochastic individual path (SIP) model for predicting the tracheobronchial deposition of pharmaceutical aerosols: Effects of transient inhalation and sampling the airways. *Journal of Aerosol Science*, *42*(11), 781–799. doi:10.1016/j.jaerosci.2011.07.005

Walenga, R. L., Tian, G., & Worth Longest, P. (2013). Development of characteristic upper tracheobronchial airway models for testing pharmaceutical aerosol delivery. *Journal of Biomechanical Engineering*, *135*(9), 091010. doi:10.1115/1.4024630 PMID:23722698

Wendell, D. C., Samyn, M. M., Cava, J. R., Krolikowski, M. M., & LaDisa, J. F. Jr. (2016). The impact of cardiac motion on aortic valve flow used in computational simulations of the thoracic aorta. *Journal of Biomechanical Engineering*, *138*(9), 091001. doi:10.1115/1.4033964 PMID:27367143

Yiallourou, T. I., Kröger, J. R., Stergiopulos, N., Maintz, D., Martin, B. A., & Bunck, A. C. (2012). Comparison of 4D phase-contrast MRI flow measurements to computational fluid dynamics simulations of cerebrospinal fluid motion in the cervical spine. *PLoS One*, *7*(12), e52284. doi:10.1371/journal.pone.0052284 PMID:23284970

Zachow, S., Steinmann, A., Hildebrandt, T., & Heppt, W. (2007). Understanding nasal airflow via CFD simulation and visualization. *Proc. Comp Aid Surg Around the Head*, 173-176.

Chapter 10
Transformative Insights:
Harnessing Artificial Intelligence for Enhanced Ovarian Cancer Prediction and Prognosis

Rita Komalasari
https://orcid.org/0000-0001-9963-2363
Yarsi University, Indonesia

ABSTRACT

This chapter explores the transformative role of artificial intelligence (AI) in ovarian cancer prediction and prognosis. It aims to analyze complex datasets, predict patient outcomes, and optimize treatment pathways. The study critically examines existing research on ovarian cancer, highlighting gaps and challenges in current prognostic methodologies. The findings demonstrate significant advancements in AI application in ovarian cancer prediction, highlighting its transformative potential in research and clinical practice. AI not only enhances understanding of ovarian cancer complexities, but also offers personalized and optimized treatment strategies, offering hope for improved patient outcomes and overall survival rates.

INTRODUCTION

The oncology field is still grappling with devastatingly poor survival rates and remarkably high incidence rates of ovarian cancer in places like Indonesia (Olatunji et al., 2023). Ovarian cancer is the fifth most common cancer in women, and human clinical decision-making is struggling to keep up with the illness because of the growing complexity of prognostic indications, treatment options, and their sequencing (Poort et al., 2021). In this light, it becomes clear that incorporating Artificial Intelligence (AI) into ovarian cancer management decision-making processes is a viable option (Akazawa & Hashimoto, 2020). The vital role of AI in ovarian cancer prognosis is examined in this chapter. The increasing amount of data accessible and the need for detailed analysis necessitate a method that is beyond what humans are capable of. The computing power of AI provides a mechanism to analyze complex patterns

DOI: 10.4018/979-8-3693-1922-2.ch010

in patient data, make predictions about outcomes, and help doctors choose the best course of therapy. The creation and implementation of AI systems for ovarian cancer patient knowledge, prediction, and possible improvement of outcomes is the focus of this chapter. Worrying patterns and statistics in ovarian cancer highlight the significance of the research. As the sixth most common cancer in women, ovarian cancer has a dismal five-year survival rate in Indonesian healthcare. (Dewi et al., 2022). Indonesia has one of the worst incidence rates in Europe, and they are becoming worse. This means that we need new solutions quickly (Reid, 2020). Because of the growing complexity of both prognostic indicators and treatment sequences, as well as the difficulties in making informed clinical decisions by humans, sophisticated computational tools are urgently needed (Degu et al., 2023). Analogies with other studies also confirm the study's usefulness. The knowledge and information yielded by this research hold significant implications for various stakeholders involved in the management of ovarian cancer, including Clinicians can leverage the predictive models developed through AI to enhance their decision-making processes in ovarian cancer treatment. The ability to analyze prognostic markers and predict outcomes, including overall survival and surgical success, equips healthcare professionals with valuable insights for personalized patient care. Patients stand to benefit directly from the improved predictive accuracy of AI models. Access to more accurate prognoses can empower patients to make informed decisions about their treatment options and actively participate in their healthcare journey. Researchers in the field of oncology and cancer biology can use the findings to explore further and refine AI applications in cancer research. The identification of novel prognostic biomarkers using AI techniques contribute to the broader understanding of ovarian cancer at the molecular level. Healthcare institutions and lawmakers can think about using AI technology in clinical practice. Ovarian cancer management healthcare policies and recommendations may be impacted by this research's findings, which show that AI may improve prediction models. Companies involved in pharmaceuticals and biotechnology may find value in the identification of novel prognostic biomarkers. This knowledge could guide the development of targeted therapies or interventions for specific subgroups of ovarian cancer patients, contributing to advancements in treatment modalities. Academic institutions can incorporate the findings into medical curricula, ensuring that the next generation of healthcare professionals is well-versed in the potential applications of AI in oncology. Organizations focused on global health, such as the World Health Organization (WHO), can use the research findings to inform strategies for addressing ovarian cancer on a broader scale, particularly in regions with high incidence rates. These stakeholders may work together to improve outcomes, make better-informed decisions, and enhance ovarian cancer care by sharing and using the information from this study.

This study makes significant contributions to the field of ovarian cancer management, offering transformative advancements that stand to improve the profession and the broader healthcare landscape. The development and validation of AI models for predicting overall survival, surgical outcomes, and HR status empower clinicians with precise tools. This enables more informed and data-driven decision-making, optimizing treatment strategies for individual patients. By identifying the most critical variables influencing ovarian cancer outcomes, such as International Federation of Gynecology and Obstetrics (FIGO) stage, surgery outcome, and biomarkers like Cancer Antigen 125 (CA125), the study facilitates a more personalized approach to patient care (Damiani et al., 2023). Clinicians can tailor treatments based on specific prognostic factors, improving the efficacy of interventions. The AI models' superior performance in predicting surgical outcomes, surpassing traditional statistical models, has direct implications for surgical planning. Surgeons can leverage this knowledge to make more accurate predictions about the success of cytoreductive procedures, leading to improved surgical outcomes and patient recovery. The identification of novel prognostic biomarkers, DNA Ligase 1 (LIG1) and DNA Polymerase Delta 3 (POLD3), associated with the Homologous Recombination (HR) pathway, contributes to the understanding of ovarian cancer at the molecular level (How et al., 2021). This finding advances cancer biology by paving the way for further studies and maybe the creation of tailored treatments. It is highly recommended that AI technologies be included in standard oncology procedures due to the proven effectiveness of AI models above conventional statistical methodologies. The way doctors approach prognosis evaluations and treatment planning will undergo a paradigm change as a result of this. Academic institutions can incorporate the study's findings into medical education, ensuring that future healthcare professionals are equipped with the knowledge and skills to harness AI in the oncology field. This contributes to the ongoing evolution of medical curricula and prepares the next generation for the integration of cutting-edge technologies. The study's findings, particularly the exploration of factors impacting ovarian cancer outcomes in a specific region like Indonesia, can have global implications. The findings have the potential to shape worldwide plans to combat ovarian cancer, leading to a more all-encompassing and universally relevant strategy for cancer treatment. This work has the potential to revolutionize ovarian cancer research and management due to its several valuable contributions, such as better clinical decision-making and the identification of new biomarkers.

BACKGROUND

Ovarian cancer ranked as the fifth most prevalent cancer in females, poses a critical healthcare challenge, particularly in regions with high incidence rates, such as Indonesia (Bukłaho et al., 2023). Indonesia has one of the lowest five-year survival rates for ovarian cancer patients (36%) (Fitriani et al., 2020). This is despite the increasing complexity of prognostic indicators, therapies, and sequencing. This alarming survival rate underscores the pressing need for advanced decision-making tools, given the intricacies that the human brain struggles to navigate (Gondhowiardjo et al., 2021). The gap in current clinical decision-making capabilities necessitates a profound solution. In order to fill this knowledge vacuum, this study will examine the feasibility of using AI systems as prediction tools for the treatment of ovarian cancer. The study seeks to determine the efficacy of AI models in predicting overall survival, surgical outcomes, and genetic biomarkers, aiming to revolutionize clinical decision-making in the challenging landscape of ovarian cancer. The dire survival rates and escalating complexities in ovarian cancer underscore the critical importance of this research. With only 36% of patients in Indonesia surviving five years post-diagnosis, the need for innovative solutions is urgent (Winarto et al., 2022). The increasing intricacies in prognostic markers and treatment sequences pose a formidable challenge to human clinical decision-making. This study addresses this pressing issue by exploring the transformative potential of Artificial Intelligence (AI) systems. By developing AI models capable of predicting overall survival, surgical outcomes, and genetic biomarkers, the research aims to revolutionize clinical decision-making in ovarian cancer. Improving patient outcomes and altering the worldwide landscape of ovarian cancer care are also possible objectives of this research. The research problem is intricately woven into the rich tapestry of existing literature on ovarian cancer, prognostic markers, and treatment modalities. Studies such as those by Falzone et al. (2021) underscore the escalating complexities in ovarian cancer management, creating a need for advanced decision-making tools. The grim survival rates in Indonesia align with global trends outlined by the Global Cancer Observatory (Singh et al., 2023), emphasizing the urgency of innovative solutions. Notably, the work of Puspitaningtyas et al. (2021) highlights the increasing incidence rates in Indonesia, reinforcing the regional specificity of the problem. Theoretical frameworks, such as those exploring decision-making in complex medical situations (Sahoo & Goswami, 2023), provide a lens to understand the challenges faced by clinicians in ovarian cancer management. The gap identified in human decision-making aligns with discussions on the limitations of traditional approaches in the face of rising complexities (Si et al., 2021). This chapter's significance lies in bridging this gap by introducing Artificial Intelligence (AI) as a transformative tool. By referencing the existing literature, the chapter positions the research problem as part of a broader discourse, emphasizing the necessity for innovative approaches. The contextualization not only showcases the authors' familiarity with the field but also sets the stage for the chapter's significant contribution to advancing ovarian cancer management through AI applications.

METHOD

The literature study conducted for this chapter critically examines existing research on ovarian cancer, highlighting gaps and challenges in current prognostic methodologies. It serves as a foundational framework, guiding the development of AI models for prediction. By synthesizing prior knowledge, the study ensures a comprehensive understanding of the field, facilitating nuanced analyses and innovative approaches. Through rigorous analysis, significant variables influencing ovarian cancer outcomes are identified, enabling the development of precise AI models. Comparative assessments validate the superiority of AI in predicting survival rates, surgical outcomes, and genetic biomarkers, reinforcing its pivotal role in ovarian cancer management.

ARTIFICIAL INTELLIGENCE FOR ENHANCED OVARIAN CANCER PREDICTION

In this section, the author discuss ovarian cancer, which accounts for 4% of all female cancer diagnoses and has 190,000 new cases every year throughout the globe (Maleki et al., 2023). Ovarian cancer is the fifth most frequent cancer in women in the United Kingdom, with a patient population that has grown by 18% in the last 30 years (Gaitskell et al., 2022). The long-term survival statistics have remained relatively unchanged over the last 20 years, even if there have been some small advances in therapies. Although survival statistics vary by stage of cancer, only 36% of patients in the UK will make it through the first five years following diagnosis (Armstrong et al., 2021). In 2020–02, the Australian government estimated a lifetime treatment cost per case of ovarian cancer at $22,000 (Lacaze et al., 2023). However, the exact cost of therapy in Indonesia and other South East Asia nations is uncertain. Our current knowledge of ovarian cancer categorizes it into four phases, each with its own prognosis and therapeutic options, according to its anatomical distribution. Aiming to tackle these issues and enhance the usual management of this malignancy, artificial intelligence is being investigated, with a focus on machine learning and feature selection approaches. Multiple variables, including grade, tumor histology, radiography, and histology, are the center of attention in ovarian cancer research. The presence of aberrant cancer cells, or grade, is a significant predictive marker in ovarian cancer. The worse the differentiation of a cancer, the worse the prognosis it is likely to have. An essential tool in the fight against ovarian cancer is CA125, a serum tumor marker (Matsas et al., 2023). Unfortunately, it is not helpful for early-stage illness because of its high sensitivity and low specificity. According to the research, a significant majority of women diagnosed with epithelial ovarian cancer had increased CA125 levels. Half to sixty percent of ovarian cancer cases are classified as severe, which impacts the course of therapy (Elorriaga et al., 2021). Ovarian cancer diagnostics rely heavily on radiography, with techniques like MRI, Computed tomography (CT), and ultrasound playing key roles (Liberto et al., 2022). While computed tomography (CT) scans provide precise and rapid measures of abdominal illness, ultrasonography offers hard pictures for evaluating therapy efficacy. In ovarian cancer, several factors are studied, including grade, tumor histology, radiography, and histology. Researchers can enhance the management and treatment of ovarian cancer by better comprehending these aspects and making these indicators more sensitive. Furthermore, the exploration of transcriptomic analysis and the identification of novel prognostic biomarkers (LIG1 and POLD3) for the Homologous Recombination (HR) pathway adds an innovative dimension to the study (Zhao et al., 2021). The AI model achieved an AUC of 0.87 in predicting HR status, demonstrating its ability to uncover genetic insights with potential implications for personalized treatment strategies (Fontaine et al., 2022). The study's importance is validated by the pressing healthcare challenge posed by ovarian cancer, supported by statistical evidence of low survival rates and rising incidence (Flasarova et al., 2023). The demonstrated improvement in predictive accuracy through AI models, especially in predicting overall survival and surgical outcomes, solidifies the study's significance in reshaping ovarian cancer management. Ovarian cancer is uncommon. However, it is more common in women and girls under the age of 30. The peak incidence occurs between the ages of 75 and 79 when the risk reaches 60.5 per 100,000 (Gaona et al., 2020). It rises with age after that. Since 70% of ovarian cancer cases are detected at an advanced stage, a thorough physical examination is essential for making a diagnosis (Stewart et al., 2019). The pelvic mass's motility, consistency, bilaterality, and cul-de-sac involvement determine if it is benign or malignant. In contrast to the statistical model's 0.62 Area under the Curve (AUC), the AI model's was 0.72. (Archila et al., 2023). This signifies a tangible advancement in predictive accuracy, reinforcing the significance of AI in enhancing prognostic capabilities. Moreover, the study addresses a crucial aspect of ovarian cancer management by predicting surgical outcomes. The AI model consistently outperformed the statistical model across various outcomes, with AUC values of 0.71, 0.73, and 0.79 for different predictions, compared to AUC values ranging from 0.47 to 0.51 for the statistical model (Damiani et al., 2023). These findings highlight the direct clinical relevance and potential impact of AI in optimizing surgical decision-making.

Surgery makes up the bulk of the detection and diagnostic process that ultimately results in an appropriate multimodal treatment plan (Charkhchi et al., 2020). By precisely defining the illness stage and removing as many tumors as feasible, surgery aims to achieve its primary purpose. Unfortunately, by the time most people seek treatment, the disease has already spread to other parts of their bodies. When no macroscopic remnant of the tumor tumor is seen after surgery, it is considered complete cytoreduction (Kepenekian et al., 2022). Diagnostic and therapeutic considerations for ovarian cancer rely heavily on prognostic and predictive variables. Prognostic criteria establish the general survival rate before to treatment or therapy, whereas predictive parameters define the response to therapy. A number of devastating consequences may develop from ovarian cancer, making it a hazardous and sometimes fatal condition. Surgery, the primary indicator of the illness stage, is the first course of therapy. A second-line surgical procedure is necessary in more advanced instances. Typically, six rounds of chemotherapy are delivered at a day center once every 21 days with the goal of targeting cancer cells that have spread to other parts of the body (Trelis et al., 2023). Blood tests are performed to make sure the bone marrow has fully recovered before each cycle. While radiation therapy has been considered for some instances of ovarian cancer, chemotherapy is the treatment of choice. The effectiveness of treatment is assessed by measurable outcomes such as cancer response rate, disease-related mortality, disease-free survival, progression-free survival, and overall survival rate. In cancer care, a group of experts called an MDT (multi-disciplinary team) must assess the patient's condition and provide treatment recommendations. Referral patients are evaluated by the MDT, who then propose treatments (surgery, chemotherapy, or inaction) depending on indicators (Pasqual et al., 2023). At the post-op MDT meetings, we go over each patient's case and think about what further they could need. One of the many obstacles to ovarian cancer research is the growing difficulty of clinical decision-making as a result of the proliferation of cancer indicators, cancer medications, and treatment sequences. More and more, it is crucial to have an expert computer system, like AI, look into the potential results for each marker, therapy, and treatment sequence (Garnier et al., 2021). A patient's overall survival rate is difficult to predict, but doing so has advantages, including allowing doctors to alter treatment plans and giving patients more accurate information about their prognosis.

By developing appropriate and high-performing models, AI approaches may aid researchers in extracting valuable insights from existing data. They find extensive use in radiation, cancer cell control, detection, diagnosis, and DNA microarray analysis and classification, among other areas of cancer research. Ovarian cancer screening, early detection, diagnosis, and serum analysis are all assisted by AI systems. The use of Machine Learning distinguishes AI systems from traditional computer algorithms; this feature enables AI systems to adapt to new data inputs by learning from old data (Yurttas et al., 2022). When making a clinical choice about ovarian cancer therapy, it is essential to look at both historical data and emerging research. One efficient way to improve computer programs' performance is via the use of previous experiences; this technique is known as machine learning (ML). Examples may be used to describe tasks, data can be explored, and hidden rules can be discovered by analyzing big datasets. When it comes to situations that are constantly changing, self-adaptive AI makes redesigning less necessary. Clustering and other forms of Unsupervised Learning are examples of ML techniques, while classification and other forms of Supervised Learning are examples of ML methods (Leigh et al., 2020). While supervised learning seeks to create indicators that lead to a result in decision-making about ovarian cancer, unsupervised learning seeks to uncover patterns and trends in the data.

Different from one another, ML uses the words "deductive" and "inductive" systems. In deductive systems, rules are provided by experts and applied to input data, but in inductive systems, rules are discovered by examining instances (Shu & Ye, 2023). General principles are introduced via inductive systems, which then train themselves using examples and put those rules into practice. To categorize objects, the Nearest Neighbour technique (also known as a decision tree) employs the k-nearest neighbor (KNN) mechanism, which is widely used in machine learning (Bansal et al., 2022). Although the KNN technique is easy to implement and works well with large training sets, finding the number of closest neighbors is computationally expensive and a challenging task. To make the most of KNN for several qualities while making clinical decisions, decision trees are a must (Mijwil & Abttan, 2021). Ovarian cancer research relies heavily on machine learning, which yields essential insights into therapy choices and results. Due to their capacity to establish comprehensible rules and pinpoint crucial prediction domains, decision trees, and closest neighbors have found extensive use in cancer research. These approaches work well for clinical decision-making and are simple enough for laypeople to employ (Boehm et al., 2022). In addition to being too sensitive to mistakes on tiny data sets, decision trees may be computationally costly to train and are not well suited for predicting continuous variables. A network of artificial neurons that mimics the way the human brain operates is called a neural network (NN). Axons, dendrites, and somas make up a biological neuron, and it is by axons that signals are sent between neurons (Worden et al., 2023). The 'feed-forward' form of ANN is the most popular and straightforward since it just requires inputs and outputs. Decision trees and KNNs are well-suited for clinical decision-making and easy enough for non-expert users to grasp, making them popular in cancer research. Cancer research also makes use of Artificial Neural Networks (ANN), which are computationally costly to train and more error-prone. Recurrent Neural Networks (RNNs) and other Artificial Neural Networks (ANNs) have been beneficial in the extensive field of cancer research due to their simplicity and broad applicability. One hidden layer links input neurons to both the hidden layer and the output neurons in a feed-forward artificial neural network (ANN) model (Abdou, 2022). The training of these models involves minimizing the Sum of Squared Error (SSE) by adjusting the weights of the network. Learning entails doing things like computing the output and error, as well as randomly initializing the network's weights. The function's threshold value, together with the values of w1 and w2, are trained into the ANN model (Ruiz et al., 2021). Numerous studies have shown that ANN is successful in predicting outcomes on an individual basis, and it has been extensively employed in several sectors of cancer research. Be that as it may, NN is not without flaws. It is able to capture non-linear correlations in the dataset and is robust enough to tolerate mistakes in the training data. Although the network may be trainable, there is no assurance that it will take a long time to train it. Furthermore, the network's weights are not easy to decipher. ANN's excellent prediction accuracy and capacity to manage dataset mistakes make it a popular choice as a prediction model in cancer research (Wang et al., 2022).

In regions such as Indonesia, grappling with alarmingly high incidence rates, the escalating challenge of ovarian cancer poses a pressing need for a transformative approach to clinical decision-making. Ovarian cancer stands as the 5th most common cancer in females, a stark reality emphasized by the sobering statistic that only 36% of Indonesian patients manage to survive for at least five years post-diagnosis (Gumilang & Hernawan, 2023). This alarming survival rate serves as a clarion call for innovative strategies, particularly as the landscape of ovarian cancer becomes increasingly intricate with a surge in prognostic markers, diverse treatment options, and intricate sequences of interventions. In response to this complexity, a compelling argument emerges for the integration of expert systems, notably Artificial Intelligence (AI), to assist clinicians in navigating the intricacies of ovarian cancer management. The primary objective is to unravel the challenging task of predicting overall survival, a crucial metric in cancer research. Impressive Area under the Curve (AUC) values of 0.72 for the AI model and 0.62 for the statistics model show that the results indicate a significant improvement in predicting accuracy (Archila et al., 2023). This not only substantiates the efficacy of AI but sets the stage for a more nuanced understanding of treatment outcomes. The identification of essential factors impacting the prediction, such as FIGO stage, surgical outcome, and CA125, enhances the comprehension of prognostic signals (Roy et al., 2022). Beyond the broad strokes of survival prediction, the study delves into the intricate aspects of treatment modalities, offering a potential pathway for tailoring interventions based on these identified variables. Furthermore, the scope of investigation extends to predicting surgical outcomes, a domain with direct implications for patient management. The AI models, when rigorously compared to the standard statistical prediction model, consistently outshine their counterparts, especially in predicting various surgical outcomes such as complete cytoreduction (Llueca et al., 2021). This marks a pivotal moment in clinical relevance, suggesting that AI can be a potent tool in pre-surgical decision support, potentially revolutionizing the surgical approach to ovarian cancer. The research's ambition does not stop at survival and surgical outcomes; it ventures into the molecular landscape through transcriptomic analysis. AI techniques unveil LIG1 and POLD3 as novel prognostic biomarkers, shedding light on the crucial DNA repair pathway, Homologous Recombination (HR) (Li et al., 2019). The AI models' ability to predict HR status with an impressive AUC of 0.87 underscores the versatility of AI in uncovering molecular insights that may redefine the understanding of ovarian cancer at the molecular level. This chapter underscores the transformative potential of AI in ovarian cancer. The evidence presented—from improved predictive models to the identification of novel biomarkers—paves the way for AI systems to be integral tools in the clinician's arsenal. The findings hold promise not only for individualized patient care but also for reshaping the landscape of ovarian cancer management. The potential of AI to inform decisions, improve prognostic accuracy, and contribute to personalized medicine emerges as a beacon of hope for a more informed and practical approach to combating ovarian cancer.

This study represents a significant advancement in the realm of ovarian cancer research. Fundamentally, the study aims to investigate how Artificial Intelligence (AI) may revolutionize ovarian cancer care by providing a forecasting tool in this complex field. The primary objective is to address the formidable challenge of predicting overall survival, a task intricately linked to the complex and multifaceted nature of ovarian cancer. The outcomes of the study, vividly illustrated by the Area under the Curve (AUC) metrics, underscore a notable improvement in predictive accuracy for the AI model. These findings show that AI is far more effective than the statistical model in prognostic modeling, with an AUC of 0.72 instead of 0.62 (Zhuhadar & Lytras, 2023). This improvement not only validates the utility of AI in predicting overall survival but also sets the stage for a paradigm shift in how clinicians approach and refine treatment strategies for ovarian cancer patients. Expert computer systems, such as Artificial Intelligence, are necessary for the investigation of potential outcomes for each marker, therapy, and treatment sequence in the complicated field of ovarian cancer management. By analyzing and forecasting potential treatment courses, these algorithms may aid physicians in learning from their mistakes and making more accurate predictions. Although it is not an easy process, predicting overall survival may alter treatment plans, study recruiting strategies, and our knowledge of patients' potential survival durations. However, recent attempts at predicting overall survival have not been successful due to the static nature of the introduced models. Additionally, GA has its limits when it comes to surgical prognosis. Surgical excision is the current standard of care for ovarian cancer after the discovery of anomalies on CT scans, increases in the CA125 biomarker, or abnormalities during a physical examination (Gumusoglu & Gunel, 2022). The purpose of referring patients for alternative treatment options in minor instances is to remove as many tumors as possible rather than undergoing surgery.

Central to the success of this study is the meticulous identification of critical variables that wield significant influence over the prediction of overall survival in ovarian cancer patients (Paoletti et al., 2020). Prognostic signs may be better understood with the addition of essential variables, including FIGO stage, surgical result, and CA125 (Furrer et al., 2019). This nuanced approach to identifying influential variables goes beyond a broad-stroke analysis, offering a detailed and comprehensive insight into the multifaceted nature of ovarian cancer prognosis. The validation of Artificial Intelligence (AI) efficacy in predicting overall survival is not merely a technological triumph but, more importantly, a groundbreaking advancement in tailoring treatment modalities. By anchoring predictions in these identified variables, the study lays the foundation for a more personalized and nuanced approach to ovarian cancer treatment. Clinicians, armed with this refined understanding of prognostic markers, are poised to make informed decisions that transcend the conventional one-size-fits-all approach, paving the way for more effective and tailored management of ovarian cancer patients. The study explores the crucial Area of surgical outcomes, which has direct and significant consequences for patient treatment, going beyond the field of overall survival prediction. When compared to the traditional statistical prediction model, Artificial Intelligence (AI) models consistently and remarkably outperform their human counterparts. This superiority is particularly evident in the prediction of various surgical outcomes, notably the achievement of complete cytoreduction (Zaidi et al., 2020). The consistent success of AI models in forecasting surgical results marks a pivotal moment in clinical relevance, signifying the potential for AI to serve as a potent and invaluable tool in pre-surgical decision support. As clinicians grapple with complex decisions regarding the extent of surgical interventions, the enhanced predictive capabilities of AI offer a guiding light, enabling more informed and precise choices that can significantly impact patient outcomes in the context of ovarian cancer treatment.

Extending its exploration beyond the realms of survival and surgery, this research boldly ventures into the intricate molecular landscape of ovarian cancer through transcriptomic analysis. Here, Artificial Intelligence (AI) techniques come to the forefront, illuminating novel insights into the disease's underlying biology. The spotlight falls on LIG1 and POLD3, identified as novel prognostic biomarkers intricately tied to the essential DNA repair pathway, Homologous Recombination (HR) (Kontizas et al., 2020). This revelation holds profound significance as disruptions in HR are pivotal in the evolution and progression of ovarian cancer. The predictive power exhibited by these AI models in determining HR status not only reaffirms their efficacy but also underscores the remarkable versatility of AI in uncovering molecular intricacies. This breakthrough in molecular profiling not only adds depth to the understanding of ovarian cancer at the molecular level but also opens new avenues for targeted interventions and personalized treatment strategies, marking a paradigm shift in how we perceive and approach the intricacies of ovarian cancer biology.

Essentially, this effort stands out as a precursor to the revolutionary power of AI in the ovarian cancer field (Breen et al., 2023). The wealth of evidence presented, spanning from the development of markedly improved predictive models to the identification of novel biomarkers, collectively points toward a future where AI systems become integral tools in the clinician's arsenal. These findings carry profound implications, not merely for the field of ovarian cancer research but for the very practice of medicine itself. The promise encapsulated in these results extends beyond the confines of statistical advancements; it heralds a paradigm shift in how clinicians perceive and approach ovarian cancer management. The newfound capability of AI to predict overall survival surgical outcomes and even delve into the molecular intricacies of DNA repair pathways, exemplified by the identification of LIG1 and POLD3, signifies a seismic leap forward (McPherson & Korzhnev, 2021). It opens avenues for a more nuanced, individualized, and targeted approach to patient care. The potential of AI to inform critical decisions, enhance prognostic accuracy, and contribute to the realm of personalized medicine emerges as a beacon of hope, promising not only improved patient outcomes but a fundamental reshaping of the ovarian cancer management landscape. We are on the cusp of a new age in healthcare, and this project's combination of artificial intelligence with cancer research paves the way for a future when cutting-edge tech does more than just help doctors; it actively changes the face of medicine. The implications are far-reaching, suggesting a trajectory toward a more informed, precise, and practical approach to combating ovarian cancer. In this synthesis of cutting-edge technology and clinical acumen, the prospect of leveraging AI as a transformative force in the fight against ovarian cancer becomes not just a possibility but a compelling reality.

SOLUTIONS AND RECOMMENDATIONS

In translating the research findings into actionable steps, several policy solutions and recommendations emerge to guide healthcare policies, intervention strategies, and therapeutic approaches in the real-world context. Policymakers should consider incorporating AI-based predictive models into clinical guidelines for ovarian cancer management. This integration ensures that clinicians have access to advanced decision support tools, fostering more informed and personalized treatment strategies. Policymakers and healthcare institutions should invest in training programs to enhance the proficiency of healthcare professionals in utilizing AI tools. This ensures that clinicians are well-equipped to interpret and integrate AI-generated insights into their decision-making processes. Healthcare systems should be more interoperable and work towards standardizing data formats. This allows for the smooth incorporation of different datasets, which improves the efficiency and applicability of AI models in various healthcare environments. Legislatively required ethical criteria must be established and continuously revised for the introduction of artificial intelligence (AI) to healthcare, especially for the treatment of ovarian cancer. To guarantee fair and ethical practices, these standards should include topics including patient consent, data protection, and the appropriate use of AI technology. Ovarian cancer treatment that makes use of artificial intelligence (AI) may be encouraged by policymakers. Artificial intelligence (AI) solutions may be developed and implemented more quickly in actual healthcare settings with the help of funding mechanisms, grants, and joint endeavors. In the context of artificial intelligence (AI) applications, policies should prioritize efforts that include and educate patients. This involves being transparent, including patients in conversations about AI in healthcare, and giving explicit information on how AI helps with decision-making. Not only can lawmakers and practitioners make use of AI's capabilities in ovarian cancer treatment, but they can also make sure that these technologies are integrated into clinical practice in a responsible, ethical, and patient-centered way by following these suggestions and solutions. This method is in step with the ever-changing medical scene and helps bring about a comprehensive change in healthcare practices. Fresh, Workable Answers: Integrate oncology practices with clinical decision support systems powered by artificial intelligence. The most recent findings in the field, along with the patient's profile, allow these systems to provide real-time support to doctors in the form of treatment suggestions. Create AI-powered, patient-focused mobile apps that update users on their treatment progress and provide additional resources to help them through difficult times. In addition to improving engagement, this gives people more agency in their healthcare. Cancer specialists, data scientists, and artificial intelligence specialists should all work together in new training programs. By working together, we can better comprehend the potential uses of AI in the fight against ovarian cancer and find ways to incorporate it into existing therapeutic processes. Get the word out about the merits and pitfalls of artificial intelligence (AI) in healthcare by launching community-based awareness campaigns. Informed communities can contribute to reducing stigma, fostering trust, and encouraging informed decision-making regarding AI applications. By embracing these innovative, practical solutions, the research not only contributes to the scientific understanding of ovarian cancer management but also paves the way for tangible improvements in clinical practice, patient engagement, and community awareness. This holistic approach ensures a seamless transition from research findings to real-world applications, fostering a positive and transformative impact on healthcare.

FUTURE RESEARCH DIRECTIONS

As we embark on the frontier of incorporating Artificial Intelligence (AI) into ovarian cancer management, the path forward is illuminated by opportunities for future research that can further refine and expand our understanding. The evolving nature of this field warrants a call for continued investigations, with a particular emphasis on three key areas: Future research should strive for clarity and precision in the terminology used to describe AI applications in ovarian cancer. Establishing standardized terms and definitions will foster a shared understanding, enabling effective communication and collaboration among researchers, clinicians, and stakeholders. The advancement of AI in the healthcare domain demands robust measurement techniques. Future studies could delve into refining and standardizing measurement methodologies for assessing the performance of AI models in predicting ovarian cancer outcomes. This includes exploring metrics that capture not only predictive accuracy but also factors like interpretability and clinical relevance. Unraveling the causal relationships and mechanisms of change underlying the impact of AI on ovarian cancer management represents a crucial avenue for future research. Investigating how AI-driven insights translate into clinical decision-making and patient outcomes will deepen our understanding of the transformative potential of these technologies. This advocacy for future research is not just a recognition of the evolving landscape but a proactive stance toward continuous improvement and innovation. By addressing these areas, future research has the potential to amplify the positive impact of AI in ovarian cancer management, ensuring that these technologies are not only adopted but optimized for the benefit of patients and clinicians alike. Highlighting Potential Research Directions: Investigate the long-term impact of integrating AI interventions in ovarian cancer care. Research could explore how sustained AI support influences patient outcomes, healthcare resource utilization, and the overall quality of care over extended periods. Explore the adaptability and performance of AI models across diverse demographic and geographic populations. Understanding how these models generalize to different patient groups ensures equitable access to advanced predictive tools. Examine the dynamics of human-AI collaboration in clinical decision-making. Research could delve into optimal models for integrating AI recommendations with clinician expertise, ensuring a harmonious and effective partnership. Investigate the impact of patient education initiatives on the acceptance and utilization of AI-generated insights. Understanding patient perceptions and preferences will be crucial for successful integration into patient-centered care.

CONCLUSION

In conclusion, this article has delved into the transformative potential of Artificial Intelligence (AI) in ovarian cancer management, showcasing a commendable intellectual exploration. To elevate its impact and relevance, the conclusion is strengthened with explicit recommendations for real-world policy solutions: Advocate for the integration of validated AI models into healthcare policies, ensuring their acknowledgment as valuable tools in guiding clinical decision-making for ovarian cancer treatment. Propose incentives for healthcare institutions to invest in training programs that enhance the proficiency of clinicians in utilizing AI tools. This ensures a seamless and effective integration of AI insights into routine clinical practice. Call for the establishment and regular updates of ethical guidelines specific to AI deployment in ovarian cancer management. These guidelines should address issues of patient consent, data privacy, and responsible AI usage. Encourage collaborations between researchers and policymakers to bridge the gap between academic insights and policy implementation. This synergy can expedite the translation of research findings into tangible improvements in healthcare policies. Recommend the development and support of patient education initiatives that demystify AI applications in healthcare. Informed patients are essential partners in the successful integration of AI into the patient care journey. By incorporating these recommendations, the article not only contributes intellectually to the discourse on AI in ovarian cancer but also positions itself as a catalyst for actionable change. Strengthening the link between research and real-world policy solutions ensures that the transformative potential of AI is not confined to academic discussions but resonates in the corridors of healthcare institutions, ultimately benefiting patients and clinicians alike. To sum up, this study is an essential step in understanding how AI might change the game for ovarian cancer treatment. The collected research highlights the revolutionary potential of AI in changing healthcare practices, with examples including improved prediction models and the identification of new biomarkers. Notably, a more sophisticated and personalized therapy paradigm may be established with the use of prognostic indicators, which have been enhanced by the discovery of critical factors impacting overall survival and surgical outcomes. Not only does the research represent a technical leap forward, but it also has the potential to revolutionize ovarian cancer management as a whole and individual patient care. Ovarian cancer clinical decision-making is set to be revolutionized by AI thanks to the complex interplay of prognostic indicators, treatment regimens, and the discovery of critical factors influencing overall survival and surgical results.

REFERENCES

Abdou, M. A. (2022). Literature review: Efficient deep neural networks techniques for medical image analysis. *Neural Computing & Applications*, *34*(8), 5791–5812. doi:10.1007/s00521-022-06960-9

Akazawa, M., & Hashimoto, K. (2020). Artificial intelligence in ovarian cancer diagnosis. *Anticancer Research*, *40*(8), 4795–4800. doi:10.21873/anticanres.14482

Archila, L. R., Smith, L., Sihvo, H. K., Koponen, V., Jenkins, S. M., O'Sullivan, D. M., & Moreira, R. K. (2023). Performance of an Artificial Intelligence Model for Recognition and Quantitation of Histologic Features of Eosinophilic Esophagitis on Biopsy Samples. *Modern Pathology*, *36*(10), 100285. doi:10.1016/j.modpat.2023.100285

Archila, L. R., Smith, L., Sihvo, H. K., Koponen, V., Jenkins, S. M., O'Sullivan, D. M., & Moreira, R. K. (2023). Performance of an Artificial Intelligence Model for Recognition and Quantitation of Histologic Features of Eosinophilic Esophagitis on Biopsy Samples. *Modern Pathology*, *36*(10), 100285. doi:10.1016/j.modpat.2023.100285

Armstrong, D. K., Alvarez, R. D., Bakkum-Gamez, J. N., Barroilhet, L., Behbakht, K., Berchuck, A., Chen, L., Cristea, M., DeRosa, M., Eisenhauer, E. L., Gershenson, D. M., Gray, H. J., Grisham, R., Hakam, A., Jain, A., Karam, A., Konecny, G. E., Leath, C. A., Liu, J., & Engh, A. M. (2021). Ovarian cancer, version 2.2020, NCCN clinical practice guidelines in oncology. *Journal of the National Comprehensive Cancer Network : JNCCN*, *19*(2), 191–226. doi:10.6004/jnccn.2021.0007

Bansal, M., Goyal, A., & Choudhary, A. (2022). A comparative analysis of K-nearest neighbor, genetic, support vector machine, decision tree, and long short term memory algorithms in machine learning. *Decision Analytics Journal*, *3*, 100071. doi:10.1016/j.dajour.2022.100071

Boehm, K. M., Aherne, E. A., Ellenson, L., Nikolovski, I., Alghamdi, M., Vázquez-García, I., Zamarin, D., Long Roche, K., Liu, Y., Patel, D., Aukerman, A., Pasha, A., Rose, D., Selenica, P., Causa Andrieu, P. I., Fong, C., Capanu, M., Reis-Filho, J. S., Vanguri, R., & Shah, S. P. (2022). Multimodal data integration using machine learning improves risk stratification of high-grade serous ovarian cancer. *Nature Cancer*, *3*(6), 723–733. doi:10.1038/s43018-022-00388-9

Breen, J., Allen, K., Zucker, K., Adusumilli, P., Scarsbrook, A., Hall, G., Orsi, N. M., & Ravikumar, N. (2023). Artificial intelligence in ovarian cancer histopathology: A systematic review. *NPJ Precision Oncology*, *7*(1), 83. doi:10.1038/s41698-023-00432-6

Bukłaho, P. A., Kiśluk, J., & Nikliński, J. (2023). Diagnostics and treatment of ovarian cancer in the era of precision medicine-opportunities and challenges. *Frontiers in Oncology*, *13*, 1227657. doi:10.3389/fonc.2023.1227657

Charkhchi, P., Cybulski, C., Gronwald, J., Wong, F. O., Narod, S. A., & Akbari, M. R. (2020). CA125 and ovarian cancer: A comprehensive review. *Cancers (Basel)*, *12*(12), 3730. doi:10.3390/cancers12123730

Damiani, C., Kalliatakis, G., Sreenivas, M., Al-Attar, M., Rose, J., Pudney, C., Lane, E. F., Cuzick, J., Montana, G., & Brentnall, A. R. (2023). Evaluation of an AI model to assess future breast cancer risk. *Radiology*, *307*(5), e222679. doi:10.1148/radiol.222679

Degu, A., Karimi, P. N., Opanga, S. A., & Nyamu, D. G. (2023). Survival outcomes among colorectal cancer patients at Kenyatta National Hospital: A retrospective cohort study. *Cancer Reports*, *6*(3), e1743. doi:10.1002/cnr2.1743

Dewi, C., Fristiohady, A., Amalia, R., Khairul Ikram, N. K., Ibrahim, S., & Muchtaridi, M. (2022). Signaling pathways and natural compounds in triple-negative breast cancer cell line. *Molecules (Basel, Switzerland)*, *27*(12), 3661. doi:10.3390/molecules27123661

Elorriaga, M. Á., Neyro, J. L., Mieza, J., Cristóbal, I., & Llueca, A. (2021). Biomarkers in ovarian pathology: From screening to diagnosis. Review of the literature. *Journal of Personalized Medicine*, *11*(11), 1115. doi:10.3390/jpm11111115

Falzone, L., Scandurra, G., Lombardo, V., Gattuso, G., Lavoro, A., Distefano, A. B., Scibilia, G., & Scollo, P. (2021). A multidisciplinary approach remains the best strategy to improve and strengthen the management of ovarian cancer. *International Journal of Oncology*, *59*(1), 1–14. doi:10.3892/ijo.2021.5233

Fitriani, N., Sutrisna, B., & Pelupessy, N. U. (2020). Survival Analysis of Ovarian Cancer Patients based on Age at The Dr. Wahidin Sudirohusodo Makassar Hospital in 2014-2018. *Indian Journal of Public Health Research & Development*, *11*, 1397.

Flasarova, D., Urban, K., Strouhal, O., Klos, D., Lemstrova, R., Dvorak, P., Soucek, P., & Mohelnikova-Duchonova, B. (2023). DNA Repair Pathway in Ovarian Cancer Patients Treated with HIPEC. *International Journal of Molecular Sciences*, *24*(10), 8868. doi:10.3390/ijms24108868

FontaineC.De BrakeleerS.TeugelsE.RenardV.Van den BulkH.VuylstekeP.De GrèveJ. (2022). Heterozygote germline mutations in Homologous Recombination core genes can predict for pathologic complete response in early triple negative breast cancer. doi:10.21203/rs.3.rs-1584056/v1

Furrer, D., Gregoire, J., Turcotte, S., Plante, M., Bachvarov, D., Trudel, D., Têtu, B., Douville, P., & Bairati, I. (2019). Performance of preoperative plasma tumor markers HE4 and CA125 in predicting ovarian cancer mortality in women with epithelial ovarian cancer. *PLoS One*, *14*(6), e0218621. doi:10.1371/journal.pone.0218621

Gaitskell, K., Hermon, C., Barnes, I., Pirie, K., Floud, S., Green, J., Beral, V., & Reeves, G. K.Million Women Study Collaborators. (2022). Ovarian cancer survival by stage, histotype, and pre-diagnostic lifestyle factors, in the prospective UK Million Women Study. *Cancer Epidemiology*, *76*, 102074. doi:10.1016/j.canep.2021.102074

Gaona-Luviano, P., Medina-Gaona, L. A., & Magaña-Pérez, K. (2020). Epidemiology of ovarian cancer. *Chinese Clinical Oncology*, *9*(4), 47. doi:10.21037/cco-20-34

Garnier, H., Murawski, M., Jastrzebski, T., Pawinska-Wasikowska, K., Balwierz, W., Sinacka, K., Gorecki, W., Izycka-Swieszewska, E., & Czauderna, P. (2021). Case report: cytoreductive surgery and hyperthermic intraperitoneal chemotherapy application in intraperitoneally disseminated inflammatory myofibroblastic tumor and in the youngest patient in the world: new indication and modification of technique. *Frontiers in Surgery*, *8*, 746700. doi:10.3389/fsurg.2021.746700

Gondhowiardjo, S., Christina, N., Ganapati, N. P., Hawariy, S., Radityamurti, F., Jayalie, V. F., Octavianus, S., Prawira Putra, A., Sekarutami, S. M., Prajogi, G. B., Giselvania, A., Adham, M., Hamid, A. R. A. H., Widyastuti, E., Prabowo, Y., Anindhita, T., Purwoto, G., Aman, R. A., Siregar, T. P., & Agianda, F. (2021). Five-year cancer epidemiology at the national referral hospital: Hospital-based cancer registry data in Indonesia. *JCO Global Oncology*, *5*(1), 190–203. doi:10.1200/GO.20.00155

Gumilang, T. J., & Hernawan, B. (2023). The Benefits of Physical Activity to Reduce Mortality in Postmenopausal Breast Cancer Patients: A Literature Review. *Indonesian Journal of Cancer*, *17*(1), 73–84. doi:10.33371/ijoc.v17i1.917

Gumusoglu-Acar, E., & Gunel, T. (2022). Ovarian Cancer Biomarkers. In *Advances in Diagnosis and Management of Ovarian Cancer* (pp. 27–42). Springer International Publishing. doi:10.1007/978-3-031-09169-8_3

How, J. A., Jazaeri, A. A., Fellman, B., Daniels, M. S., Penn, S., Solimeno, C., Yuan, Y., Schmeler, K., Lanchbury, J. S., Timms, K., Lu, K. H., & Yates, M. S. (2021). Modification of homologous recombination deficiency score threshold and association with long-term survival in epithelial ovarian cancer. *Cancers (Basel)*, *13*(5), 946. doi:10.3390/cancers13050946

Kepenekian, V., Péron, J., You, B., Bonnefoy, I., Villeneuve, L., Alyami, M., Bakrin, N., Rousset, P., Benzerdjeb, N., & Glehen, O. (2022). Non-resectable malignant peritoneal mesothelioma treated with pressurized intraperitoneal aerosol chemotherapy (PIPAC) plus systemic chemotherapy could lead to secondary complete cytoreductive surgery: A cohort study. *Annals of Surgical Oncology*, *29*(3), 1–10. doi:10.1245/s10434-021-10983-2

Kontizas, E., Tastsoglou, S., Karamitros, T., Karayiannis, Y., Kollia, P., Hatzigeorgiou, A. G., & Sgouras, D. N. (2020). Impact of Helicobacter pylori infection and its major virulence factor CagA on DNA damage repair. *Microorganisms*, *8*(12), 2007. doi:10.3390/microorganisms8122007

Lacaze, P., Marquina, C., Tiller, J., Brotchie, A., Kang, Y. J., Merritt, M. A., Green, R. C., Watts, G. F., Nowak, K. J., Manchanda, R., Canfell, K., James, P., Winship, I., McNeil, J. J., & Ademi, Z. (2023). Combined population genomic screening for three high-risk conditions in Australia: A modelling study. *EClinicalMedicine*, *66*, 102297.. doi:10.1016/j.eclinm.2023.102297

Leigh, N., Solomon, D., Pletcher, E., Labow, D. M., Magge, D. R., Sarpel, U., & Golas, B. J. (2020). Is cytoreductive surgery and hyperthermic intraperitoneal chemotherapy indicated in hepatobiliary malignancies? *World Journal of Surgical Oncology*, *18*(1), 1–8. doi:10.1186/s12957-020-01898-5

Li, J., Sun, H., Huang, Y., Wang, Y., Liu, Y., & Chen, X. (2019). Pathways and assays for DNA double-strand break repair by homologous recombination. *Acta Biochimica et Biophysica Sinica*, *51*(9), 879–889. doi:10.1093/abbs/gmz076

Liberto, J. M., Chen, S. Y., Shih, I. M., Wang, T. H., Wang, T. L., & Pisanic, T. R. II. (2022). Current and emerging methods for ovarian cancer screening and diagnostics: A comprehensive review. *Cancers (Basel)*, *14*(12), 2885. doi:10.3390/cancers14122885

Llueca, A., Climent, M. T., Escrig, J., Carrasco, P., Serra, A., Gomez-Quiles, L., Játiva, R., Cebrian, G., Bosso, V., Villarin, A., Maiocchi, K., Delgado-Barriga, K., Rodrigo-Aliaga, M., Ruiz, N., Herrero, C., Frances, A., Beato, I., Ferrer, C., Aracil, J. P., & Adell, R. (2021). Validation of three predictive models for suboptimal cytoreductive surgery in advanced ovarian cancer. *Scientific Reports*, *11*(1), 8111. doi:10.1038/s41598-021-86928-2

Maleki, Z., Vali, M., Nikbakht, H. A., Hassanipour, S., Kouhi, A., Sedighi, S., Farokhi, R., & Ghaem, H. (2023). Survival rate of ovarian cancer in Asian countries: A systematic review and meta-analysis. *BMC Cancer*, *23*(1), 1–11. doi:10.1186/s12885-023-11041-8

Matsas, A., Stefanoudakis, D., Troupis, T., Kontzoglou, K., Eleftheriades, M., Christopoulos, P., Panoskaltsis, T., Stamoula, E., & Iliopoulos, D. C. (2023). Tumor Markers and Their Diagnostic Significance in Ovarian Cancer. *Life (Basel, Switzerland)*, *13*(8), 1689. doi:10.3390/life13081689

McPherson, K. S., & Korzhnev, D. M. (2021). Targeting protein-protein interactions in the DNA damage response pathways for cancer chemotherapy. *RSC Chemical Biology*, *2*(4), 1167–1195. doi:10.1039/D1CB00101A

Mijwil, M. M., & Abttan, R. A. (2021). Utilisation of machine learning techniques in testing and training of different medical datasets. *Asian Journal of Computer and Information Systems*, *9*(4). doi:10.24203/ajcis.v9i4.6765

Olatunji, G., Aderinto, N., Adefusi, T., Kokori, E., Akinmoju, O., Yusuf, I., Olusakin, T., & Muzammil, M. A. (2023). Efficacy of tumour-treating fields therapy in recurrent glioblastoma: A narrative review of current evidence. *Medicine*, *102*(48), e36421. doi:10.1097/MD.0000000000036421

Paoletti, X., Lewsley, L. A., Daniele, G., Cook, A., Yanaihara, N., Tinker, A., Kristensen, G., Ottevanger, P. B., Aravantinos, G., Miller, A., Boere, I. A., Fruscio, R., Reyners, A. K. L., Pujade-Lauraine, E., Harkin, A., Pignata, S., Kagimura, T., Welch, S., Paul, J., & Glasspool, R. M. (2020). Assessment of progression-free survival as a surrogate end point of overall survival in first-line treatment of ovarian cancer: A systematic review and meta-analysis. *JAMA Network Open*, *3*(1), e1918939–e1918939. doi:10.1001/jamanetworkopen.2019.18939

Pasqual, E. M., Londero, A. P., Robella, M., Tonello, M., Sommariva, A., De Simone, M., Bacchetti, S., Baiocchi, G., Asero, S., Coccolini, F., De Cian, F., Guaglio, M., Cinquegrana, A., Cenzi, C., Scaringi, S., & Macrì, A. (2023). Repeated Cytoreduction Combined with Hyperthermic Intraperitoneal Chemotherapy (HIPEC) in Selected Patients Affected by Peritoneal Metastases: Italian PSM Oncoteam Evidence. *Cancers (Basel)*, *15*(3), 607. doi:10.3390/cancers15030607

Poort, H., Fenton, A. T., Thompson, E., Dinardo, M. M., Liu, J. F., Arch, J. J., & Wright, A. A. (2021). Lived experiences of women reporting fatigue during PARP inhibitor maintenance treatment for advanced ovarian cancer: A qualitative study. *Gynecologic Oncology*, *160*(1), 227–233. doi:10.1016/j.ygyno.2020.10.034

Puspitaningtyas, H., Espressivo, A., Hutajulu, S. H., Fuad, A., & Allsop, M. J. (2021). Mapping and visualization of cancer research in Indonesia: A scientometric analysis. *Cancer Control*, *28*, 10732748211053464. doi:10.1177/10732748211053464

Reid, F. (2020). World Ovarian Cancer Coalition Atlas. 2020. *Global Trends in Incidence, Mortality, and Survival*, 1-42.

Roy, S., Das, A., Vernekar, M., Mandal, S., & Chatterjee, N. (2022). Understanding the correlation between metabolic regulator SIRT1 and exosomes with CA-125 in ovarian cancer: A clinicopathological study. *BioMed Research International*, *2022*, 1–16. doi:10.1155/2022/5346091

Ruiz-Aguilar, J. J., Turias, I., González-Enrique, J., Urda, D., & Elizondo, D. (2021). A permutation entropy-based EMD-ANN forecasting ensemble approach for wind speed prediction. *Neural Computing & Applications*, *33*(7), 2369–2391. doi:10.1007/s00521-020-05141-w

Sahoo, S. K., & Goswami, S. S. (2023). A comprehensive review of multiple criteria decision-making (MCDM) Methods: Advancements, applications, and future directions. *Decision Making Advances*, *1*(1), 25–48. doi:10.31181/dma1120237

Shu, X., & Ye, Y. (2023). Knowledge Discovery: Methods from data mining and machine learning. *Social Science Research*, *110*, 102817. doi:10.1016/j.ssresearch.2022.102817

Si, A., Das, S., & Kar, S. (2021). Picture fuzzy set-based decision-making approach using Dempster-Shafer theory of evidence and grey relation analysis and its application in COVID-19 medicine selection. *Soft Computing*, 1–15. doi:10.1007/s00500-021-05909-9

Singh, D., Vignat, J., Lorenzoni, V., Eslahi, M., Ginsburg, O., Lauby-Secretan, B., Arbyn, M., Basu, P., Bray, F., & Vaccarella, S. (2023). Global estimates of incidence and mortality of cervical cancer in 2020: A baseline analysis of the WHO Global Cervical Cancer Elimination Initiative. *The Lancet. Global Health*, *11*(2), e197–e206. doi:10.1016/S2214-109X(22)00501-0

Stewart, C., Ralyea, C., & Lockwood, S. (2019, April). Ovarian cancer: an integrated review. In Seminars in oncology nursing (Vol. 35, No. 2, pp. 151-156). WB Saunders. https://doi.org/ doi:10.1016/j.soncn.2019.02.001

Trelis, A., Lago, V., Padilla, P., Martínez, R. P., Ripollés, V. B., Matute, L., & Domingo, S. (2023). Optimal cytoreduction: Is a CT's picture worth a surgeon's word? *Surgical Oncology*, *101948*, 101948. Advance online publication. doi:10.1016/j.suronc.2023.101948

Wang, Y., Zhang, M., Chen, Y., & Qu, H. (2022). Signed neuron with memory: Towards simple, accurate and high-efficient ann-snn conversion. In *International Joint Conference on Artificial Intelligence*. IEEE. 10.24963/ijcai.2022/347

Winarto, H., Welladatika, A., Habiburrahman, M., Purwoto, G., Kusuma, F., Utami, T. W., & Nuryanto, K. H. (2022). Overall survival and related factors of advanced-stage epithelial ovarian cancer patients underwent debulking surgery in Jakarta, Indonesia: A single-center experience. *Open Access Macedonian Journal of Medical Sciences*, *10*(B), 265-280. https://doi.org/ doi:10.3889/oamjms.2022.8296

Worden, K., Tsialiamanis, G., Cross, E. J., & Rogers, T. J. (2023). Artificial neural networks. In *Machine Learning in Modeling and Simulation: Methods and Applications* (pp. 85–119). Springer International Publishing. doi:10.1007/978-3-031-36644-4_2

Yurttas, C., Überrück, L., Nadiradze, G., Königsrainer, A., & Horvath, P. (2022). Limitations of laparoscopy to assess the peritoneal cancer index and eligibility for cytoreductive surgery with HIPEC in peritoneal metastasis. *Langenbeck's Archives of Surgery*, *407*(4), 1667–1675. doi:10.1007/s00423-022-02455-2

Zaidi, M. Y., Lee, R. M., Gamboa, A. C., Speegle, S., Cloyd, J. M., Kimbrough, C., Grotz, T., Leiting, J., Fournier, K., Lee, A. J., Dineen, S., Dessureault, S., Kelly, K. J., Kotha, N. V., Clarke, C., Gamblin, T. C., Patel, S. H., Lee, T. C., Hendrix, R. J., & Staley, C. A. III. (2020). Preoperative risk score for predicting incomplete cytoreduction: A 12-institution study from the US HIPEC collaborative. *Annals of Surgical Oncology*, *27*(1), 156–164. doi:10.1245/s10434-019-07626-y

Zhao, S., Wei, C., Tang, H., Ding, H., Han, B., Chen, S., Song, X., Gu, Q., Zhang, Y., Liu, W., & Wang, J. (2021). Elevated DNA polymerase delta 1 expression correlates with tumor progression and immunosuppressive tumor microenvironment in hepatocellular carcinoma. *Frontiers in Oncology*, *11*, 736363. doi:10.3389/fonc.2021.736363

Zhuhadar, L. P., & Lytras, M. D. (2023). The Application of AutoML Techniques in Diabetes Diagnosis: Current Approaches, Performance, and Future Directions. *Sustainability (Basel)*, *15*(18), 13484. doi:10.3390/su151813484

ADDITIONAL READINGS

Komalasari, R. (2022). A Social Ecological Model (SEM) to Manage Methadone Programmes in Prisons. In Handbook of Research on Mathematical Modeling for Smart Healthcare Systems (pp. 374-382). IGI Global. doi:10.4018/978-1-6684-4580-8.ch020

Komalasari, R. (2023). The Relationship Between Cybersecurity and Public Health. In *Handbook of Research on Current Trends in Cybersecurity and Educational Technology* (pp. 78–91). IGI Global., doi:10.4018/978-1-6684-6092-4.ch005

Komalasari, R. (2023). Postnatal Mental Distress: Exploring the Experiences of Mothers Navigating the Healthcare System. In Perspectives and Considerations on Navigating the Mental Healthcare System (pp. 159-181). IGI Global. doi:10.4018/978-1-6684-5049-9.ch007

Komalasari, R. (2023). Designing Health Systems for Better, Faster, and Less Expensive Treatment. In *Exploring the Convergence of Computer and Medical Science Through Cloud Healthcare* (pp. 1–13). IGI Global. doi:10.4018/978-1-6684-5260-8.ch001

Komalasari, R. (2023). The Ethical Consideration of Using Artificial Intelligence (AI) in Medicine. In Advanced Bioinspiration Methods for Healthcare Standards, Policies, and Reform (pp. 1-16). IGI Global. doi:10.4018/978-1-6684-5656-9.ch001

Komalasari, R. (2023). Healthcare for the Elderly With Digital Twins. In Digital Twins and Healthcare: Trends, Techniques, and Challenges (pp. 145-156). IGI Global. doi:10.4018/978-1-6684-5925-6.ch010

Komalasari, R. (2023). Cloud Computing's Usage in Healthcare. In Recent Advancements in Smart Remote Patient Monitoring, Wearable Devices, and Diagnostics Systems (pp. 183-194). IGI Global. doi:10.4018/978-1-6684-6434-2.ch009

Komalasari, R. (2023). History and Legislative Changes Governing Medical Cannabis in Indonesia. In Medical Cannabis and the Effects of Cannabinoids on Fighting Cancer, Multiple Sclerosis, Epilepsy, Parkinson's, and Other Neurodegenerative Diseases (pp. 274-284). IGI Global. doi:10.4018/978-1-6684-5652-1.ch012

Komalasari, R. (2023). Ambient Assisted Living (AAL) Systems to Help Older People. In Exploring Future Opportunities of Brain-Inspired Artificial Intelligence (pp. 84-99). IGI Global. doi:10.4018/978-1-6684-6980-4.ch006

Komalasari, R. (2023). Telemedicine in Pandemic Times in Indonesia: Healthcare Professional's Perspective. In Health Informatics and Patient Safety in Times of Crisis (pp. 138-153). IGI Global. doi:10.4018/978-1-6684-5499-2.ch008

Komalasari, R. (2023). Cloud Computing's Usage in Healthcare. In Recent Advancements in Smart Remote Patient Monitoring, Wearable Devices, and Diagnostics Systems (pp. 183-194). IGI Global. doi:10.4018/978-1-6684-6434-2.ch009

Komalasari, R. (2023). The Relationship Between Cybersecurity and Public Health. In *Handbook of Research on Current Trends in Cybersecurity and Educational Technology* (pp. 78–91). IGI Global., doi:10.4018/978-1-6684-6092-4.ch005

Komalasari, R. (2023). Treatment of Menstrual Discomfort in Young Women and a Cognitive Behavior Therapy (CBT) Program. In Perspectives on Coping Strategies for Menstrual and Premenstrual Distress (pp. 194-211). IGI Global. doi:10.4018/978-1-6684-5088-8.ch011

Komalasari, R. (2023). Healthcare for the Elderly With Digital Twins. In Digital Twins and Healthcare: Trends, Techniques, and Challenges (pp. 145-156). IGI Global. doi:10.4018/978-1-6684-5925-6.ch010

Komalasari, R. (2023). Telemedicine in Pandemic Times in Indonesia: Healthcare Professional's Perspective. In Health Informatics and Patient Safety in Times of Crisis (pp. 138-153). IGI Global. doi:10.4018/978-1-6684-5499-2.ch008

Komalasari, R. (2023). Regulatory Shift Healthcare Applications in Industry 5.0. In Advanced Research and Real-World Applications of Industry 5.0 (pp. 149-165). IGI Global. doi:10.4018/978-1-7998-8805-5.ch008

Komalasari, R. (2023). Exploring Pedagogies of Affect in Secondary School Physical Education for Enhanced Emotional Well-Being. *Pedagogik: Jurnal Pendidikan Guru Sekolah Dasar*, *11*(2), 117–127. doi:10.33558/pedagogik.v11i2.7374

Komalasari, R. (2023). Integrating sport education model and the athletics challenges approach for transformative physical education in Indonesian Middle Schools. *Motion: Jurnal Riset Physical Education*, *13*(2), 118–135. doi:10.33558/motion.v13i2.7372

Komalasari, R. (2023). Efek Ganja Medis pada Pasien Parkinson: A Literature Review of Clinical Evidence. *Journal of Islamic Pharmacy*, *8*(1), 44–48. doi:10.18860/jip.v8i1.17832

Komalasari, R. (2023). Rancangan Program Bagi Keberhasilan Gerakan Mencuci Tangan Yang Sehat Bagi Para Ibu Dan Anak. *Sulolipu: Media Komunikasi Sivitas Akademika dan Masyarakat*, *23*(1), 55-60. doi:10.32382/sulolipu.v23i1.3103

Komalasari, R. (2024). Unravelling the Veil: Exploring the Nexus of Insecure Attachment and Functional Somatic Disorders in Adults. In *Discourses, Inquiries, and Case Studies in Healthcare, Social Sciences, and Technology*. IGI Global.

Komalasari, R. (2024). Medical Educators (ME) and their Continuous Professional Development (CPD). In The Lifelong Learning Journey of Health Professionals: Continuing Education and Professional Development. IGI Global.

Komalasari, R. (2024). Advancing Healthcare: Economic Implications of Immediate MRI in Suspected Scaphoid Fractures - A Comprehensive Exploration. In *Future of AI in Medical Imaging*. IGI Global.

Komalasari, R. (2024). Women's perceptions of midlife mothering during perimenopause: the impact on health and well-being through life's transitions. In *Utilizing AI Techniques for the Perimenopause to Menopause Transition*. IGI Global.

Komalasari, R. (2024). Machine Learning in Health Information Security: Unraveling Patterns, Concealing Secrets, and Mitigating Vulnerabilities. In *Enhancing Steganography Through Deep Learning Approaches*. IGI Global.

Komalasari, R. (2024). Navigating Complexity: Unraveling the IVHM Requirements Puzzle in Unmanned Aerial Systems Through Innovative IVHM-RD Methodology and Rigorous Data Analysis. In *Ubiquitous Computing and Technological Innovation for Universal Healthcare*. IGI Global.

Komalasari, R. (2024). Trust Dynamics in Remote Patient-Expert Communication: Unraveling the Role of ICT in Indonesia's Private Healthcare Sector. In *Improving Security, Privacy, and Connectivity Among Telemedicine Platforms*. IGI Global.

Komalasari, R. (2024). Cyberbullying in the Healthcare Workplace: How to Find Your Way Through the Digital Maze. In M. Aslam, Y. Kim, & Q. Linchao (Eds.), *Workplace Cyberbullying and Behavior in Health Professions* (pp. 84–112). IGI Global., doi:10.4018/979-8-3693-1139-4.ch004

Komalasari, R. (2024). Harmonizing Minds and Machines: A Transformative AI-Based Mental Health Care Framework in Medical Tourism. In *Impact of AI and Robotics on the Medical Tourism Industry*. IGI Global.

Komalasari, R. (2024). Navigating the Future of Healthcare: A User-Centric Approach to Designing Lucrative Business Models for the IoMT. In *Lightweight Digital Trust Architectures in the Internet of Medical Things (IoMT)*. IGI Global.

Komalasari, R. (2024). Biospheric Reverie: Unraveling Indoor Air Quality through Bio-Inspired Textiles, Awareness, and Decision-Making. In *Intelligent Decision Making Through Bio-Inspired Optimization*. IGI Global.

Komalasari, R. (2024). An Integrated Approach to Next-Generation Telemedicine and Health Advice Systems through AI Applications in Disease Diagnosis. In *AI-Driven Innovations in Digital Healthcare: Emerging Trends, Challenges, and Applications*. IGI Global.

Komalasari, R. (2024). Transformative Insights: Harnessing Artificial Intelligence for Enhanced Ovarian Cancer Prediction and Prognosis. In *Biomedical Research Developments for Improved Healthcare*. IGI Global.

Komalasari, R. (2024). Bridging the Gap: Theory of Change Guided Digital Health Implementation in Indonesian Primary Care. In *Analyzing Current Digital Healthcare Trends Using Social Networks*. IGI Global.

Komalasari, R. (2024). Navigating the Digital Frontier: A Socio-Technical Review of Indonesia's NHS E-Health Strategy and the Path to Seamless Healthcare Transformation. In *Inclusivity and Accessibility in Digital Health*. IGI Global.

Komalasari, R. (2024). *Pursuit of Excellence: A Comprehensive Framework for Ensuring Quality in Digital Health Apps. In Multi-Sector Analysis of the Digital Healthcare Industry*. IGI Global.

Komalasari, R. (2024). Fostering Transparency in AI for Disease Detection: Exploring Meaningful Explanations and Interactive Modalities in Healthcare. In *Federated Learning and Privacy-Preserving in Healthcare AI*. IGI Global.

Komalasari, R. (2024). Predictive Pioneers: Bridging the Mental Health Gap in Older Adults through Advanced Digital Technologies. In *Using Machine Learning to Detect Emotions and Predict Human Psychology*. IGI Global.

Komalasari, R. (2024). Unravelling the Veil: Exploring the Nexus of Insecure Attachment and Functional Somatic Disorders in Adults. In *Discourses, Inquiries, and Case Studies in Healthcare, Social Sciences, and Technology*. IGI Global.

Komalasari, R. (2024). Cyberbullying in the Health Care Workplace: How to Find Your Way Through the Digital Maze. In *Workplace Cyberbullying and Behavior in Health Professions*. IGI Global. doi:10.4018/979-8-3693-1139-4.ch004

Komalasari, R. (2024). Navigating Mental Health Crises: Understanding Pivotal Phases, Health Campaigns, and Community Resilience for Recovery and Reform. In *The Role of Health Literacy in Major Healthcare Crises*. IGI Global.

Komalasari, R. (2024). Decoding the Genomic and Proteomic Landscape: Dissemination and Mechanisms of Antibiotic Resistance in Environmental Staphylococci. In *Contemporary Approaches to Mitigating Antibacterial Drug Resistance*. IGI Global.

Komalasari, R. (2024). Unveiling the Future: Blockchain-Powered Digital Twins for Personalized Privacy Preservation in Metaverse Healthcare Data. In D. Burrell (Ed.), *Innovations, Securities, and Case Studies Across Healthcare, Business, and Technology* (pp. 321–342). IGI Global., doi:10.4018/979-8-3693-1906-2.ch017

Komalasari, R. (2024). Safeguarding the Future: Cutting-Edge Time-Stamping Services and Blockchain Technologies for Maintaining EHR Data Integrity Over the Long Term. In *Blockchain and IoT Approaches for Secure Electronic Health Records (EHR)*. IGI Global.

Komalasari, R. (2024). Empowering Informed Healthcare Choices in Rural Areas: SIKDA - A Novel Model for Enhancing Patient Access to Reliable Consumer Health Information in Indonesia. In *Emerging Technologies for Health Literacy and Medical Practice*. IGI Global.

Komalasari, R. (2024). Advancing Trauma Care: Contemporary Cranio-Maxillo-Facial Surgery in Indonesia and Beyond. In *Contemporary Cranio-Maxillo-Facial Surgery: From Trauma to Reconstruction*. IGI Global.

Komalasari, R. (2024). Innovations in Neurodegenerative Disease Diagnosis: Unraveling the Transformative Potential of Deep Learning and MRI Integration. In *Deep Learning Approaches for Early Diagnosis of Neurodegenerative Diseases*. IGI Global.

Komalasari, R. (2024). Secure and privacy-preserving federated learning with explainable artificial intelligence for smart healthcare systems. In *Federated Learning and AI for Healthcare 5.0*. IGI Global.

Komalasari, R. (2024) Optimizing Prostate Cancer Radiotherapy: Advanced Machine Learning in Virtual Patient-Specific Plan Verification. *Indonesian Journal of Cancer.*

Komalasari, R. (2024). Bridging Uncertainties: Exploring the Lived Experiences of Pediatric Long-COVID-19 Through ME/CFS Caregiver Narratives. In G. Rao & S. Dhamdhere-Rao (Eds.), *Clinical Practice and Post-Infection Care for COVID-19 Patients* (pp. 207–227). IGI Global. doi:10.4018/978-1-6684-6855-5.ch009

Komalasari, R., & Mustafa, C. (2023). Combating International Cyber Conflict: A Healthy Just War and International Law Analysis of NATO and Indonesian Policies. Jurnal Pertahanan: Media Informasi tentang Kajian & Strategi Pertahanan yang Mengedepankan Identity. *Nasionalism & Integrity*, *9*(3), 559–570. doi:10.33172/jp.v9i2.16794

Komalasari, R., & Mustafa, C. (2023). Analysis Of the Propulsion and Maneuvering Characteristics of Autonomous Underwater Vehicles and Their Strategic Defense. *Propeller Jurnal Permesinan*, *1*(2).

Komalasari, R., & Mustafa, C. (2023). Amphibious Forces in the Total War Age: Exploring Indonesia's Multifaceted Contributions to Statecraft in the Asia-Pacific Region. *Jurnal Strategi Pertahanan Laut*, *9*(2), 62–77. doi:10.33172/spl.v9i2.12606

Komalasari, R., & Mustafa, C. (2023). The Evolution and Institutionalization of the Naval Partnership between Indonesia and ASEAN. *Jurnal Strategi Pertahanan Laut*, *9*(2), 1–18. doi:10.33172/spl.v9i2.12576

Komalasari, R., & Mustafa, C. (2023). A Healthy Game-Theoretic Evaluation of NATO and Indonesia's Policies in the Context of International Law. *Jurnal Pertahanan: Media Informasi tentang Kajian & Strategi Pertahanan yang Mengedepankan Identity, Nasionalism & Integrity*, *9*(2), 333–349. doi:10.33172/jp.v9i2.16794

Komalasari, R., & Mustafa, C. (2024). Strengthening National Security Through Integrated Farming Systems: A Comprehensive Assessment and Strategic Outlook. *Jurnal Pertanian Terpadu*, *1*(1), 19–33.

Komalasari, R., & Mustafa, C. (2024). Dimensi Sosial-Ekologi Pengelolaan Danau Tropis Berkelanjutan: Pelajaran Dari Danau Toba, Indonesia. *Wrasse Jurnal Perikanan dan Kelautan Nusantara*, *1*(1).

Komalasari, R., Nurhayati, N., & Mustafa, C. (2022). Enhancing the Online Learning Environment for Medical Education: Lessons From COVID-19. In Policies and procedures for the implementation of safe and healthy educational environments: Post-COVID-19 perspectives (pp. 138-154). IGI Global. doi:10.4018/978-1-7998-9297-7.ch009

Komalasari, R., Nurhayati, N., & Mustafa, C. (2022). Insider/outsider issues: Reflections on qualitative research. *The Qualitative Report*, *27*(3), 744–751. doi:10.46743/2160-3715/2022.5259

Komalasari, R., Nurhayati, N., & Mustafa, C. (2022). Professional Education and Training in Indonesia. In *Public Affairs Education and Training in the 21st Century* (pp. 125–138). IGI Global. doi:10.4018/978-1-7998-8243-5.ch008

Komalasari, R., Nurhayati, N., & Mustafa, C. (2022). Professional Education and Training in Indonesia. In *Public Affairs Education and Training in the 21st Century* (pp. 125–138). IGI Global. doi:10.4018/978-1-7998-8243-5.ch008

Komalasari, R., Wilson, S., & Haw, S. (2021). A systematic review of qualitative evidence on barriers to and facilitators of the implementation of opioid agonist treatment (OAT) programmes in prisons. *The International Journal on Drug Policy*, *87*, 102978. doi:10.1016/j.drugpo.2020.102978

Komalasari, R., Wilson, S., & Haw, S. (2021). A social ecological model (SEM) to exploring barriers of and facilitators to the implementation of opioid agonist treatment (OAT) programmes in prisons. *International Journal of Prisoner Health*, *17*(4), 477–496. doi:10.1108/IJPH-04-2020-0020

Komalasari, R., Wilson, S., Nasir, S., & Haw, S. (2020). Multiple burdens of stigma for prisoners participating in Opioid Antagonist Treatment (OAT) programmes in Indonesian prisons: A qualitative study. *International Journal of Prisoner Health*, *17*(2), 156–170. doi:10.1108/IJPH-03-2020-0018

Mustafa, C. (2020). The influence of sunni islamic values on rehabilitation as judicial decision for minor drug users in Indonesian court. *Ijtihad: Jurnal Wacana Hukum Islam dan Kemanusiaan*, *20*(1), 79-96. doi:10.18326/ijtihad.v20i1.79-96

Mustafa, C. (2021). Qualitative Method Used in Researching the Judiciary: Quality Assurance Steps to Enhance the Validity and Reliability of the Findings. *The Qualitative Report*, *26*(1), 176–185. doi:10.46743/2160-3715/2021.4319

Mustafa, C. (2021). Key Finding: Result of a Qualitative Study of Judicial Perspectives on the Sentencing of Minor Drug Offenders in Indonesia: Structural Inequality. *The Qualitative Report*, *26*(5), 1678–1692. doi:10.46743/2160-3715/2021.4436

Mustafa, C. (2021). The view of judicial activism and public legitimacy. *Crime, Law, and Social Change*, *76*(1), 23–34. doi:10.1007/s10611-021-09955-0

Mustafa, C. (2021). The News Media Representation of Acts of Mass Violence in Indonesia. In *Mitigating Mass Violence and Managing Threats in Contemporary Society* (pp. 127–140). IGI Global. doi:10.4018/978-1-7998-4957-5.ch008

Mustafa, C. (2021). The Challenges to Improving Public Services and Judicial Operations: A unique balance between pursuing justice and public service in Indonesia. In Handbook of research on global challenges for improving public services and government operations (pp. 117-132). IGI Global. doi:10.4018/978-1-7998-4978-0.ch007

Mustafa, C., Malloch, M., & Hamilton Smith, N. (2020). Judicial perspectives on the sentencing of minor drug offenders in Indonesia: Discretionary practice and compassionate approaches. *Crime, Law, and Social Change*, *74*(3), 297–313. doi:10.1007/s10611-020-09896-0

Suhariyanto, B. (2023). Contradiction Over the Application Of Corporate Liability In Corruption Court Decisions In Indonesia. *Indonesia Law Review*, *13*(1), 8.

Suhariyanto, B., Mustafa, C., & Santoso, T. (2021). Liability incorporate between transnational corruption cases Indonesia and the United States of America. *J. Legal Ethical & Regul. Isses*, *24*, 1.

KEY TERMS AND DEFINITIONS

Artificial Intelligence: A computer system is said to have artificial intelligence (AI) if it can learn and perform tasks typically associated with a human brain. Visual perception, speech recognition, decision-making, and language translation are all within the capabilities of these computers. Clinical decision-making, outcome forecasting, and complex data analysis are all areas where artificial intelligence (AI) is finding utility in healthcare.

Biomarkers: When it comes to biological processes, the existence of illness, or the response to therapy, there are measurable signs or chemicals in the body that may be used as biomarkers. Biomarkers are characteristics of a disease that may be measured chemically, genetically, or molecularly to help researchers understand the nature and course of cancer.

Ovarian Cancer: This malignancy develops in the female reproductive organs that lay the eggs. The fact that symptoms can not appear until the illness has progressed significantly has earned it the nickname "silent killer" among medical professionals. Because of its high death rates and difficulty in diagnosing, ovarian cancer is a significant public health issue and the subject of much study.

Predictive Models: The term "predictive model" refers to a class of mathematical and computer tools that may foretell the future by analyzing past data and using statistical techniques. Using pertinent clinical data and characteristics, predictive models are used in healthcare to foretell the course of illness, the efficacy of treatments, or the final results for patients.

Treatment Personalization: Treatment customization refers to the process of tailoring medical interventions, such as medications and therapies, to the specific characteristics of each patient. Personalization of cancer treatment ensures optimal efficacy with minimal side effects by considering factors such as genetic makeup, biomarker profiles, and disease characteristics.

Chapter 11
Nanoparticles for the Prevention and Treatment of Bacterial Biofilms on Orthopedic Implants

Priyanka Prabhu
https://orcid.org/0000-0001-6163-4128
Shobhaben Pratapbhai Patel School of Pharmacy and Technology Management, SVKM's NMIMS, Mumbai, India

Vishvesh Joshi
https://orcid.org/0000-0003-3744-1507
Chartwell Pharmaceuticals LLC, USA

ABSTRACT

Orthopedic implants are a boon for devastating bone diseases such as osteoarthritis and bone fractures. However, implantation of an orthopedic device is associated with the risk of bacterial biofilm formation on the surface of the implant. Treating biofilm-associated infections is extremely challenging owing to the acidic environment, enzymatic degradation, presence of sessile bacterial cells, and hypoxic environment in the biofilm. Biofilms are highly resistant to both antibacterials and the human immune system. Present antibiotic therapies have limited success due to low blood flow, high density, and poor permeability of the bone causing insufficient bone permeation of the antibiotic. Nanoparticles have a high surface-to-volume ratio and superior penetration across cell membranes, and their nanostructure provides flexibility in designing effective strategies to tackle biofilm-associated infection. This chapter summarizes the various research endeavours in which nanoparticles have been explored for the prevention and or/treatment of bacterial biofilms on orthopedic implants.

DOI: 10.4018/979-8-3693-1922-2.ch011

INTRODUCTION

Orthopedic implants have indubitably enhanced the quality of life for patients with incapacitating bone diseases such as osteoarthritis and bone fractures (Connaughton et al., 2014). Recent times have witnessed substantial advances in the design of orthopedic implants with a focus on enhancing their performance and durability (Caplin & García, 2019). It is estimated that the global market size of orthopedic implants will reach 79.5 billion USD by 2030 (Wu et al., 2022). However, implantation is associated with a risk of infection due to bacterial adherence on the surface of implants (Abouaitah et al., 2021). This is attributed to the fact that ~ 10,000 times lower bacterial load is necessary to colonize an implant as compared to native tissue (Khatoon et al., 2018). The risk of implant-associated infection is enhanced in patients having a compromised immune system (Abouaitah et al., 2021). The rate of implant-associated infections also depends on the condition and lifestyle of the patient such as patients having diabetes mellitus, obesity, smokers, etc. (Billings & Anderson, 2022).

Biofilms are defined as organized clusters of bacteria enclosed in a self-generated matrix. The composition of this matrix varies with the type of species, strain, and environmental conditions. These biofilm bacteria are resistant to conventional antibiotics, immune mechanisms of destruction such as phagocytosis, and adaptive immune responses (Arciola et al., 2012). It is imperative to preclude bacterial adherence and propagation at 6 h after implantation to achieve an aseptic environment at the site of implantation for a longer time, as bacterial adherence at this stage can culminate in biofilm formation on the implant surface (Pawar et al., 2018). Bacterial biofilm-associated infections account for 65–80% of orthopedic implant-associated infections (Wei et al., 2022). Thus, biofilm formation on orthopedic implants remains a grave concern threatening their clinical success (Caplin & García, 2019).

The skin barrier is disrupted during the surgical procedure of implantation and the incorporation of foreign material in the body renders the body vulnerable to infection (Connaughton et al., 2014). Post-implantation there is a race between the host cells and the microorganisms to adhere and colonize the implantation area (Chopra et al., 2021). Implantation of an orthopedic device results in the formation of an immunosuppressed environment termed the immune-competent fibroinflammatory zone which triggers bacterial infection and biofilm formation. Micromovement of the orthopedic device within the host tissue causes suppression of the immune response to bacterial infection. When the orthopedic device moves within the tissue, it releases fragments detrimental to the neighbouring tissue further exacerbating the bacterial infection (Moure et al., 2017).

Staphylococcus, *Pseudomonas,* and *Enterobacteriaceae* are the notorious bacteria commonly responsible for serious implant-associated bacterial infections such as prosthetic joint infections. These bacteria are capable of biofilm formation on the surface of implants (Abouaitah et al., 2021). *Staphylococcus aureus* and *Staphylococcus epidermidis* account for almost 70% of orthopedic implant infections and *Pseudomonas aeruginosa* accounts for 8% of orthopedic implant infections (Raphel et al., 2016).

Implant-related factors influencing the rate of infection include surface hydrophilicity, surface porosity, surface charge, surface energy, and composition of the implanted material (Chopra et al., 2021; Khatoon et al., 2018). Bacteria-related factors influencing adhesion onto implants include the type of pathogen whether Gram + ve or Gram –ve, genus/species, and shape (Chopra et al., 2021). Bacterial adhesion is hindered on smooth implant surfaces and surface imperfections or roughness encourage bacterial adhesion (Rodríguez-Merchán et al., 2021). Ceramic implants are reported to resist biofilm formation compared to other implant materials which is probably attributed to the nature of ceramic implants to undergo minimal surface degradation compared to their metal or polymeric counterparts, thus preserving the implant surface smoothness for a longer time (Rodríguez-Merchán et al., 2021).

Implant-associated infections are both a health concern and an enormous economic burden for low-income countries (Wu et al., 2022). Orthopedic implant-associated infections finally culminate in osteomyelitis with catastrophic consequences on bone and surrounding soft tissues (Arciola et al., 2012). Chronic implant-associated infections lead to implant mobilization and failure. Such cases require the substitution of the implant to exterminate the infection which is associated with a greater risk of relapse (Arciola et al., 2012). Osteomyelitis is an inflammatory condition that leads to bone destruction and gradual loss of bone. *Staphylococcus aureus* and *Staphylococcus epidermidis* are the main bacteria causing osteomyelitis (Billings & Anderson, 2022). Osteomyelitis is common in 2 to 5% of surgical interventions with internal fixation devices (Caplin & García, 2019).

Orthopedic implants are intended to have a functional period of 20 years in the patient and hence are at a potential risk of getting infected during that entire period. Hence, long-term strategies are desired to prevent infection of orthopedic implants (Raphel et al., 2016). Present therapeutic strategies for implant-associated bacterial infections involve long-lasting systemic antimicrobial therapy, corrective surgeries, and implant removal (Billings & Anderson, 2022). Systemic antimicrobial therapy with high doses causes toxicities such as hepatotoxicity and nephrotoxicity (Nie et al., 2022). Presently, antibiotic therapy has limited success owing to delayed diagnosis of orthopedic implant-associated infection; and low blood flow, high density, and poor permeability of the bone causing insufficient permeation of the antibiotic into the bone (Nie et al., 2022). There has been a trend of using antimicrobial drug-eluting implants over systemic antimicrobial agents. Bone cements containing antibiotics have been developed to prevent orthopedic implant-associated infections. However, these implants suffer from issues of burst release and only provide the release of antimicrobial agents for a short duration (Y. Wang et al., 2021).

Corrective surgery includes local debridement of affected tissues followed by removal of the initial implant and re-implantation. It is important to clear the infection before re-implanting the new implant. Corrective surgery increases the cost for the patient (Billings & Anderson, 2022). Corrective surgery may be done as a one-stage procedure or a two-stage procedure. The one-stage procedure, having a success rate of 80%, includes surgical removal of the first device followed by debridement and immediate reimplantation of a new device (Moure et al., 2017). Antibiotic-impregnated bone cements may be utilized during a single-stage surgical procedure. However, these allow limited loading of antibiotics, show burst release of the antibiotic, and decrease bone repair causing slow and painful patient recovery (Min et al., 2016). The two-stage procedure, having a success rate of > 90% includes surgical removal of the first implant and debridement. A temporary joint spacer is then placed and parenteral antibiotics are given to the patient before the implantation of a new device (Moure et al., 2017). However, the formation of bacterial biofilms on these joint spacers hinders sterility in the intraarticular space before the second stage of corrective surgery (Indelli et al., 2021).

Therefore, the need of the hour is to develop innovative strategies to prevent biofilm formation on orthopedic implants and also destroy the biofilm-associated bacteria in case of biofilm formation on orthopedic implants.

This chapter outlines the process of bacterial biofilm formation on orthopedic implants, highlights the challenges associated with the treatment of bacterial biofilm-associated orthopedic implant infections, and showcases the immense potential of nanoparticles in fighting the peril of bacterial biofilm formation on orthopedic implants.

PROCESS OF BIOFILM FORMATION ON ORTHOPEDIC IMPLANTS

The process of biofilm formation on orthopedic implants involves 4 steps:

Initiation

This step involves the initial passive attachment of the bacteria onto the implant surface fostered via hydrophobic, electrostatic, and van der Waals interactions. Certain proteins called autolysins are responsible for this attachment. The autolysin in the case of *Staphylococcus aureus* is AtlA and in the case of *Staphylococcus epidermidis* is AtlE (Arciola et al., 2012). Surface roughness and surface hydrophobicity have been found to play an important role in this step. Rough surfaces offer greater surface area and thus augment bacterial attachment. Bacteria have been found to adhere more easily to hydrophobic surfaces as compared to hydrophilic surfaces (Behbahani et al., 2022).

Accumulation

This step involves accumulation to form multiple layers and is actively mediated by Microbial Surface Components Recognizing Adhesive Matrix Molecules (MSCRAMMs) and intercellular bonding (Arciola et al., 2012).

Maturation

In this step, the biofilm further matures with the formation of various matrix components such as exopolysaccharides, proteins, and extracellular DNA (eDNA) (Arciola et al., 2012). The final structure is a 3-dimensional stack of bacterial cells. This stack has fluid-filled channels in its structure enabling the supply of nutrients to the bacteria which are deep-seated within the biofilm (Behbahani et al., 2022).

Dispersal

This step entails the detachment of bacterial cells from the biofilm matrix into a planktonic state (Arciola et al., 2012). The planktonic bacteria are ready to carry the infection further to newer sites (Behbahani et al., 2022).

CHALLENGES IN THE TREATMENT OF BIOFILM-ASSOCIATED ORTHOPEDIC IMPLANT INFECTIONS

Bacterial Resistance

The biggest challenge in the treatment of biofilm-associated implant infections is the resistance of the biofilm-forming bacteria (Behbahani et al., 2022). Sessile bacterial cells present in biofilms are 10- to 1000 times less vulnerable to antimicrobial therapy as compared to planktonic bacterial cells (Berini et al., 2022). Numerous mechanisms confer the biofilm-forming bacteria with resistance (Behbahani et al., 2022).

- Poor antimicrobial penetration into the biofilm matrix (Behbahani et al., 2022)
- Acidic microenvironment inside the biofilm: The environment inside the biofilm is acidic and oxygen-deficient which hampers the effectiveness of antibiotics (Dhar & Han, 2020).
- Enzymatic degradation of antimicrobial agents by the biofilm: β-lactamase produced by the bacteria degrades the antibiotics and precludes their ingress into the biofilm (Khatoon et al., 2018).
- Stress responses induced by biofilm-forming bacteria on encountering a challenge: These stress responses may cause the expression of efflux transporters which will expel the antimicrobial agent out of the bacterial cell. Stress responses could also push the bacterial cells into a dormant phase thus inactivating those antimicrobial agents that work against fast-growing bacteria (Behbahani et al., 2022).
- Generation of highly tolerant phenotype called persisters: These persisters are highly tolerant to antibiotics, stay in a dormant state, and get activated once the antibiotic therapy is terminated (Khatoon et al., 2018).
- Hypoxia in biofilms causes bacteria to undergo anaerobic metabolism thereby compromising antibiotic effectiveness (Khatoon et al., 2018).
- eDNA present in the exopolysaccharide of the biofilm chelates cations and renders the biofilm tolerant to zinc, copper, and lead (Khatoon et al., 2018).
- Fungal coinfection: Polymicrobial biofilms containing a mixture of bacteria and fungi may show enhanced antimicrobial resistance. The presence of *Candida albicans* along with *Staphylococcus epidermidis* was more resistant to vancomycin due to the fungal matrix acting as a barrier (Khatoon et al., 2018).
- Biofilm-associated bacteria can enhance the frequency of mutations to escape the host immune system, augment the activity of efflux pumps, and exchange plasmids for the transfer of genes for antimicrobial resistance (Q. Hong et al., 2021).
- The usage of systemic antimicrobial agents for an extended period results in systemic adverse effects and also enhances the potential for antimicrobial resistance (Billings & Anderson, 2022). Using antibiotics in large doses to tackle the menace of biofilms further contributes to bacterial resistance (Khatoon et al., 2018).

Immunosuppression

Biofilm-linked infections can tilt the primary macrophages (M0) towards an anti-inflammatory phenotype (M2) having reduced migration and compromised bactericidal ability, which impairs biofilm clearance and prolongs biofilm presence on the implant. The eDNA, which is a part of the extracellular polymeric substance (EPS) of the biofilm, plays a cardinal role in the formation, maturation, and structural stability of the biofilm. eDNA is responsible for the local innate immune response and bacterial resistance (Guo et al., 2020).

STRATEGIES TO TACKLE BIOFILM-ASSOCIATED IMPLANT INFECTIONS

As the famous phrase, 'Prevention is better than cure', it is advisable to prevent biofilm formation rather than its treatment, as the former approach is easier, safer, and more economical (Khatoon et al., 2018). However, it is difficult to completely prevent bacterial adhesion on the surface of implants. This necessitates the establishment of a strategy to tackle the bacteria adhered to the surface of the implant. This may be achieved by immobilizing antibacterial agents such as antibiotics and antimicrobial peptides on the implant surface to contact the adhered bacteria and kill them to prevent biofilm formation. Vancomycin is a wide-spectrum antibiotic that is generally immobilized onto the surface of orthopedic implants. It can be easily anchored onto the implant surface via C-terminal carboxylic acid since the acid is not responsible for antibacterial activity. If both prevention of bacterial adhesion and killing of adherent bacteria fail, then the adhered bacteria would eventually form a biofilm on the surface of the implant. In such cases, enzymes may be employed to destroy the biofilm matrix and help the antibiotic to get access to the bacteria. Two enzymes that break the biofilm matrix include dispersin B which degrades polysaccharides and deoxyribonuclease I (DNAse I) which breaks eDNA (M. Wang & Tang, 2019).

Biodegradable materials are more favourable than non-degradable metallic implants fabricated from titanium alloys and stainless steel as they preclude the removal of implants post-bone healing (Z. Lin et al., 2019). An ideal orthopedic implant is expected to prevent bacterial adhesion onto its surface to hinder biofilm formation and also facilitate the formation of new bone. Certain metals, polymers, and ceramics are known to exert both antibacterial and osteogenic activities. Copper containing stainless steel (317 L Cu SS) was found to possess antibacterial activity against *Staphylococcus aureus* and *Escherichia coli* and hamper biofilm formation. Further, it was also found to be osteogenic. Commonly used orthopedic implants may be endowed with antibacterial and osteogenic potential via techniques such as alloying, doping, etc. Cu^{2+} and Zn^{2+} are known to exert both antibacterial activity and osteogenic activity. Ag, Cu, and Zn are antibacterial. Cu, Si, and Mg are known to facilitate angiogenesis for new bone formation. However, controlled release of these metal ions from modified implants is essential to prevent damage to neighbouring tissues and organs. Surface topography of the orthopedic implant is also crucial to prevent bacterial adhesion and biofilm formation. It has been shown that bacterial adhesion and biofilm formation is facilitated by an increase in surface roughness (X. Lin et al., 2017).

A polymer brush strategy has also been used to prevent bacterial adhesion on the implant surface. Polyethylene glycol (PEG) was altered with cell-adhesive arginine-glycine-aspartic acid (RGD) ligand to recover interactions with osteoblasts and the modified coating was found to substantially decrease *Staphylococcus aureus* adhesion. Dextran was also used to coat titanium-based implants and was found to decrease the adhesion of *S. aureus* and *S. epidermidis* by around 50% compared to the uncoated titanium implant. However, the polymer brush technique is associated with the drawbacks of difficulty in achieving a uniform polymer coating and degradation of the polymer brush with time. Further, these polymer brush coatings can only inhibit bacterial adhesion and do not have any antimicrobial activity of their own (Raphel et al., 2016).

Antibiotic-releasing coatings may be applied on the surface of implants to destroy the bacteria. However, these suffer from the problem of the initial burst release of the antibiotic. Release of the antibiotic in high concentrations is detrimental to host cells as well as causes exhaustion of the antimicrobial agent thereby rendering the implant susceptible to infection. There is a need for the on-demand release of antimicrobials when the implant comes in contact with the bacterial load. This will help to curb antimicrobial resistance as well as prevent the exhaustion of the antimicrobial agent (Raphel et al., 2016).

NANOPARTICLES FOR THE PREVENTION AND TREATMENT OF BACTERIAL BIOFILMS ON ORTHOPEDIC IMPLANTS

Nanotechnology involves matter that has at least one of its dimensions in the size range of 1 to 100 nm (Gavaskar et al., 2018). Nanoparticles have promising properties of high surface-to-volume ratio and penetration across cell membranes as compared to their bulk counterparts (Mahamuni-Badiger et al., 2020).

Application of nanotechnology to the field of orthopedics is promising since the bone and its components such as hydroxyapatite and collagen fibrils are nano compounds. Nanoparticles have diverse applications in orthopedics such as surface coating to preclude biofilm formation and facilitate bone formation, controlled drug release to tackle implant-associated infections, etc. Nanotextured surfaces and nanoengineered implants enhance osteoblast activity. The decreased grain size due to the nanoengineering of implants translates into a huge surface area enabling better interactions between the host bone and the implant surface. This facilitates osseointegration and enhances the lifespan of the implant. Nanoscaffolds can also be fabricated to resorb over time (Gavaskar et al., 2018). The ultra-fine size facilitates penetration of bacterial cell membranes and the biofilm extracellular polysaccharide layer (Sahoo et al., 2022). Nanoparticles can be used to deliver both hydrophilic and hydrophobic antibacterial drugs at the site of infection (Indelli et al., 2021). Nanoparticles can provide controlled release of the drug thus overcoming the toxicity associated with the bolus release profile of antibiotics (Indelli et al., 2021). The chemical composition, size, and surface charge of nanoparticles can be tuned as per the need (Q. Hong et al., 2021). Smart nanoparticles can be designed to respond to certain triggers present inside the biofilm microenvironment (pH, enzymes) or external triggers (temperature, ultrasound, light, magnetic field) and release the antimicrobial at the required site at the required time (Q. Hong et al., 2021; Wu et al., 2022). Biomimetic nanoparticles such as nanowires, nanopillars, and nano blades can be designed to mimic the architecture of cicada and dragonfly wings and achieve inherent bacterial destruction (Wu et al., 2022).

Factors affecting nanoparticle performance include particle size and surface charge. Antibacterial activity is mostly inversely related to the particle size of the nanoparticle because a smaller size facilitates penetration across the bacterial cell membrane. Further, the release of the metal ion is quicker when the size of the metallic nanoparticles is small. Zinc oxide nanoparticles having a size < 10 nm were found to be more effective against *Staphylococcus aureus* as compared to their larger counterparts. Similarly, silver nanoparticles having a size of 7 nm showed lower minimum inhibitory concentration (MIC) than 29 nm and 89 nm silver nanoparticles (Berini et al., 2022). Positively charged nanoparticles are known to be more efficacious owing to their electrostatic bonding with negatively charged bacterial cell membranes and biofilms (Berini et al., 2022).

The usage of antibacterial orthopedic implants alone is associated with the risk of bacteria remaining in the bone. Bacteria race to the implant surface before the mesenchymal cells in the absence of an osteoinductive agent. An osteoinductive agent helps to promote osteoblast proliferation and ensures that bone-implant integration ensues before bacterial colonization of the implant surface. Incorporation of both antibacterial agents and osteoinductive growth factors within the implant surface is beneficial to avoid biofilm formation on implants (Min et al., 2016). Nanoparticles can load both antibacterial agents and osteoinductive growth factors within a single platform thereby facilitating osteointegration and infection control (Min et al., 2016).

The following subsections summarize the various research endeavours in which nanoparticles have been explored for the prevention and or/treatment of bacterial biofilms on orthopedic implants.

Use of Nanoparticles with Intrinsic Antibacterial Activity

These include nanoparticles such as silver, copper, and zinc oxide (Indelli et al., 2021). Advantages of nano silver include efficacy against *Escherichia coli* and *Staphylococcus aureus*, its anti-corrosive nature, and its ability to be incorporated into hydroxyapatite and chitosan to yield an antibacterial coating with osteo-integrative characteristics (Indelli et al., 2021). Silver nanoparticles have also been shown to act in synergy with a few antibiotics (Khatoon et al., 2018). Silver nanoparticles are also effective against methicillin-resistant *Staphylococcus aureus* (MRSA), methicillin-resistant *Staphylococcus epidermidis* (MRSE), and *Pseudomonas aeruginosa* (Sahoo et al., 2022). The positive charge of silver nanoparticles enables interaction between silver nanoparticles and the negatively charged bacterial cell membrane through electrostatic binding. Once they get attached to the bacterial cell membrane and enter the bacterial cell, they interact with ribosomes leading to ribosomal denaturation. Ribosomal denaturation hinders translation and protein synthesis. Silver nanoparticles also generate huge amounts of reactive oxygen species (ROS) and free radicals like hydrogen peroxide, superoxide anion, hydroxyl radical, and singlet oxygen. Silver nanoparticles further downregulate the expression of antioxidants like glutathione, superoxide dismutase, and catalase which promotes the accumulation of ROS (Qing et al., 2018). Silver nanoparticles bind to the bacterial cell membrane disturbing the cell membrane and causing the build-up of peroxides which oxidize the cell wall. Silver nanoparticles attack the respiratory chain of the bacteria and disrupt the bacterial cell via the generation of ROS (Khatoon et al., 2018). Silver nanoparticles are biocompatible since they can be phagocytosed by mammalian cells and processed via lysosomal fusion (Khatoon et al., 2018). Using metallic ions is also associated with decreased propensity to develop resistance (Y. Wang et al., 2021). The smaller the size of the silver nanoparticles, the greater the antimicrobial activity owing to the larger surface area enabling better interaction with the bacterial cell membrane (Qing et al., 2018). An important drawback of silver nanoparticles is that they are susceptible to agglomeration and thus require surface modification before usage (Mahamuni-Badiger et al., 2020).

Copper catalyzes the formation of ROS, and disrupts the bacterial cell membrane permeability causing leakage of reducing sugars and proteins from the bacterial cell (J. Liu et al., 2020). Copper is reported to possess antibacterial activity against *Staphylococcus aureus* and *Escherichia coli* which are the typical microbes usually responsible for bone infections (Zou et al., 2020). Copper has an affinity for sulfhydryl groups present in thiols of proteins and thus is bactericidal. Copper is less expensive compared to silver (Sahoo et al., 2022). Cu^{2+} is also known to facilitate angiogenesis and vascular maturation which promotes new bone formation (Wei et al., 2022). Copper is known to expedite bone regeneration via the maturation of collagen through crosslinking lysyl oxidase and by stimulating osteogenic differentiation of mesenchymal stem cells (Yan et al., 2020).

Zinc oxide nanoparticles have shown antibacterial and antibiofilm activity against *Pseudomonas aeruginosa*, *Streptococcus pneumoniae*, and *Escherichia coli* (Mahamuni-Badiger et al., 2020). Zinc oxide is a safe material recognized by the US FDA (Mahamuni-Badiger et al., 2020). The physical interaction of zinc oxide with the bacterial cell also contributes to antibacterial activity. Positively charged zinc oxide undergoes electrostatic interaction with negatively charged bacterial cell membranes (Mahamuni-Badiger et al., 2020). Zinc oxide nanoparticles interfere with the biofilm formation of *Escherichia coli* and *Staphylococcus aureus* via ROS production. Zinc is also involved in osteoblast mineralization and the regulation of bone growth (M. Wang & Tang, 2019).

van Hengel et al. developed porous titanium implants using Ti-6Al-4V. The implants had interconnected pores resulting in an enormous increase in surface area as compared to solid implants. However, the increased surface area also fosters adherence of biofilm-forming bacteria onto the implant surface. The solid implants were surface-modified with silver nanoparticles. The silver nanoparticles were embedded in a surface-grown oxide layer on the implants. This immobilizes the silver nanoparticles and precludes their release in the bloodstream thus avoiding any toxicity, as well as provides sustained release of silver ions. The implants were found to release silver ions for 28 days. The developed implants demonstrated good antibacterial activity against MRSA and prevented biofilm formation. The porous implants were superior as compared to their solid counterparts: four times greater amount of released silver ions, two times larger zone of inhibition, and one additional order of magnitude of reduction in numbers of colony forming units in an *ex vivo* mouse infection model (van Hengel et al., 2017).

Hu et al. developed TaON–Ag nanocomposites and coated them onto titanium nanorods in the form of a thin film. The group prepared TaN films using reactive co-sputtering with a Ta target in plasma with argon and nitrogen. Silver is immiscible with TaN and hence doping TaN with silver gives rise to thin nanocomposite films. The modified titanium nanorods showed good antibacterial activity against *Staphylococcus aureus*, coagulase-negative *Staphylococcus* and Gram-negative *Escherichia coli,* and *Pseudomonas aeruginosa*. Implantation of these titanium nanorods in a rat femur fracture model revealed good fracture healing. The modified titanium nanorods also inhibited both *Staphylococcus aureus* and *Escherichia coli* biofilm formation for 14 days *in vivo* (C. C. Hu et al., 2020).

Li et al. superimposed titanium nanotubes onto the titanium implant surface followed by the loading of silver nanoparticles into the titanium nanotubes via centrifugation and irradiation. This modified implant possessed self-adjusting bactericidal activity. During the early stages, silver nanoparticles are loosely held onto titanium nanotubes and hence silver ions are readily released. However, with time the silver nanoparticles get recrystallized onto the walls of titanium nanotubes thus obstructing the release of silver ions, but bacteria would still be killed on coming in contact with the implant surface. Such a self-adjusting system is beneficial since the early phase post-implantation is highly vulnerable to bacterial infection, and hence necessitates a strong release of antibacterial agent. Later on, contact bactericidal action is sufficient to protect the implant against bacterial adhesion. The developed titanium surface showed good antibacterial activity against both *Staphylococcus aureus* and *Escherichia coli* (B. Li et al., 2019).

Cheng et al. developed strontium (Sr) and silver (Ag) loaded nanotubes on titanium surfaces for osseointegration and prevention of infection in orthopedic implants. They fabricated nanotubular titanium surfaces by the hydrothermal method of immersion in Ag and Sr solutions. The release of Ag and Sr from the nanotubes could be programmed by modifying the nanotube diameter and time of hydrothermal treatment. The nanotubular structures on the titanium surface provide prolonged release of Sr which facilitates the filling of bone defects. The implants showed good spreading and proliferation of human fibroblasts on their surface. The nanotubular surfaces showed antibacterial activity against MRSA, methicillin-sensitive *Staphylococcus aureus,* and *Escherichia coli.* This inhibited colonization and biofilm formation on the implant surface (Cheng et al., 2016).

Ghilini et al. developed multifunctional titanium surfaces for combined antibacterial activity and osseointegration. Lactoferrin is osteogenic and also has antimicrobial activity. Lactoferrin and silver nanoparticles were adsorbed onto the titanium surface by immersion of titanium disks in lactoferrin solution and silver nanoparticle dispersion respectively. Adsorption of lactoferrin occurs via electrostatic interaction between lactoferrin and the negatively charged titanium surface and silver nanoparticles. Adsorption of lactoferrin and silver nanoparticles on titanium surface renders the surface highly hydrophilic and rough thereby hindering the adhesion of macrophages and promoting preosteoblast cell adhesion. The multifunctional titanium surface also decreased bacterial colonization by 97.7% (Ghilini et al., 2021).

Bee et al. used natural hydroxyapatite extracted from chicken bone waste as a support for the deposition of silver nanoparticles to form nanocomposites. Hydroxyapatite has hydroxyl groups which serve as nucleating sites for attachment of silver nanoparticles. The developed nanocomposites showed sustained release of Ag^+ ion for 42 days and excellent antibacterial activity against *Staphylococcus aureus*. The nanocomposites also showed good bioactivity with increased formation of apatite on the surface of the nanocomposites. The nanocomposites can thus be explored to tackle infection on orthopedic implants (Bee et al., 2021). Reduced graphene oxide comprises a dense honeycomb-like framework including monolayers of carbon atoms and oxygen-containing functional groups at the edges. Reduced graphene oxide is known to produce ROS which exerts antibacterial activity. Further, it also acts like a nano knife and physically damages the bacteria. The huge surface area of reduced graphene oxide enables adsorption of other antibacterial compounds thus making it ideal for combination therapy. San et al. combined reduced graphene oxide nanosheets with silver nanoparticles to generate a multifunctional titanium-based orthopedic implant having both antibacterial and osteogenic potential. Reduced graphene oxide nanosheets were parallel to the implant surface. The developed multifunctional implants showed antibacterial activity against MRSA and enhanced osteogenic differentiation of preosteoblasts (San et al., 2022).

Wang et al. developed a trilogy strategy based on hydrothermal reaction and biomimetic mineralization. They applied the trilogy strategy-based coating on standard orthopedic implants like locking compression plates and pedicle screws. The upper layer of the coating comprised calcium phosphate and the lower layer was sodium titanate. Na^+ ions in the sodium titanate porous structure of the Ti6Al4V implant were replaced by Sr^{2+} and Zn^{2+} ions. The nano topography in the top calcium phosphate layer exerts bacteriostatic activity during the early stage of implant infection. The calcium phosphate coating degrades slowly and subsequently releases Zn^{2+} which helps to tackle the implant infection during the middle stage. Once the calcium phosphate layer is exhausted completely, the underlying zinc titanate layer exerts long-term antibacterial action. The mechanical properties and structural stability of the orthopedic implants were intact after the surface modification. *In vivo* studies using *Staphylococcus aureus* and *Escherichia coli* revealed that the strategy could exterminate the established bacterial biofilm on Sr–Zn2@CaP implant after 20 weeks. The strategy enabled the sustained release of Sr^{2+} and Zn^{2+} during calcium phosphate coating degradation which facilitated implant osseointegration even under an infectious environment (Y. Wang et al., 2021).

Elimination of *Staphylococcus aureus* from the implant surface depends on neutrophil density surrounding the implant. Delayed recruitment of neutrophils fosters the creation of biofilms and attenuates host defence. Recruitment of neutrophils around the infected implant kills bacteria by phagocytosis and via the formation of neutrophil extracellular traps which trap and kill bacteria extracellularly. Li et al. developed zinc-doped ferric oxyhydroxide nanolayers on an Mg substrate for antibacterial activity and osseointegration. The group used a simple green method. The modification using zinc-doped ferric oxyhydroxide nanolayers enhanced the biocompatibility, corrosion resistance, antibacterial activity, and osteoblast differentiation activity of the Mg implant. The zinc-doped ferric oxyhydroxide nanolayer inhibited immune evasion-related gene expression, facilitated the formation of neutrophil extracellular traps, and promoted osteointegration in *Staphylococcus*-infected rat femurs (M. Li et al., 2022).

Lin et al. developed a hybrid coating of chitosan and zinc oxide nanoparticles onto porous titanium implant surfaces. Pretreatment of the titanium surface was done with alkali to induce a porous morphology on the titanium implant surface which would facilitate contact between chitosan and titanium. The incorporation of zinc oxide with chitosan enhanced the adhesion of the coating onto the titanium implant. The hybrid coating inhibited the growth of *Escherichia coli* and adhesion onto the titanium oxide implant (M. H. Lin et al., 2021).

Magnesium silicate is reported to stimulate the formation of apatite on its surface in simulated body fluid. Poly (ether ether ketone) (PEEK) has good biocompatibility and mechanical strength to form orthopedic implants. However, it is biologically inert and cannot bond with the host bone tissue. Hence, Tang et al. incorporated nano zinc-magnesium silicate into the PEEK matrix. This enhanced the mechanical properties, surface roughness, and hydrophilicity of the resulting bio composite. The developed composite showed antibacterial activity against *Escherichia coli* due to the release of Zn ions. The incorporation of bioactive nano zinc-magnesium silicate with PEEK resulted in enhanced attachment, proliferation, and differentiation of MC3T3-E1 cells on the bio composite (Tang et al., 2019).

Zinc-based zeolitic imidazolate framework (ZIF-8) has excellent antibacterial activity due to the presence of zinc sites. Zou and the group prepared copper-loaded ZIF-8 nanoparticles by adsorption. Cu (I) @ZIF-8 nanoparticles were then combined with poly lactic-co-glycolic acid (PLGA) using 3D printing to obtain porous scaffolds. The scaffolds showed good porosity and mechanical properties. The scaffolds substantially reduced the bacterial count in infected rats implanted with the scaffolds. The scaffolds also facilitated the proliferation of murine mesenchymal stem cells, promoted cell adhesion and spreading, and induced osteoblastic differentiation of mesenchymal stem cells. The scaffolds thus hold promising potential for infected bone repair (Zou et al., 2020).

Titanium dioxide exerts a bactericidal effect via the production of ROS and oxidation of lipids in the bacterial cell wall. γ-Fe_2O_3 nanoparticles exert antibacterial action via the generation of ROS, superoxide radicals, and singlet oxygen. Calabrese et al. fabricated titanium biomimetic scaffolds functionalized with either titanium dioxide (TiO_2) nanoparticles (300 nm–800 nm) or iron oxide (γ-Fe_2O_3) nanoparticles (10–20 nm) using wet chemistry. The implants coated with titanium dioxide nanoparticles showed a 99.4% reduction in the colony count of *Staphylococcus aureus* whereas the implants coated with γ-Fe_2O_3 nanoparticles showed a 99.9% reduction in the colony count of *Staphylococcus aureus.* Both implants also showed good osseointegration (Calabrese et al., 2021).

Gallium nitrate [Ga $(NO_3)_3$] is reported to have wide-spectrum antibacterial activity against *Staphylococcus aureus*, MRSA, *Pseudomonas aeruginosa,* and *Escherichia coli.* Dong et al. modified the titanium surface with titanium dioxide nanotubes and a coating of biodegradable poly D, L-lactic acid (PDLLA) loaded with gallium nitrate. Incorporating the gallium nitrate into the PDLLA matrix results in controlling the release of gallium ions causing an initial burst release followed by a gradual release. The modified implants were implanted in rat spine followed by the instillation of mixed *Staphylococcus aureus-Escherichia coli* bacteria at the site. The developed implants showed good anti-inflammatory and antibacterial activity, and compatibility with osteoblasts (Dong et al., 2019).

Selenium nanoparticles are promising for coating orthopedic implants due to their stability, ability to kill microorganisms via numerous mechanisms, and ease of immobilization onto surfaces. Tran et al. applied selenium nanoparticles having a size of 30–70 nm on titanium implants and studied the antibacterial activity against MRSA and MRSE in an infected femur model in rats. The selenium nanoparticles possessed antibacterial activity at 0.5 ppm concentration. The selenium nanoparticle coating on titanium implants inhibited biofilm formation on implants and decreased the number of viable bacteria in neighbouring tissues after the implants were inoculated with biofilm-forming doses of MRSA and MRSE. The nano topography formed by the selenium nanoparticles is also beneficial to facilitate bone-forming osteoblast cell functions via enhanced adsorption of specific proteins on the nano rough surface (Tran et al., 2019).

Nano Topography

Some naturally occurring surface morphologies such as the wings of insects have been reported to have strong antimicrobial activity. This may be due to the mechanical interactions between the adhering bacteria and the nano-dimensional microstructure of the surface. Orthopedic implants can be endowed with such nanoscale microstructure to generate bioinspired surface-modified implants with anti-adhesion characteristics (M. Wang & Tang, 2019). Lotus leaves, shark skin, and taro leaves repel bacterial adhesion and cell attachment because of micro and nano superhydrophobic structures and surface patterns. Dragonflies, cicada wings, and geko skin are bactericidal surfaces because they kill bacteria on contact. The surface of taro leaves has a structure endowed with bumps which render the surface superhydrophobic with a contact angle of 90° to 150 °. Lotus leaves also have microscale bumps covered by nanosized crystals on their surface which makes them superhydrophobic. Dragonfly wings also show a superhydrophobic surface with a contact angle of 153 ° (Jaggessar et al., 2017). Fabrication of nanoscale structures on the implant surface has been reported to exert antibacterial activity. These nanostructures may be designed to mechanically damage the bacterial membranes by the 'bed of nails' mechanism or to load antimicrobial agents (Chopra et al., 2021). An advantage of this approach is the low probability of resistance (Wu et al., 2022). However, the snag associated with this approach is that the anti-adhesion activity is due to mere physical interactions, and hence adhesion of osteoblasts is also obstructed. Therefore, it is necessary to include bioactive molecules like bone morphogenetic protein-2 (BMP-2) to facilitate the interaction of the implant with the neighbouring tissue. It is also challenging to achieve a stable bioinspired surface-modified implant (M. Wang & Tang, 2019). Further, the antibacterial activity is only restricted to bacteria that adhere to the implant surface. Planktonic bacteria would be spared (Wu et al., 2022).

Titanium nanotubes may be coated onto orthopedic titanium implants via anodization. These nanotube arrays covering the implant surface can be used to deliver antibacterial drugs and/or growth factors. Titanium nanotubes themselves also promote osteoblast adhesion to the implant surface, this effect is enhanced with smaller-diameter nanotubes. Titanium nanotubes have been found to inhibit the adhesion and colonization of *Staphylococcus epidermidis* (M. Wang & Tang, 2019). The creation of titanium dioxide nanotubes on the titanium surface also enhances bone mineralization 3-fold compared to non-modified titanium surfaces (E. P. Su et al., 2018).

Electrochemical anodization is an economical method that has been extensively used to create controlled nano-topographies of titania nanotubes on titanium implants. Electrochemical anodization involves a bare titanium cathode and the target titanium implant as the anode. Both the anode and cathode are immersed in an electrolyte solution containing fluoride ions and water and connected to a direct current power supply. When conditions such as voltage, current, and time are optimized, the nanotubes of titanium undergo self-assembly onto the titanium implant. These implants modified via electrochemical anodization offer numerous benefits for antibacterial activity. They are biocompatible, amenable to surface modification to alter crystallinity and wettability, favourable for loading antibacterial agents and tuning their localized release profile, show good osseointegration and immunomodulation, enable chemical modification to achieve on-demand release of actives, and have established *in vivo* potential. Titania nanotubes have been loaded with both antibiotics and metal nanoparticles to exert antibacterial activity. The open pores in the nanotubes may be obstructed with a polymer coating to sustain the release of antibiotics. Polymers like chitosan and PLGA have been explored to avoid the high local release in the initial few hours of implantation. Using chitosan has the advantage of being antibacterial and osteogenic, thus exerting a synergistic effect with the antibiotic. However, long-term toxicity studies and functional studies under continuous load-bearing conditions are essential to enable the successful clinical translation of these systems (Chopra et al., 2021).

Jia and the group developed a porous coating of titanium dioxide on titanium implants. This was followed by a coating of polydopamine which would entrap Ag^+ ions via chelation. The strategy enabled the successful immobilization of silver nanoparticles on the implant surface and entrapment within the nano porous cavities of titanium dioxide coating. The nano topography of titanium dioxide coating mimics the extracellular matrix thus facilitating osteogenesis. The coating shielded titanium against corrosion and also provided sustained release of Ag^+ for 28 days. Interactions between the catechol groups of polydopamine and Ag^+ help to control the release of Ag^+. The implant was capable of osteoblast adhesion, spreading, proliferation, and differentiation. The modified implants were effective in reducing the adhesion of *Staphylococcus aureus* on the implant, trapping the adhered *Staphylococcus aureus* within the pores and subsequently killing the bacteria via dual physical entrapment within the pores and the chemical antibacterial effect of Ag^+ (Jia et al., 2016).

Silk-based implants can avoid the problem of non-biodegradability associated with titanium implants. Silk has good mechanical strength to be used as a bone implant. Mehrjou et al. fabricated nano cones on *Bombyx mori* silk films using oxygen plasma etching. This treatment increased the surface energy by 176%. The hydrophilic surface nanostructure decreased bacterial adhesion by > 90% for *Staphylococcus aureus* and *Escherichia coli* and also enhanced osteoblast proliferation by 30% (Mehrjou et al., 2019).

Wang et al. developed a multifunctional coating for antibacterial activity and osseointegration. Silver nanoparticles wrapped with silk fibroin were entrapped in strontium-loaded titanium dioxide nanotubes. Nano porous strontium titanate imitates the size and background of bone tissue. Silk fibroin helps to control the release of Ag^+ ions, thereby reducing their cytotoxicity. It also enhances the biocompatibility of the coating. The multifunctional coating simultaneously inhibited the growth of *Escherichia coli* and *Staphylococcus aureus*. The multifunctional coating fostered the attachment, proliferation, and differentiation of MC3T3-E1 cells and promoted early osteointegration between the titanium implant and bone tissue *in vivo* (B. Wang et al., 2021).

Yang et al. developed magnesium-containing titania nanotube-modified titanium implants combining the osteogenic and antibacterial potential of magnesium with the antimicrobial effect of titania nanotubes. Titanium nanotube coatings of size 80 nm were generated via electrochemical anodization and hydrothermally treated to incorporate magnesium. The 80 nm diameter titania nanotubes showed promising antibacterial activity, predictable release patterns, and excellent osseointegration. The modified implants exhibited continuous release of Mg ions from the titania nanotubes. *In vivo* studies in rats using MRSA showed that the modified implants substantially avoided bone infection and osteolysis, and also resulted in good osseointegration between implant and bone tissue (Yang et al., 2019).

Cirsium arvense is a flower containing a characteristic structure of a stack of petals wrapping the buds. This structure functions to gradually open the petals and sepals and expose the buds in response to the stimuli of temperature and humidity to promote the buds to transfer pollen. Drawing inspiration from this flower, Zhang et al. developed capsule-integrated, pH-responsive biomimetic implant nanocoatings using the layer-by-layer self-assembly method to deliver an immune-tuning antimicrobial agent interleukin–12 (IL-12) and a clinically approved growth factor recombinant human BMP-2 to combat infection and foster wound healing in open fractures. Biocompatible polypeptides such as poly-L-lysine or poly-L-glutamic acid were used to prepare capsule reservoirs for loading actives. These capsule reservoirs were interspersed in multilayer films via self-assembly. The multilayer nanofilms are similar to petals and sepals which cover the drug-loaded capsules and help to achieve sustained release of the payload within the capsules on demand. IL-12 triggers naïve CD4 T cells to differentiate into Th1 cells and secrete interferon-γ thereby activating macrophages for bacterial destruction. The biomimetic structure helped to enhance the loading of both IL-12 and BMP-2 by > 7-fold. The developed system released IL-12 for 10 days and BMP-2 for 6 weeks. The nano coating had good hydrophilicity and surface roughness thus indicating its ability to foster cell adhesion and proliferation. The developed nanocoating also exhibited 3-fold higher bacterial killing efficacy against *Staphylococcus aureus* and 5-fold enhanced bone healing in a rat femur fracture model (S. Zhang et al., 2021).

Use of Nanoparticles as Carriers of Antibacterial Agents

Nanoparticles may be used as carriers of antibacterial agents to combat orthopedic implant-associated bacterial infections. They help to provide sustained release of the antibacterial agent (Pawar et al., 2018).

Liu et al. developed a coating containing composite polyelectrolyte multilayer and mesoporous bioactive glass nanoparticles on the surface of 316L stainless steel implant using the spin coating technique. Bioactive glass is osteoconductive and promotes bone formation. The polyelectrolyte layer contained collagen as the cationic layer, γ-poly glutamic acid as the anionic layer, and chitosan as the barrier layer. Tetracycline-loaded 57S mesoporous bioactive glass nanoparticles were incorporated in the γ-poly glutamic acid layer. Chitosan serves to postpone the degradation and release of tetracycline from the γ-poly glutamic acid layer. The mesoporous bioactive glass nanoparticles released tetracycline in a sustained manner for 7 days. They showed a greater release of tetracycline at acidic pH. The surface-modified implants also promoted the mineralization of bone marrow mesenchymal stem cells. The surface-modified implants showed a surface roughness of 384 nm which helps in stimulating osteoblast proliferation and differentiation. Coating 316L stainless steel decreased the mechanical properties of stainless steel and mimicked the mechanical properties of natural bone. The surface-modified implants showed good antibacterial activity against *Staphylococcus aureus* and *Escherichia coli* (X. Liu et al., 2018).

In situ hydrogel is defined as a liquid that after being injected into the body is transformed into a gel which prolongs retention at the site of injection. *In situ* gels help to deliver the drug directly to the target site, thus enhancing its efficacy and mitigating the side effects. Pawar et al. developed an *in-situ* alginate gel containing vancomycin-loaded chitosan nanoparticles and povidone-iodine. Chitosan was used as a carrier since its polycationic nature helps in binding the antibiotic to bacteria. The developed formulation showed sustained release of vancomycin and povidone-iodine for 10–15 days. The formulation showed antibacterial and antibiofilm activity against *Staphylococcus aureus* indicating its promise for the prevention of orthopedic implant-associated infections. The gel showed good hemocompatibility and cell compatibility. Vancomycin exerts antibacterial activity and povidone-iodine shows anti-biofilm activity. The anti-biofilm activity of povidone-iodine is due to the suppression of the icaA gene in bacteria which encodes polysaccharide intercellular adhesin molecule involved in biofilm formation (Pawar et al., 2018).

AbouAitah et al. developed PEEK implants coated with a layer of hydroxyapatite nanoparticles using the sonocoating method. The sonocoating method involves immersion of the implant to be coated into a suspension containing nanoparticles of the coating agent in water followed by the application of high-power ultrasound. The PEEK implants were coated with a hydroxyapatite layer of thickness in the range of 219 nm to 764 nm. The nanohydroxyapatite layer serves the dual function of facilitating bone growth as well as loading the antibiotic cefuroxime sodium. The nanohydroxyapatite coating showed good porosity which is essential for both osseointegration as well as drug absorption. The implant showed an initial burst release of around 86 to 96% of the drug within 24 hours and then a stable release. The developed implant also showed antibacterial activity against *Staphylococcus aureus* (Abouaitah et al., 2021).

Min et al. developed PEEK implants with a nanolayer coating showing the sequential release of gentamicin sulphate and BMP-2 for eradicating biofilm and accelerating bone repair respectively. The implant was 1.3 mm in diameter and 4 mm in length and had lateral channels having a diameter of 250 μm, which is ideal for facilitating vascular invasion and bone mineralization. The top part of the implant coating released gentamicin and then the underlying part released BMP-2. The top part of the implant coating contained alternate layers of gentamicin sulphate with polymer 1 and poly acrylic acid. The underlying part of the implant coating contained alternate layers of BMP-2 with polymer 2 and poly acrylic acid. Polymer 1 has a half-life of < 5 h and polymer 2 has a half-life of ~ 40 h. Release studies of the implant showed first-order release of gentamicin sulphate with minimal burst release on the first day and sustained release remaining above MIC, and slow and continuous release of BMP-2. The implants showed effective osteointegration with the host tissue (80% bone coverage within 3 weeks post-implantation) in rat tibia with induced osteomyelitis. Long-term bone-implant integration was observed with 15-fold higher interfacial strength compared to uncoated implants. *Staphylococcus aureus* Xen 29 was used to induce osteomyelitis since the strain is 20 times more resistant to gentamicin than *Staphylococcus aureus* (Min et al., 2016).

Dwivedi and group developed a layer-by-layer coating containing vancomycin onto a stainless-steel implant. The implant was subjected to 40 coating cycles to obtain a transparent and flexible coating and also achieve the desired concentration of vancomycin required to inhibit *Staphylococcus aureus*. Each coating cycle comprised 3 layers in which the implant was dipped: vancomycin (innermost layer), poly lactic acid (PLA) as the middle layer, and vancomycin-loaded niosomes (outermost layer). The first layer applied on the stainless-steel implant was vancomycin since the surface of the implant was hydrophilic post-plasma treatment. Vancomycin-loaded niosomes could bind well with the porous layer of PLA, hence, PLA was used as the middle layer in the coating. The incorporation of vancomycin in niosomes prolonged its release above the MIC against *Staphylococcus aureus* up to 28 h. When incubated with *Staphylococcus aureus*, the implants coated with niosomes showed minimal bacterial adhesion and inhibition of bacterial growth (Dwivedi et al., 2018).

Liu et al. developed a bifunctional titanium alloy with a porous nanohydroxyapatite coating having a thickness of 20–30 μm and a size of 100–150 nm. The porous nature enabled the loading of antimicrobial peptide human β-defensin 3(HBD-3) and BMP-2. The implants showed sustained release of both HBD-3 and BMP-2 for 20 days. The implant inhibited the growth of *Staphylococcus aureus* and *Escherichia coli* in the incubation medium and the surface of the implant was free from living bacteria. The implants also showed adhesion, proliferation, and osteogenic differentiation of human mesenchymal stromal cells (H. W. Liu et al., 2018).

Ashbaugh et al. developed an electrospun composite coating of PLGA nanofibers embedded in a poly (Ɛ-caprolactone) (PCL) film for locally sustained codelivery of antibiotics from the titanium implant surface. The PLGA:PCL ratio can be varied to change the release kinetics of the antibiotics. The system allows the loading of more than one set of PLGA fibers thus enabling delivery of two different antibiotics at different release rates. The release rate of the drug can also be adjusted by loading it either in the PLGA coating or in the PCL film. Three different sets of PLGA: PCL composite coatings containing different antibiotic combinations namely: vancomycin in PLGA: rifampicin in PCL; linezolid in PLGA: linezolid and rifampicin in PCL film; and daptomycin in PLGA: daptomycin and rifampicin in PCL film were prepared. The objective was to achieve a faster release of rifampicin as compared to another antibiotic in the combination to ensure that rifampicin is never present as a single antibiotic during the complete release period to avoid resistance. In an *in vivo* mouse model of *Staphylococcus aureus* prosthetic joint infection, the combination of linezolid with rifampicin and the combination of daptomycin with rifampicin prevented bacterial infection and biofilm formation on the implant surface (Ashbaugh et al., 2016).

Use of Stimulus-Responsive Nanoparticles

One of the major drawbacks associated with the usage of nanoparticles is the problem of burst release of the drug in less than 5 h. Typically, an orthopedic implant surgery lasts for 1–2 h, hence, some amount of antibacterial drug would be released during the surgery which gets wasted. Further, the low amount of antibacterial drug retained in the system would not be sufficient to kill the bacteria fostering antimicrobial resistance. Stimulus-responsive nanoparticles can be designed to release the antibacterial payload only in response to internal stimuli present within the infection site matrix (enzymes, pH) or external stimuli such as light, magnetic field, and temperature (Wu et al., 2022). Antimicrobial surface coatings have been explored for the treatment of implant-associated infections. However, they suffer from compromised effectiveness owing to two reasons: They slowly lose their antimicrobial load over time and are ineffective at the time of infection on the implant; secondly, the long-term low release of antibiotics fosters antibiotic resistance. The usage of stimulus-responsive nanoparticles is highly beneficial to overcome these problems associated with antimicrobial surface coatings. Temperature or low pH in the biofilm may be used to stimulate the release of antimicrobial agents (Quan et al., 2021). The biofilm microenvironment has the following features: low pH (4.5 – 6.5) and an increased concentration of H_2O_2 (Guo et al., 2020).

Temperature-Responsive Nanoparticles

Wickramasinghe developed a thermo-responsive nanocomposite hydrogel containing a blend of D-amino acids (D-tyrosine, D-tryptophan, and D-phenylalanine) and gold nanorods to promote biofilm disruption and also achieve photothermal destruction of residual bacterial cells. The hydrogel undergoes a sol-to-gel conversion at physiological temperatures. The developed hydrogel was effective in destroying *Staphylococcus aureus* biofilm on the surface of the Ti-6Al-4V implant without causing any harm to the surrounding soft tissues around the implant. The implant was incubated with the system for 2 h followed by a 20 min photothermal treatment using an 808 nm laser. The group also studied the effectiveness of vancomycin and a mixture of Hibiclens and Bacitracin to disrupt the biofilm. Vancomycin showed 30% eradication of biofilm and a mixture of Hibiclens and Bacitracin showed 20% eradication of biofilm (Wickramasinghe et al., 2020).

pH-Responsive Nanoparticles

pH of the bloodstream is 7.4, however, infected tissues show an acidic microenvironment (Intravaia et al., 2023). Biofilm shows an acidic microenvironment due to the accumulation of acid metabolites post microbial anaerobic fermentation (Y. Hu et al., 2022). This differential pH can be exploited to deliver the antimicrobial drug selectively at the site of infection in response to the acidic pH (Intravaia et al., 2023).

Zeolitic imidazolate framework-67 (ZIF-67) is endowed with favourable features such as large surface area, excellent biodegradability, and controllable pore size. ZIF-67 also shows pH-responsive release of payload thereby rendering it useful for delivery in the acidic bacterial infection microenvironment. Tao et al. designed a multifunctional coating on a titanium implant that displays drug release in an acidic infection environment, suppresses the inflammatory response, and promotes osseointegration. ZIF-67 nanoparticles encapsulating the osteogenic growth peptide were fabricated on titanium dioxide nanotube substrates. The ZIF-67 nanoparticles could eradicate MRSA infection due to the synergism between the rapid release of cobalt ions and the created alkaline environment. pH of 8.5 is considered ideal for osteoblast proliferation. The released osteogenic growth peptide and cobalt ions regulate the macrophages to M2 type, create an anti-inflammatory milieu, and enhance cross-talk between macrophages and mesenchymal stem cells for osseointegration. *In vivo* studies revealed the potential of the system in facilitating MRSA-infected femoral defect osseointegration (Tao et al., 2023).

Chen et al., immobilized dopamine onto titanium nanotubes and then conjugated pH-responsive polymer poly methacrylic acid (PMMA) onto titanium nanotubes via bonding between the carboxyl group of PMMA and the amino group of dopamine. At the physiological pH of 7.4, PMMA swells and in the acidic environment of infection (pH < 6), it collapses. Thus, it helps to achieve on-demand delivery of antimicrobial peptide HHC36 at the infection site. The developed nanotubes showed excellent bactericidal activity against *Staphylococcus aureus*, MRSA, *Escherichia coli,* and *Pseudomonas aeruginosa.* The nanotubes also exhibited osteogenic activity with human bone mesenchymal stem cells. *In vivo* studies in New Zealand rabbits having osteomyelitis revealed inhibition of *Staphylococcus infection* and biocompatibility during the acute infection period post-implantation (Chen et al., 2020).

Zhou et al. fabricated a multifunctional platform for infection-responsive antibacterial activity and osseointegration. The group designed antimicrobial peptides with osteogenic fragments and complexed them on the surface of silver nanoparticles via hydrogen bonding. Silk fibroin was then applied to these modified nanoparticles. This system was then coated onto titanium implants. Antimicrobial peptide complexation facilitated the dispersion and size regulation of silver nanoparticles. Silk fibroin coating showed a pH-responsive release of silver ions in response to an acidic pH of 5, which is the pH at the site of infection. Silk fibroin shares structural similarities with type II collagen. Hence, the silk fibroin coating promoted adhesion, diffusion and proliferation of bone marrow stem cells on the implant surface and also helped in collagen secretion. The developed system showed good anti-adhesion, planktonic killing for up to 21 days, and biofilm formation inhibition against *Staphylococcus aureus*. *In vivo* implantation in rats revealed excellent osseointegration at week 4 and 8. The usage of antimicrobial peptides in combination with silver nanoparticles offers a synergistic effect with dose reduction in silver nanoparticles. The phosphorothioate group in the biofilm of *Staphylococcus aureus* is endowed with many negatively charged phosphate groups and can bond with positively charged antimicrobial peptides leading to the collapse of the bacterial cell membrane. This further enhances the ingress of silver nanoparticles into the bacterial cell (Zhou et al., 2023).

Silk fibroin possesses diverse isoelectric points on various parts of its backbone thus enabling its usage in pH-responsive release of the payload. It can release a low amount of metal ions required to initiate osteointegration at physiological pH and a high concentration of metal ions in the event of an infection (low pH). Yan et al. developed a pH-responsive nanocoating onto a porous PEEK surface for combined antibacterial activity, osseointegration, and angiogenesis. Copper oxide microspheres decorated with silver nanoparticles were deposited onto a porous PEEK surface with the help of silk fibroin. The resulting coating released large amounts of Cu^{2+} and Ag^{+} at pH 5 killing 99.9% of planktonic bacteria and eliminating sessile bacteria thus preventing biofilm formation. *In vivo* studies in rabbit tibia model showed good new bone formation. Copper and silver act synergistically owing to their distinct mechanisms of action. Copper catalyzes Fenton chemistry and hydroxyl radical formation to increase ROS generation whereas silver depletes antioxidant supply to increase intracellular ROS. Copper causes lipid peroxidation and affects bacterial membrane function whereas silver compromises the integrity of the bacterial membrane or disrupts the bacterial electron transport chain function (Yan et al., 2020).

Enzyme-Responsive Nanoparticles

Enzyme-responsive nanoparticles are appropriate for biomedical uses as compared to other stimulus-responsive nanoparticles due to their selectivity and specificity. Enzymes present in the infection microenvironment cleave the nanoparticle to release the payload (Q. Hong et al., 2021).

Mesoporous silica nanoparticles (MSNs) have multiple features that make them ideal drug delivery vehicles. These include a large surface area, tunable pore size, and the ability to control drug release (Moure et al., 2017). Ding et al. fabricated an enzyme-responsive nanoplatform to simultaneously eradicate bacteria and facilitate tissue regeneration in osteomyelitis. The group encapsulated silver nanoparticles in MSNs and then poly glutamic acid (anionic) and polyallylamine hydrochloride (cationic) were assembled on the MSNs containing silver nanoparticles using the layer-by-layer assembly method. The resulting nanoparticles were then deposited on the surface of a polydopamine-modified titanium substrate. Poly glutamic acid contains an amide linkage which can be cleaved by the enzyme glutamyl endonuclease secreted by *Staphylococcus aureus*. The titanium implants were effective in treating bacterial infection in a bacterium-infected femur-defect rat model. The modified titanium implants also enhanced the formation of new bone tissue post-implantation for 4 weeks. When the modified system is implanted *in vivo* and encounters *Staphylococcus aureus* infection, glutamyl endonuclease which is overexpressed in the bacterial infection microenvironment destroys the poly glutamic acid–polyallylamine hydrochloride capping present around MSNs. This causes the release of silver nanoparticles entrapped within the MSNs. Polyglutamic acid is a polypeptide that enhances the growth of cells and tissue. It is a biomimetic polypeptide that resembles the natural extracellular matrix of subchondral bone. Silver nanoparticles have a positive charge and thus can readily interact with the negative charge of the bacterial cell membrane. Accumulation of silver nanoparticles in the cell membrane alters the bacterial membrane structure and enhances bacterial cell permeability. This facilitates the transport of Ag^+ ions across the bacterial cell membrane to combine with the negatively charged groups present in the DNA and protein causing DNA condensation and cell membrane distortion. Silver nanoparticles further cause an increase in intracellular ROS levels (Ding et al., 2020).

Staphylococcus aureus secretes hyaluronidase enzyme. Shen et al. developed hyaluronidase-responsive multi-layered titanium nanotubes onto titanium surfaces for combating *Staphylococcus aureus* infection. Titanium dioxide nanotubes were loaded with the antimicrobial peptide cecropin B followed by coating with hyaluronidase-responsive multiple layers of chitosan/sodium hyaluronic-cecropin B. Cecropin B loaded in the nanotubes would be released in a sustained fashion. Sodium hyaluronic-conjugated cecropin B enabled cecropin B to be released early in the event of an infection with *Staphylococcus aureus*. Further, the conjugate of sodium hyaluronic-cecropin B helped to reduce the hydrophilicity of sodium hyaluronic thereby enhancing its cytocompatibility with osteoblasts. The developed nanotubes released cecropin B when exposed to exogenous hyaluronidase enzyme and also when incubated with *Staphy-*

lococcus aureus. The developed nanotubes showed good early (4 h) and long-term (72 h) antibacterial activity against *Staphylococcus aureus* (Shen et al., 2016).

Magnetic Field-Responsive Nanoparticles

A magnetic demand trigger is desirable over other demand triggers since fever or other disease symptoms will not interfere with a magnetic demand trigger (Quan et al., 2021).

Quan et al. designed a magnetic nanoparticle coating from which the magnetic nanoparticles can be pulled off using a magnetic field whenever there is biofilm formation on the implant surface. The group hypothesized that the path along which magnetic nanoparticles have been pulled through the biofilm will produce artificial by-pass channels via which antibiotics can better penetrate the biofilm to destroy bacteria over the entire thickness of the biofilm. The group found that larger magnetic nanoparticles having a 70 nm diameter could create artificial bypass channels without the dispersal of bacteria. Dispersal of bacteria from the infected biofilm can cause bacteremia if the dispersed bacteria are not sufficiently destroyed by the host immune system or antibiotic treatment. The group reported the greatest efficacy using a coating of around 0.64 mg/cm^2 magnetic nanoparticles. Higher concentrations showed less efficacy of antibiotics probably attributed to increased EPS production through more contact with the sharp edges of cubic magnetic nanoparticles. Magnetically pulled-off magnetic nanoparticles created additional by-pass channels in adhering biofilms of *Staphylococcus aureus* or *Pseudomonas aeruginosa*, resulting in increased antibiotic penetration and bacterial killing by gentamicin and tobramycin (Quan et al., 2021).

Bone-Targeted Therapy with Nanoparticles

Targeted delivery of antibiotics to the infected bone tissue can enhance the efficacy of antibiotics and reduce their toxicity (Nie et al., 2022).

Peptide D_6 helps to target the bone tissue since it has aspartic acid which has a high affinity for the bone tissue. Peptide UBI_{29-41} helps in targeting bacteria since it has six positively charged residues which facilitate targeting the higher negative charge of *S. aureus*. Nie et al. fabricated bone-and-bacteria dual-targeted MSNs for selective delivery of vancomycin to the infected bone. The developed nanoparticles when injected intravenously in rats inhibited bacterial growth around the titanium implant infected with MRSA. They were biocompatible and also resulted in decreased bone destruction and maintained bone integrity (Nie et al., 2022).

Yao et al. developed MSNs for the prevention of implant-associated bacterial infections and subsequent bone loss. They prepared MSNs containing enoxacin, coated them with PEG, and conjugated them with bone-targeting D-Asp8. PEG coating helps to prevent the premature release of enoxacin before reaching the target tissue. The bone-targeting ligand D-Asp8 can chelate hyaluronic acid and result in accumulation of the nanoparticles in bone tissue. Enoxacin exerts both antibacterial and osteoclast-inhibitory effects. The developed nanoparticles showed good antibacterial activity *in vivo* against *Staphylococcus aureus* as evidenced by the decreased biofilm formation on titanium rods implanted in rats (Yao et al., 2021).

Multimodal Therapy

Using multiple modes of destruction is a promising way to tackle the menace of resistant bacterial biofilms on the surface of orthopedic implants (Skłodowski et al., 2023). This sub-section highlights the various multimodal therapies that have been explored to combat bacterial biofilms on orthopedic implants.

Multimodal Therapy using Metallic Nanoparticles and Antibiotics

The combination of metallic nanoparticles with antibiotics is useful to augment the efficacy of antibiotics and also hinder biofilm formation (Altun et al., 2021). It further helps to decrease the dose of antibiotics; tackle antimicrobial resistance; and achieve synergistic antimicrobial efficacy against both planktonic and biofilm-embedded bacteria, and multidrug-resistant bacteria strains (Skłodowski et al., 2023).

Xiang et al. developed a surface coating for prolonging the drug release from titanium nanotubes and preventing implant-associated bacterial infections. Folic acid was conjugated onto the surface of zinc oxide quantum dots. Titania nanotubes containing vancomycin were capped with folic acid-conjugated zinc oxide quantum dots. These nanotubes when exposed to acidic conditions in bacterial infection release Zn^{2+} and vancomycin via dissolution of the zinc oxide capping. Both vancomycin and Zn^{2+} act synergistically to destroy the bacteria. The system showed good antibacterial activity against *Staphylococcus aureus* (Xiang et al., 2018). Zhou et al. developed a collagen-mimicking silk fibroin-based coating loaded with silver nanoparticles and gentamicin wherein Ag^{+} was reduced to silver nanoparticles by silk fibroin under UV radiation. Polydopamine coating assisted in anchoring silk fibroin-based coating onto the titanium implant surface. The coating effectively inhibited biofilm formation by *Staphylococcus aureus* attributed to the synergy between gentamicin and silver nanoparticles. The coating also enhanced the attachment and proliferation of MC3T3 cells revealing its osteogenic potential (Zhou et al., 2017).

Multimodal Therapy using Metallic Nanoparticles and Immunotherapy

The immune system is of paramount importance in eradicating bacterial infection from the body. Hence, it is beneficial to activate the immune system to combat bacterial infection on orthopedic implants. The usage of metallic nanoparticles in combination with immunomodulatory agents can provide superior efficacy as compared to monotherapy with metallic nanoparticles alone (W. Liu et al., 2019).

Liu et al. fabricated a porous PEEK implant loaded with Cu ions for antibacterial therapy and immunotherapy of implant-associated infections. The group treated PEEK with sulphuric acid to create a porous surface. This was followed by the deposition of Cu onto the porous implant surface via a magnetron sputtering apparatus. The PEEK implants showed a sustained release of Cu^{++} ions for 15 days. Cu exerts a dual function of killing the microorganisms as well as modulating the immune response thereby inhibiting biofilm formation. The implant showed good antibacterial activity against sessile MRSA *in vitro* owing to the porous surface and Cu. Cu^{++} ions destroy the permeability of MRSA membranes causing leakage of proteins and sugars. Cu^{++} ions then enter the MRSA cell and increase ROS generation. Cu^{++} ions are also genotoxic and interfere with DNA and RNA replication. Culturing of macrophages on the PEEK implants showed that the implants were able to generate an inflammatory microenvironment characterized by the generation of pro-inflammatory cytokines such as IL-6 and TNF-α (Fig. 1). These macrophages also showed enhanced phagocytosis of MRSA. Thus, the implants could polarize the macrophages towards the M1 phenotype. The implants also showed good antibacterial activity in the implant-associated infection model in mice and the osteomyelitis model in rats (W. Liu et al., 2019).

Figure 1. Schematic representation of a porous PEEK implant loaded with Cu ions for antibacterial therapy and immunotherapy of implant-associated infections

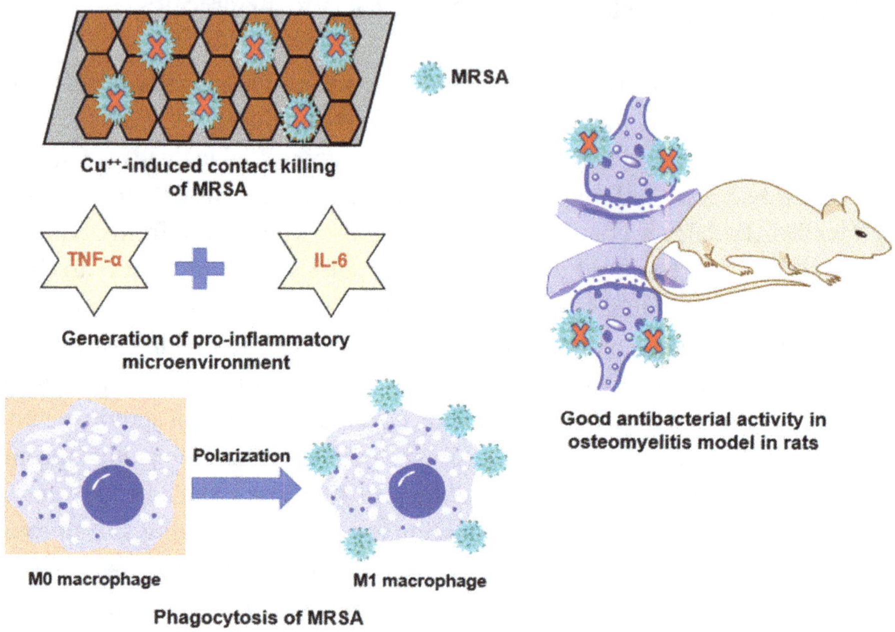

Zinc oxide nanoparticles also show immunomodulatory activity. They can activate human monocytes and generate a pro-inflammatory response. Wang et al. immobilized zinc oxide nanoparticles on the titanium implant surface. Coculturing of bacteria and innate immunocytes with the developed implant showed that the zinc oxide nanoparticles enhanced phagocytosis and secretion of inflammatory cytokines by macrophages and polymorphonuclear leukocytes. The developed zinc oxide nanoparticles showed excellent antibacterial activity against sessile bacteria namely: *Staphylococcus epidermidis*, *Staphylococcus aureus*, *Pseudomonas aeruginosa*, and *Escherichia coli*. This was attributed to the release of Zn^{2+} and ROS. The implants also showed excellent antibacterial effect against implant-associated soft tissue infection in mouse model (J. Wang et al., 2017).

Multimodal Therapy using Photothermal Therapy (PTT) and Photodynamic Therapy (PDT)

Photodynamic therapy (PDT) involves the generation of ROS using substances that absorb light energy in the presence of oxygen (Khatoon et al., 2018). However, PDT suffers from the drawback of poor penetration of the generated ROS into bacterial cells (Naskar & Kim, 2023). Photothermal therapy (PTT) involves using a probe followed by exposure to light which causes the conversion of light into heat to destroy the bacteria (Mosselhy et al., 2021). Near-infrared (NIR) light is more beneficial than UV and visible light since it is endowed with deep tissue penetration and minimum damage to living cells (L. Hong et al., 2019). PTT based on NIR radiation is beneficial for eradicating bacterial biofilms owing to its remote and minimally invasive nature, good tissue penetration, and preventing the emergence of drug-resistant bacteria. The downside of using PTT for the eradication of bacterial biofilms on implants is the usage of high laser power and local high temperature which can cause damage to neighbouring healthy tissues. Further, mature biofilms formed on implant surfaces are highly resistant and cannot be eliminated via PTT alone (J. Hu et al., 2022). A potential solution for both these problems is to reduce the temperature of PTT to < 48 °C by using the technique in combination with other strategies (Wu et al., 2022). The combined use of PDT and PTT merges the benefits of both approaches and compensates for the drawbacks of each (Naskar & Kim, 2023). Hyperthermia caused by PTT enhances the penetration ability of ROS into the bacterial cells by making them more susceptible to ROS. Further, the PDT-derived ROS destroy the bacteria thereby reducing the high-temperature requirement associated with the single use of PTT (Naskar & Kim, 2023).

Copper sulphide (CuS) nanoparticles can convert light energy into heat energy and thus can be used in PTT of infections. Wang et al. synthesized PEG-modified dandelion-shaped Cu_9S_8 nanoparticles. Under NIR radiation (808 nm, 1 W/cm^2), the developed nanoparticles showed excellent antibacterial activity and antibiofilm activity against clinically collected *Staphylococcus aureus* proliferating on titanium surfaces. The promising antibacterial effect was due to the dual action of photothermal and photodynamic effects as ROS were also formed. Scanning electron microscopy revealed that the structure of the biofilm was strongly destroyed after treatment and the remaining bacteria exhibited a distorted shape (W. Wang et al., 2019).

Yuan et al. developed mesoporous polydopamine nanoparticles loaded with indocyanine green (ICG) for PTT and PDT of implant-associated bacterial biofilms. Mesoporous polydopamine nanoparticles were immobilized onto titanium implants. The mesoporous polydopamine nanoparticles were functionalized with osteogenic RGD peptides to endow them with biocompatibility. Polydopamine has many aromatic rings and a mesoporous structure which promotes significant loading of ICG by π-π stacking interaction. The mesoporous polydopamine surface is also endowed with many dihydroxy indole/indolequinone groups which enable conjugation of RGD peptide through the Michael addition reaction or Schiff base reaction. Hyperthermia (50 °C) caused due to photothermal action enhances the diffusion of ICG into the biofilm matrix (Fig. 2). The ROS produced by PDT would further destroy the bacterial membranes and make them more vulnerable to the hyperthermia effect. The titanium implants modified with the developed system were implanted in rats after the establishment of *Staphylococcus aureus* biofilm on the implant surface. They showed good antibiofilm activity (95.4% efficiency) and also showed osseointegration on exposure to NIR light of 808 nm. The developed strategy offers a remote and non-invasive approach to tackle the menace of implant-associated bacterial biofilms (Yuan et al., 2019).

Figure 2. Schematic diagram of mesoporous polydopamine nanoparticles loaded with indocyanine green for photothermal and photodynamic therapy of implant-associated bacterial biofilms

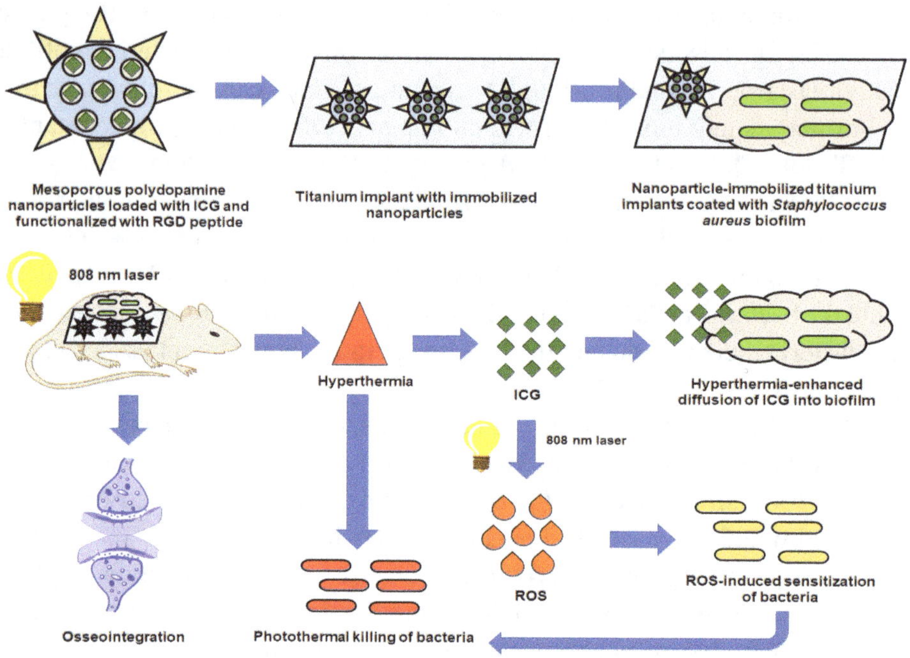

Inorganic semiconductors have high stability and hence are better ROS generators for PDT than certain photosensitizers. Hong et al. developed a NIR-activated inorganic hybrid semiconductor coating for surface modification of orthopedic implants to accomplish biofilm elimination via combined PDT and PTT. Bismuth sulphide (Bi_2S_3) is biocompatible and has good photothermal efficacy. Trisilver phosphate (Ag_3PO_4) is a photocatalyst that shows bacteriostatic characteristics and is non-toxic to mammalian cells. Hong et al. grew Bi_2S_3 nanorod arrays on Ti implants through hydrothermal methods, and Ag_3PO_4 nanoparticles were loaded onto Bi_2S_3 nanorod arrays through stepwise electrostatic adsorption. The developed system was capable of achieving biofilm elimination within 15 min after exposure to 808 nm light in a *Staphylococcus aureus* biofilm-infected rat model (L. Hong et al., 2019).

Multimodal Therapy Using Magnetothermal Therapy and Immunotherapy

Magnetic hyperthermia is better than PTT due to better tissue penetration depth thus enabling its usage for the treatment of deep orthopedic implant-associated infections. Magnetic hyperthermia penetrates the biofilm matrix and dissociates the latter. Nitric oxide penetrates deep inside the biofilm matrix to destroy bacteria via the generation of reactive nitrogen species. S-nitrosothiol is a thermoresponsive nitric oxide donator that can be stimulated to release nitric oxide under hyperthermia. The synergistic effect of magnetic hyperthermia and nitric oxide may not be able to completely eradicate pathogens, hence it is necessary to activate the immune response. Magnetic nanoparticles are reported to act as immunomodulators and recruit macrophages thereby preventing relapse of implant-associated infection. Wang et al. developed a system for synergistic magnetothermal therapy and immunotherapy of orthopedic implant-associated bacterial infections. They synthesized magnetic nanoparticles ($CoFe_2O_4$

core/$MnFe_2O_4$ shell) which were further coated with nitrosothiol. The developed system on exposure to an alternating magnetic field could produce a significant amount of heat and quickly release nitric oxide (Fig. 3). The system showed good activity against *Staphylococcus aureus* and *Escherichia coli* biofilms. Further, *in vivo* evaluation using PEEK implants infected with *Staphylococcus aureus* biofilms revealed good anti-biofilm activity with the recruitment of macrophages and their polarization to the M1 phenotype. Polarization of macrophages to the M1 phenotype exerts a local innate immune response against the biofilm (J. Wang et al., 2021).

Figure 3. Schematic representation of a system for synergistic magnetothermal therapy and immunotherapy of orthopedic implant-associated bacterial infections

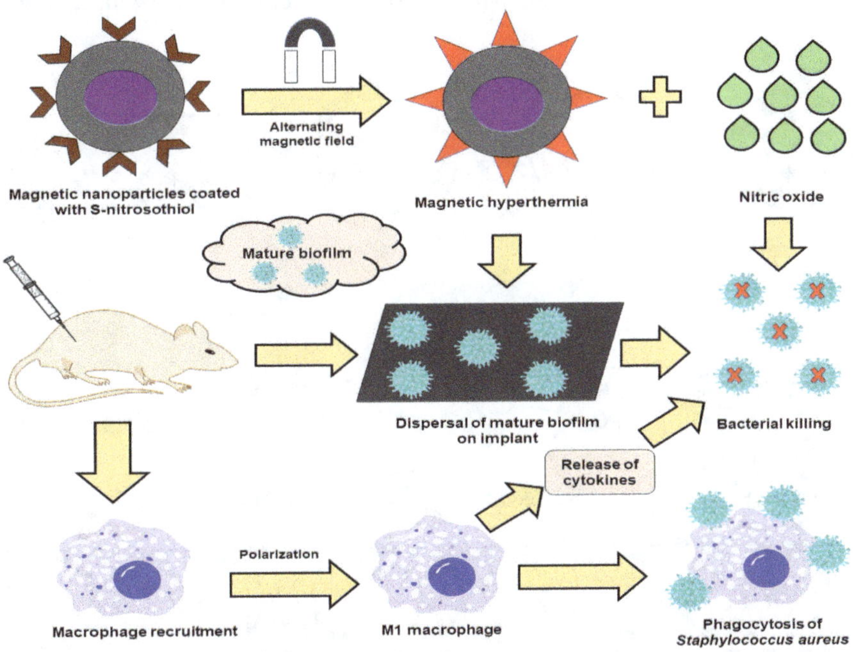

Multimodal Therapy Using Quorum Sensing Inhibitors and PTT

A problem associated with using titanium materials for bone tissue repair and replacement is their inability to form a tight and long-lasting bond with the neighbouring tissues post-implantation. This culminates in the gradual loosening of the orthopedic implant. Hu et al. utilized quorum-sensing inhibitors in combination with PTT to eradicate bacterial biofilms on titanium-based orthopedic implants. Luteolin is a quorum-sensing inhibitor that causes the dispersion of a bacterial biofilm. It is challenging for quorum-sensing inhibitors to reach the interface between the implant and bone tissue after systemic metabolism and blood circulation. Hu et al. loaded luteolin inside mesoporous polydopamine nanoparticles which were used as a photothermal agent. Calcium phosphate coating was applied onto the mesoporous polydopamine nanoparticles and the coated nanoparticles were deposited onto the surface of the titanium substrate. When the modified implant encounters the acidic environment of the bacterial biofilm, the calcium phosphate coating gets degraded to release the entrapped luteolin and Ca^{2+} and PO_4^{3-} (Fig.

4). Release of Ca^{2+} and PO_4^{3-} facilitates osseointegration. Luteolin causes the dispersal of biofilm and renders the bacteria sensitive, irradiation of the implants with NIR light at 808 nm kills the bacteria via PTT. The modified implants showed antibacterial efficacy of 96.1% *in vivo* in the knee joint humeral implant *Staphylococcus aureus* infection model and also resulted in large areas of newly formed bone tissue around the implant (J. Hu et al., 2022).

Figure 4. Schematic diagram of mesoporous polydopamine nanoparticles for multimodal photothermal therapy and inhibition of quorum sensing. Panel A: Mesoporous polydopamine nanoparticles; Panel B: Mesoporous polydopamine nanoparticles loaded with luteolin; Panel C: Mesoporous polydopamine nanoparticles loaded with luteolin and coated with calcium phosphate; Panel D: Deposition of mesoporous polydopamine nanoparticles loaded with luteolin and coated with calcium phosphate on the titanium implant

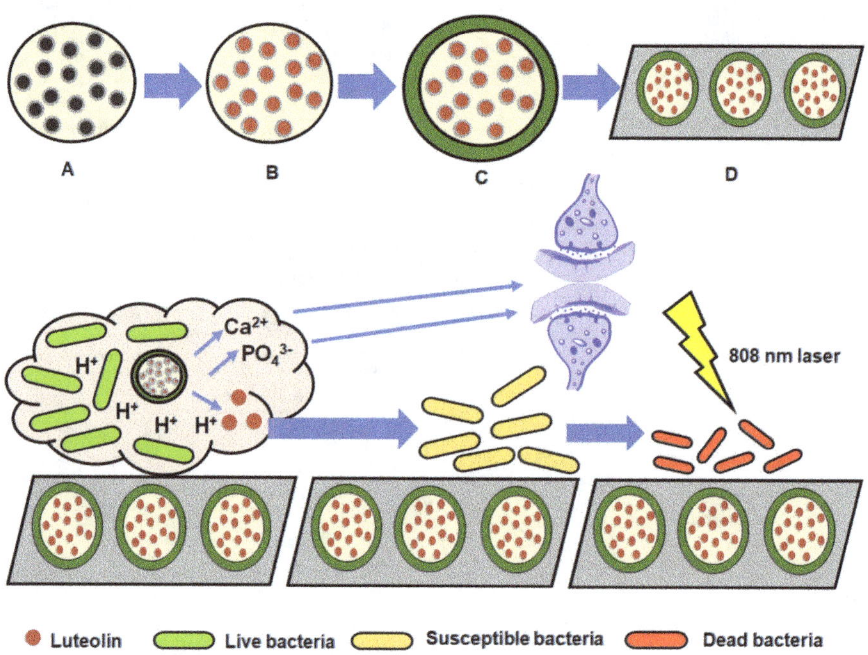

Multimodal Therapy Using PDT and Metallic Nanoparticles

The usage of PDT in combination with metallic nanoparticles would help to decrease the dose of photosensitizer and light required for bacterial destruction, thereby curbing tissue toxicity. Hydroxy apatite is commonly used for the surface coating of orthopedic implants because it is biocompatible and biodegradable. However, it lacks antibacterial activity. Lotfali and Meshkani developed a redox-responsive hydroxyapatite-based nanocomposite having photocatalytic activity. The group first synthesized hydroxyapatite–zinc oxide nanocomposites, coated them with gallic acid, and deposited silver nanoparticles on the surface of the nanocomposites. Lysozyme was then conjugated on the surface of the nanocomposite via a disulfide linkage. The incorporation of zinc oxide with hydroxyapatite confers antibacterial activity to the nanocomposite and also generates a UV-responsive surface. When the nanocomposite is exposed to UV radiation, zinc oxide generates ROS. The developed system showed superior antibacterial activity (> 5 times) against *Escherichia coli* as compared to plain hydroxyapatite attributed to the synergistic antibacterial activity of zinc oxide nanoparticles, silver nanoparticles, gallic acid, and lysozyme. The nanocomposite also eradicated *Escherichia coli* biofilm. The cellular uptake of the nanocomposite and ROS generation by the developed nanocomposite was augmented with the use of UV A irradiation. The developed nanocomposite showed promising potential to be used as a coating on orthopedic devices (Lotfali & Meshkini, 2021).

Multimodal Therapy Using PTT and Metallic Nanoparticles

Fang et al. designed a multifunctional implant by deposition of polydopamine–black phosphorus nanosheets and zinc oxide (ZnO) nanowires onto the titanium substrate. The implant was developed to exert both antibacterial activity and phototherapy. Black phosphorus nanosheets and polydopamine can transform NIR light into heat. When exposed to NIR radiation, the photothermal effect dissipates the biofilm and also enhances the Zn^{2+} releasing ability of the ZnO nanowires, thereby killing the biofilm-producing bacteria. The synergistic effect of PTT and antibacterial ZnO resulted in 99.5% eradication of *Staphylococcus aureus* biofilm *in vivo* in rat model, which is superior to ZnO or PTT used alone. Black phosphorus nanosheets are highly biocompatible and biodegradable corroborating their use for surface modification of orthopedic implants (Fang et al., 2022).

Multimodal Therapy Using PTT and Immunotherapy

Li et al. developed a strategy to eradicate biofilm formation on orthopedic implants via dual PTT and immunotherapy. The group developed a hydrogel coating comprising polyvinyl alcohol modified with chitosan, polydopamine, and NO donor on a red phosphorus nanofilm deposited on a titanium implant. Polydopamine acts as a photothermal agent. Chitosan was used since it has antibacterial activity. NIR irradiation of the system at 808 nm led to peroxynitrite ($.ONOO^-$) formation through a reaction between NO and superoxide ($.O_2^-$) generated via the red phosphorus nanofilm. The release of NO facilitated inflammatory polarization by upregulating the expression of *Opn* and *Ocn* genes and TNF-α. *In vivo* studies in rats revealed excellent bone formation and eradication of MRSA biofilm (Y. Li et al., 2020).

Figure 5. Schematic representation of a strategy combining hydrogen sulfide (H_2S)-sensitized hyperthermia-based bacterial killing with macrophage-linked immune modulation

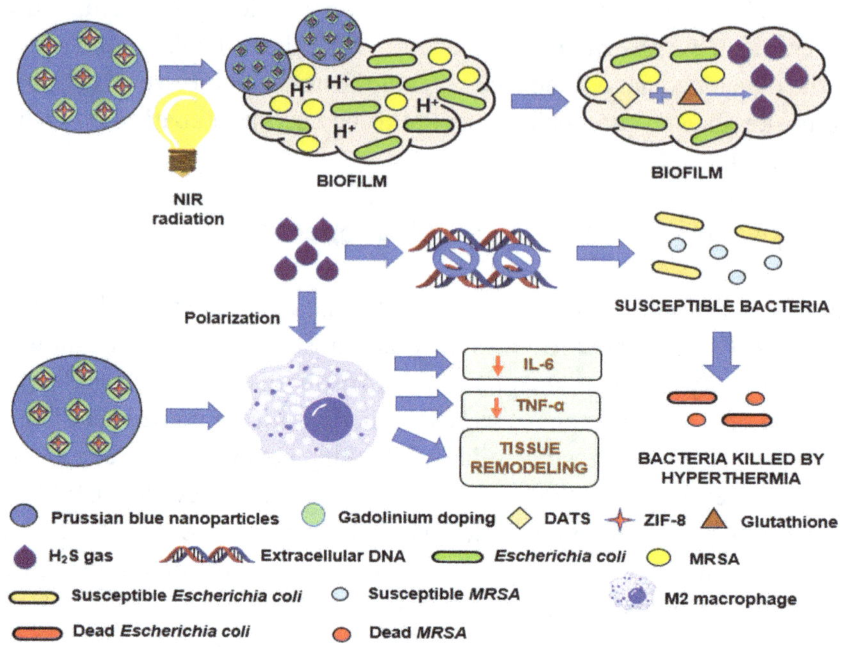

Su et al. developed a strategy combining hydrogen sulfide (H_2S)-sensitized hyperthermia-based bacterial killing with macrophage-linked immune modulation to combat recalcitrant implant-associated bacterial infections. Diallyl trisulphide (DATS) was used for glutathione-responsive generation of H_2S. Prussian blue is a promising substance for NIR-activated hyperthermia owing to its biocompatibility. Doping of Prussian blue with gadolinium improves the magnetic resonance imaging contrast of the former translating into excellent spatial resolution, high sensitivity, and tissue permeability. These features aid in the exact location of biofilm and monitoring of real-time images. Gadolinium-doped Prussian blue can encapsulate drugs in its porous network. However, gatekeepers that can respond to certain triggers present in the biofilm environment or external triggers and release the payload are required to preclude off-target burst release of the encapsulated payload. Prussian blue nanoparticles were synthesized followed by gadolinium doping. DATS was then loaded into these Prussian blue nanoparticles. Metal-organic framework (MOF) ZIF-8 was used as the gatekeeper to seal the pores of gadolinium-doped Prussian blue. MOF remains stable in neutral conditions and is degraded in the acidic biofilm microenvironment. Degradation of MOF releases DATS which then interacts with overexpressed glutathione in the biofilm microenvironment to liberate H_2S (Fig. 5). The liberated H_2S damages the eDNA. This collapses the tough biofilm structure and enhances the sensitivity of the bacteria released from the biofilm to hyperthermia resulting from NIR-activated Prussian blue. H_2S also facilitates the polarization of macrophages toward the M2 phenotype. Due to stimulation by MOF-gated Prussian blue nanoparticles containing DATS, M2 macrophages secrete cytokines to foster tissue remodeling. The group investigated the antibiofilm effects of the developed strategy in a mouse model of implant-associated bacterial infection. Immunomodulatory effects and tissue remodeling were also studied in mice. MRSA-infested PEEK implants were implanted inside the mouse. The developed strategy in combination with NIR radiation was found to reverse bacterial infection-induced pro-inflammatory microenvironment and facilitate tissue remodeling (Z. Su et al., 2022).

Multimodal Therapy Using PDT, PTT, and Quorum Sensing Inhibitor

Preventing quorum sensing in bacterial biofilms via the usage of analogs of molecules involved in quorum sensing is a promising strategy to tackle biofilms (Khatoon et al., 2018). Curcumin is known to inhibit quorum sensing. Zhang et al. developed multifunctional titanium nanorods for biofilm eradication via PDT, PTT, and quorum sensing; and bone regeneration. The nanorods were doped with F, Yb, and Ho to enhance photocatalytic activity. Curcumin was immobilized onto the nanorods. When the multifunctional nanorods are illuminated with a 1060 nm laser, they can eradicate *Staphylococcus aureus* biofilms via combined PDT, PTT, quorum sensing, and physical puncturing at a mild temperature of 45 °C. BMP-2 was also grafted onto the titanium nanorods. BMP-2 confers osteogenic differentiation capacity to the system (G. Zhang et al., 2021).

Miscellaneous

Guo et al. explored space-selective chemodynamic therapy using metastable $CuFe_5O_8$ nanocubes to tackle implant-associated infections. A high concentration of hydroxyl radials (.OH) cleaves the eDNA in the biofilm. This cleavage is catalyzed by $CuFe_5O_8$ nanocubes. The environment surrounding the biofilm has a low concentration of H_2O_2 and higher pH. ROS such as hydroxyl radicals, when present at low levels trigger the expression of pro-inflammatory cytokines thus facilitating the inflammatory response against bacteria and their elimination. Destroying the intricate biofilm architecture requires massive cleavage of eDNA. Therefore, a Fenton-like reaction catalyst is required to increase the concentration of hydroxyl radicals and catalyze the cleavage of eDNA within the biofilm. Further, disruption of the biofilm matrix would liberate biofilm fragments and planktonic bacteria which need to be eliminated from the body. These biofilm fragments and planktonic bacteria can otherwise lead to acute infection at other sites. Using smart catalysts like $CuFe_5O_8$ nanocubes results in low levels of hydroxyl radicals outside the biofilm thus activating the pro-inflammatory arm of the immune system. Thus, the $CuFe_5O_8$ nanocubes exert a dual function of the destruction of eDNA and remodeling the immune response. The developed nanocubes demonstrated excellent activity against titanium implant-associated *Staphylococcus aureus* biofilm infection in a mouse model (Guo et al., 2020).

CONCLUSION

There will be an increased demand for orthopedic implants in the coming years, given the global rise in the elderly population. Orthopedic implant-associated infections can result in devastating consequences of joint deformity, amputation, and even death. The need of the hour is to prevent biofilm formation on the surface of orthopedic implants and improve their function. Treatment of orthopedic implant-associated bacterial infections requires simultaneous eradication of bacterial load and promotion of tissue regeneration in the vicinity of implants. The idea is to combine materials having biomimetic and osseointegration properties with antimicrobial substances to expedite recovery and mobility in patients.

Currently, available options for antibiotic use and surgery are not reliable owing to their cost and morbidity coupled with the grave issue of bacterial resistance.

The eradication of biofilms on orthopedic implants will require multifunctional systems addressing various aspects of biofilm formation. Combination therapies are required to tackle both metabolically active and inactive bacteria present within biofilms. Nanoparticles used as a coating on implants should release the drug in a slow and sustained manner to eradicate the biofilm. When designing therapies to combat bacterial biofilms on orthopedic implants, one needs to consider host-pathogen interactions as well and not just focus on the biofilm matrix.

Nanoparticles undoubtedly possess immense potential in the battle against tough biofilm-forming orthopedic implant-associated infections. However, certain issues need to be addressed in detail before nanoparticles can make their way into the clinic. Fabrication of nanoparticles should provide reproducible batches. Further, these nanoparticles need to retain their structural and functional stability during sterilization and storage. Results obtained from *in vivo* studies in animal models may not necessarily guarantee similar efficacy in humans owing to the physiological differences between animals and humans. Further, toxicity studies of nanoparticles in humans are lacking. It is necessary to investigate the interaction of nanoparticles with cells, tissues, and organs in the human body to establish their safety and foster their clinical translation. Fundamental studies addressing the interactions of nanoparticles with biofilm are important. Successful collaboration between formulation scientists, clinicians, microbiologists, immunologists, toxicologists, and material science engineers would foster the development and evaluation of nanoparticle-modified orthopedic implants.

It is important to study the performance of the modified orthopedic implants in a combined model of bone defect and exogenous infection to get a true picture of the efficacy of the system. Also, the dose of bacteria used to induce infection is crucial.

REFERENCES

Abouaitah, K., Bil, M., Pietrzykowska, E., Szałaj, U., Fudala, D., Woźniak, B., Nasiłowska, J., Swiderska-Sroda, A., Lojkowski, M., Sokołowska, B., Swieszkowski, W., & Lojkowski, W. (2021). Drug-releasing antibacterial coating made from nano-hydroxyapatite using the sonocoating method. *Nanomaterials (Basel, Switzerland)*, *11*(7), 1690. doi:10.3390/nano11071690 PMID:34203218

Altun, E., Aydogdu, M. O., Chung, E., Ren, G., Homer-Vanniasinkam, S., & Edirisinghe, M. (2021). Metal-based nanoparticles for combating antibiotic resistance. *Applied Physics Reviews*, *8*(4), 041303. doi:10.1063/5.0060299

Arciola, C. R., Campoccia, D., Speziale, P., Montanaro, L., & Costerton, J. W. (2012). Biofilm formation in *Staphylococcus* implant infections. A review of molecular mechanisms and implications for biofilm-resistant materials. *Biomaterials*, *33*(26), 5967–5982. doi:10.1016/j.biomaterials.2012.05.031 PMID:22695065

Ashbaugh, A. G., Jiang, X., Zheng, J., Tsai, A. S., Kim, W. S., Thompson, J. M., Miller, R. J., Shahbazian, J. H., Wang, Y., Dillen, C. A., Ordonez, A. A., Chang, Y. S., Jain, S. K., Jones, L. C., Sterling, R. S., Mao, H. Q., & Miller, L. S. (2016). Polymeric nanofiber coating with tunable combinatorial antibiotic delivery prevents biofilm-associated infection *in vivo*. *Proceedings of the National Academy of Sciences of the United States of America*, *113*(45), E6919–E6928. doi:10.1073/pnas.1613722113 PMID:27791154

Bee, S. L., Bustami, Y., Ul-Hamid, A., Lim, K., & Abdul Hamid, Z. A. (2021). Synthesis of silver nanoparticle-decorated hydroxyapatite nanocomposite with combined bioactivity and antibacterial properties. *Journal of Materials Science. Materials in Medicine*, *32*(9), 106. doi:10.1007/s10856-021-06590-y PMID:34426879

Behbahani, S. B., Kiridena, S. D., Wijayaratna, U. N., Taylor, C., Anker, J. N., & Tzeng, T. R. J. (2022). pH variation in medical implant biofilms: Causes, measurements, and its implications for antibiotic resistance. *Frontiers in Microbiology*, *13*(October), 1–28. doi:10.3389/fmicb.2022.1028560 PMID:36386694

Berini, F., Orlandi, V., Gornati, R., Bernardini, G., & Marinelli, F. (2022). Nanoantibiotics to fight multidrug resistant infections by Gram-positive bacteria: Hope or reality? *Biotechnology Advances*, *57*, 107948. doi:10.1016/j.biotechadv.2022.107948 PMID:35337933

Billings, C., & Anderson, D. E. (2022). Role of Implantable Drug Delivery Devices with Dual Platform Capabilities in the Prevention and Treatment of Bacterial Osteomyelitis. *Bioengineering (Basel, Switzerland)*, *9*(2), 65. doi:10.3390/bioengineering9020065 PMID:35200418

Calabrese, G., Franco, D., Petralia, S., Monforte, F., Condorelli, G. G., Squarzoni, S., Traina, F., & Conoci, S. (2021). Dual-functional nano-functionalized titanium scaffolds to inhibit bacterial growth and enhance osteointegration. *Nanomaterials (Basel, Switzerland)*, *11*(10), 2634. doi:10.3390/nano11102634 PMID:34685075

Caplin, J. D., & García, A. J. (2019). Implantable antimicrobial biomaterials for local drug delivery in bone infection models. *Acta Biomaterialia*, *93*, 2–11. doi:10.1016/j.actbio.2019.01.015 PMID:30654212

Chen, J., Shi, X., Zhu, Y., Chen, Y., Gao, M., Gao, H., Liu, L., Wang, L., Mao, C., & Wang, Y. (2020). On-demand storage and release of antimicrobial peptides using Pandora's box-like nanotubes gated with a bacterial infection-responsive polymer. *Theranostics*, *10*(1), 109–122. doi:10.7150/thno.38388 PMID:31903109

Cheng, H., Xiong, W., Fang, Z., Guan, H., Wu, W., Li, Y., Zhang, Y., Alvarez, M. M., Gao, B., Huo, K., Xu, J., Xu, N., Zhang, C., Fu, J., Khademhosseini, A., & Li, F. (2016). Strontium (Sr) and silver (Ag) loaded nanotubular structures with combined osteoinductive and antimicrobial activities. *Acta Biomaterialia*, *31*, 388–400. doi:10.1016/j.actbio.2015.11.046 PMID:26612413

Chopra, D., Gulati, K., & Ivanovski, S. (2021). Understanding and optimizing the antibacterial functions of anodized nano-engineered titanium implants. *Acta Biomaterialia*, *127*, 80–101. doi:10.1016/j.actbio.2021.03.027 PMID:33744499

Connaughton, A., Childs, A., Dylewski, S., & Sabesan, V. J. (2014). Biofilm disrupting technology for orthopedic implants: What's on the horizon? *Frontiers in Medicine*, *1*(AUG), 1–4. doi:10.3389/fmed.2014.00022 PMID:25705632

Dhar, Y., & Han, Y. (2020). Current developments in biofilm treatments: Wound and implant infections. *Engineered Regeneration*, *1*, 64–75. doi:10.1016/j.engreg.2020.07.003

Ding, Y., Hao, Y., Yuan, Z., Tao, B., Chen, M., Lin, C., Liu, P., & Cai, K. (2020). A dual-functional implant with an enzyme-responsive effect for bacterial infection therapy and tissue regeneration. *Biomaterials Science*, *8*(7), 1840–1854. doi:10.1039/C9BM01924C PMID:31967110

Dong, J., Fang, D., Zhang, L., Shan, Q., & Huang, Y. (2019). Gallium-doped titania nanotubes elicit anti-bacterial efficacy *in vivo* against *Escherichia coli* and *Staphylococcus aureus* biofilm. *Materialia*, *5*, 100209. doi:10.1016/j.mtla.2019.100209

Dwivedi, A., Mazumder, A., & Nasongkla, N. (2018). Layer-by-layer nanocoating of antibacterial niosome on orthopedic implant. *International Journal of Pharmaceutics*, *547*(1–2), 235–243. doi:10.1016/j.ijpharm.2018.05.075 PMID:29864515

Fang, J., Wan, Y., Sun, Y., Sun, X., Qi, M., Cheng, S., Li, C., Zhou, Y., Xu, L., Dong, B., & Wang, L. (2022). Near-infrared-activated nanohybrid coating with black phosphorus/zinc oxide for efficient biofilm eradication against implant-associated infections. *Chemical Engineering Journal*, *435*(P1), 134935. doi:10.1016/j.cej.2022.134935

Gavaskar, A., Rojas, D., & Videla, F. (2018). Nanotechnology: The scope and potential applications in orthopedic surgery. *European Journal of Orthopaedic Surgery & Traumatology*, *28*(7), 1257–1260. doi:10.1007/s00590-018-2193-z PMID:29602957

Ghilini, F., Fagali, N., Pissinis, D. E., Benítez, G., & Schilardi, P. L. (2021). Multifunctional Titanium Surfaces for Orthopedic Implants: Antimicrobial Activity and Enhanced Osseointegration. *ACS Applied Bio Materials*, *4*(8), 6451–6461. doi:10.1021/acsabm.1c00613 PMID:35006865

Guo, G., Zhang, H., Shen, H., Zhu, C., He, R., Tang, J., Wang, Y., Jiang, X., Wang, J., Bu, W., & Zhang, X. (2020). Space-selective chemodynamic therapy of $CuFe_5O_8$ nanocubes for implant-related infections. *ACS Nano*, *14*(10), 13391–13405. doi:10.1021/acsnano.0c05255 PMID:32931252

Hong, L., Liu, X., Tan, L., Cui, Z., Yang, X., Liang, Y., Li, Z., Zhu, S., Zheng, Y., Yeung, K. W. K., Jing, D., Zheng, D., Wang, X., & Wu, S. (2019). Rapid Biofilm Elimination on Bone Implants Using Near-Infrared-Activated Inorganic Semiconductor Heterostructures. *Advanced Healthcare Materials*, *8*(19), 1–11. doi:10.1002/adhm.201900835 PMID:31464096

Hong, Q., Huo, S., Tang, H., Qu, X., & Yue, B. (2021). Smart Nanomaterials for Treatment of Biofilm in Orthopedic Implants. *Frontiers in Bioengineering and Biotechnology*, *9*(September), 694635. doi:10.3389/fbioe.2021.694635 PMID:34589470

Hu, C. C., Chang, C. H., Chang, Y., Hsieh, J. H., & Ueng, S. W. N. (2020). Beneficial effect of TaON-Ag nanocomposite titanium on antibacterial capacity in orthopedic application. *International Journal of Nanomedicine*, *15*, 7889–7900. doi:10.2147/IJN.S264303 PMID:33116507

Hu, J., Ding, Y., Tao, B., Yuan, Z., Yang, Y., Xu, K., Li, X., Liu, P., & Cai, K. (2022). Surface modification of titanium substrate via combining photothermal therapy and quorum-sensing-inhibition strategy for improving osseointegration and treating biofilm-associated bacterial infection. *Bioactive Materials*, *18*, 228–241. doi:10.1016/j.bioactmat.2022.03.011 PMID:35387171

Hu, Y., Ruan, X., Lv, X., Xu, Y., Wang, W., Cai, Y., Ding, M., Dong, H., Shao, J., Yang, D., & Dong, X. (2022). Biofilm microenvironment-responsive nanoparticles for the treatment of bacterial infection. *Nano Today*, *46*, 101602. doi:10.1016/j.nantod.2022.101602

Indelli, P. F., Ghirardelli, S., Iannotti, F., Indelli, A. M., & Pipino, G. (2021). Nanotechnology as an anti-infection strategy in periprosthetic joint infections (PJI). *Tropical Medicine and Infectious Disease*, *6*(2), 1–6. doi:10.3390/tropicalmed6020091 PMID:34071727

Intravaia, J. T., Graham, T., Kim, H. S., Nanda, H. S., Kumbar, S. G., & Nukavarapu, S. P. (2023). Smart orthopedic biomaterials and implants. *Current Opinion in Biomedical Engineering*, *25*, 100439. doi:10.1016/j.cobme.2022.100439 PMID:36642994

Jaggessar, A., Shahali, H., Mathew, A., & Yarlagadda, P. K. D. V. (2017). Bio-mimicking nano and micro-structured surface fabrication for antibacterial properties in medical implants. *Journal of Nanobiotechnology*, *15*(1), 1–20. doi:10.1186/s12951-017-0306-1 PMID:28969628

Jia, Z., Xiu, P., Li, M., Xu, X., Shi, Y., Cheng, Y., Wei, S., Zheng, Y., Xi, T., Cai, H., & Liu, Z. (2016). Bioinspired anchoring AgNPs onto micro-nanoporous TiO_2 orthopedic coatings: Trap-killing of bacteria, surface-regulated osteoblast functions and host responses. *Biomaterials*, *75*, 203–222. doi:10.1016/j.biomaterials.2015.10.035 PMID:26513414

Khatoon, Z., McTiernan, C. D., Suuronen, E. J., Mah, T. F., & Alarcon, E. I. (2018). Bacterial biofilm formation on implantable devices and approaches to its treatment and prevention. *Heliyon*, *4*(12), e01067. doi:10.1016/j.heliyon.2018.e01067 PMID:30619958

Li, B., Ma, J., Wang, D., Liu, X., Li, H., Zhou, L., Liang, C., & Wang, H. (2019). Self-adjusting antibacterial properties of Ag-incorporated nanotubes on micro-nanostructured Ti surfaces. *Biomaterials Science*, *7*(10), 4075–4087. doi:10.1039/C9BM00862D PMID:31355387

Li, M., Zhang, D., Peng, F., Xie, J., Zhang, X., Qian, S., Zhang, Y., Liu, X., & Yu, B. (2022). Zinc-doped ferric oxyhydroxide nano-layer enhances the bactericidal activity and osseointegration of a magnesium alloy through augmenting the formation of neutrophil extracellular traps. *Acta Biomaterialia*, *152*, 575–592. doi:10.1016/j.actbio.2022.08.066 PMID:36070834

Li, Y., Liu, X., Li, B., Zheng, Y., Han, Y., Chen, D. F., Yeung, K. W. K., Cui, Z., Liang, Y., Li, Z., Zhu, S., Wang, X., & Wu, S. (2020). Near-Infrared Light Triggered Phototherapy and Immunotherapy for Elimination of Methicillin-Resistant *Staphylococcus aureus* Biofilm Infection on Bone Implant. *ACS Nano*, *14*(7), 8157–8170. doi:10.1021/acsnano.0c01486 PMID:32585104

Lin, M. H., Wang, Y. H., Kuo, C. H., Ou, S. F., Huang, P. Z., Song, T. Y., Chen, Y. C., Chen, S. T., Wu, C. H., Hsueh, Y. H., & Fan, F. Y. (2021). Hybrid ZnO/chitosan antimicrobial coatings with enhanced mechanical and bioactive properties for titanium implants. *Carbohydrate Polymers*, *257*, 117639. doi:10.1016/j.carbpol.2021.117639 PMID:33541664

Lin, X., Yang, S., Lai, K., Yang, H., Webster, T. J., & Yang, L. (2017). Orthopedic implant biomaterials with both osteogenic and anti-infection capacities and associated *in vivo* evaluation methods. *Nanomedicine; Nanotechnology, Biology, and Medicine*, *13*(1), 123–142. doi:10.1016/j.nano.2016.08.003 PMID:27553074

Lin, Z., Zhao, Y., Chu, P. K., Wang, L., Pan, H., Zheng, Y., Wu, S., Liu, X., Cheung, K. M. C., Wong, T., & Yeung, K. W. K. (2019). A functionalized TiO_2/Mg_2TiO_4 nano-layer on biodegradable magnesium implant enables superior bone-implant integration and bacterial disinfection. *Biomaterials, 219*, 119372. doi:10.1016/j.biomaterials.2019.119372 PMID:31362176

Liu, H. W., Wei, D. X., Deng, J. Z., Zhu, J. J., Xu, K., Hu, W. H., Xiao, S. H., & Zhou, Y. G. (2018). Combined antibacterial and osteogenic in situ effects of a bifunctional titanium alloy with nanoscale hydroxyapatite coating. *Artificial Cells, Nanomedicine and Biotechnology, 46*(sup3), S460–S470. doi: 10.1080/21691401.2018.1499662

Liu, J., Liu, J., Attarilar, S., Wang, C., Tamaddon, M., Yang, C., Xie, K., Yao, J., Wang, L., Liu, C., & Tang, Y. (2020). Nano-Modified Titanium Implant Materials: A Way Toward Improved Antibacterial Properties. *Frontiers in Bioengineering and Biotechnology, 8*(November), 576969. doi:10.3389/fbioe.2020.576969 PMID:33330415

Liu, W., Li, J., Cheng, M., Wang, Q., Qian, Y., Yeung, K. W. K., Chu, P. K., & Zhang, X. (2019). A surface-engineered polyetheretherketone biomaterial implant with direct and immunoregulatory antibacterial activity against methicillin-resistant *Staphylococcus aureus*. *Biomaterials, 208*, 8–20. doi:10.1016/j.biomaterials.2019.04.008 PMID:30986611

Liu, X., Chen, H. H., Lin, Y. C., Nabilla, S. C., Liu, W. C., Wang, W. C., Shih, S. J., Li, Y., Lin, C. P., Zhao, G., & Chung, R. J. (2018). Composite polyelectrolyte multilayer and mesoporous bioactive glass nanoparticle coating on 316L stainless steel for controlled antibiotic release and biocompatibility. *Journal of Biomedical Nanotechnology, 14*(4), 725–735. doi:10.1166/jbn.2018.2531 PMID:31352946

Lotfali, H., & Meshkini, A. (2021). Synthesis and characterization of lysozyme-conjugated Ag.ZnO@HA nanocomposite: A redox and pH-responsive antimicrobial agent with photocatalytic activity. *Photodiagnosis and Photodynamic Therapy, 35*, 102418. doi:10.1016/j.pdpdt.2021.102418 PMID:34197967

Mahamuni-Badiger, P. P., Patil, P. M., Badiger, M. V., Patel, P. R., Thorat-Gadgil, B. S., Pandit, A., & Bohara, R. A. (2020). Biofilm formation to inhibition: Role of zinc oxide-based nanoparticles. *Materials Science and Engineering C, 108*, 110319. doi:10.1016/j.msec.2019.110319 PMID:31923962

Mehrjou, B., Mo, S., Dehghan-Baniani, D., Wang, G., Qasim, A. M., & Chu, P. K. (2019). Antibacterial and Cytocompatible Nanoengineered Silk-Based Materials for Orthopedic Implants and Tissue Engineering. *ACS Applied Materials & Interfaces, 11*(35), 31605–31614. doi:10.1021/acsami.9b09066 PMID:31385497

Min, J., Choi, K. Y., Dreaden, E. C., Padera, R. F., Braatz, R. D., Spector, M., & Hammond, P. T. (2016). Designer Dual Therapy Nanolayered Implant Coatings Eradicate Biofilms and Accelerate Bone Tissue Repair. *ACS Nano, 10*(4), 4441–4450. doi:10.1021/acsnano.6b00087 PMID:26923427

Mosselhy, D. A., Assad, M., Sironen, T., & Elbahri, M. (2021). Nanotheranostics: A possible solution for drug-resistant *Staphylococcus aureus* and their biofilms? *Nanomaterials (Basel, Switzerland), 11*(1), 1–36. doi:10.3390/nano11010082 PMID:33401760

Moure, J. S. F., Evangelopoulos, M., Colvill, K., Van Eps, J. L., & Tasciotti, E. (2017). Nanoantibiotics: A new paradigm for the treatment of surgical infection. *Nanomedicine (London), 12*(11), 1319–1334. doi:10.2217/nnm-2017-0401 PMID:28520517

Naskar, A., & Kim, K. S. (2023). Friends against the Foe: Synergistic Photothermal and Photodynamic Therapy against Bacterial Infections. *Pharmaceutics, 15*(4), 1116. doi:10.3390/pharmaceutics15041116 PMID:37111601

Nie, B., Huo, S., Qu, X., Guo, J., Liu, X., Hong, Q., Wang, Y., Yang, J., & Yue, B. (2022). Bone infection site targeting nanoparticle-antibiotics delivery vehicle to enhance treatment efficacy of orthopedic implant related infection. *Bioactive Materials, 16*(February), 134–148. doi:10.1016/j.bioactmat.2022.02.003 PMID:35386313

Pawar, V., Topkar, H., & Srivastava, R. (2018). Chitosan nanoparticles and povidone iodine containing alginate gel for prevention and treatment of orthopedic implant associated infections. *International Journal of Biological Macromolecules, 115*, 1131–1141. doi:10.1016/j.ijbiomac.2018.04.166 PMID:29727653

Qing, Y., Cheng, L., Li, R., Liu, G., Zhang, Y., Tang, X., Wang, J., Liu, H., & Qin, Y. (2018). Potential antibacterial mechanism of silver nanoparticles and the optimization of orthopedic implants by advanced modification technologies. *International Journal of Nanomedicine, 13*, 3311–3327. doi:10.2147/IJN.S165125 PMID:29892194

Quan, K., Zhang, Z., Ren, Y., Busscher, H. J., van der Mei, H. C., & Peterson, B. W. (2021). On-demand pulling-off of magnetic nanoparticles from biomaterial surfaces through implant-associated infectious biofilms for enhanced antibiotic efficacy. *Materials Science and Engineering C, 131*, 112526. doi:10.1016/j.msec.2021.112526 PMID:34857305

Raphel, J., Holodniy, M., Goodman, S. B., & Heilshorn, S. C. (2016). Multifunctional coatings to simultaneously promote osseointegration and prevent infection of orthopaedic implants. *Biomaterials, 84*, 301–314. doi:10.1016/j.biomaterials.2016.01.016 PMID:26851394

Rodríguez-Merchán, E. C., Davidson, D. J., & Liddle, A. D. (2021). Recent strategies to combat infections from biofilm-forming bacteria on orthopaedic implants. *International Journal of Molecular Sciences, 22*(19), 10243. doi:10.3390/ijms221910243 PMID:34638591

Sahoo, J., Sarkhel, S., Mukherjee, N., & Jaiswal, A. (2022). Nanomaterial-Based Antimicrobial Coating for Biomedical Implants: New Age Solution for Biofilm-Associated Infections. *ACS Omega, 7*(50), 45962–45980. doi:10.1021/acsomega.2c06211 PMID:36570317

San, H., Paresoglou, M., Minneboo, M., van Hengel, I. A. J., Yilmaz, A., Gonzalez-Garcia, Y., Fluit, A. C., Hagedoorn, P. L., Fratila-Apachitei, L. E., Apachitei, I., & Zadpoor, A. A. (2022). Fighting Antibiotic-Resistant Bacterial Infections by Surface Biofunctionalization of 3D-Printed Porous Titanium Implants with Reduced Graphene Oxide and Silver Nanoparticles. *International Journal of Molecular Sciences, 23*(16), 9204. doi:10.3390/ijms23169204 PMID:36012467

Shen, X., Zhang, F., Li, K., Qin, C., Ma, P., Dai, L., & Cai, K. (2016). Cecropin B loaded TiO_2 nanotubes coated with hyaluronidase sensitive multilayers for reducing bacterial adhesion. *Materials & Design, 92*, 1007–1017. doi:10.1016/j.matdes.2015.12.126

Skłodowski, K., Chmielewska-Deptuła, S. J., Piktel, E., Wolak, P., Wollny, T., & Bucki, R. (2023). Metallic Nanosystems in the Development of Antimicrobial Strategies with High Antimicrobial Activity and High Biocompatibility. *International Journal of Molecular Sciences*, *24*(3), 2104. doi:10.3390/ijms24032104 PMID:36768426

Su, E. P., Justin, D. F., Pratt, C. R., Sarin, V. K., Nguyen, V. S., Oh, S., & Jin, S. (2018). Effects of titanium nanotubes on the osseointegration, cell differentiation, mineralisation and antibacterial properties of orthopaedic implant surfaces. *The Bone & Joint Journal*, *100B*(1), 9–16. doi:10.1302/0301-620X.100B1.BJJ-2017-0551.R1 PMID:29292334

Su, Z., Kong, L., Dai, Y., Tang, J., Mei, J., Qian, Z., Ma, Y., Li, Q., Ju, S., Wang, J., Fan, W., & Zhu, C. (2022). Bioresponsive nano-antibacterials for H_2S-sensitized hyperthermia and immunomodulation against refractory implant–related infections. *Science Advances*, *8*(14), 1–20. doi:10.1126/sciadv.abn1701 PMID:35394829

Tang, X., Dai, J., Sun, H., Nabanita, S., Petr, S., Wang, D., Cheng, Q., & Wei, J. (2019). Mechanical Strength, Surface Properties, Cytocompatibility and Antibacterial Activity of Nano Zinc-Magnesium Silicate/Polyetheretherketone Biocomposites. *Journal of Nanoscience and Nanotechnology*, *19*(12), 7615–7623. doi:10.1166/jnn.2019.16727 PMID:31196268

Tao, B., Yi, W., Qin, X., Wu, J., Li, K., Guo, A., Hao, J., & Chen, L. (2023). Improvement of antibacterial, anti-inflammatory, and osteogenic properties of OGP loaded Co-MOF coating on titanium implants for advanced osseointegration. *Journal of Materials Science and Technology*, *146*, 131–144. doi:10.1016/j.jmst.2022.11.013

Tran, P. A., O'brien-Simpson, N., Palmer, J. A., Bock, N., Reynolds, E. C., Webster, T. J., Deva, A., Morrison, W. A., & O'connor, A. J. (2019). Selenium nanoparticles as anti-infective implant coatings for trauma orthopedics against methicillin-resistant *Staphylococcus aureus* and *epidermidis: In vitro* and *in vivo* assessment. *International Journal of Nanomedicine*, *14*, 4613–4624. doi:10.2147/IJN.S197737 PMID:31308651

van Hengel, I. A. J., Riool, M., Fratila-Apachitei, L. E., Witte-Bouma, J., Farrell, E., Zadpoor, A. A., Zaat, S. A. J., & Apachitei, I. (2017). Selective laser melting porous metallic implants with immobilized silver nanoparticles kill and prevent biofilm formation by methicillin-resistant *Staphylococcus aureus*. *Biomaterials*, *140*, 1–15. doi:10.1016/j.biomaterials.2017.02.030 PMID:28622569

Wang, B., Wu, Z., Lan, J., Li, Y., Xie, L., Huang, X., Zhang, A., Qiao, H., Chang, X., Lin, H., Zhang, H., Li, T., & Huang, Y. (2021). Surface modification of titanium implants by silk fibroin/Ag co-functionalized strontium titanate nanotubes for inhibition of bacterial-associated infection and enhancement of *in vivo* osseointegration. *Surface and Coatings Technology*, *405*, 126700. doi:10.1016/j.surfcoat.2020.126700

Wang, J., Wang, L., Pan, J., Zhao, J., Tang, J., Jiang, D., Hu, P., Jia, W., & Shi, J. (2021). Magneto-Based Synergetic Therapy for Implant-Associated Infections via Biofilm Disruption and Innate Immunity Regulation. *Advancement of Science*, *8*(6), 1–12. doi:10.1002/advs.202004010 PMID:33747747

Wang, J., Zhou, H., Guo, G., Tan, J., Wang, Q., Tang, J., Liu, W., Shen, H., Li, J., & Zhang, X. (2017). Enhanced anti-infective efficacy of ZnO nanoreservoirs through a combination of intrinsic anti-biofilm activity and reinforced innate defense. *ACS Applied Materials & Interfaces*, *9*(39), 33609–33623. doi:10.1021/acsami.7b08864 PMID:28884578

Wang, M., & Tang, T. (2019). Surface treatment strategies to combat implant-related infection from the beginning. *Journal of Orthopaedic Translation*, *17*, 42–54. doi:10.1016/j.jot.2018.09.001 PMID:31194031

Wang, W., Cheng, X., Liao, J., Lin, Z., Chen, L., Liu, D., Zhang, T., Li, L., Lu, Y., & Xia, H. (2019). Synergistic Photothermal and Photodynamic Therapy for Effective Implant-Related Bacterial Infection Elimination and Biofilm Disruption Using Cu_9S_8 Nanoparticles. *ACS Biomaterials Science & Engineering*, *5*(11), 6243–6253. doi:10.1021/acsbiomaterials.9b01280 PMID:33405531

Wang, Y., Teng, W., Zhang, Z., Zhou, X., Ye, Y., Lin, P., Liu, A., Wu, Y., Li, B., Zhang, C., Yang, X., Li, W., Yu, X., Gou, Z., & Ye, Z. (2021). A trilogy antimicrobial strategy for multiple infections of orthopedic implants throughout their life cycle. *Bioactive Materials*, *6*(7), 1853–1866. doi:10.1016/j.bioactmat.2020.11.030 PMID:33336116

Wei, H., Song, X., Liu, P., Liu, X., Yan, X., & Yu, L. (2022). Antimicrobial coating strategy to prevent orthopaedic device-related infections: Recent advances and future perspectives. *Biomaterials Advances*, *135*, 212739. doi:10.1016/j.bioadv.2022.212739 PMID:35929213

Wickramasinghe, S., Ju, M., Milbrandt, N. B., Tsai, Y. H., Navarreto-Lugo, M., Visperas, A., Klika, A., Barsoum, W., Higuera-Rueda, C. A., & Samia, A. C. S. (2020). Photoactivated Gold Nanorod Hydrogel Composite Containing D-Amino Acids for the Complete Eradication of Bacterial Biofilms on Metal Alloy Implant Materials. *ACS Applied Nano Materials*, *3*(6), 5862–5873. doi:10.1021/acsanm.0c01018

Wu, Z., Chan, B., Low, J., Chu, J. J. H., Hey, H. W. D., & Tay, A. (2022). Microbial resistance to nanotechnologies: An important but understudied consideration using antimicrobial nanotechnologies in orthopaedic implants. *Bioactive Materials*, *16*, 249–270. doi:10.1016/j.bioactmat.2022.02.014 PMID:35415290

Xiang, Y., Liu, X., Mao, C., Liu, X., Cui, Z., Yang, X., Yeung, K. W. K., Zheng, Y., & Wu, S. (2018). Infection-prevention on Ti implants by controlled drug release from folic acid/ZnO quantum dots sealed titania nanotubes. *Materials Science and Engineering C*, *85*, 214–224. doi:10.1016/j.msec.2017.12.034 PMID:29407150

Yan, J., Xia, D., Zhou, W., Li, Y., Xiong, P., Li, Q., Wang, P., Li, M., Zheng, Y., & Cheng, Y. (2020). pH-responsive silk fibroin-based CuO/Ag micro/nano coating endows polyetheretherketone with synergistic antibacterial ability, osteogenesis, and angiogenesis. *Acta Biomaterialia*, *115*, 220–234. doi:10.1016/j.actbio.2020.07.062 PMID:32777292

Yang, Y., Liu, L., Luo, H., Zhang, D., Lei, S., & Zhou, K. (2019). Dual-Purpose Magnesium-Incorporated Titanium Nanotubes for Combating Bacterial Infection and Ameliorating Osteolysis to Realize Better Osseointegration. *ACS Biomaterials Science & Engineering*, *5*(10), 5368–5383. doi:10.1021/acsbiomaterials.9b00938 PMID:33464078

Yao, C., Zhu, M., Han, X., Xu, Q., Dai, M., Nie, T., & Liu, X. (2021). A Bone-Targeting Enoxacin Delivery System to Eradicate *Staphylococcus aureus*-Related Implantation Infections and Bone Loss. *Frontiers in Bioengineering and Biotechnology*, *9*(November), 1–17. doi:10.3389/fbioe.2021.749910 PMID:34869262

Yuan, Z., Tao, B., He, Y., Mu, C., Liu, G., Zhang, J., Liao, Q., Liu, P., & Cai, K. (2019). Remote eradication of biofilm on titanium implant via near-infrared light triggered photothermal/photodynamic therapy strategy. *Biomaterials*, *223*, 119479. doi:10.1016/j.biomaterials.2019.119479 PMID:31520885

Zhang, G., Yang, Y., Shi, J., Yao, X., Chen, W., Wei, X., Zhang, X., & Chu, P. K. (2021). Near-infrared light II - assisted rapid biofilm elimination platform for bone implants at mild temperature. *Biomaterials*, *269*, 120634. doi:10.1016/j.biomaterials.2020.120634 PMID:33421708

Zhang, S., Vaida, J., Parenti, J., Lindsey, B. A., Xing, M., & Li, B. (2021). Programmed Multidrug Delivery Based on Bio-Inspired Capsule-Integrated Nanocoatings for Infected Bone Defect Treatment. *ACS Applied Materials & Interfaces*, *13*(10), 12454–12462. doi:10.1021/acsami.0c20332 PMID:33683872

Zhou, W., Bai, T., Wang, L., Cheng, Y., Xia, D., Yu, S., & Zheng, Y. (2023). Biomimetic AgNPs@ antimicrobial peptide/silk fibroin coating for infection-trigger antibacterial capability and enhanced osseointegration. *Bioactive Materials*, *20*(5), 64–80. doi:10.1016/j.bioactmat.2022.05.015 PMID:35633877

Zhou, W., Jia, Z., Xiong, P., Yan, J., Li, Y., Li, M., Cheng, Y., & Zheng, Y. (2017). Bioinspired and Biomimetic AgNPs/Gentamicin-Embedded Silk Fibroin Coatings for Robust Antibacterial and Osteogenetic Applications. *ACS Applied Materials & Interfaces*, *9*(31), 25830–25846. doi:10.1021/acsami.7b06757 PMID:28731325

Zou, F., Jiang, J., Lv, F., Xia, X., & Ma, X. (2020). Preparation of antibacterial and osteoconductive 3D-printed PLGA/Cu(I)@ZIF-8 nanocomposite scaffolds for infected bone repair. *Journal of Nanobiotechnology*, *18*(1), 1–14. doi:10.1186/s12951-020-00594-6 PMID:32103765

KEY TERMS AND DEFINITIONS

Angiogenesis: Physiological process for formation of new blood vessels from existing ones.

Bacteremia: Presence of living bacteria in the blood.

Biocompatible: Able to be in contact with living tissue without causing any harm.

Biodegradable: Able to be broken down in the biological system into harmless products.

Biomimetic: Mimicking models from nature or biological processes.

Chemodynamic Therapy: A type of therapy using a Fenton-like reaction to produce toxic hydroxyl radicals.

Debridement: Process of removing dead or damaged tissue to enable wound healing.

Hydrogel: A three-dimensional network capable of taking up large amounts of water.

Hyperthermia: Abnormally high body temperature > 100 °F.

Hypoxia: A condition in which there is a lack of adequate oxygen to sustain body functions.

Inorganic Semiconductors: Conductors having conductivity between conductors and insulators and made of non-carbon material.

Minimum Inhibitory Concentration: The lowest concentration of an antimicrobial substance that would inhibit visible microbial growth.

Mesenchymal stem cells: Multipotent stem cells present in bone marrow which can differentiate into various cells.

Mesoporous Silica Nanoparticles: Silica nanoparticles with pore size from 2 to 50 nm.

Nanocomposites: Multicomponent materials containing different phases in which one phase is continuous and at least one phase has nanoscale dimensions.

Nanofibers: Fibers with diameter in the nanoscale.

Nanorods: Rod-like nanoscale materials.

Nanosheets: Two-dimensional nanostructures with nanoscale thickness.

Nanotubes: Nanoscale material with a tube-like structure.

Nanowires: Nanostructures in the form of a wire.

Niosomes: Vesicular structures made of non-ionic surfactants.

Osseointegration: Bone growth into a metal implant.

Osteoblasts: Cells that form new bones and grow and heal existing bones.

Osteogenic: Relating to the development and formation of bones.

Osteoinductive: Able to naturally form bone with the skeleton.

Persisters: Dormant forms of bacterial cells highly tolerant to antibiotics.

Planktonic: Bacteria: Free-flowing bacteria in suspension.

Quorum Sensing: Process of cell-cell communication between bacteria to share information.

Quantum Dots: Ultra-small particles of a semiconducting material having size 2 to 10 nm.

Sessile Bacteria: Bacteria attached to a substrate that cannot move freely.

Chapter 12
Evolutionary Global Neuroscience for Cognition and Brain Health:
Strengthening Innovation in Brain Science

Bhupinder Singh
https://orcid.org/0009-0006-4779-2553
Sharda University, India

ABSTRACT

The evolutionary foundations of brain development represent point of departure. It delves into the depths of evolutionary history, unveiling the intricate journey that has shaped the human brain. Comparative neuroanatomy and phylogeny provide the canvas upon which we paint the portrait of our cognitive evolution. It contemplates the constraints, both genetic and epigenetic, that have sculpted the human brain's architecture. These insights serve as the bedrock upon which Evolutionary Global Neuroscience stands. In the quest to unlock the enigmatic complexities of the human brain, Evolutionary Global Neuroscience emerges as a transformative paradigm that transcends disciplinary boundaries. This expansive field explores the profound interplay between our evolutionary heritage and contemporary challenges in cognition, brain health, and the ever-evolving landscape of neurotechnology. Cognition, that quintessential hallmark of human existence, is next to claim our attention. It interrogate the adaptive value of cognition across epochs, deciphering its role in survival and prosperity. Yet, in this age of rapid societal transformation, the concept of "evolutionary mismatch" comes to the fore. It explore how this mismatch underlies a burgeoning epidemic of brain disorders, underscoring the urgent need for a new paradigm in addressing cognitive health. This paper illuminates the grim landscape of modernity's brain disorders. Alzheimer's disease and autism spectrum disorders serve as poignant exemplars of the evolutionary puzzle that engulfs us. From genetic predispositions to environmental influences, it contemplates the multifaceted origins of these disorders, ever mindful of the shadows cast by evolutionary

DOI: 10.4018/979-8-3693-1922-2.ch012

history. The promise of Evolutionary Global Neuroscience, however, is not confined to diagnosis and treatment. It extends to innovative approaches that safeguard cognitive well-being. Drawing inspiration from our evolutionary past, it uncovers preventive strategies, therapeutic interventions, and personalized medicine approaches. Lifestyle modifications and environmental adaptations emerge as potent tools in preserving cognitive resilience. Amidst this transformative journey, neurotechnology emerges as the catalyst of innovation. Neuroimaging, brain-machine interfaces, neuromodulation, and artificial intelligence converge to redefine the boundaries of human brain exploration. The fusion of biology and technology holds the key to unlocking new frontiers in brain science, offering hope to individuals grappling with brain disorders. Yet, as it propels ourselve toward these novel discoveries, it confront an array of ethical challenges. Questions of cognitive enhancement, informed consent, privacy, and equitable access to neurotechnological advancements loom large. Our abstract engages with these ethical considerations, advocating for responsible innovation that harmonizes with our ethical and societal values. Evolutionary Global Neuroscience is not merely a scientific endeavor; it is a call to action. It beckons us to unite disciplines, fostering interdisciplinary collaboration that transcends borders and cultures. It urges us to invest in longitudinal and cross-cultural research that expands our understanding of the human brain's exquisite diversity. It demands a commitment to brain health literacy and thoughtful policy recommendations to ensure that the fruits of neurotechnological advancement are accessible to all. This paper journeyed through the realm of Evolutionary Global Neuroscience, a journey that begins in our evolutionary past and charts a course toward an enlightened, innovative, and equitable future for brain science, brain health, and the human mind.

INTRODUCTION

Evolutionary Global Neuroscience for Cognition and Brain Health represents a groundbreaking and interdisciplinary approach to understanding the intricate workings of the human brain. As our world continues to advance at an unprecedented pace, our understanding of the brain and its fundamental role in shaping cognition, behavior, and overall well-being becomes increasingly critical. This holistic perspective, often referred to as "neuroscience," seeks to unravel the complexities of the brain's evolution and its impact on brain disorders while fostering innovation in neurotechnology, thereby opening doors to novel discoveries that hold the potential to revolutionize the field of brain science (Thompson, 2020).

The human brain, with its vast network of neurons and intricate neural circuits, remains one of the most enigmatic and fascinating organs in the human body. It has been the subject of scientific inquiry for centuries, and yet, we are only scratching the surface of its immense capabilities. Evolutionary Global Neuroscience delves deep into the evolutionary history of the brain, tracing its development through millennia of biological change (Kramer, 2007). This perspective not only provides us with insights into how the brain has adapted to various environmental pressures over time but also sheds light on the potential for further adaptation in our rapidly changing world.

The paramount goals of Evolutionary Global Neuroscience is to advance our understanding of cognition and brain health. Cognition, the set of mental processes that encompasses perception, attention, memory, language, problem-solving, and decision-making, underpins all human endeavors. By delving into the neural underpinnings of these processes, we can unlock new strategies for enhancing cognitive abilities, improving educational techniques, and optimizing mental health interventions (Kasai, 2015). Moreover, a deeper understanding of brain health empowers us to develop targeted interventions and therapies for a wide range of brain disorders, from neurodegenerative diseases like Alzheimer's and Parkinson's to neuropsychiatric conditions like schizophrenia and depression.

In the pursuit of innovation, Evolutionary Global Neuroscience acts as a catalyst, propelling the field of brain science forward. This involves not only cutting-edge research but also the integration of technology into the study of the brain. Neurotechnology, encompassing techniques such as functional magnetic resonance imaging (fMRI), electroencephalography (EEG), and brain-computer interfaces (BCIs), provides us with unprecedented access to the brain's inner workings. These tools enable researchers to map neural activity, decipher the neural code, and even manipulate brain functions, offering vast potential for developing novel therapies and interventions (Stein, 2015). The fusion of neuroscience and technology opens up exciting avenues for novel discoveries. As we gain a deeper understanding of the brain's evolution, its role in brain disorders, and the potential of neurotechnology, we can explore uncharted territories. This includes the development of personalized treatments, the enhancement of human capabilities through brain-machine interfaces, and the creation of innovative tools for education and cognitive enhancement.

This paper represents a comprehensive and forward-thinking approach to understanding the brain's evolution, its impact on brain disorders, and the potential for innovation through neurotechnology. By embracing this holistic perspective, we are poised to make groundbreaking discoveries that will not only advance our knowledge of the brain but also pave the way for transformative advancements in brain science and the enhancement of cognitive function and brain health for individuals around the world.

Background and Significance

The background and significance of Evolutionary Global Neuroscience for Cognition and Brain Health lie at the intersection of critical scientific inquiry and pressing societal needs. Historically, our understanding of the brain and its evolution has been fragmented across various disciplines, making it imperative to establish a comprehensive framework that unites biology, psychology, and technology. This convergence addresses the pressing need for insights into cognition and brain health, given the rising global burden of neurological and neuropsychiatric disorders (Chiao, 2009). The significance of this approach becomes even more apparent as it offers the potential to drive innovations that can revolutionize healthcare, education, and human-machine interactions. By bringing together diverse fields and harnessing cutting-edge technology, Evolutionary Global Neuroscience promises to unlock profound discoveries and transformative solutions for the complex challenges facing humanity in the realm of brain science.

Scope and Objectives

The scope of this paper on Evolutionary Global Neuroscience for Cognition and Brain Health is expansive and multifaceted. Firstly, it encompasses a comprehensive exploration of the evolutionary history of the human brain, spanning from its origins to its present state, providing insights into how it has adapted over millennia and continues to adapt in response to changing environments (Rossell, 2003). Secondly, it delves into the intricate neural mechanisms that underlie cognition, shedding light on the fundamental processes that govern human thought, emotion, and behavior. This paper discusses the pressing issue of brain disorders, elucidating their neurological basis and exploring novel avenues for treatment and prevention. It also underscores the transformative potential of neurotechnology, detailing its application in mapping neural activity, understanding the neural code, and developing innovative interventions. It addresses the broader societal implications, emphasizing the importance of enhancing cognitive abilities, promoting brain health, and improving our educational and healthcare systems through insights gained from this multidisciplinary approach. Lastly, it highlights the global relevance of Evolutionary Global Neuroscience, emphasizing its capacity to transcend geographical and cultural boundaries, fostering international collaboration and sharing of knowledge to advance brain science on a global scale. In essence, this paper offers a holistic perspective that not only elucidates the intricate mysteries of the brain but also presents a roadmap for harnessing this knowledge to address some of the most critical challenges facing humanity in the 21st century.

The objectives of the paper on Evolutionary Global Neuroscience for Cognition and Brain Health are as follows-

To Explore Brain Evolution: Investigate the evolutionary history of the human brain, tracing its development over millions of years to gain a deeper understanding of how it has adapted to changing environments and pressures.

To Uncover Neural Mechanisms: Examine the neural mechanisms underpinning cognition, including perception, memory, decision-making, and emotion, to unravel the mysteries of human thought processes and behavior.

To Address Brain Disorders: Investigate the neurological basis of brain disorders such as Alzheimer's, Parkinson's, schizophrenia, and depression, with the aim of identifying novel therapeutic approaches and preventive strategies.

To Harness Neurotechnology: Explore the potential of cutting-edge neurotechnologies like functional MRI, EEG, and brain-computer interfaces in mapping neural activity, deciphering the neural code, and developing innovative interventions for brain-related conditions.

To Promote Global Collaboration: Emphasize the significance of international collaboration and knowledge-sharing in advancing Evolutionary Global Neuroscience, transcending geographical and cultural boundaries to foster a global community of researchers and practitioners dedicated to enhancing brain science and its applications.

EVOLUTIONARY FOUNDATIONS OF BRAIN DEVELOPMENT

The evolutionary foundations of brain development are a testament to the remarkable adaptive power of life on Earth. Over millions of years, the vertebrate brain has undergone an intricate process of development and expansion, shaped by the relentless forces of natural selection. At its core, this process reflects an exquisite interplay between genetic factors and environmental pressures, resulting in the highly complex and specialized brains we find in modern species (Lubans, 2016).

The earliest vertebrate ancestors possessed rudimentary neural structures geared primarily towards basic sensory and motor functions. However, as species faced diverse ecological challenges and niches, the brain evolved to become increasingly sophisticated. The emergence of complex social behaviors, the need for advanced problem-solving abilities, and the development of intricate communication systems all contributed to the expansion of the brain. The pivotal milestone in the evolution of brain development was the emergence of the neocortex, a region responsible for higher cognitive functions in mammals. This evolutionary leap allowed for advanced learning, memory, and decision-making capabilities, ultimately shaping the adaptive success of various species. It's fascinating to observe how different lineages took distinct evolutionary paths, resulting in the diverse array of brain structures seen in mammals today (Singh, 2023). The evolutionary foundations of brain development remind us that our own human brains are a product of this ancient process. Our cognitive abilities, complex emotions, and capacity for innovation are deeply rooted in the evolutionary history of the brain. Understanding these foundations not only provides insights into our own species but also offers a broader perspective on the importance of brain development in shaping the diversity of life on Earth. It underscores the incredible plasticity and adaptability of the brain, which continues to evolve in response to the ever-changing challenges of our dynamic world (Bherer, 2013).

The Evolutionary History of the Brain

The evolutionary history of the brain is an awe-inspiring journey that stretches back hundreds of millions of years, beginning with the simplest neural structures in the earliest multicellular organisms. Over eons of time, this humble beginning has culminated in the complexity and sophistication of the human brain. This remarkable journey is marked by a series of key milestones, including the emergence of the first nervous systems in primitive marine organisms, the development of more elaborate brains in early vertebrates, and the evolutionary expansion of specific brain regions to support increasingly complex behaviors and adaptations.

As life on Earth diversified and species encountered varied ecological challenges, natural selection acted upon the brain, favoring traits that enhanced survival and reproduction. This ongoing process led to the evolution of diverse brain structures tailored to meet the specific needs of different species. For instance, the olfactory centers of the brain in mammals evolved to process complex scents, while the cerebellum developed to coordinate fine motor skills (Schumann, 2023). In primates, including humans, the evolution of the neocortex brought forth a higher level of cognitive complexity, enabling advanced problem-solving, social interactions, and the development of culture.

The evolutionary history of the brain not only highlights the incredible adaptability and plasticity of this organ but also underscores its central role in shaping the trajectory of life on our planet. It serves as a testament to the power of evolution to sculpt and refine a structure as intricate and vital as the brain, providing the biological foundation for the remarkable cognitive and behavioral diversity seen across species today (Kyriazis, 2015). Understanding this evolutionary journey is not only a scientific endeavor but also a profound exploration of our own place within the continuum of life's history, offering insights into the origins of human consciousness and intelligence.

Comparative Neuroanatomy and Phylogeny

Comparative neuroanatomy and phylogeny are fields of study that offer profound insights into the evolution of the nervous system and the diversity of brain structures across different species. By examining the anatomical and functional differences and similarities among brains of various organisms, scientists can trace the evolutionary history of neural structures, shedding light on how different species have adapted to their ecological niches over millions of years. The comparative neuroanatomy explores the intricate neural architecture of different organisms, from the simplest nervous systems of invertebrates to the highly developed brains of vertebrates. This discipline allows researchers to identify common neural structures and specialized adaptations, providing clues about the evolutionary relationships between species (Keating, 2004). By comparing the organization of brain regions, the presence of specific neural circuits, and the development of sensory and motor systems, scientists can unravel the evolutionary trajectories that have shaped the diversity of nervous systems on Earth.

Phylogeny focuses on reconstructing the evolutionary history of organisms based on genetic and morphological data. When applied to the study of nervous systems, phylogenetic analyses can reveal the ancestral relationships between different species and help place them within the context of the tree of life. By integrating phylogenetic data with neuroanatomical studies, researchers can elucidate how changes in brain structure and function have occurred throughout evolutionary history, offering insights into the selective pressures that have shaped neural adaptations in response to environmental challenges.

Together, comparative neuroanatomy and phylogeny provide a holistic perspective on the evolution of the nervous system, enriching our understanding of the relationship between form and function in brains across the animal kingdom (McDonough, 2022). These disciplines not only deepen our appreciation for the remarkable diversity of neural adaptations but also highlight the shared ancestry that unites all living creatures, ultimately contributing to a comprehensive picture of life's evolutionary history.

Evolutionary Constraints on Brain Structure

Evolutionary constraints on brain structure are fundamental principles that highlight both the remarkable adaptability and inherent limitations of the brain as it evolves across different species. While evolution can shape neural structures in response to environmental pressures, certain constraints imposed by genetic, developmental, and structural factors can limit the extent of change. These constraints are essential in understanding why brains of different organisms exhibit both commonalities and divergences in their structures.

The prominent constraint is the trade-off between brain size and metabolic demands. Enlarging the brain comes at a significant energy cost, as neurons require substantial energy resources to function optimally. Therefore, evolutionary changes in brain size must strike a balance between cognitive advantages and energy efficiency (Papini, 2020). This constraint often leads to variations in brain size and complexity among species, with some having larger brains to support complex cognitive functions, while others prioritize energy conservation. The constraint relates to the architectural organization of the brain. The development of neural connections and structures during embryonic growth is highly conserved across species due to genetic and developmental constraints. This conservation leads to similarities in basic brain organization, such as the presence of regions responsible for sensory processing, motor control, and higher-order cognitive functions, despite differences in overall brain size and complexity.

Also, constraints can also arise from historical evolutionary legacies. The brain's evolutionary history has left behind vestiges of structures that were once adaptive but have become functionally obsolete in some species. These remnants, known as exaptations, can constrain further neural evolution, as they are not easily altered due to their ancestral origins (Chiao, 2017). So, understanding these evolutionary constraints on brain structure is essential for unraveling the intricate web of adaptations that have shaped neural diversity across species. It emphasizes that while evolution can drive remarkable innovations in brain function, it does so within the confines of genetic and developmental limitations, leading to a fascinating mosaic of neural structures and functions across the animal kingdom (Jackson, 2009).

Genetic and Epigenetic Influences on Brain Evolution

Genetic and epigenetic influences play pivotal roles in the complex process of brain evolution. Genetic factors provide the blueprint for the development of neural structures, while epigenetic mechanisms act as dynamic regulators, modulating gene expression in response to environmental cues. Together, they contribute to the remarkable diversity and adaptability of brains across species.

The genetic influences are deeply rooted in the evolutionary history of the brain. The genetic code encodes the proteins and molecular pathways necessary for brain development and function. Mutations and genetic variations can lead to changes in neural architecture, neural circuitry, and even brain size, which can have profound implications for an organism's cognitive abilities and behavior. Over generations, natural selection acts upon these genetic changes, favoring adaptations that enhance survival and reproduction. This interplay between genetics and natural selection has shaped the unique features of brain structures in different species (Deshpande, 2017).

Epigenetic influences, on the other hand, offer a layer of complexity to brain evolution. Epigenetic modifications, such as DNA methylation and histone modification, can alter gene expression patterns without changing the underlying DNA sequence. These modifications are highly responsive to environmental factors, allowing organisms to adapt to changing conditions. In the context of brain development and function, epigenetic mechanisms can influence synaptic plasticity, learning, and memory. They also enable the brain to respond dynamically to environmental cues, such as early life experiences and exposure to stressors, which can shape neural circuits and behavior over time.

The interplay between genetics and epigenetics in brain evolution highlights the intricate relationship between nature and nurture. While genetic factors provide the foundation for neural development, epigenetic mechanisms allow the brain to fine-tune its responses to the environment. This dynamic interplay enables organisms to adapt to diverse ecological niches and challenges, resulting in the incredible diversity of neural structures and functions seen across species (Best, 2015). Understanding these genetic and epigenetic influences on brain evolution is crucial for unraveling the complexity of brain development and its role in shaping the diversity of life on Earth.

COGNITION AND BRAIN HEALTH: AN EVOLUTIONARY PERSPECTIVE

Cognition and Brain Health: An Evolutionary Perspective represents a paradigm shift in how we approach the understanding and promotion of cognitive function and mental well-being. This perspective takes us on a journey through time, tracing the intricate interplay between human cognition and the evolutionary history of the brain. It reminds us that the cognitive abilities we cherish today are the product of millions of years of biological refinement, shaped by the relentless forces of natural selection (Gillan, 2021). This perspective acknowledges that our cognitive capacities, including memory, attention, problem-solving, and social cognition, have evolved as adaptive traits. These cognitive abilities conferred significant advantages in our ancestral environments, enabling our species to navigate complex social hierarchies, solve ecological challenges, and communicate effectively. As such, they are not just abstract constructs of the mind but are deeply intertwined with our biology and the history of our species. Moreover, an evolutionary perspective highlights that cognition and brain health are not static entities but dynamic processes continually influenced by our modern environment. The rapid pace of technological and societal changes in recent centuries has created a cognitive dissonance, as our brains grapple with novel stimuli and demands that differ significantly from those for which they were originally adapted. This dissonance underlies the prevalence of modern mental health challenges, such as stress, anxiety, and depression, as our evolved cognitive mechanisms often struggle to cope with the fast-paced and interconnected world.

In terms of brain health, an evolutionary perspective emphasizes the importance of aligning our contemporary lifestyles with the needs of our ancient brains. Diet, exercise, sleep, and social connections all play pivotal roles in brain health, as they are intricately tied to the neural mechanisms that have evolved to sustain our cognitive functions (Paul, 2007). By recognizing these connections, we can make informed choices about our lifestyles and health habits, ultimately promoting not only cognitive vitality but also overall well-being. This perspective underscores the value of interventions and therapies that harness our innate cognitive processes, leveraging evolutionary insights to develop effective strategies for cognitive enhancement and mental health support. By understanding the cognitive strengths and limitations that have evolved in us, we can design interventions that align with our evolutionary heritage, making them more intuitive and efficacious. This perspective that bridges cognition and brain health with evolution offers a profound understanding of the human mind and its vulnerabilities in the modern world. It calls us to respect the ancestral wisdom encoded in our brains, adapt our lifestyles to its requirements, and develop innovative strategies that harmonize with our evolutionary heritage. Through this lens, we gain valuable insights into the complex interplay between our cognition and brain health, paving the way for a more holistic and effective approach to mental well-being in our ever-changing world.

The Adaptive Value of Cognition

The adaptive value of cognition stands as a cornerstone of human evolution and survival. Cognition, encompassing a wide range of mental processes such as memory, problem-solving, decision-making, and social intelligence, has been honed by natural selection over countless generations. These cognitive abilities conferred significant advantages to our ancestors as they navigated a challenging and ever-changing environment. Memory allowed for the retention of valuable information about food sources, predators, and social interactions. Problem-solving skills enabled the development of tools and strategies to overcome environmental obstacles. Social cognition facilitated cooperation, communication, and the formation of complex social structures (Zheng, 2019). The cognition is not a one-size-fits-all phenomenon but rather a dynamic and adaptable set of mental tools that can be fine-tuned to suit various ecological niches. This adaptability has been crucial in enabling humans to thrive in diverse environments across the globe. Our cognitive flexibility allows us to adjust to new challenges, innovate in response to changing circumstances, and learn from both success and failure. It also underpins our ability to create and transmit culture, passing down knowledge and traditions from one generation to the next.

In the modern context, understanding the adaptive value of cognition has implications for education, mental health, and overall well-being. By appreciating the evolutionary significance of cognitive processes, we can design educational systems that align with our cognitive strengths and foster critical thinking, problem-solving, and creativity (Singh, 2022). Additionally, recognizing the evolutionary roots of cognition helps us appreciate the importance of mental health and provides insights into addressing cognitive challenges and mental disorders effectively. The adaptive value of cognition underscores its fundamental role in human survival and success. It is a testament to the power of natural selection in shaping the human mind and a reminder of the deep connection between our cognitive abilities and our evolutionary heritage. Understanding this connection can inform and enrich various aspects of our lives, from education to mental health, and guide us in harnessing the full potential of our cognitive capacities.

Evolutionary Mismatch and Brain Disorders

Evolutionary mismatch, a concept rooted in evolutionary biology, sheds light on the complex relationship between our ancestral past and the emergence of brain disorders in the modern world. It posits that many of the health challenges we face today, including various brain disorders, result from a misalignment between our evolved biological traits and our contemporary lifestyles. Our brains, shaped by millions of years of adaptation to a vastly different environment, are not always well-suited to the demands of our fast-paced, technology-driven society.

This mismatch becomes particularly apparent in the context of mental health. Conditions such as anxiety, depression, and attention deficit hyperactivity disorder (ADHD) are increasingly prevalent in modern societies (Chiao, 2016). They are theorized to be, in part, a consequence of our brains grappling with stressors, information overload, and sedentary lifestyles that diverge significantly from the conditions in which our ancestors evolved. Evolutionary mismatch highlights the dissonance between our ancient neural mechanisms, which evolved to respond to immediate physical threats and process information at a slower pace, and the contemporary world's chronic stressors and constant cognitive demands. Recognizing the role of evolutionary mismatch in brain disorders opens doors to novel approaches for prevention and treatment. It underscores the importance of lifestyle factors, including diet, exercise, sleep, and social connections, in maintaining optimal brain health. Additionally, it emphasizes the value of interventions that align with our evolutionary heritage, such as mindfulness practices that help manage stress or outdoor activities that promote physical and mental well-being.

So, understanding the concept of evolutionary mismatch provides a valuable lens through which we can view the rising incidence of brain disorders in the modern era. It encourages us to reevaluate our lifestyles, prioritize strategies that are in harmony with our evolutionary past, and explore innovative approaches to enhance mental health and cognitive well-being in a rapidly changing world (Gomes-Osman, 2018).

Cognitive Resilience and Brain Health

Cognitive resilience stands as a pivotal component of brain health, representing the brain's capacity to adapt, withstand adversity, and recover from cognitive challenges. It encompasses a range of mental processes that enable individuals to maintain optimal cognitive functioning despite encountering stressors, aging, or various neurological conditions. This resilience reflects the brain's inherent plasticity, allowing it to reorganize neural networks, form new connections, and adapt to changing circumstances throughout life.

Cognitive resilience has far-reaching implications for brain health, particularly in the context of aging and neurodegenerative diseases. As individuals grow older, the brain naturally undergoes changes that can impact cognitive abilities (Amit, 1989). However, those with higher levels of cognitive resilience exhibit a greater ability to stave off cognitive decline and maintain their mental faculties as they age. Moreover, cognitive resilience plays a crucial role in mitigating the impact of brain disorders like Alzheimer's and Parkinson's disease, allowing individuals to retain functional independence and quality of life.

The promotion of cognitive resilience involves various factors, including intellectual engagement, social interactions, physical exercise, and stress management. Engaging in lifelong learning, maintaining an active and socially connected lifestyle, and practicing cognitive exercises can bolster cognitive resilience and contribute to brain health. By recognizing and harnessing the brain's remarkable capacity for resilience, we can take proactive steps to preserve and enhance cognitive function, promoting not only individual well-being but also the collective cognitive vitality of our aging population.

BRAIN DISORDERS: AN EPIDEMIC OF MODERNITY

Brain disorders have emerged as a concerning epidemic of modernity, reflecting the intricate interplay between our evolving lifestyles and the vulnerabilities of the human brain. The rapid pace of technological advancement, changes in dietary habits, increased stressors, and altered sleep patterns in contemporary societies have contributed to the rising prevalence of brain disorders. Conditions such as anxiety, depression, attention deficit disorders, and neurodegenerative diseases have become increasingly common, casting a shadow on mental health worldwide.

This modern epidemic underscores the importance of understanding the complex relationship between our environment and brain health. Many of these disorders, rooted in genetics and neurobiology, are exacerbated by factors such as chronic stress, sedentary lifestyles, and the pervasive use of digital devices. The resulting neural and psychological strain highlights the profound disconnect between the brains we have evolved and the challenges of our modern world (Erickson, 2019).

Addressing this epidemic necessitates a multifaceted approach that combines advances in neuroscience, mental health awareness, and lifestyle interventions. By recognizing the role of modernity in the emergence of brain disorders, we can develop strategies to mitigate their impact. These strategies may include promoting mental health education, fostering a culture of mindfulness and stress management, and encouraging lifestyle choices that align with our brain's evolutionary heritage. In essence, the epidemic of brain disorders in the modern era serves as a poignant reminder of the importance of preserving and prioritizing brain health in our rapidly changing world (Dauncey, 2009). It calls for a holistic approach that encompasses not only medical treatments but also societal awareness and lifestyle adaptations to safeguard the cognitive and emotional well-being of individuals and communities.

Prevalence and Economic Burden

The prevalence and economic burden of brain disorders have reached alarming proportions, impacting individuals, families, and societies at large. These disorders, which include conditions like Alzheimer's disease, depression, and schizophrenia, have become increasingly prevalent in our modern world. The statistics are staggering, with millions of people affected worldwide. Beyond the personal suffering and diminished quality of life experienced by those with these conditions, there is a substantial economic toll as well.

The economic burden of brain disorders encompasses a range of costs, from direct medical expenses for diagnosis and treatment to indirect costs like lost productivity, caregiving expenses, and the strain on healthcare systems. These conditions often require long-term care and can lead to disability, resulting in substantial economic consequences for individuals and their families. Moreover, the societal impact is significant, as brain disorders strain healthcare resources, reduce workforce productivity, and place an additional burden on social support systems (Singh, 2022). Recognizing the extent of the economic burden underscores the urgent need for investment in research, prevention, and improved access to mental healthcare services. By addressing brain disorders comprehensively, we can alleviate the personal suffering endured by affected individuals and families while also mitigating the substantial economic costs associated with these conditions. Furthermore, a proactive approach that emphasizes early intervention, mental health awareness, and destigmatization can lead to more efficient and cost-effective strategies for managing and reducing the burden of brain disorders on both individuals and society as a whole.

Evolutionary Mismatch Hypothesis in Brain Disorders

The Evolutionary Mismatch Hypothesis in brain disorders provides a thought-provoking perspective on the origins and prevalence of neurological and neuropsychiatric conditions in our modern world. This hypothesis posits that many brain disorders result from a misalignment between the environment in which our brains evolved and the rapidly changing conditions of contemporary society (Gordon, 2008). Our brains have evolved over millions of years to adapt to the challenges and conditions of our ancestral environment, but the pace of societal and technological change has far outstripped the capacity of our brains to adjust.

This mismatch between our evolved neural mechanisms and the demands of modern life has far-reaching implications for brain health. Conditions like anxiety, depression, and attention deficit disorders are theorized to be, in part, a consequence of our brains grappling with chronic stressors, information overload, sedentary lifestyles, and social disconnection that diverge significantly from the conditions for which they were originally adapted (Collins, 2011). Additionally, the modern prevalence of neurodegenerative diseases like Alzheimer's may, in part, be influenced by factors such as dietary changes and decreased physical activity, which are at odds with the evolutionary history of our species.

So, understanding the Evolutionary Mismatch Hypothesis underscores the importance of adopting a holistic approach to brain health. It highlights the significance of lifestyle factors such as diet, exercise, sleep, and social connections in mitigating the impact of brain disorders and promoting mental well-being. Furthermore, it emphasizes the value of interventions that align with our evolutionary heritage, such as mindfulness practices and outdoor activities that reduce stress and support cognitive health. By recognizing and addressing the challenges posed by the evolutionary mismatch in brain disorders, we can strive for a better alignment between our biology and the demands of modern life, ultimately enhancing brain health and well-being.

Alzheimer's Disease and Evolutionary Insights

Alzheimer's disease, a devastating neurodegenerative condition that affects millions worldwide, offers a unique lens through which we can gain evolutionary insights into the human brain. While Alzheimer's is a modern affliction, the study of this condition unveils crucial aspects of our brain's evolutionary history. This disease primarily affects older adults, prompting questions about its prevalence in an evolutionary context. Some researchers suggest that Alzheimer's may be a result of our extended lifespan, as it disproportionately affects individuals who have outlived the typical reproductive age.

From an evolutionary perspective, Alzheimer's disease challenges our understanding of brain function and aging. It raises questions about the trade-offs between cognitive decline in later life and the benefits of longevity. Additionally, the study of Alzheimer's offers insights into the importance of genes involved in brain health and the potential role of genetic variations in disease susceptibility. By examining how Alzheimer's intersects with our evolutionary history, scientists may uncover valuable clues about the genetic and environmental factors that contribute to the condition, ultimately guiding efforts for prevention and treatment (Patel, 2018). Alzheimer's disease provides a window into the intricate relationship between brain aging, evolution, and genetic vulnerability. While it is a heartbreaking affliction, the pursuit of evolutionary insights into this condition holds the promise of advancing our understanding of both brain health and the complex interplay between genetics and aging in the human species.

Autism Spectrum Disorders: A Complex Evolutionary Puzzle

Autism Spectrum Disorders (ASD) present a multifaceted and intricate evolutionary puzzle that continues to challenge researchers and scientists. This group of neurodevelopmental disorders, characterized by difficulties in social communication and repetitive behaviors, raises questions about their origins in the context of human evolution. While the precise etiology of ASD remains unclear, evolutionary theories suggest that genetic variants associated with these conditions may have been maintained in human populations due to their potential benefits in specific contexts. The theory posits that some traits associated with ASD, such as enhanced attention to detail and heightened pattern recognition, may have conferred advantages in ancestral human environments. These traits could have been advantageous in activities such as hunting, foraging, or tool-making. However, the same genetic variants that provided benefits in certain contexts might lead to the manifestation of ASD when they interact with modern environmental factors or when they are inherited in certain combinations (Basar, 2012). Understanding the evolutionary puzzle of ASD is not only a scientific endeavor but also holds potential implications for treatment and support. By elucidating the genetic and environmental factors that contribute to ASD, researchers can develop more targeted interventions and therapies. Additionally, recognizing the complexity of ASD through an evolutionary lens reminds us of the diversity of human neurodevelopment and the importance of embracing and supporting individuals with ASD in our modern, rapidly changing society. Ultimately, the study of ASD serves as a testament to the intricate interplay between genetics, evolution, and the human brain's remarkable variability.

INNOVATIVE APPROACHES TO BRAIN HEALTH

Innovative approaches to brain health represent a dynamic frontier in the field of neuroscience and mental well-being. In a world where cognitive challenges and neurological disorders are on the rise, these approaches offer novel strategies to protect and enhance brain function, thereby improving the quality of life for individuals across the lifespan (Biswal, 2010). The most promising innovations in brain health is the burgeoning field of neurotechnology. Advanced brain imaging techniques, such as functional magnetic resonance imaging (fMRI) and electroencephalography (EEG), provide unprecedented insights into neural activity, allowing researchers to map brain function with exceptional precision. Brain-computer interfaces (BCIs) take this a step further, enabling direct communication between the brain and external devices. BCIs have immense potential for individuals with paralysis, offering restored mobility and independence. Additionally, neurofeedback technologies enable individuals to train their brains for optimal performance, enhancing focus, reducing stress, and improving cognitive abilities.

Another innovative avenue in brain health involves personalized medicine and genetic insights. Advances in genomics have revealed the role of genetics in neurological conditions, allowing for more tailored treatments and interventions. Precision medicine approaches can identify genetic markers associated with conditions like Alzheimer's disease, enabling earlier diagnosis and targeted therapies. Furthermore, the emerging field of epigenetics explores how lifestyle factors can influence gene expression, opening doors to strategies for promoting brain health through diet, exercise, and stress management (Bellugi, 2013).

In the mental health perspective, digital therapeutics and telehealth services are revolutionizing access to care. Mobile apps and online platforms offer evidence-based interventions for conditions such as anxiety, depression, and substance abuse, making mental health support more accessible and convenient. Telehealth services extend this reach, providing remote consultations with mental health professionals, reducing barriers to treatment, and addressing the shortage of mental health providers in many regions.

Besides, innovative approaches to brain health are increasingly emphasizing the importance of a holistic lifestyle. Diet, exercise, sleep, and mindfulness practices are recognized as powerful tools for promoting cognitive well-being. Nutritional psychiatry explores the impact of diet on mental health, while physical exercise has been shown to enhance neuroplasticity and cognitive resilience. Mindfulness and meditation techniques contribute to stress reduction and emotional regulation, nurturing a positive mental state.

The innovative approaches to brain health represent a multifaceted and dynamic frontier in healthcare and neuroscience (Chiao, 2020). These approaches harness cutting-edge technologies, genetic insights, and holistic lifestyle strategies to protect and enhance brain function, offering hope for a future where cognitive challenges are met with resilience, and mental well-being is optimized for all individuals. By embracing these innovations, we can strive for a world where brain health is a fundamental component of overall health and quality of life.

Evolution-Inspired Preventive Strategies

Evolution-inspired preventive strategies represent a forward-thinking approach to promoting health and well-being by recognizing the deep connection between our biology and the environments in which we evolved. Drawing insights from our evolutionary history, these strategies emphasize aligning our modern lifestyles with the needs of our ancient genes. For instance, our ancestors lived in environments where physical activity was integral to survival, so modern preventive strategies encourage regular exercise as a means to combat sedentary lifestyles and associated health issues. Additionally, dietary choices are influenced by evolutionary principles, emphasizing whole foods over highly processed ones. This approach acknowledges that our genes have adapted to diets rich in natural, unprocessed foods, and that many modern health problems arise from a mismatch between our ancestral diet and the highly processed, sugar-laden options prevalent today. The stress management techniques are rooted in evolutionary biology as recognizing that our stress response mechanisms evolved to cope with acute threats, these strategies advocate mindfulness, meditation, and relaxation exercises to counteract chronic stress, which is prevalent in our fast-paced modern world (Hillman, 2008). By embracing evolution-inspired preventive strategies, individuals can make informed choices about their lifestyles, promoting physical and mental health and reducing the risk of chronic diseases that have become prevalent in the modern era. These strategies serve as a testament to the timeless wisdom embedded in our genetic heritage, offering a roadmap to a healthier and more resilient future.

Therapeutic Approaches Informed by Evolutionary Biology

Therapeutic approaches informed by evolutionary biology represent a groundbreaking paradigm in the field of medicine and healthcare. These approaches recognize that many health conditions and diseases have roots deeply embedded in our evolutionary history. By understanding the evolutionary origins of certain traits and vulnerabilities, researchers and clinicians can develop more targeted and effective treatments. The notable area where evolutionary insights have been invaluable is in infectious disease management. Understanding how our immune system has evolved to combat pathogens has led to the development of vaccines, antibiotics, and antiviral drugs. These treatments exploit our immune system's capacity to recognize and defend against pathogens, offering critical tools for disease prevention and treatment (Stangl, 2023).

The mental health therapies informed by evolutionary biology delve into the fundamental aspects of human behavior and psychology. They explore how our brains have evolved to cope with stress, form social bonds, and navigate complex social dynamics. By aligning therapeutic interventions with our innate cognitive and emotional processes, these approaches offer innovative strategies for addressing conditions such as anxiety, depression, and addiction. In personalized medicine, genetic insights derived from evolutionary biology are transforming how we treat and manage various diseases. Tailored treatments based on an individual's genetic makeup are becoming increasingly common, optimizing therapeutic outcomes while minimizing side effects. The therapeutic approaches informed by evolutionary biology are revolutionizing healthcare by providing a deeper understanding of the human condition and its biological underpinnings. They offer a path toward more effective treatments, prevention strategies, and personalized care, ultimately improving the well-being and health outcomes of individuals and populations alike.

Personalized Medicine and the Role of Genomics

Personalized medicine, driven by the ever-expanding field of genomics, represents a transformative shift in healthcare. It recognizes that each individual's genetic makeup plays a pivotal role in their susceptibility to diseases, response to treatments, and overall health. Genomics, the study of an individual's complete set of genes, enables healthcare professionals to tailor medical interventions to the specific genetic profile of a patient, optimizing both the efficacy and safety of treatments (Basar, 2012).

The role of genomics in personalized medicine is multi-faceted. Genetic testing can reveal an individual's genetic predisposition to certain diseases, allowing for early intervention and prevention strategies. It can also identify specific genetic markers that influence drug metabolism, guiding healthcare providers in prescribing medications that are most likely to be effective and well-tolerated by the patient. Additionally, genomics plays a crucial role in cancer treatment by identifying genetic mutations that drive tumor growth, enabling the development of targeted therapies that precisely attack cancer cells while sparing healthy tissue. The genomics has a profound impact on rare and genetic diseases, offering hope to individuals and families affected by these conditions. By pinpointing the genetic basis of rare diseases, personalized medicine can facilitate accurate diagnoses and the development of tailored treatments.

The integration of genomics into healthcare is revolutionizing the way we approach disease prevention, diagnosis, and treatment. It holds the promise of more effective, patient-centered care by harnessing the power of individual genetic information to improve health outcomes, reduce adverse effects, and ultimately enhance the quality of life for countless individuals. As genomics continues to advance, the era of personalized medicine is poised to transform healthcare into a more precise, proactive, and patient-centric field.

Lifestyle and Environmental Interventions

Lifestyle and environmental interventions represent a cornerstone of preventive healthcare, offering powerful tools to mitigate the risk of chronic diseases and promote overall well-being. These interventions recognize that our daily choices, surroundings, and behaviors have a profound impact on our health outcomes. Lifestyle interventions often encompass diet, exercise, sleep, stress management, and social interactions. They emphasize the importance of a balanced, nutritious diet that supports physical and cognitive health, regular physical activity that strengthens the body and mind, and adequate sleep to facilitate recovery and cognitive function (Morris, 2019). Stress management techniques, such as mindfulness and relaxation exercises, equip individuals with tools to navigate life's challenges, reducing the risk of stress-related health problems. Furthermore, fostering social connections and a supportive community environment contributes to emotional resilience and mental health.

Environmental interventions extend beyond personal choices to address broader societal and ecological factors. They involve initiatives to improve air and water quality, reduce exposure to environmental toxins, and create healthier built environments that encourage physical activity and community interaction. These interventions recognize that a healthy environment is essential for human well-being and play a pivotal role in disease prevention (Walsh, 2011).

Collectively, lifestyle and environmental interventions empower individuals to take an active role in safeguarding their health and preventing disease. They underscore the interconnectedness of individual choices and the world in which we live, highlighting the importance of fostering a holistic approach to health that considers both personal behaviors and the broader environment. By embracing these interventions, we can move closer to a future where health is not solely a matter of medical treatment but is primarily a product of proactive, health-promoting choices and supportive surroundings.

NEUROTECHNOLOGY: A CATALYST FOR DISCOVERY

Neurotechnology stands as a remarkable catalyst for discovery in the realm of neuroscience and brain research. It represents a convergence of cutting-edge technologies, ranging from advanced brain imaging to brain-computer interfaces, and offers unprecedented access to the inner workings of the human brain. This transformative field has revolutionized our understanding of brain structure and function, paving the way for groundbreaking discoveries that span a spectrum of scientific disciplines.

One of the most notable contributions of neurotechnology is its role in decoding the mysteries of the brain's intricate architecture. High-resolution brain imaging techniques, such as functional magnetic resonance imaging (fMRI) and diffusion tensor imaging (DTI), enable researchers to map neural activity and connectivity with unparalleled precision (Debette, 2009). These tools have revealed the functional specialization of various brain regions, shedding light on how the brain processes information, regulates emotions, and controls bodily functions. Neuroimaging has also been instrumental in elucidating the neural signatures of various neurological and neuropsychiatric conditions, providing valuable insights for diagnosis and treatment.

Additionally, neurotechnology has unleashed the potential of brain-computer interfaces (BCIs), which facilitate direct communication between the brain and external devices. BCIs have proven transformative for individuals with paralysis, enabling them to control prosthetic limbs or communicate through computers simply by thinking. This technology not only restores mobility and independence but also blurs the lines between human and machine, offering exciting prospects for human augmentation and neuroprosthetics.

The neurotechnology plays a central role in advancing our understanding of cognitive processes and the mind. Functional neuroimaging studies have explored the neural underpinnings of memory, decision-making, and problem-solving, unraveling the complex interplay of brain regions involved in these functions. Such insights have practical applications in fields like education, cognitive enhancement, and neuropsychology, opening doors to innovative approaches for improving cognitive performance and mental well-being (Tamminga, 2014).

Neurotechnology's role as a catalyst for discovery is profound and multifaceted. It empowers researchers to explore the brain's intricate mechanisms, unravel the mysteries of cognition and consciousness, and develop cutting-edge interventions for neurological disorders. As this field continues to advance, the possibilities for new discoveries and innovations in neuroscience are limitless, promising a future where our understanding of the brain is more profound than ever, and where the application of this knowledge transforms the landscape of healthcare and human potential.

Neuroimaging and Brain Mapping

Neuroimaging and brain mapping represent a revolution in our ability to explore the complex terrain of the human brain. These techniques utilize advanced technologies to visualize and characterize the brain's structure, function, and connectivity with remarkable precision. Magnetic resonance imaging (MRI), functional MRI (fMRI), positron emission tomography (PET), and diffusion tensor imaging (DTI) are among the key tools that have enabled us to map the brain's intricate neural networks and understand how they underlie cognition, emotion, and behavior (Bell, 2011).

Neuroimaging has unveiled the brain's functional topography, revealing specialized regions responsible for tasks as diverse as language processing, motor control, and memory formation. It has also illuminated the dynamic interplay of these regions during complex cognitive processes, offering insights into the neural basis of human experience. Moreover, brain mapping has played a pivotal role in elucidating the structural and functional abnormalities associated with neurological and neuropsychiatric disorders, providing valuable diagnostic and therapeutic insights.

The implications of neuroimaging and brain mapping extend far beyond the realm of research. They have practical applications in clinical medicine, aiding in the early detection and treatment of brain disorders. Additionally, these techniques hold promise for advancing brain-computer interfaces and neuroprosthetics, enhancing our ability to restore lost sensory or motor functions. In essence, neuroimaging and brain mapping have become indispensable tools in neuroscience, empowering us to explore the intricate landscape of the brain, unlock its secrets, and harness its potential for the benefit of both individual health and our collective understanding of the human mind.

Brain-Machine Interfaces and Neuroprosthetics

Brain-machine interfaces (BMIs) and neuroprosthetics represent groundbreaking innovations in the field of neuroscience and medical technology. These transformative technologies bridge the gap between the human brain and external devices, offering remarkable prospects for individuals with neurological impairments and providing valuable insights into the brain's functioning.

BMIs, often referred to as brain-computer interfaces, allow direct communication between the brain and computers or external devices. By translating neural activity into actionable commands, BMIs empower individuals with paralysis to control robotic limbs, computer software, or even communication devices simply by thinking. This not only restores a sense of agency and independence but also blurs the boundaries between the human mind and artificial systems (Baars, 2005). Neuroprosthetics take BMIs a step further, encompassing devices implanted directly into the nervous system. These devices can replace lost sensory or motor functions, such as cochlear implants for hearing-impaired individuals or retinal implants for the visually impaired. By interfacing with the nervous system, neuroprosthetics have the potential to restore a level of function and quality of life previously unimaginable for those with disabilities.

Beyond their immediate clinical applications, BMIs and neuroprosthetics offer profound insights into the brain's plasticity and adaptability. Researchers are increasingly utilizing these technologies to study neural circuits and decode the brain's intricate workings. This knowledge has far-reaching implications, extending to fields like neurology, psychology, and cognitive science, and holds the promise of advancing our understanding of the human brain and consciousness. BMIs and neuroprosthetics exemplify the transformative power of interdisciplinary research and technological innovation. They not only enhance the lives of individuals living with neurological conditions but also broaden our horizons in neuroscience and expand the boundaries of what is possible in human-computer interaction. As these fields continue to advance, the future holds exciting prospects for the integration of these technologies into healthcare, augmenting human capabilities and deepening our understanding of the brain's remarkable complexity.

Advancements in Neuromodulation Techniques

Advancements in neuromodulation techniques have ushered in a new era of precision and effectiveness in the treatment of neurological and neuropsychiatric conditions. Neuromodulation, which involves the targeted stimulation or modulation of neural circuits, encompasses a diverse range of therapies, including deep brain stimulation (DBS), transcranial magnetic stimulation (TMS), and spinal cord stimulation (SCS), among others. These techniques have evolved rapidly, offering innovative solutions for conditions that were once challenging to manage. The most notable advancements is in DBS, where surgically implanted electrodes are used to deliver electrical pulses to specific brain regions. Originally developed to treat Parkinson's disease, DBS has expanded its applications to include conditions like epilepsy, obsessive-compulsive disorder (OCD), and even depression. The precision of DBS allows for symptom relief with fewer side effects compared to traditional therapies, enhancing the quality of life for many patients (Cattaneo, 2018).

Transcranial magnetic stimulation (TMS) has emerged as a non-invasive neuromodulation technique with promising applications in treating depression, anxiety, and various neuropsychiatric disorders. TMS involves the use of magnetic fields to stimulate specific brain areas, offering a safer and more targeted alternative to medications or electroconvulsive therapy.

Spinal cord stimulation (SCS) has also seen significant advancements, providing relief to individuals suffering from chronic pain conditions. Miniaturized devices with improved programming capabilities allow for highly customized pain management, reducing the reliance on opioids and improving patients' overall well-being.

These advancements reflect the growing recognition of the central role of neuromodulation in addressing neurological and neuropsychiatric disorders. As technology continues to evolve, we can anticipate even more sophisticated and precise neuromodulation techniques that offer new hope to individuals facing the challenges of these conditions (Demnitz, 2017). Moreover, the expanding applications of neuromodulation underscore the interdisciplinary nature of modern healthcare, where neuroscience, engineering, and clinical expertise converge to improve the lives of countless individuals.

Artificial Intelligence and Machine Learning in Neurotechnology

Artificial intelligence (AI) and machine learning are revolutionizing the field of neurotechnology, offering powerful tools to decode the complexities of the brain and develop innovative solutions for neurological disorders. These technologies have significantly enhanced our ability to analyze vast datasets generated by neuroimaging, electrophysiological recordings, and genetic studies, allowing researchers to uncover patterns and insights that were once hidden.

In neuroimaging, AI algorithms are employed to improve the accuracy of brain scans, aiding in early disease detection and precise localization of abnormalities. Machine learning models can predict neurological outcomes, such as the progression of Alzheimer's disease or the risk of epileptic seizures, by analyzing diverse patient data over time. AI-driven approaches are transforming brain-computer interfaces (BCIs) and neuroprosthetics. Advanced algorithms enable BCIs to interpret neural signals more efficiently, facilitating smoother control of external devices by individuals with paralysis. In the realm of neuroprosthetics, machine learning algorithms are optimizing the performance of implanted devices, such as prosthetic limbs or cochlear implants, by adapting to the user's needs and preferences (Holmes, 2014). Additionally, AI has found applications in mental health, where natural language processing and sentiment analysis can provide insights into individuals' emotional states and psychological well-being. Chatbots and virtual mental health assistants powered by AI offer support and interventions, reducing the stigma associated with seeking help for mental health concerns.

The integration of AI and machine learning in neurotechnology represents a synergistic partnership between human ingenuity and computational power. These technologies hold immense potential for the development of more effective treatments, personalized interventions, and a deeper understanding of brain function and dysfunction. As AI continues to advance, it promises to reshape the landscape of neuroscience and neurology, ushering in a new era of precision, accessibility, and innovation in brain research and healthcare.

CHALLENGES AND ETHICAL CONSIDERATIONS

While the advancements in neurotechnology are undeniably promising, they also bring forth a set of profound challenges and ethical considerations that demand careful scrutiny. One of the foremost concerns is privacy. As neuroimaging and brain-computer interfaces become more sophisticated, the potential for intrusion into an individual's inner thoughts and emotions raises questions about data security and the protection of sensitive neural information. Safeguarding the privacy of individuals and ensuring the responsible use of neurodata is paramount. Another pressing challenge relates to the potential misuse of neurotechnology. Brain-computer interfaces, for example, could be susceptible to hacking or unauthorized access, leading to potential risks and ethical dilemmas. Additionally, the prospect of cognitive enhancement using neurotechnology raises questions about equity and access, as these enhancements may only be available to those who can afford them, exacerbating existing societal inequalities.

Ethical dilemmas also emerge in the realm of neuroethics, particularly concerning issues of consent and agency. Individuals with neurological conditions may have limited capacity to provide informed consent for certain neurotechnological interventions, raising concerns about autonomy and the potential for coercion. Ethical guidelines and regulatory frameworks must evolve in tandem with technological advancements to address these complex issues (Small, 2022). The blurring of lines between the human brain and external devices in brain-computer interfaces raises profound questions about identity, agency, and the nature of consciousness. As neurotechnology continues to advance, society must grapple with questions about the boundaries of the self and the implications of integrating human cognition with artificial systems.

While neurotechnology holds immense promise for advancing our understanding of the brain and improving healthcare outcomes, it also poses formidable challenges and ethical dilemmas. Addressing these concerns requires a multidisciplinary approach that involves scientists, ethicists, policymakers, and the public. By proactively addressing these challenges and upholding ethical principles, we can harness the full potential of neurotechnology while ensuring its responsible and equitable use in an increasingly complex and interconnected world.

Ethical Implications of Enhancing Cognitive Abilities

The ethical implications of enhancing cognitive abilities are both profound and multifaceted. While cognitive enhancement techniques, such as pharmaceuticals or brain-computer interfaces, hold the potential to improve memory, attention, and problem-solving skills, they also raise concerns about fairness, safety, and the potential for unintended consequences. Questions about equitable access to cognitive enhancements, the risks of long-term use, and the potential for social pressure to enhance cognitive abilities are paramount. Additionally, ethical considerations extend to issues of informed consent, individual autonomy, and the impact of cognitive enhancements on personal identity. Striking a balance between the desire to augment human cognition and the need to navigate these complex ethical challenges is a central concern in the ongoing discourse surrounding cognitive enhancement.

Informed Consent and Privacy in Neurotechnology

Informed consent and privacy are crucial ethical considerations in the realm of neurotechnology. Obtaining informed consent from individuals participating in neurotechnology research or receiving neurotechnological interventions is essential to respect their autonomy and ensure they understand the potential risks and benefits. Moreover, the sensitive nature of neural data and the potential for invasive access to an individual's thoughts and emotions raise significant privacy concerns. Protecting the privacy of neural information is paramount to prevent unauthorized access, misuse, or the inadvertent exposure of deeply personal data. Balancing the advancement of neurotechnology with robust informed consent practices and stringent privacy safeguards is essential to uphold ethical standards and protect individuals' rights and well-being.

Equity and Access to Neurotechnological Advancements

Equity and access to neurotechnological advancements are pressing ethical considerations. As these technologies progress, ensuring that they are accessible and affordable to all individuals, regardless of socioeconomic status or geographical location, becomes essential. Without careful attention to equity, there is a risk of exacerbating existing disparities in healthcare and cognitive enhancement. Policymakers, researchers, and stakeholders must work collaboratively to develop strategies that promote equitable access to the benefits of neurotechnology, thereby fostering a more just and inclusive society where these advancements are harnessed for the well-being of all.

Balancing Innovation with Ethical Responsibility

Balancing innovation with ethical responsibility is a fundamental imperative in the development and deployment of emerging technologies like neurotechnology. While innovation drives progress and offers transformative solutions, it must always be tempered by ethical considerations. Responsible innovation in neurotechnology requires rigorous ethical scrutiny to ensure that these advancements prioritize the well-being, autonomy, and privacy of individuals (Dickinson, 2009). Ethical frameworks and guidelines should be integrated into every stage of research, development, and deployment, with a commitment to transparency and accountability. Striking this delicate balance ensures that the potential benefits of neurotechnology are maximized while minimizing the risks and unintended consequences, fostering a society that harnesses innovation for the greater good while upholding ethical principles and values.

FUTURE PROSPECTS AND RECOMMENDATIONS

The future prospects of neurotechnology are both exciting and promising, offering the potential to unlock new frontiers in neuroscience, healthcare, and human potential. As these technologies continue to advance, it is imperative that we proactively address the ethical, privacy, and accessibility challenges they present. Recommendations for navigating this evolving landscape include the development of comprehensive regulatory frameworks that promote responsible innovation and safeguard individual rights. Increased collaboration among scientists, ethicists, policymakers, and the public is crucial to ensure that ethical considerations remain at the forefront of neurotechnological advancements. Moreover, efforts should be made to ensure equitable access to these technologies, preventing the exacerbation of societal disparities. By fostering a multidisciplinary and inclusive approach to neurotechnology, we can harness its full potential while upholding the ethical principles and values that guide us toward a future where brain health and cognitive enhancement are accessible, safe, and ethically sound for all.

Advancing Interdisciplinary Collaboration

Advancing interdisciplinary collaboration is essential in the pursuit of knowledge and innovation, particularly in complex fields like neuroscience and neurotechnology. Bringing together experts from diverse disciplines, such as biology, engineering, ethics, and psychology, fosters a rich exchange of ideas and methodologies. This collaboration enables a more comprehensive understanding of the brain's intricacies and its applications in healthcare and technology. Moreover, it enhances problem-solving by approaching challenges from various angles, leading to innovative solutions and breakthroughs. As the boundaries between disciplines blur, interdisciplinary collaboration is poised to play a pivotal role in shaping the future of neuroscience and neurotechnology, driving progress and addressing complex ethical and scientific questions.

Investment in Longitudinal and Cross-Cultural Research

Investing in longitudinal and cross-cultural research is an imperative for gaining deeper insights into the complexities of human cognition, behavior, and health. Longitudinal studies, tracking individuals over extended periods, provide invaluable data on developmental trajectories and age-related changes in various populations. Such research is vital for understanding how factors like genetics, environment, and lifestyle impact health outcomes over time. Additionally, cross-cultural studies reveal the influence of culture on cognition, emotion, and behavior, shedding light on the universality and cultural specificity of certain phenomena. These investments in research not only advance scientific knowledge but also inform policies and interventions that are more sensitive to individual and cultural diversity, promoting better health and well-being for people across the globe.

Promoting Brain Health Literacy

Promoting brain health literacy is essential in an era where neuroscience and neurotechnology are rapidly advancing. Brain health literacy empowers individuals to understand the complexities of their own brains and make informed decisions about their mental well-being. It also encourages critical thinking about the ethical and societal implications of emerging neurotechnologies. By fostering brain health literacy through education and awareness campaigns, we can equip the public with the knowledge and tools to maintain cognitive health, reduce stigma associated with mental disorders, and engage in meaningful discussions about the responsible use of neurotechnology. Ultimately, brain health literacy is a crucial step toward ensuring that advancements in brain science benefit society as a whole.

Policy Recommendations for Neurotechnology Governance

Policy recommendations for neurotechnology governance are pivotal in managing the ethical, legal, and societal challenges posed by these advancing technologies. Policymakers should prioritize the development of comprehensive and adaptable regulatory frameworks that keep pace with technological advancements. These frameworks must safeguard individual privacy, ensure equitable access, and establish clear guidelines for informed consent, data security, and responsible research conduct. Collaboration among governments, academia, industry, and civil society is essential to address neuroethical concerns effectively. Additionally, ongoing dialogue and public engagement should be promoted to include diverse perspectives in shaping neurotechnology policies. In a rapidly evolving field like neurotechnology, robust and forward-thinking governance is crucial to harness the benefits while mitigating potential risks and ethical dilemmas.

CONCLUDING OBSERVATIONS

It is evident that the intersection of neuroscience, neurotechnology, and ethics presents a fascinating and complex landscape. As our understanding of the brain deepens and technology continues to advance, the potential for transformative breakthroughs in healthcare, neurology, and human enhancement is unparalleled. However, these advancements also bring forth profound ethical considerations, ranging from issues of consent, privacy, and equity to questions about the boundaries of human identity and autonomy.

The ethical responsibility of researchers, policymakers, and society as a whole is to ensure that neurotechnological progress is guided by principles of respect, justice, and beneficence. This necessitates robust regulatory frameworks that balance innovation with safeguards, ensuring that individual rights and well-being are upheld. It also calls for equitable access to neurotechnological advancements, preventing the creation of disparities in cognitive enhancement and healthcare. Moreover, interdisciplinary collaboration and public engagement are essential in navigating this complex terrain. By fostering dialogue among scientists, ethicists, policymakers, and the public, we can collectively shape a future where neurotechnology serves as a force for good, enhancing human lives while respecting fundamental ethical principles.

In this evolving landscape, the pursuit of knowledge, innovation, and ethical responsibility must go hand in hand. By upholding ethical values in the development and deployment of neurotechnology, we can aspire to a future where the potential of the human brain is harnessed for the betterment of individuals and society as a whole.

Synthesis of Key Findings

The synthesis of key findings underscores the intricate interplay between neuroscience, neurotechnology, and ethics. Advances in brain research and technology offer transformative potential for healthcare, cognition, and human capabilities. However, ethical considerations loom large, encompassing issues of privacy, consent, equity, and the very essence of what it means to be human. Striking a balance between innovation and ethical responsibility is paramount. Robust governance frameworks, interdisciplinary collaboration, and public engagement are vital to navigate this complex landscape. As we peer into the future of neurotechnology, we must do so with a commitment to uphold ethical principles, ensuring that these advancements benefit society as a whole while respecting the dignity and rights of individuals.

The Promise of Evolutionary Global Neuroscience

The promise of evolutionary global neuroscience lies in its potential to unite diverse perspectives, knowledge, and expertise from across the globe to advance our understanding of the brain and its evolution. By drawing upon insights from evolutionary biology, neurology, psychology, and other fields, we can uncover the fundamental principles that underpin brain development, cognition, and behavior in a wide range of species, including humans. This interdisciplinary approach not only deepens our scientific understanding but also holds the potential to inform novel interventions for brain disorders, enhance cognitive health, and explore the ethical implications of emerging neurotechnologies. Ultimately, evolutionary global neuroscience offers a promising path to unlock the mysteries of the brain and improve the well-being of individuals and societies worldwide.

The Way Forward: Uniting Science and Society

The way forward in the realm of neuroscience, neurotechnology, and ethics hinges on the unity of science and society. As we embark on a journey to unravel the mysteries of the brain and harness the potential of neurotechnology, it is imperative that we do so with the active participation and understanding of the broader community. Bridging the gap between scientific advancements and public awareness, engagement, and consent is vital. Together, we can navigate the ethical complexities, ensure equitable access, and make informed decisions about the responsible use of neurotechnologies. In this united endeavor, we can harness the power of science and innovation to improve brain health, enhance human potential, and address the profound ethical questions that lie at the intersection of neuroscience and society.

REFERENCES

Amit, D. J., & Amit, D. J. (1989). *Modeling brain function: The world of attractor neural networks.* Cambridge university press. doi:10.1017/CBO9780511623257

Baars, B. J. (2005). Global workspace theory of consciousness: Toward a cognitive neuroscience of human experience. *Progress in Brain Research, 150*, 45–53. doi:10.1016/S0079-6123(05)50004-9 PMID:16186014

Başar, E. (2012). Brain function and oscillations: volume I: brain oscillations. Principles and approaches. Springer Science & Business Media.

Başar, E. (2012). Brain function and oscillations: volume ii: integrative brain function. Neurophysiology and cognitive processes. Springer Science & Business Media.

Bell, B., Lin, J. J., Seidenberg, M., & Hermann, B. (2011). The neurobiology of cognitive disorders in temporal lobe epilepsy. *Nature Reviews. Neurology, 7*(3), 154–164. doi:10.1038/nrneurol.2011.3 PMID:21304484

Bellugi, U., Bihrle, A., Neville, H., Doherty, S., & Jernigan, T. (2013). Language, cognition, and brain organization in a neurodevelopmental disorder. In *Developmental behavioral neuroscience* (pp. 201–232). Psychology Press.

Best, J. R., Chiu, B. K., Hsu, C. L., Nagamatsu, L. S., & Liu-Ambrose, T. (2015). Long-term effects of resistance exercise training on cognition and brain volume in older women: Results from a randomized controlled trial. *Journal of the International Neuropsychological Society, 21*(10), 745–756. doi:10.1017/S1355617715000673 PMID:26581787

Bherer, L., Erickson, K. I., & Liu-Ambrose, T. (2013). A review of the effects of physical activity and exercise on cognitive and brain functions in older adults. *Journal of Aging Research, 2013*, 2013. doi:10.1155/2013/657508 PMID:24102028

Biswal, B. B., Mennes, M., Zuo, X. N., Gohel, S., Kelly, C., Smith, S. M., & Milham, M. P. (2010). Toward discovery science of human brain function. *Proceedings of the National Academy of Sciences of the United States of America, 107*(10), 4734–4739. doi:10.1073/pnas.0911855107 PMID:20176931

Cattaneo, G., Bartrés-Faz, D., Morris, T. P., Sánchez, J. S., Macià, D., Tarrero, C., Tormos, J. M., & Pascual-Leone, A. (2018). The Barcelona brain health initiative: A cohort study to define and promote determinants of brain health. *Frontiers in Aging Neuroscience, 10*, 321. doi:10.3389/fnagi.2018.00321 PMID:30405394

Chiao, J. Y. (Ed.). (2009). Cultural neuroscience: Cultural influences on brain function.

Chiao, J. Y., Li, S. C., Seligman, R., & Turner, R. (Eds.). (2016). *The Oxford handbook of cultural neuroscience*. Oxford University Press.

Chiao, J. Y., Li, S. C., Turner, R., & Lee-Tauler, S. Y. (2020). Cultural neuroscience and the research domain criteria: Implications for global mental health. *Neuroscience and Biobehavioral Reviews, 116*, 109–119. doi:10.1016/j.neubiorev.2020.06.005 PMID:32540352

Chiao, J. Y., Li, S. C., Turner, R., Lee-Tauler, S. Y., & Pringle, B. A. (2017). Cultural neuroscience and global mental health: Addressing grand challenges. *Culture and Brain*, *5*(1), 4–13. doi:10.1007/s40167-016-0045-4 PMID:28642836

Collins, P. Y., Patel, V., Joestl, S. S., March, D., Insel, T. R., Daar, A. S., & Walport, M. (2011). Grand challenges in global mental health. *Nature*, *475*(7354), 27–30. doi:10.1038/475027a PMID:21734685

Dauncey, M. J. (2009). New insights into nutrition and cognitive neuroscience: symposium on 'Early nutrition and later disease: current concepts, research and implications'. *The Proceedings of the Nutrition Society*, *68*(4), 408–415. doi:10.1017/S0029665109990188 PMID:19698201

Debette, S., Wolf, P. A., Beiser, A., Au, R., Himali, J. J., Pikula, A., Auerbach, S., DeCarli, C., & Seshadri, S. (2009). Association of parental dementia with cognitive and brain MRI measures in middle-aged adults. *Neurology*, *73*(24), 2071–2078. doi:10.1212/WNL.0b013e3181c67833 PMID:20007524

Demnitz, N., Zsoldos, E., Mahmood, A., Mackay, C. E., Kivimäki, M., Singh-Manoux, A., Dawes, H., Johansen-Berg, H., Ebmeier, K. P., & Sexton, C. E. (2017). Associations between mobility, cognition, and brain structure in healthy older adults. *Frontiers in Aging Neuroscience*, *9*, 155. doi:10.3389/fnagi.2017.00155 PMID:28588477

Deshpande, S. K., Hasegawa, R. B., Rabinowitz, A. R., Whyte, J., Roan, C. L., Tabatabaei, A., Baiocchi, M., Karlawish, J. H., Master, C. L., & Small, D. S. (2017). Association of playing high school football with cognition and mental health later in life. *JAMA Neurology*, *74*(8), 909–918. doi:10.1001/jamaneurol.2017.1317 PMID:28672325

Dickinson, D., & Elvevåg, B. (2009). Genes, cognition and brain through a COMT lens. *Neuroscience*, *164*(1), 72–87. doi:10.1016/j.neuroscience.2009.05.014 PMID:19446012

Erickson, K. I., Hillman, C., Stillman, C. M., Ballard, R. M., Bloodgood, B., Conroy, D. E., & Powell, K. E. (2019). Physical activity, cognition, and brain outcomes: A review of the 2018 physical activity guidelines. *Medicine and Science in Sports and Exercise*, *51*(6), 1242–1251. doi:10.1249/MSS.0000000000001936 PMID:31095081

Gillan, C. M., & Rutledge, R. B. (2021). Smartphones and the neuroscience of mental health. *Annual Review of Neuroscience, 44*, 129-151.

Gomes-Osman, J., Cabral, D. F., Morris, T. P., McInerney, K., Cahalin, L. P., Rundek, T., & Pascual-Leone, A. (2018). Exercise for cognitive brain health in aging: A systematic review for an evaluation of dose. *Neurology. Clinical Practice*, *8*(3), 257–265. doi:10.1212/CPJ.0000000000000460 PMID:30105166

Gordon, E., Barnett, K. J., Cooper, N. J., Tran, N., & Williams, L. M. (2008). An" integrative neuroscience" platform: Application to profiles of negativity and positivity bias. *Journal of Integrative Neuroscience*, *7*(03), 345–366. doi:10.1142/S0219635208001927 PMID:18988297

Hillman, C. H., Erickson, K. I., & Kramer, A. F. (2008). Be smart, exercise your heart: Exercise effects on brain and cognition. *Nature Reviews. Neuroscience*, *9*(1), 58–65. doi:10.1038/nrn2298 PMID:18094706

Holmes, E. A., Craske, M. G., & Graybiel, A. M. (2014). Psychological treatments: A call for mental-health science. *Nature*, *511*(7509), 287–289. doi:10.1038/511287a PMID:25030152

Jackson, J. C., Mitchell, N., & Hopkins, R. O. (2009). Cognitive functioning, mental health, and quality of life in ICU survivors: An overview. *Critical Care Clinics*, *25*(3), 615–628. doi:10.1016/j.ccc.2009.04.005 PMID:19576534

Kasai, K., Fukuda, M., Yahata, N., Morita, K., & Fujii, N. (2015). The future of real-world neuroscience: Imaging techniques to assess active brains in social environments. *Neuroscience Research*, *90*, 65–71. doi:10.1016/j.neures.2014.11.007 PMID:25433093

Keating, D. P. (2004). Handbook of adolescent psychology: Vol. 45-84. *Cognitive and brain development*.

Kramer, A. F., & Erickson, K. I. (2007). Capitalizing on cortical plasticity: Influence of physical activity on cognition and brain function. *Trends in Cognitive Sciences*, *11*(8), 342–348. doi:10.1016/j.tics.2007.06.009 PMID:17629545

Kyriazis, M. (2015). Systems neuroscience in focus: From the human brain to the global brain? *Frontiers in Systems Neuroscience*, *9*, 7. doi:10.3389/fnsys.2015.00007 PMID:25705180

Lubans, D., Richards, J., Hillman, C., Faulkner, G., Beauchamp, M., Nilsson, M., Kelly, P., Smith, J., Raine, L., & Biddle, S. (2016). Physical activity for cognitive and mental health in youth: A systematic review of mechanisms. *Pediatrics*, *138*(3), e20161642. doi:10.1542/peds.2016-1642 PMID:27542849

McDonough, I. M., Erwin, H. B., Sin, N. L., & Allen, R. S. (2022). Pet ownership is associated with greater cognitive and brain health in a cross-sectional sample across the adult lifespan. *Frontiers in Aging Neuroscience*, *14*, 953889. doi:10.3389/fnagi.2022.953889 PMID:36337704

Morris, T. P., Tormos Muñoz, J. M., Cattaneo, G., Solana-Sánchez, J., Bartrés-Faz, D., & Pascual-Leone, A. (2019). Traumatic brain injury modifies the relationship between physical activity and global and cognitive health: Results from the barcelona brain health initiative. *Frontiers in Behavioral Neuroscience*, *13*, 135. doi:10.3389/fnbeh.2019.00135 PMID:31275124

Papini, C., Palaniyappan, L., Kroll, J., Froudist-Walsh, S., Murray, R. M., & Nosarti, C. (2020). Altered cortical gyrification in adults who were born very preterm and its associations with cognition and mental health. *Biological Psychiatry: Cognitive Neuroscience and Neuroimaging*, *5*(7), 640–650. doi:10.1016/j.bpsc.2020.01.006 PMID:32198001

Patel, V., Saxena, S., Lund, C., Thornicroft, G., Baingana, F., Bolton, P., Chisholm, D., Collins, P. Y., Cooper, J. L., Eaton, J., Herrman, H., Herzallah, M. M., Huang, Y., Jordans, M. J. D., Kleinman, A., Medina-Mora, M. E., Morgan, E., Niaz, U., Omigbodun, O., & UnÜtzer, J. Ü. (2018). The Lancet Commission on global mental health and sustainable development. *Lancet*, *392*(10157), 1553–1598. doi:10.1016/S0140-6736(18)31612-X PMID:30314863

Paul, R. H., Gunstad, J., Cooper, N., Williams, L. M., Clark, C. R., Cohen, R. A., Lawrence, J. J., & Gordon, E. (2007). Cross-cultural assessment of neuropsychological performance and electrical brain function measures: Additional validation of an international brain database. *The International Journal of Neuroscience*, *117*(4), 549–568. doi:10.1080/00207450600773665 PMID:17365135

Rossell, S. L., Coakes, J., Shapleske, J., Woodruff, P. W. R., & David, A. S. (2003). Insight: Its relationship with cognitive function, brain volume and symptoms in schizophrenia. *Psychological Medicine*, *33*(1), 111–119. doi:10.1017/S0033291702006803 PMID:12537042

Schumann, G., Andreassen, O. A., Banaschewski, T., Calhoun, V. D., Clinton, N., Desrivieres, S., Brandlistuen, R. E., Feng, J., Hese, S., Hitchen, E., Hoffmann, P., Jia, T., Jirsa, V., Marquand, A. F., Nees, F., Nöthen, M. M., Novarino, G., Polemiti, E., Ralser, M., & Ogoh, G. (2023). Addressing Global Environmental Challenges to Mental Health Using Population Neuroscience: A Review. *JAMA Psychiatry*, *80*(10), 1066. doi:10.1001/jamapsychiatry.2023.2996 PMID:37610741

Sharma, A., & Singh, B. (2022). Measuring Impact of E-commerce on Small Scale Business: A Systematic Review. *Journal of Corporate Governance and International Business Law*, *5*(1).

Singh, B. (2019). Profiling Public Healthcare: A Comparative Analysis Based on the Multidimensional Healthcare Management and Legal Approach. *Indian Journal of Health and Medical Law*, *2*(2), 1–5.

Singh, B. (2020). Global Science and Jurisprudential Approach Concerning Healthcare and Illness. *Indian Journal of Health and Medical Law*, *3*(1), 7–13.

Singh, B. (2022). Understanding Legal Frameworks Concerning Transgender Healthcare in the Age of Dynamism. *Electronic Journal of Social and Strategic Studies*, *3*(1), 56–65. doi:10.47362/EJSSS.2022.3104

Singh, B. (2022). Relevance of Agriculture-Nutrition Linkage for Human Healthcare: A Conceptual Legal Framework of Implication and Pathways. *Justice and Law Bulletin*, *1*(1), 44–49.

Singh, B. (2022). COVID-19 Pandemic and Public Healthcare: Endless Downward Spiral or Solution via Rapid Legal and Health Services Implementation with Patient Monitoring Program. *Justice and Law Bulletin*, *1*(1), 1–7.

Singh, B. (2023). Blockchain Technology in Renovating Healthcare: Legal and Future Perspectives. In Revolutionizing Healthcare Through Artificial Intelligence and Internet of Things Applications (pp. 177-186). IGI Global.

Singh, B. (2023). Federated Learning for Envision Future Trajectory Smart Transport System for Climate Preservation and Smart Green Planet: Insights into Global Governance and SDG-9 (Industry, Innovation and Infrastructure). *National Journal of Environmental Law*, *6*(2), 6–17.

Singh, B. (2024). Legal Dynamics Lensing Metaverse Crafted for Videogame Industry and E-Sports: Phenomenological Exploration Catalyst Complexity and Future. *Journal of Intellectual Property Rights Law*, *7*(1), 8–14.

Small, G. W., Lee, J., Kaufman, A., Jalil, J., Siddarth, P., Gaddipati, H., & Bookheimer, S. Y. (2022). Brain health consequences of digital technology use. *Dialogues in Clinical Neuroscience*. PMID:32699518

Stangl, M., Maoz, S. L., & Suthana, N. (2023). Mobile cognition: Imaging the human brain in the 'real world'. *Nature Reviews. Neuroscience*, *24*(6), 1–16. doi:10.1038/s41583-023-00692-y PMID:37046077

Stein, D. J., He, Y., Phillips, A., Sahakian, B. J., Williams, J., & Patel, V. (2015). Global mental health and neuroscience: Potential synergies. *The Lancet. Psychiatry*, *2*(2), 178–185. doi:10.1016/S2215-0366(15)00014-0 PMID:26359754

Tamminga, C. A. (2014). Approaching human neuroscience for disease understanding. *World Psychiatry; Official Journal of the World Psychiatric Association (WPA)*, *13*(1), 41–43. doi:10.1002/wps.20099 PMID:24497244

Thompson, P. M., Jahanshad, N., Ching, C. R., Salminen, L. E., Thomopoulos, S. I., Bright, J., & Sønderby, I. E. (2020). ENIGMA and global neuroscience: A decade of large-scale studies of the brain in health and disease across more than 40 countries. *Translational Psychiatry*, *10*(1), 100. doi:10.1038/s41398-020-0705-1 PMID:32198361

Walsh, R. (2011). Lifestyle and mental health. *The American Psychologist*, *66*(7), 579–592. doi:10.1037/a0021769 PMID:21244124

World Health Organization. (2001). [Mental health: new understanding, new hope.]. *World Health Report*. WHO.

Zheng, G., Ye, B., Zheng, Y., Xiong, Z., Xia, R., Qiu, P., Tao, J., & Chen, L. (2019). The effects of exercise on the structure of cognitive related brain regions: A meta-analysis of functional neuroimaging data. *The International Journal of Neuroscience*, *129*(4), 406–415. doi:10.1080/00207454.2018.1508135 PMID:30073877

Chapter 13
Understanding the Multifactorial Background of Attention-Deficit/Hyperactivity Disorder

Payal Kapoor
Lovely Professional University, India

Simran Sunil Khutarkar
Lovely Professional University, India

ABSTRACT

Attention-Deficit/Hyperactivity Disorder (ADHD), a common neuro-behavioral disorder, affects children at an early age and its symptoms may continue in adults. The disorder impacts the individual's social as well as academic development. The multifactorial nature of ADHD is likely due to the combination of genetic, environmental, and neurological factors. Family history poses a great risk for ADHD development and is known to involve genes that regulate neurotransmitters associated with attention, impulses, and behavior. Maternal smoking during pregnancy, metal ion and toxin exposure and traumatic brain injury are some of the environmental factors that increase the risk of ADHD. Further, brain imaging studies in ADHD have shown altered structure and brain activity specifically in the prefrontal cortex. Reduced prefrontal cortex volume and altered brain connections in ADHD patients result in impulsivity, poor behavioral control, and difficulty in maintaining attention. Although with no cure, ADHD can be successfully managed, and some symptoms may improve as the child ages.

INTRODUCTION AND BACKGROUND

A neurodevelopmental condition known as Attention-Deficit/Hyperactivity Disorder (ADHD) is most typically diagnosed in childhood, however its symptoms can last into adulthood. The disorder is marked by recurrent patterns of impulsivity, hyperactivity, and inattentiveness that severely impede day-to-day

DOI: 10.4018/979-8-3693-1922-2.ch013

operations, personal development and may have detrimental effects on a variety of aspects of a person's life including relationships, profession, academic success and overall quality of life (Wilens & Spencer, 2010). Though the actual etiology of ADHD is unknown, research indicates that it is most likely the result of a complex interaction of genetic, environmental, and neurological variables and the proportional importance of each element varies between individuals. Furthermore, studies are being conducted to better understand the complex interplay between these factors (Thapar, Cooper, Eyre, Langley, & Psychiatry, 2013).

Earlier ADHD was believed to impact children with symptoms progressively disappearing throughout adolescence or the early stages of adulthood. However, follow-up studies showed the persistence of ADHD from childhood to adulthood (Sibley, Mitchell, & Becker, 2016) rather the disorder affected a significant population of adults (Song et al., 2021). The prevalence of ADHD in childhood has been reported to be 5%-7% (Polanczyk, De Lima, Horta, Biederman, & Rohde, 2007) and 1/3 of the patients are positive for persistent ADHD (Barbara Franke et al., 2018). A recent review indicated the pooled prevalence of ADHD in adults to be 3.10% where ADHD-I (inattentive type of ADHD) remained the commonest type of ADHD, followed by ADHD-HI (hyperactive type) and ADHD-C (combined type) (Ayano et al., 2023). Further, the prevalence rate of ADHD in adult population is higher than other well-known mental disorders including bipolar disorders (1%) and some anxiety disorders including post-traumatic stress disorder (PTSD), Obsessive-compulsive disorder (OCD), Generalized anxiety disorder (GAD) and panic disorders (Clemente et al., 2015). Several factors such as drug usage, harsh childhood experiences, the severity of the disorder and/or untreated ADHD throughout childhood may be responsible for the persistence of ADHD in adults (Kessler et al., 2005). ADHD treatment necessitates a thorough and customized strategy that takes into account the particular requirements and difficulties of every patient. A successful and satisfying life may be achieved by those with ADHD by combining several treatment approaches and offering continuous support.

This chapter seeks to provide succinct information on the variables that are important in the development of ADHD, assisting us in comprehending the complex elements that lead to this crippling condition (Figure 1).

Figure 1. Schematic representation of the factors contribution to ADHD

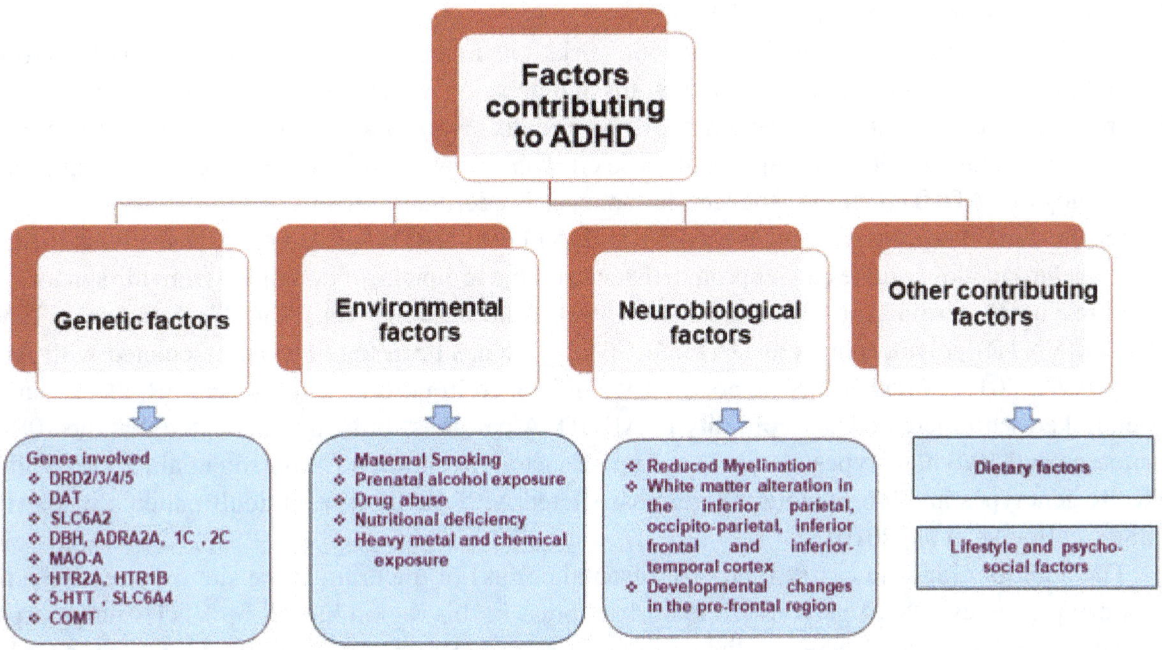

FACTORS CONTRIBUTING TO ADHD

1. GENETIC FACTORS

ADHD is known to be familial and is a highly heritable disorder (Chen et al., 2017). ADHD has a high heredity rate when compared rates reported in autism spectrum disorder (ASD), bipolar disorder, and schizophrenia (Sullivan, Daly, & O'donovan, 2012). ADHD heritability has been estimated to be 77-88% based on family and twin studies conducted in recent decades (Faraone & Larsson, 2019). Further findings imply that the relative risk of ADHD for first-degree relatives is 5- to 10-fold (B Franke et al., 2012) which applies to core symptoms and is gender-independent (Martin et al., 2018)

Several studies have indicated many genes to be involved in the development of ADHD in children. (Ogdie et al., 2004) (Klein et al., 2017) (Neale et al., 2010). These include dopamine (DA) genes such as dopamine receptor genes (DRD2, DRD3, DRD4, DRD5), dopamine transporter gene (DAT, SLC6A3), norepinephrine transporter gene (SLC6A2) (DiMaio, Grizenko, Joober, & Neuroscience, 2003), dopamine-beta-hydroxylase (DBH) gene (Zhang, Wang, Li, Wang, & Yang, 2005) noradrenergic receptor genes such as ADRA2A, 2C and 1C, monoamine oxidase-A (MAO-A) gene, 5-hydroxytryptamine receptor

HTR2A, HTR1B, serotonin transporter (5-HTT) and SLC6A4, catechol-O-methyltransferase (COMT) gene (Sun, Yuan, Shen, Xiong, & Wu, 2014) and at least one Gamma-aminobutyric acid (GABA) gene GABRB3. Studies indicate that the genes contributing to the development of child ADHD may contribute differently in the development of ADHD in adults (B Franke et al., 2012). A longitudinal twin study showed that the persistence of ADHD in adults is associated with higher genetic loading (Z. Chang, Lichtenstein, Asherson, & Larsson, 2013).

Existing research has revealed that genetic variations as well as pre- and perinatal risk factors are related to the presentation of ADHD symptoms and appear to be connected with a variety of neurodevelopmental and psychiatric consequences. Brain circuits involved with DA regulate the processes such as motivation and attention. Disruption of these circuits or any or both of these processes contributes to the etiology of ADHD (Uchida, Spencer, Faraone, & Biederman, 2018).

Of the candidate genes linked with ADHD, DAT1 and DRD4 genes are extensively studied. The DAT1 gene encodes a solute carrier protein that causes the re-uptake of dopamine from the synaptic cleft into presynaptic terminals (Amara & Kuhar, 1993). A 40 base-pair (bp) Variable Number of Tandem Repeat (VNTR) polymorphism of DAT1/SLC6A3 gene has been found to be associated with ADHD (Šerý et al., 2015). Another VNTR polymorphism located in intron8 with 5R and 6R alleles, has also been linked with increased susceptibility to ADHD. A previous study also indicated that the 10R/10R genotype and 10/6 haplotypes are likely to be risk factors in children (Asherson et al., 2007) while the 9R/9R genotypes and 9/6 haplotypes were associated with the disease in adult patients with ADHD (Barbara Franke et al., 2010).

The anterior cingulate cortex (and orbitofrontal cortex) of the brain is the site for attention and inhibitory processes. DRD4 gene which majorly express in this region has 48-bp VNTR polymorphism in exon3, the 7R allele of which is linked to an increased risk of ADHD in the Caucasian population (Wu, Xiao, Sun, Zou, & Zhu, 2012). However earlier studies reported that the patients with persistent ADHD symptoms were less likely to have DRD4 7R allele (Shaw, Gornick, et al., 2007). Emergence of new genes might contribute to changes in symptoms at a later age (Z. Chang et al., 2013). Further, genetic regions with ADHD-related endophenotypes (neurobiological biomarkers associated with an illness) such as Latrophilin 3 gene, Fibroblast growth factor 1 gene, Cholinergic receptor, nicotinic and alpha 4, have also been identified (Mastronardi et al., 2016).

Genome-wide association studies (GWAS) have revealed that common genetic variants such as cholinergic receptor nicotinic alpha 7 subunit (CHRNA7), Charged Multivesicular Body Protein 7(CHMP7), 5'-nucleotidase cytosolic II(NT5C2), and Ribosomal Protein L13a Pseudogene 3 (RPL13AP3) are linked to phenotypes shared by multiple psychiatric disorders, including schizophrenia, bipolar disorder, anxiety, major depressive disorder and ADHD but none of these genetic variants are robustly associated with ADHD (Lancet, 2013) (Xu et al., 2017).

2. ENVIRONMENTAL FACTORS

Prenatal and Post-Natal Risk Factors for ADHD

Several studies have looked at how inherited genetic variables interact with environmental risk factors to impact the development and effects of ADHD (Nigg, Nikolas, Burt, & Psychiatry, 2010). The DAT 1 gene has been linked to maternal smoking and alcohol use during pregnancy (Brookes et al., 2006) and has been shown to increase the risk for combined-type ADHD by nine-fold (Neuman et al., 2007).

A study showed that the gene-environment interaction was responsible for 59% of ADHD diagnoses in children with the DRD4 GG genotype who had been exposed to dimethyl phosphate. (C.-H. Chang et al., 2018). Toxin exposure during pregnancy and after birth has also been linked to an increased incidence of ADHD (Thapar et al., 2013). Damage to the brain after birth caused by traumatic brain injury has also been identified as a risk factor for ADHD (Adeyemo et al., 2014).

The period of pregnancy being the critical time for brain development, exposure to certain hazardous environment during this period can increase the risk of ADHD development in children. Some of the most well-studied prenatal exposures linked to ADHD include:

Maternal Smoking

Studies have consistently shown a strong association between maternal smoking and increased ADHD risk in offspring as it increases the risk of hyperactivity in offsprings (Pauly & Slotkin, 2008). Exposure to cigarette smoke during pregnancy negatively impacts both prenatal and postnatal growth, damages neural pathways, cause abnormalities in cellular proliferation and differentiation, inhibits the development of catecholaminergic and cholinergic systems and increases the risk of behavioral issues and cognitive developmental defects in children and adolescents (Ernst, Moolchan, Robinson, & Psychiatry, 2001). Reduced uterine blood flow caused by smoking disrupts the normal placental processes (Milberger, Biederman, Faraone, Chen, & Jones, 1996).The resulting hypoxia-ischemia and malnutrition (Suzuki, Minei, Johnson, & gynecology, 1980) causes retardation in intrauterine development (Milberger, Biederman, Faraone, & Jones, 1998).

Prenatal Alcohol Exposure

Exposure to alcohol during pregnancy raises the child's risk of mental illnesses, impulsive or destructive conduct and hyperactivity. This impacts the cognitive processes such as the overall intellectual performance, learning and memory, language, attention, response time, visual-spatial ability, executive functions, fine and gross motor skills, adaptive and forming and maintaining social relationships (Huizink, Mulder, & Reviews, 2006). Although some studies have demonstrated that prenatal alcohol exposure increases the risk for ADHD, (Knopik et al., 2006), several studies found no correlation between the two.(Rodriguez et al., 2009) (D'Onofrio et al., 2007)

Drug Abuse

Cocaine exposure during pregnancy has been linked to behavioral issues and attention impairments in preschoolers and school-aged children (Whitaker, 2011). These effects are generated by disruptions in the development in the monoaminergic system (Yolton et al., 2014). An ideal environment after birth was found to bring good influence on the intellectual ability, but all school-aged children born to heroin-dependent parents, even those reared at home and those adopted, had a high risk for the development of ADHD (Ornoy, 2003).

Nutritional Deficiencies

Deficiencies in certain nutrients, particularly iron, zinc, and omega-3 fatty acids during pregnancy may contribute to ADHD risk. These nutrients are essential for healthy brain development and function, and deficiencies can lead to structural and functional changes in the brain (Matsudaira & health, 2007). Understanding these potential risk factors can help healthcare professionals provide better prenatal care and early intervention for children at risk of ADHD. Animal studies have shown the impact of iron deficiency on dopamine receptor density in brain (Erikson et al., 2001). Iron plays a crucial role in dopamine synthesis wherein it acts as a cofactor of tyrosine hydroxylase enzyme (JM & aspects, 1988) A study indicated a link between hyperactive children with ADHD and low ferritin levels, but not between hyperactivity and cognitive skills (Oner, Alkar, & Oner, 2008).

Heavy Metal and Chemical Exposure

Exposure to heavy metals and certain chemicals has emerged as a potential environmental factor contributing to the development of ADHD. ADHD has been linked to exposure to chemicals like colors and preservatives in food, as well as poisons like lead, mercury, and manganese(Collipp, Chen, Maitinsky, & metabolism, 1983).

Lead Toxicity

Lead is a neurotoxin that readily crosses the blood-brain barrier, particularly in children whose brains are still developing. Lead exposure can disrupt neurotransmitter function, impair cognitive development and negatively impact impulse control and attention. Even low levels of lead exposure can be detrimental and efforts to minimize children's exposure are crucial (Donzelli et al., 2019). According to Braun *et al.* there is a substantial association between ADHD and higher blood lead concentrations in children. Children who have blood lead levels >2.0 μg/dL are at a 4.1-fold greater risk of developing ADHD (Braun, Kahn, Froehlich, Auinger, & Lanphear, 2006).

Mercury Toxicity

Mercury can damage brain cells and interfere with neurotransmitter signalling, affecting the cognitive function and behavior. Mercury, particularly methylmercury, is a potent neurotoxin that can accumulate in the body through fish consumption and industrial exposure. According to a meta-analysis, a significant correlation between environmental perinatal mercury exposure and an increased risk of ADHD has been found (Yoshimasu, Kiyohara, Takemura, & Nakai, 2014).

Pesticide Toxicity

Organophosphates are a class of pesticides widely used in agriculture and pest control. Exposure to organophosphates, especially during prenatal or early childhood development, has been linked to an increased risk of ADHD. These chemicals can disrupt the nervous system and affect brain development, leading to cognitive and behavioral impairments (Banerjee, Middleton, & Faraone, 2007). Rodríguez (2012) found a link between ADHD in females and urine levels of 3-phenoxy-benzoic acid (3-PBA), a metabolite of pyrethroid insecticide(Rodríguez, 2012). However, QuirŌs-Alcalá *et al.* demonstrated that postnatal pyrethroid exposure did not correlate with ADHD in children (Quirós-Alcalá, Mehta, & Eskenazi, 2014).

Phthalates Toxicity

Phthalates are chemicals commonly found in plastics and personal care product. Research suggests that exposure to phthalates during pregnancy may be associated with an increased risk of ADHD in children. Phthalates can interfere with the endocrine system and disrupt hormone function, potentially impacting brain development and behavior (Chopra, Harley, Lahiff, & Eskenazi, 2014).

Other Metals and Chemicals

Studies are also investigating the potential link between ADHD and exposure to other metals such as cadmium and manganese, as well as chemicals like polychlorinated biphenyls (PCBs) and bisphenol A (BPA).While the evidence is still emerging for these substances, further research is needed to elucidate their potential role in ADHD development (Grova, Schroeder, Olivier, & Turner, 2019).

3. NEUROBIOLOGICAL FACTORS

Neuroimaging technique has resulted in some significant improvements enhancing the knowledge of ADHD, demonstrating a definite brain foundation for the condition in areas involved in attention, executive and inhibitory control. An expanding corpus of studies have indicated that the pathophysiology of ADHD likely involves the fronto-striatal network which includes the caudate nucleus and putamen, the dorsal anterior cingulate cortex, and the lateral prefrontal cortex. Total brain volume, the prefrontal cortex, the basal ganglia (striatum), the dorsal anterior cingulate cortex, the corpus callosum, and the cerebellum have all been shown to have decreased volume in ADHD individuals (Emond, Joyal, & Poissant, 2008). A delay in cortical maturation has been observed in ADHD children where the grey matter was seen to peak with a delay of 3 years as compared to healthy controls (Shaw & Rabin, 2009).

Prefrontal cortex region which controls the cognitive functions was found to be most affected regions (Shaw et al., 2006) (Shaw, Eckstrand, et al., 2007). As compared with controls, the ADHD patients showed reduction in the volume of white matter (WM) (Castellanos et al., 2002), reduced midsagittal corpus callosum areas (Krain & Castellanos, 2006), and reduced cortical thickness. Studies have also revealed the existence of alterations in white matter in children, adolescent and adults with ADHD (Makris et al., 2008). WM abnormalities has been found in several distinct regions of the brain including the inferior parietal, occipito-parietal, inferior frontal, and inferior temporal cortex (Silk, Vance, Rinehart, Bradshaw, & Cunnington, 2009). Reduced myelination and hence reduced neuronal communication in ADHD patients has been revealed indicating developmental changes in cortical WM pathways in prefrontal regions and in pathways surrounding the basal ganglia and cerebellum (D'Agati, Casarelli, Pitzianti, Pasini, & Psychiatry, 2010).

OTHER CONTRIBUTING FACTORS

Dietary Factors and ADHD: A Complex Relationship

Several studies have investigated the potential role of dietary factors in ADHD development and management. While there is no single "ADHD diet," understanding the relationship between nutrition and ADHD can help manage symptoms and support overall well-being (Lange, Lange, Nakamura, & Reissmann, 2023)

Food Additives

Artificial food additives, such as color, preservatives and sweeteners, have been suspected of worsening ADHD symptoms in some children. Studies have shown mixed results, with some suggesting a link between specific additives and increased hyperactivity and impulsivity, while others demonstrate no significant association. More research is needed to fully understand the potential impact of food additives on ADHD symptoms (Stevens, Kuczek, Burgess, Hurt, & Arnold, 2011).

Diet Quality

A diet high in processed foods, refined carbohydrates, and unhealthy fats may contribute to ADHD symptoms. These foods can lead to fluctuations in blood sugar levels, which can impact mood, focus, and energy levels. Conversely a diet rich in fruits, vegetables, whole grains, lean protein, and healthy fats may offer potential benefits for ADHD management. These foods provide essential nutrients that support brain function and neurotransmitter activity, potentially improving focus, attention, and behavior (Lange, Nakamura, & Reissmann, 2022).

Specific Nutrients

Deficiencies in certain nutrients, such as iron, zinc, omega-3 fatty acids, and vitamin D, have been linked to an increased risk of ADHD and worsening of symptoms. Iron and zinc are essential for brain development and function, while omega-3 fatty acids are crucial for neurotransmitter signalling and brain health. Vitamin D has also been shown to play a role in brain development and regulation, and deficiencies may contribute to ADHD symptoms. Ensuring adequate intake of these nutrients through dietary sources or supplementation, if needed, can potentially support ADHD management (Lange et al., 2017).

Sugar and ADHD

The link between sugar and ADHD is a complex one with conflicting evidence about its impact on symptoms. Some studies suggest that excessive sugar intake may lead to increased hyperactivity and impulsivity, while others have found no significant association (Farsad-Naeimi et al., 2020). It's important to note that the sugar intake interacts with other dietary factors and individual differences, making it difficult to establish a definitive link with ADHD. Limiting added sugars and focusing on whole, unprocessed foods remains a healthy recommendation for overall well-being, regardless of ADHD status (Kanarek, 1994).

LIFESTYLE AND PSYCHOSOCIAL FACTORS

Sleep deprivation: Poor sleep quality can worsen ADHD symptoms. Children with ADHD often have difficulty falling asleep, staying asleep, and getting enough sleep (Um, Hong, Jeong, & Neuroscience, 2017).

Stress and anxiety: Chronic stress and anxiety can exacerbate ADHD symptoms and make it difficult for children to manage their behavior (Saccaro, Schilliger, Perroud, & Piguet)

Family environment: A chaotic or dysfunctional family environment can contribute to ADHD symptoms. Children in such environments may experience more stress, instability and lack of structure, which can worsen their ADHD (Pires, da Silva, & de Assis, 2013).

Socioeconomic status: Children living in poverty or low-income families may be at a higher risk of developing ADHD due to limited access to healthcare, healthy food, and educational resources (Holton & Nigg, 2020)

CONCLUSION

Attention-Deficit/Hyperactivity Disorder, or ADHD, is a multifaceted disorder that is a confluence of neurological, environmental, and hereditary variables. Research shows a strong hereditary component associated with ADHD. Environmental factors such as early childhood experiences, premature birth and chemical exposure during fetal development can also lead to the development of ADHD. Further it is important to understand that several genetic and environmental variables interact, rather than one genetic component being the cause of ADHD. The variety of ways that ADHD presents itself highlights just how complex its causes are. In order to effectively treat ADHD in individuals, a thorough approach to diagnosis and treatment is necessary, one that considers both environmental factors and genetic predispositions. Deviations from normal brain structure and function, particularly in the domains of executive, impulse control and attention also impact the manifestation of symptoms. Further study in this area is crucial for a better understanding of the disorder and the development of more tailored therapies to assist individuals with ADHD.

REFERENCES

Adeyemo, B. O., Biederman, J., Zafonte, R., Kagan, E., Spencer, T. J., Uchida, M., & Faraone, S. V. J. J. o. a. d. (2014). Mild traumatic brain injury and ADHD: a systematic review of the literature and meta-analysis. *Journal of Attention Disorders, 18*(7), 576-584.

Amara, S. G., & Kuhar, M. (1993). Neurotransmitter transporters: recent progress. *Annual Review of Neuroscience, 16*(1), 73-93.

Asherson, P., Brookes, K., Franke, B., Chen, W., Gill, M., Ebstein, R. P., & Eisenberg, J. J. A. J. o. P. (2007). Confirmation that a specific haplotype of the dopamine transporter gene is associated with combined-type ADHD. *American Journal of Psychiatry, 164*(4), 674-677.

Ayano, G., Tsegay, L., Gizachew, Y., Necho, M., Yohannes, K., Abraha, M., & Alati, R. J. P. R. (2023). *Prevalence of attention deficit hyperactivity disorder in adults: Umbrella review of evidence generated across the globe.*

Banerjee, T. D., Middleton, F., & Faraone, S. V. J. A. p. (2007). Environmental risk factors for attention-deficit hyperactivity disorder. *Acta Paeddiatra, 96*(9), 1269-1274.

Braun, J. M., Kahn, R. S., Froehlich, T., Auinger, P., & Lanphear, B. P. J. E. h. p. (2006). Exposures to environmental toxicants and attention deficit hyperactivity disorder in US children. *Environmental Health Perspective, 114*(12), 1904-1909.

Brookes, K.-J., Mill, J., Guindalini, C., Curran, S., Xu, X., Knight, J., & Taylor, E. J. A. o. g. p. (2006). A common haplotype of the dopamine transporter gene associated with attention-deficit/hyperactivity disorder and interacting with maternal use of alcohol during pregnancy. *Arc. Gen. Psychology, 63*(1), 74-81.

Castellanos, F. X., Lee, P. P., Sharp, W., Jeffries, N. O., Greenstein, D. K., Clasen, L. S., & Walter, J. M. J. J. (2002). Developmental trajectories of brain volume abnormalities in children and adolescents with attention-deficit/hyperactivity disorder. *JAMA, 288*(14), 1740-1748.

Chang, C.-H., Yu, C.-J., Du, J.-C., Chiou, H.-C., Chen, H.-C., Yang, W., & Mao, I.-F. J. E. R. (2018). The interactions among organophosphate pesticide exposure, oxidative stress, and genetic polymorphisms of dopamine receptor D4 increase the risk of attention deficit/hyperactivity disorder in children. *Environmental Resources, 160*, 339-346.

Chang, Z., Lichtenstein, P., Asherson, P. J., & Larsson, H. J. J. p. (2013). Developmental twin study of attention problems: high heritabilities throughout development. *JAMA, 70*(3), 311-318.

Chen, Q., Brikell, I., Lichtenstein, P., Serlachius, E., Kuja-Halkola, R., Sandin, S., & Psychiatry. (2017). Familial aggregation of attention-deficit/hyperactivity disorder. *Journal of Childhood Psychology, 58*(3), 231-239.

Chopra, V., Harley, K., Lahiff, M., & Eskenazi, B. J. E. r. (2014). Association between phthalates and attention deficit disorder and learning disability in US children, 6–15 years. *Environmental Resources, 128*, 64-69.

Clemente, A. S., Diniz, B. S., Nicolato, R., Kapczinski, F. P., Soares, J. C., Firmo, J. O., & Castro-Costa, É. J. B. J. o. P. (2015). Bipolar disorder prevalence: a systematic review and meta-analysis of the literature. *Braz Journal of Psychology, 37*, 155-161.

Collipp, P., Chen, S., Maitinsky, S. J. A. o. n., & metabolism. (1983). Manganese in infant formulas and learning disability. *Ann Nutr Metab, 27*(6), 488-494.

D'Agati, E., Casarelli, L., Pitzianti, M. B., Pasini, A. J. P. i. N.-P., & Psychiatry, B. (2010). Overflow movements and white matter abnormalities in ADHD. *Prog Neuropsychopharmacol Biol Psychiatry, 34*(3), 441-445.

D'Onofrio, B. M., Van Hulle, C. A., Waldman, I. D., Rodgers, J. L., Rathouz, P. J., & Lahey, B. B. J. A. o. g. p. (2007). Causal inferences regarding prenatal alcohol exposure and childhood externalizing problems. *Archive of General Psychiatry, 64*(11), 1296-1304.

DiMaio, S., Grizenko, N., Joober, R. J. J. o. P., & Neuroscience. (2003). Dopamine genes and attention-deficit hyperactivity disorder: a review. *Journal of Psychiatry and Neuroscience, 28*(1), 27-38.

Donzelli, G., Carducci, A., Llopis-Gonzalez, A., Verani, M., Llopis-Morales, A., Cioni, L., & health, p. (2019). The association between lead and attention-deficit/hyperactivity disorder: a systematic review. *Int J Environ Res Public Health, 16*(3), 382.

Emond, V., Joyal, C., & Poissant, H. J. L. e. (2008). Structural and functional neuroanatomy of attention-deficit hyperactivity disorder (ADHD). *35*(2), 107-114.

Erikson, K. M., Jones, B. C., Hess, E. J., Zhang, Q., Beard, J. L. J. P. b., & behavior. (2001). Iron deficiency decreases dopamine D1 and D2 receptors in rat brain. *69*(3-4), 409-418.

Ernst, M., Moolchan, E. T., Robinson, M. L. J. J. o. t. A. A. o. C., & Psychiatry, A. (2001). Behavioral and neural consequences of prenatal exposure to nicotine. *40*(6), 630-641.

Faraone, S. V., & Larsson, H. J. M. p. (2019). Genetics of attention deficit hyperactivity disorder. *24*(4), 562-575.

Farsad-Naeimi, A., Asjodi, F., Omidian, M., Askari, M., Nouri, M., Pizarro, A. B., & Daneshzad, E. J. C. t. i. m. (2020). Sugar consumption, sugar sweetened beverages and Attention Deficit Hyperactivity Disorder: A systematic review and meta-analysis. *53*, 102512.

Franke, B., Faraone, S., Asherson, P., Buitelaar, J., Bau, C., Ramos-Quiroga, J. A., & Haavik, J. J. M. p. (2012). The genetics of attention deficit/hyperactivity disorder in adults, a review. *17*(10), 960-987.

Franke, B., Michelini, G., Asherson, P., Banaschewski, T., Bilbow, A., Buitelaar, J. K., & Haavik, J. J. E. N. (2018). Live fast, die young? A review on the developmental trajectories of ADHD across the lifespan. *28*(10), 1059-1088.

Franke, B., Vasquez, A. A., Johansson, S., Hoogman, M., Romanos, J., Boreatti-Hümmer, A., & Casas, M. J. N. (2010). Multicenter analysis of the SLC6A3/DAT1 VNTR haplotype in persistent ADHD suggests differential involvement of the gene in childhood and persistent ADHD. *35*(3), 656-664.

Grova, N., Schroeder, H., Olivier, J.-L., & Turner, J. D. J. I. j. o. g. (2019). Epigenetic and neurological impairments associated with early life exposure to persistent organic pollutants. *2019*.

Holton, K. F., & Nigg, J. T. J. J. o. a. d. (2020). The association of lifestyle factors and ADHD in children. *24*(11), 1511-1520.

Huizink, A. C., Mulder, E. J. J. N., & Reviews, B. (2006). Maternal smoking, drinking or cannabis use during pregnancy and neurobehavioral and cognitive functioning in human offspring. *30*(1), 24-41.

JM, W. J. B. i. n., & aspects, b. (1988). Iron-dependent enzymes in the brain. 25-66.

Kanarek, R. B. J. N. r. (1994). Does sucrose or aspartame cause hyperactivity in children?, *52*(5), 173-175.

Kessler, R. C., Adler, L. A., Barkley, R., Biederman, J., Conners, C. K., Faraone, S. V., & Spencer, T. J. B. p. (2005). Patterns and predictors of attention-deficit/hyperactivity disorder persistence into adulthood: results from the national comorbidity survey replication. *57*(11), 1442-1451.

Klein, M., Onnink, M., van Donkelaar, M., Wolfers, T., Harich, B., Shi, Y., & Reviews, B. (2017). Brain imaging genetics in ADHD and beyond–mapping pathways from gene to disorder at different levels of complexity. *80*, 115-155.

Knopik, V. S., Heath, A. C., Jacob, T., Slutske, W. S., Bucholz, K. K., Madden, P. A., & Martin, N. G. J. P. m. (2006). Maternal alcohol use disorder and offspring ADHD: disentangling genetic and environmental effects using a children-of-twins design. *36*(10), 1461-1471.

Krain, A. L., & Castellanos, F. X. J. C. p. r. (2006). Brain development and ADHD. *26*(4), 433-444.

Lancet, C.-D. G. o. t. P. G. C. J. T. (2013). Identification of risk loci with shared effects on five major psychiatric disorders: a genome-wide analysis. *381*(9875), 1371-1379.

Lange, K. W., Hauser, J., Lange, K. M., Makulska-Gertruda, E., Nakamura, Y., Reissmann, A., & Takeuchi, Y. J. C. p. r. (2017). The role of nutritional supplements in the treatment of ADHD: what the evidence says. *19*, 1-9.

Lange, K. W., Lange, K. M., Nakamura, Y., & Reissmann, A. J. C. N. R. (2023). Nutrition in the Management of ADHD: A Review of Recent Research. *12*(3), 383-394.

Lange, K. W., Nakamura, Y., & Reissmann, A. J. J. o. F. F. (2022). Diet and food in attention-deficit hyperactivity disorder. *2*(2), 112-118.

Makris, N., Buka, S. L., Biederman, J., Papadimitriou, G. M., Hodge, S. M., Valera, E. M., & Caviness, V. S. J. C. c. (2008). Attention and executive systems abnormalities in adults with childhood ADHD: a DT-MRI study of connections. *18*(5), 1210-1220.

Martin, J., Walters, R. K., Demontis, D., Mattheisen, M., Lee, S. H., Robinson, E., & Lichtenstein, P. J. B. p. (2018). A genetic investigation of sex bias in the prevalence of attention-deficit/hyperactivity disorder. *83*(12), 1044-1053.

Mastronardi, C. A., Pillai, E., Pineda, D. A., Martinez, A. F., Lopera, F., Velez, J. I., & Acosta, M. T. J. M. p. (2016). Linkage and association analysis of ADHD endophenotypes in extended and multigenerational pedigrees from a genetic isolate. *21*(10), 1434-1440.

Matsudaira, T. J. N., & health. (2007). Attention deficit disorders–drugs or nutrition?, *19*(1-2), 57-60.

Milberger, S., Biederman, J., Faraone, S. V., Chen, L., & Jones, J. J. T. A. j. o. p. (1996). Is maternal smoking during pregnancy a risk factor for attention deficit hyperactivity disorder in children?, *153*(9), 1138-1142.

Milberger, S., Biederman, J., Faraone, S. V., & Jones, J. J. J. o. c. c. p. (1998). Further evidence of an association between maternal smoking during pregnancy and attention deficit hyperactivity disorder: findings from a high-risk sample of siblings. *27*(3), 352-358.

Neale, B. M., Medland, S. E., Ripke, S., Asherson, P., Franke, B., Lesch, K.-P., & Psychiatry, A. (2010). Meta-analysis of genome-wide association studies of attention-deficit/hyperactivity disorder. *49*(9), 884-897.

Neuman, R. J., Lobos, E., Reich, W., Henderson, C. A., Sun, L.-W., & Todd, R. D. J. B. p. (2007). Prenatal smoking exposure and dopaminergic genotypes interact to cause a severe ADHD subtype. *61*(12), 1320-1328.

Nigg, J., Nikolas, M., Burt, S. A. J. J. o. t. A. A. o. C., & Psychiatry, A. (2010). Measured gene-by-environment interaction in relation to attention-deficit/hyperactivity disorder. *49*(9), 863-873.

Ogdie, M. N., Fisher, S. E., Yang, M., Ishii, J., Francks, C., Loo, S. K., & Smalley, S. L. J. T. A. J. o. H. G. (2004). Attention deficit hyperactivity disorder: fine mapping supports linkage to 5p13, 6q12, 16p13, and 17p11. *75*(4), 661-668.

Oner, O., Alkar, O. Y., & Oner, P. J. P. I. (2008). Relation of ferritin levels with symptom ratings and cognitive performance in children with attention deficit–hyperactivity disorder. *50*(1), 40-44.

Ornoy, A. J. T. l. (2003). The impact of intrauterine exposure versus postnatal environment in neurodevelopmental toxicity: long-term neurobehavioral studies in children at risk for developmental disorders. *140*, 171-181.

Pauly, J. R., & Slotkin, T. A. J. A. P. (2008). Maternal tobacco smoking, nicotine replacement and neurobehavioural development. *97*(10), 1331-1337.

Pires, T. d. O., da Silva, C. M. F. P., & de Assis, S. G. J. B. p. (2013). Association between family environment and attention deficit hyperactivity disorder in children–mothers' and teachers' views. *13*(1), 1-9.

Polanczyk, G., De Lima, M. S., Horta, B. L., Biederman, J., & Rohde, L. A. J. A. j. o. p. (2007). The worldwide prevalence of ADHD: a systematic review and metaregression analysis. *164*(6), 942-948.

Quirós-Alcalá, L., Mehta, S., & Eskenazi, B. J. E. h. p. (2014). Pyrethroid pesticide exposure and parental report of learning disability and attention deficit/hyperactivity disorder in US children: NHANES 1999–2002. *122*(12), 1336-1342.

Rodriguez, A., Olsen, J., Kotimaa, A., Kaakinen, M., Moilanen, I., Henriksen, T., & Psychiatry. (2009). Is prenatal alcohol exposure related to inattention and hyperactivity symptoms in children? Disentangling the effects of social adversity. *50*(9), 1073-1083.

Rodríguez, T. (2012). *Environmental pesticide exposure and neurobehavioral effects among children of Nicaraguan agricultural workers.* Acta Universitatis Upsaliensis, Saccaro, L., Schilliger, Z., Perroud, N., & Piguet, C. Inflammation, Anxiety, and Stress in Attention-Deficit/Hyperactivity Disorder. Biomedicines 2021, 9, 1313. In.

Šerý, O., Paclt, I., Drtílková, I., Theiner, P., Kopečková, M., Zvolský, P., & Functions, B. (2015). A 40-bp VNTR polymorphism in the 3′-untranslated region of DAT1/SLC6A3 is associated with ADHD but not with alcoholism. *11*(1), 1-8.

Shaw, P., Eckstrand, K., Sharp, W., Blumenthal, J., Lerch, J., Greenstein, D., & Rapoport, J. J. P. o. t. n. a. o. s. (2007). Attention-deficit/hyperactivity disorder is characterized by a delay in cortical maturation. *104*(49), 19649-19654.

Shaw, P., Gornick, M., Lerch, J., Addington, A., Seal, J., Greenstein, D., & Castellanos, F. X. J. A. o. g. p. (2007). Polymorphisms of the dopamine D4 receptor, clinical outcome, and cortical structure in attention-deficit/hyperactivity disorder. *64*(8), 921-931.

Shaw, P., Lerch, J., Greenstein, D., Sharp, W., Clasen, L., Evans, A., & Rapoport, J. J. A. o. g. p. (2006). Longitudinal mapping of cortical thickness and clinical outcome in children and adolescents with attention-deficit/hyperactivity disorder. *63*(5), 540-549.

Shaw, P., & Rabin, C. J. C. p. r. (2009). New insights into attention-deficit/hyperactivity disorder using structural neuroimaging. *11*(5), 393-398.

Sibley, M. H., Mitchell, J. T., & Becker, S. P. J. T. L. P. (2016). Method of adult diagnosis influences estimated persistence of childhood ADHD: a systematic review of longitudinal studies. *3*(12), 1157-1165.

Silk, T. J., Vance, A., Rinehart, N., Bradshaw, J. L., & Cunnington, R. J. H. b. m. (2009). White-matter abnormalities in attention deficit hyperactivity disorder: A diffusion tensor imaging study. *30*(9), 2757-2765.

Song, P., Zha, M., Yang, Q., Zhang, Y., Li, X., & Rudan, I. J. J. o. g. h. (2021). The prevalence of adult attention-deficit hyperactivity disorder: A global systematic review and meta-analysis. *11*.

Stevens, L. J., Kuczek, T., Burgess, J. R., Hurt, E., & Arnold, L. E. J. C. p. (2011). Dietary sensitivities and ADHD symptoms: thirty-five years of research. *50*(4), 279-293.

Sullivan, P. F., Daly, M. J., & O'donovan, M. J. N. R. G. (2012). Genetic architectures of psychiatric disorders: the emerging picture and its implications. *13*(8), 537-551.

Sun, H., Yuan, F., Shen, X., Xiong, G., & Wu, J. J. M. n. (2014). Role of COMT in ADHD: a systematic meta-analysis. *49*, 251-261.

Suzuki, K., Minei, L. J., Johnson, E. E. J. A. j. o. o., & gynecology. (1980). Effect of nicotine upon uterine blood flow in the pregnant rhesus monkey. *136*(8), 1009-1013.

Thapar, A., Cooper, M., Eyre, O., Langley, K. J. J. o. C. P., & Psychiatry. (2013). Practitioner review: what have we learnt about the causes of ADHD?, *54*(1), 3-16.

Uchida, M., Spencer, T. J., Faraone, S. V., & Biederman, J. J. J. o. a. d. (2018). Adult outcome of ADHD: an overview of results from the MGH longitudinal family studies of pediatrically and psychiatrically referred youth with and without ADHD of both sexes. *22*(6), 523-534.

Um, Y. H., Hong, S.-C., Jeong, J.-H. J. C. P., & Neuroscience. (2017). Sleep problems as predictors in attention-deficit hyperactivity disorder: causal mechanisms, consequences and treatment. *15*(1), 9.

Whitaker, R. (2011). *Anatomy of an epidemic: Magic bullets, psychiatric drugs, and the astonishing rise of mental illness in America*. Crown.

Wilens, T. E., & Spencer, T. J. J. P. m. (2010). Understanding attention-deficit/hyperactivity disorder from childhood to adulthood. *122*(5), 97.

Wu, J., Xiao, H., Sun, H., Zou, L., & Zhu, L.-Q. J. M. n. (2012). Role of dopamine receptors in ADHD: a systematic meta-analysis. *45*, 605-620.

Xu, B., Jia, T., Macare, C., Banaschewski, T., Bokde, A. L., Bromberg, U., & Psychiatry, A. (2017). Impact of a common genetic variation associated with putamen volume on neural mechanisms of attention-deficit/hyperactivity disorder. *56*(5), 436-444. e434.

Yolton, K., Cornelius, M., Ornoy, A., McGough, J., Makris, S., Schantz, S. J. N., & teratology. (2014). Exposure to neurotoxicants and the development of attention deficit hyperactivity disorder and its related behaviors in childhood. *44*, 30-45.

Yoshimasu, K., Kiyohara, C., Takemura, S., & Nakai, K. J. N. (2014). A meta-analysis of the evidence on the impact of prenatal and early infancy exposures to mercury on autism and attention deficit/hyperactivity disorder in the childhood. *44*, 121-131.

Zhang, H.-B., Wang, Y.-F., Li, J., Wang, B., & Yang, L. J. Z. e. k. z. z. C. J. o. P. (2005). Association between dopamine beta hydroxylase gene and attention deficit hyperactivity disorder complicated with disruptive behavior disorder. *43*(1), 26-30.

Chapter 14
Precision Medicine and Epigenetics:
Personalized Diabetes Care

N. L. Swathi
https://orcid.org/0000-0002-3695-0732
Jawaharlal Nehru Technological University, India

Ishwar Basavan
https://orcid.org/0009-0005-1805-6971
Jawaharlal Nehru Medical College, India

Muhammad Muzzamil
https://orcid.org/0000-0001-9871-5837
Health Services Academy, Pakistan

Aparna Venkatesan
https://orcid.org/0009-0005-2703-8695
Independent Researcher, India

Gorthy Shanmukha Rama
https://orcid.org/0009-0006-6965-4944
Dr. N.T.R. University of Health Sciences, India

Akhil Arun
https://orcid.org/0000-0003-1787-054X
Amrita School of Pharmacy, India

ABSTRACT

With diabetes presenting as a multifactorial and heterogeneous metabolic disorder, the imperative for precision in therapeutic strategies is underscored by an in-depth analysis of its genetic and environmental determinants. This chapter explores precision medicine for diabetes, focusing on genetic and environmental factors influencing the condition. It emphasizes the importance of accuracy in therapeutic approaches and the role of epigenetics in tailoring therapies based on a person's unique epigenetic profile. Technological advancements in genetic and epigenetic analysis provide empirical evidence for the viability and effectiveness of these approaches.

DOI: 10.4018/979-8-3693-1922-2.ch014

INTRODUCTION

Setting the Stage for Precision Medicine in Diabetes

The emergence of Precision Medicine has signified a paradigm shift in the field of diabetes care. This transformative approach is a shift from traditional, broad treatments to a more individualised, nuanced approach. With the use of cutting-edge research and scientific ideas, our goal in this exploration is to establish the foundation for the application of precision medicine to the treatment of diabetes.

Diabetes, characterized by its heterogeneity among individuals, necessitates a departure from traditional, uniform treatment methodologies. Precision medicine emerges as a promising avenue, recognizing the inherent diversity in diabetes manifestations and tailoring interventions to the specific genetic and epigenetic characteristics of each patient.

Technological developments are essential to this trip because they offer instruments for the detailed examination of genetic and epigenetic components. By enabling healthcare professionals to understand the genetic landscape, these state-of-the-art technologies enable a more customised and focused approach to treating diabetes.

As we continue our investigation, we explore the real-world uses of precision medicine, including lifestyle modifications, real-time monitoring, and diagnostics. The symposium concludes with a discussion of the critical role that patient engagement plays, acknowledging that informed and involved patients are essential to attaining the best possible results.

Figure 1. Applications and principles of precision medicine

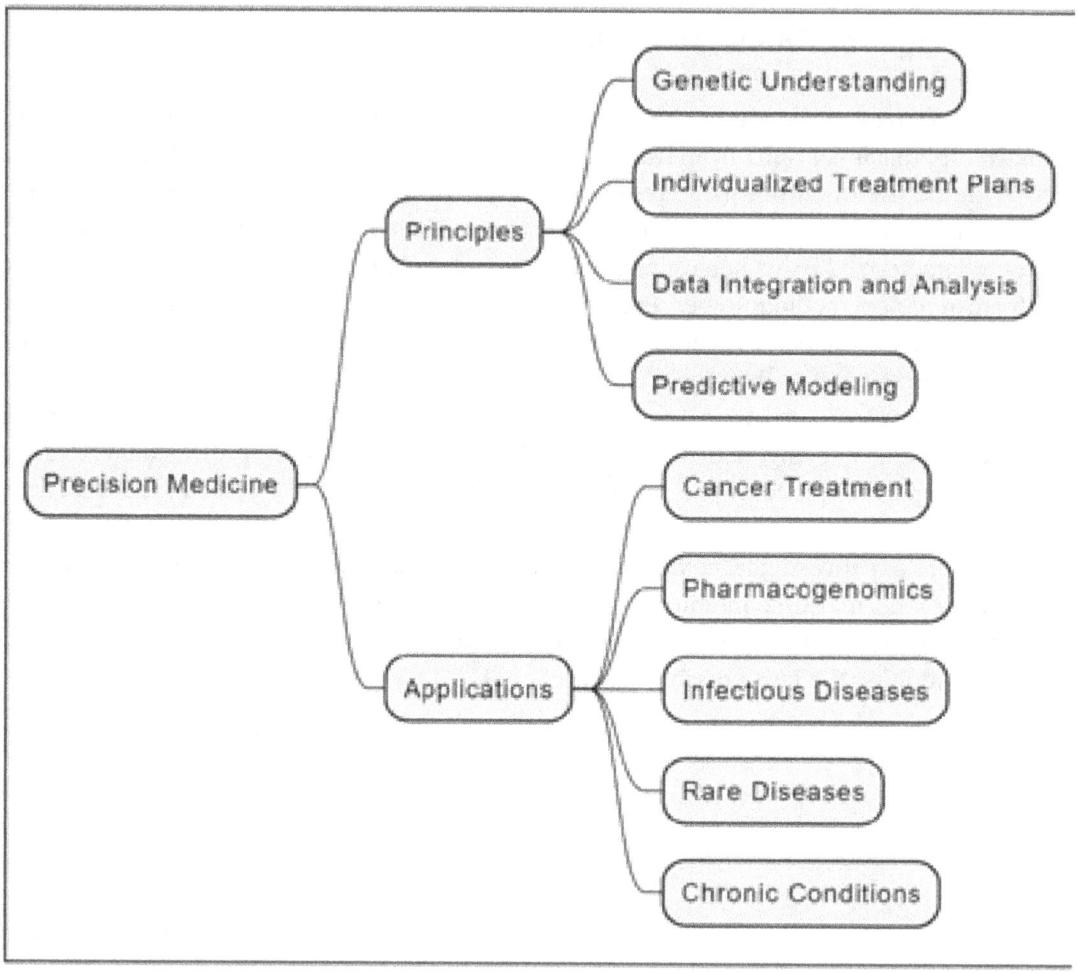

UNDERSTANDING DIABETES

Hyperglycaemia is a defining feature of diabetes mellitus (DM), commonly referred to as diabetes. A chronic and diverse condition characterized by persistently increased blood glucose levels is a physiologically abnormal state that presents as metabolic dysfunctions of fat, protein, and carbohydrates (Banday Mz et al.,2020).

The American Diabetes Association defines diabetes as a collection of metabolic disorders marked by elevated blood sugar levels brought on by deficiencies in insulin secretion, insulin action, or both. The World Health Assembly approved five worldwide goals for diabetes treatment and coverage in May 2022, with a target date of 2030(American Diabetes Association,2010). The World Health Organization (WHO) 2023 estimates that 422 million people worldwide have diabetes, with the majority of those affected living in low- and middle-income nations.

A glycated haemoglobin (A1c) of 6.0% to 6.4%, impaired glucose tolerance (IGT), or impaired fasting glucose (IFG) combined put a person at high risk of developing diabetes and its complications. These conditions are collectively referred to as prediabetes, a phrase that is both practical and convenient.

The distinction between type 1 and type 2 diabetes was established in 1936.The most prevalent kind of diabetes mellitus (DM) is type 2, also referred to as non-insulin-dependent DM (Olokoba et al., 2012). If diabetes is not managed, it can lead to major health problems such as heart disease, stroke, chronic kidney disease, ulcers on the feet, damage to the eyes, and chronic kidney disease. Diabetes sufferers must rely on a healthy lifestyle and appropriate medicine as there is currently no permanent solution for the condition. Type 1, type 2, and gestational diabetes—diabetes during pregnancy—are the three primary forms of the disease.

It is believed that a combination of multiple lifestyle factors, medical conditions, genetics, psychosocial factors, and demographic risk factors—such as elevated serum uric acid levels, sleep patterns, smoking, depression, cardiovascular disease, dyslipidaemia, hypertension, aging, ethnicity, diabetes in the family history, physical inactivity, and obesity (Ismail et al.,2021).

Figure 2. Illustrating etiopathogenesis of Diabetes Mellitus

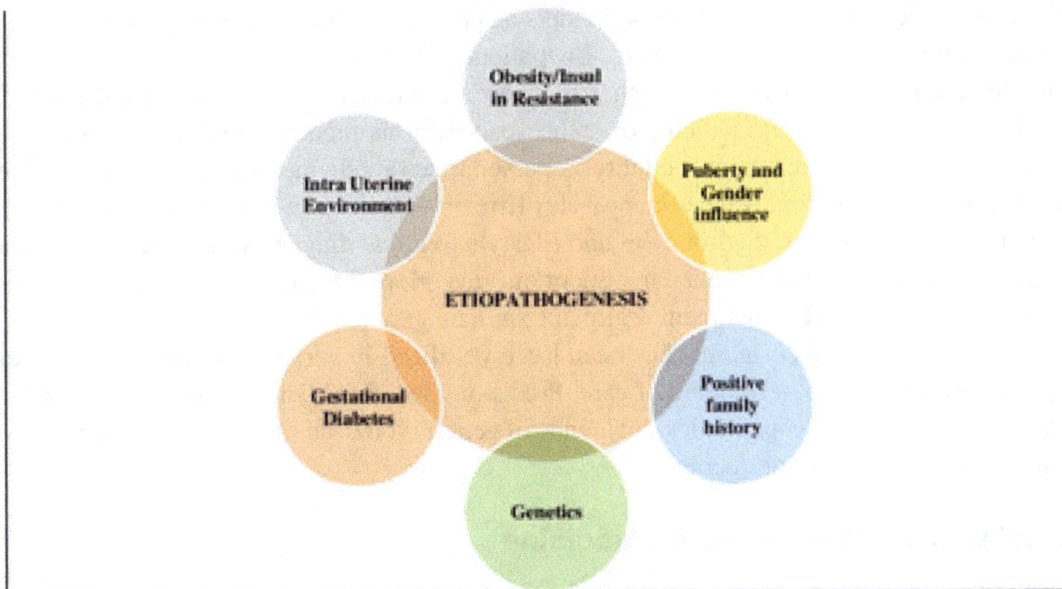

UNDERSTANDING THE COMPLEXITIES OF DIABETES MILLETUS

Accurate understanding of diabetes management goals is linked to enhanced diabetes self-care and better glycemic control, according to Trivedi et al.'s 2017 study. Still, previous research indicates that as few as 25% of diabetics can correctly define the hemoglobin a1c value (HBA1C) or recollect their most recent value.

According to Lee Y et al., providers must understand the roles that self-efficacy and self-care activities play in patients' diabetes control models. Providers' recommendations and feedback can be more in line with patients' perspectives and priorities if they focus on building self-efficacy and supporting and reinforcing self-care in their communications with patients, hopefully increasing the effectiveness of these communications.

Nielsen (2011) stated that previous research has found associations between patient-reported symptoms and symptom severity and diabetes outcomes, such as patients' self-rated health status and measured glycemic control.

In 2014, Holt ri et al., Stated notably, malaise and fatigue may be symptoms of depression, a condition that affects diabetic patients disproportionately, is underdiagnosed in the population, and may contribute to poor diabetes outcomes. Second, the absence of symptoms may provide patients with false reassurance about their current diabetes control. Many patients with blood glucose levels well above target have no symptoms, despite an increased risk of future diabetes-related complications.

Exploring the Principles and Applications of Precision Medicine

Medicine and treatment of patients has been an ever-evolving area wherein new innovations and breakthrough therapies pave the way for treating the patients optimally, traditionally medicine used a "one-size fits all approach" where the treatment choices were made largely based on the existing knowledge at given pharmacological doses or intervention models as per pre-existing treatment protocols.

This approach however does not consider how patients, their bodies differ in composition and how many people respond to medications and that is where the lacunae of treating the patient optimally surfaces. With advancements in medicine over the years various modalities of treatment have gained a lot of attention one such treatment modality is personalized medicine synonymously referred to as precision medicine is a more advanced approach of treating patients which utilizes patient's minute details including their genetics, environment and lifestyle to guide diagnosis, treatment and prevention of disease. Factors that can influence the treatment or prevention of the diseases are taken into to make a more accurate predictions for treatment regimen (Naithani et al., 2021) .

Precision medicine and its principle has been largely utilized in various specialties of medicines to ensure that patients receive best standard of care. Precision medicine blends the use of genetic information using genomics, Pharmacogenomics (influence of genetics on drug metabolism) along with the use of existing standard treatment protocol.

Historical Evolution of Precision Medicine

The development and progress in field of modern personalized medicine began with the Human Genome project 2003, wherein significant measures were taken to understand the genetics behind human diseases and health(Powledge, 2003).The domain saw further development after the US President Barack Obama embarked on the journey of introducing precision medicine in America enrolling nearly over a million under a project called "All of US" .With recent years the amalgamation of artificial intelligence and other healthcare technologies including Digital Twin, tele-health has helped in further expediting the process of incorporating precision medicine into mainstream healthcare system, these approaches have been particularly in the field of oncology, endocrinology, psychiatry (Ginsburg & Phillips, 2018) .

Epigenetics in Diabetes

While genetic factors are known to play a crucial role in diabetes susceptibility, emerging research has shed light on the significant influence of epigenetic factors. Epigenetic modifications, which involve heritable changes in gene expression without altering the underlying DNA sequence, are increasingly recognized as key contributors to the development and progression of both type 1 diabetes (T1D) and type 2 diabetes (T2D).

DNA Methylation in Diabetes

DNA methylation is mediated by DNA methyltransferase enzymes (DNMTs), primarily DNMT1, DNMT3A, and DNMT3B. DNMT1 is responsible for maintaining existing methylation patterns during DNA replication, while DNMT3A and DNMT3B are involved in de novo DNA methylation, establishing methylation patterns in previously unmethylated regions.(Jin B et al., 2013) The addition of a methyl group to a cytosine residue in CpG dinucleotide results in the formation of 5-methylcytosine (5mC) (Kurdyukov S et al., 2016). This modification typically leads to gene silencing when it occurs within gene promoter regions by preventing the binding of transcription factors and RNA polymerase (Jones PA et al., 2001)

Type 1 diabetes is characterized by the autoimmune destruction of pancreatic beta cells, leading to insulin deficiency. While genetic susceptibility plays a significant role in T1D, epigenetic factors, including DNA methylation, have gained attention in recent research. Hypermethylation in these regions may contribute to immune dysregulation and the autoimmune response against beta cells

In T2D, DNA methylation alterations have been observed in various metabolic tissues. (Yang et al. 2011) reported hypermethylation of key metabolic genes, such as PGC-1α, PDX1, and IRS1, in pancreatic islets and peripheral tissues of individuals with T2D. These changes can impair insulin sensitivity and beta cell function.

DNA methylation changes in genes involved in the insulin signaling pathway contribute to insulin resistance. Hypermethylation of genes like PPARγ and GLUT4 in adipose tissue is associated with reduced gene expression and impaired glucose uptake (Dayeh et al., 2014).Moreover, the influence of environmental factors, such as diet and physical activity, on DNA methylation underscores the dynamic nature of epigenetic regulation in diabetes (Bouchard et al., 2010).

The concept of “metabolic memory” suggests that early glycemic control can influence long-term diabetes outcomes. DNA methylation is believed to play a role in this phenomenon. For instance, persistent hypermethylation of specific genes, even after glucose levels are normalized, has been observed (Miao et al., 2014). These epigenetic changes may contribute to the progression of diabetes complications.

Figure 3. Multifactorial links between epigenetics factors and diabetes (Tajudeen, Y. O., & Shemishere, U. B., 2019)

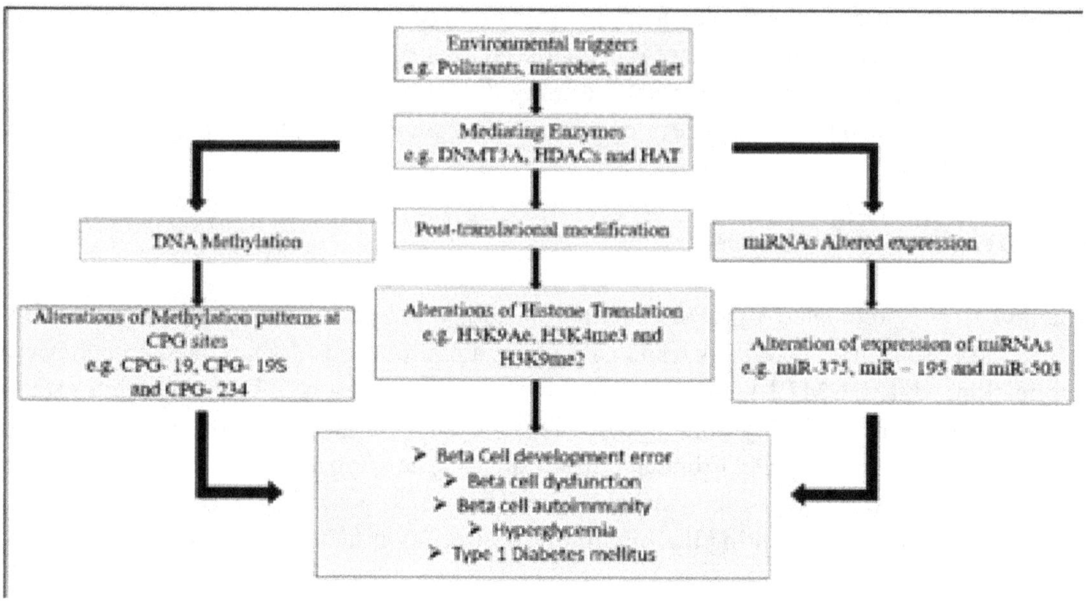

Histone Modifications in Diabetes

Histones are proteins that package and condense DNA within the cell nucleus. Post-translational modifications of histones, such as acetylation and methylation, can influence chromatin structure and gene accessibility. Histone modifications, including acetylation, methylation, and phosphorylation, play a pivotal role in chromatin remodeling and gene expression regulation Dysregulation of histone modifications have been implicated in insulin resistance, a hallmark of T2D. Histone deacetylases (HDACs) remove acetyl groups from histones, resulting in gene repression. Increased HDAC activity has been associated with insulin resistance, as it reduces the expression of genes involved in insulin signaling (Villeneuve et al., 2010). Inhibition of HDACs has shown promise as a potential therapeutic approach for T2D by restoring normal histone acetylation levels. Altered histone methylation patterns have been observed in tissues affected by diabetic nephropathy and retinopathy, suggesting a potential link between epigenetic modifications and diabetic complications (Reddy et al., 2015).

Non-Coding RNAs in Diabetes

Non-coding RNAs, particularly microRNAs (miRNAs) and long non-coding RNAs (lncRNAs), have emerged as important regulators in diabetes. MiRNAs can post-transcriptionally modulate gene expression by binding to messenger RNAs, while lncRNAs can influence chromatin structure and gene transcription. Dysregulated miRNA expression has been observed in both T1D and T2D, affecting insulin secretion, beta cell survival, and insulin signaling (Sebastiani et al., 2015). miRNAs play a vital role in the regulation of beta-cell function and insulin secretion. Aberrant expression of specific miRNAs can lead to impaired insulin production and glucose homeostasis (Poy et al., 2004).In T2D, lncRNAs such as HOTAIR and MALAT1 have been linked to insulin resistance and inflammation (Yan et al., 2016).

Circulating miRNAs have shown promise as diagnostic biomarkers for diabetes. Altered levels of specific miRNAs have been detected in the blood of individuals with diabetes, suggesting their potential utility for disease detection and monitoring (Kong et al., 2011).

Environmental Factors and Epigenetics

Environmental factors, including maternal diet during pregnancy and exposure to toxins, can impact epigenetic modifications associated with diabetes risk. Lilly crop et al. (2008) demonstrated that maternal diet can induce altered epigenetic regulation of the hepatic glucocorticoid receptor in offspring, increasing their susceptibility to metabolic diseases later in life. Gestational exposures can also have lasting effects on DNA methylation. Maternal obesity and gestational diabetes have been linked to altered methylation patterns in offspring, potentially predisposing them to diabetes later in life (Bouchard et al., 2010). Understanding the intricate interplay between genetics, epigenetics, and the environment is essential for tailoring personalized prevention and treatment strategies for diabetes.

Therapeutic Implications

The recognition of epigenetic contributions to diabetes has opened new avenues for therapeutic intervention. Targeting epigenetic mechanisms presents promising opportunities for the development of novel therapeutic strategies in diabetes management. Epigenetic drugs, such as DNA methyltransferase inhibitors and HDAC inhibitors, are under investigation for their potential to modify disease progression and improve glucose homeostasis (Keating et al., 2019). Additionally, miRNA-based therapies and lncRNA targeting strategies offer exciting prospects for restoring normal gene expression patterns in diabetic individuals. The recognition of epigenetic factors in diabetes may pave the way for personalized medicine approaches. Tailoring treatments based on an individual's epigenetic profile could improve therapeutic outcomes and reduce side effects (Ling et al., 2019). Epigenetic changes resulting from environmental exposures can be reversible. Lifestyle modifications, such as a healthy diet and regular physical activity, may help counteract detrimental epigenetic alterations associated with diabetes risk (Nilsson et al., 2015).

Advanced Technologies: Cutting-Edge Tools for Genetic and Epigenetic Analysis

The swift progressions in the field of molecular biology have fostered the creation of an array of methodologies and instruments for scrutinizing genetic and epigenetic aspects. Genomic, epigenomic, and integrative tools, coupled with functional assays and multi-omics approaches, offer a holistic understanding of the molecular landscape in diabetes. These cutting-edge tools not only unravel the complexities of genetic and epigenetic regulation but also pave the way for personalized diagnostics and therapeutic interventions

GENOMIC TOOLS

Genome-Wide Association Studies (GWAS)

GWAS is a potent method for identifying prevalent genetic variants associated with diabetes. Recent investigations have utilized extensive GWAS to reveal new loci connected to both type 1 and type 2 diabetes, providing insights into the genetic framework of the disease. (Mahajan et al., 2018; Morris et al., 2012).

Single-Cell Genomics

Single-cell RNA sequencing (scRNA-seq) has emerged as a breakthrough tool, allowing researchers to explore cellular heterogeneity in pancreatic islets and beta cells. This technology provides insights into gene expression at the single-cell level, offering a more granular understanding of the molecular basis of diabetes The application of single-cell transcriptomics enabled the creation of transcriptional profiles for individual cells within the pancreatic endocrine and exocrine regions. (Segerstolpe et al., 2016; Baron et al., 2016).

Next-Generation Sequencing (NGS)

The emergence of Next-Generation Sequencing (NGS) technologies has transformed genomic research in diabetes. NGS facilitates DNA sequencing, enabling thorough genomic profiling and the detection of uncommon genetic variants linked to susceptibility to diabetes. (Metzker, 2010).

EPIGENOMIC TOOLS

Bisulfite Sequencing

BS-Seq remains a cornerstone in DNA methylation analysis. Recent advancements, such as the Bismark tool, enhance the efficiency and accuracy of bisulfite sequencing data. The integration of bisulfite treatment of DNA with high-throughput sequencing (BS-Seq) provides a means to capture a snapshot of a cell's epigenomic state. This approach unveils the genome-wide cytosine methylation at single-base resolution, offering detailed insights into the epigenetic landscape. (Krueger & Andrews, 2011).

Chromatin Immunoprecipitation Sequencing (ChIP-Seq)

ChIP-Seq has emerged as a potent methodology for mapping histone modifications and identifying transcription factor binding sites in the context of diabetes research. Utilizing tools like MACS (Model-based Analysis of ChIP-Seq), researchers can effectively discern epigenetic signatures that serve as indicators of disease progression. (Zhang et al., 2008).

Nanopore Sequencing

In nanopore sequencing devices, the electrical current signals produced during electrolysis demonstrate sensitivity to changes in DNA bases, notably 5-methylcytosine (5-mC). It offers a real-time, long-read sequencing approach that holds promise for detecting DNA methylation patterns directly. This innovation provides a unique advantage in capturing dynamic epigenetic changes associated with diabetes as the epigenetic state is highly mutable, changing more easily and rapidly than a genetic mutation (Simpson et al., 2017).

Histone Peptide Array Assays allow high-throughput screening of histone modification patterns, aiding in the identification of specific histone marks associated with diabetes-related genes (Ravnskjaer et al., 2013). Advanced microscopy techniques, such as super-resolution microscopy, provide spatial information about histone modifications, revealing subcellular epigenetic alterations in diabetes (Bintu et al., 2018).

Functional Tools

The advent of CRISPR/Cas9 technology has revolutionized functional genomics research by enabling precise genome editing. Within the field of diabetes research, CRISPR/Cas9 plays a crucial role in generating genetically modified cellular models. These models facilitate the exploration of functional consequences associated with genetic variants linked to the disease, offering valuable insights into its underlying mechanisms. (Doudna & Charpentier, 2014; Cong et al., 2013).

Integrated analysis platforms like the UCSC Genome Browser and the ENCODE project play a pivotal role in cross-disciplinary research. By consolidating a wide range of genomic and epigenomic datasets, these platforms serve as centralized resources, enabling seamless access and interpretation of genomic and epigenomic information. (Rosenbloom et al., 2015; Consortium et al., 2012).The integration of diverse omics datasets, spanning genomics, epigenomics, transcriptomics, and metabolomics, provides a comprehensive insight into the molecular landscape of diabetes.

Utilizing tools such as OmicsNet proves instrumental in seamlessly integrating and interpreting multi-omics data. This integration aids in the identification of key regulatory networks, contributing to a more thorough understanding of the complex molecular mechanisms underlying diabetes. (Zhou & Xia, 2018). Utilizing techniques like mass spectrometry and nuclear magnetic resonance spectroscopy, metabolomic profiling offers a snapshot of the metabolic state linked to diabetes.

The integration of metabolomic data with genomic and epigenomic information enriches our comprehension of the intricate interplay among genetics, epigenetics, and metabolism, providing a more holistic view of the factors contributing to diabetes. (Bain et al., 2009).The application of pharmacogenomic tools facilitates the identification of genetic variants that impact drug responses in the context of diabetes treatment. A comprehension of individual variations in drug metabolism, as demonstrated by genes like CYP2C9 and VKORC1, informs the selection of personalized pharmacological interventions, enhancing the effectiveness of treatment strategies. (Roden et al., 2019).

Personalized Diagnostics: Harnessing Genetic Insights for Tailored Diagnostics

Personalised diagnostics is an emerging field with great potential to transform the landscape of patient care in chronic diseases. While the term was first introduced into the scientific literature in April 1999 in the form of an article published by the Wall Street Journal titled "New Era of Personalized Medicine: Targeting Drugs for Each Unique Genetic Profile", it has taken on a multitude of meanings throughout the following decades branching into fields such as precision and stratified diagnostic approaches. Over the following pages, we will review the overall definitions, the various advances made in the fields, examples of this in action, and the future directions that this branch of medicine can take.

Personalised medicine generally refers to the practice of studying a person's genotype and extrapolating information to use in the process of risk identification, diagnostics, and therapeutics (Klonoff, 2008). A stratified diagnostic approach, on the other hand, refers to the division of the patient population into groups according to the risk of either developing diabetes in the first place or if already diagnosed, the odds of developing associated complications and providing more targeted therapeutic options (Pearson, 2016). Precision diagnostics, the latest stage in the evolution of personalised medicine, takes it a step further and analyses not only genetic data but also epigenetics and environmental influences to better diagnose, profile and treat the patient with the best plan of management. This is why predictive accuracy is an important parameter to account for in precision diagnostics. In recent decades, precision medicine has also led to the creation of subgroups in diabetes (Chung et al., 2020).

A Look Into Current Opportunities for Precision Diagnostics in Diabetes

Figure 4. Precision medicine in diabetes: From challenges to opportunities

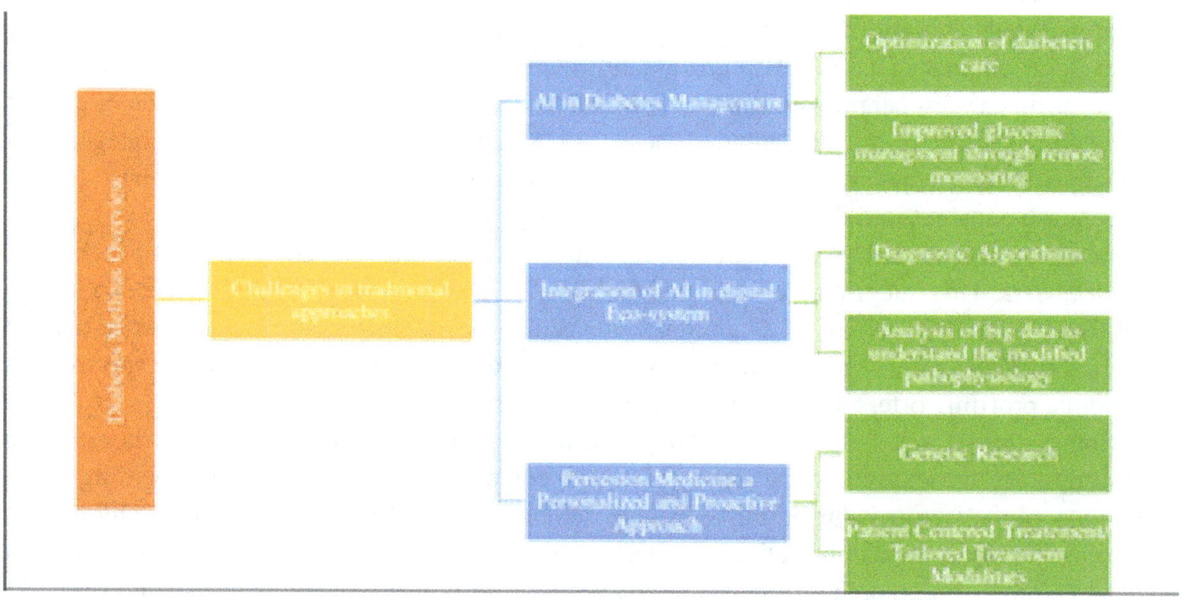

The most well-established field that has been studied with respect to precision diagnostics is its applications in the creation of monogenic diabetes subtypes, such as MODY. (Kleinberger & Pollin, 2015; Pearson, 2016). This is because, unlike polygenic forms of the disorder (T1DM and T2DM), monogenic diabetes has a single identifiable genetic cause. This in turn results in the ability to determine characteristics of the manifestation (phenotype) of the disorder ranging from the stage of diagnosis to prognosis on an individual level (Chung et al., 2020). On the other hand, precision diagnostics often take different approaches when discussed in the context of T1DM and T2DM.

T1DM

Type 1 diabetes mellitus is a disorder with complex etiology. The current extent of precision diagnosis involves the identification of patient characteristics (particularly the age of onset) and laboratory testing (islet autoantibody and less commonly, C-peptide levels). The avenues being explored involve using this collected data and analyzing them in order to detect "clusters" which could then potentially lead to the classification of T1DM by subtypes with well-defined features and prognosis. (Michalek et al., 2023). At present, T1D can be divided into T1a and T1b based on the presence or absence of islet autoantibodies (Unnikrishnan et al., 2021). While the disease's genetic background has been studied extensively and is well-established, there is still much information that needs to be added regarding the multiple individual genes that may be responsible, and in addition, how they may interplay in the manifestation of the phenotype as well as how they could influence the prognosis. At present, thus, there is no sole genetic-based diagnostic method that can be used without also considering patient history, clinical features, and relevant laboratory data. (Chung et al., 2020).

T2DM

T1DM is considered to be a homogenous disease, whereas T2DM is heterogenous in nature. While there is some genetic basis to the etiologies of this disorder group, there is a larger role played by the environment and lifestyle of the individual; how they interlace with and influence the expression of genes is indeed an emerging field of study with great promise. Thus, there is a greater emphasis placed on the identification of risk factors and predictors for patients with T2DM, in order to facilitate early diagnosis (Kleinberger & Pollin, 2015). The use of cluster analysis to determine various subtypes of T2DM would allow for "decoding" the heterogeneity of the disorder; this would effectively offer the opportunity for targeted therapeutics and precise monitoring for complications that are most likely to occur according to the respective subtypes (Misra et al., 2023). At present, there are five such clusters identified: SAID, SIDD, SIRD, MOD, and MARD (Unnikrishnan et al., 2021). Like with T1DM, there is currently no genetic-based diagnostic method that can be utilized; the majority of diagnostic burden falls upon clinical features and laboratory tests of blood glucose levels (Chung et al., 2020).

Monogenic Diabetes

Monogenic Diabetes is the third category of diabetes that is distinct from T1DM and T2DM. The present research is focused on finding novel gene variations that could be causal, however, the genetic basis for this category has for the most part been well-studied. It is thus in this category that precision diagnosis has already been implemented, particularly in MODY, which stands for Maturity Onset Diabetes of the Young, subclassified into about 13 forms based on the genetic etiology (Kleinberger & Pollin, 2015). Individuals are then provided therapeutics that the particular subtype would respond better to. As the phenotype for MD overlaps with the clinical presentation for T1DM and T2DM, there is some difficulty in identifying patients suitable for precision diagnosis, given the time, cost and technology involved in genetic testing (R et al., 2023).

Barriers, Challenges, and Future Directions

While precision medicine has begun its foray into applications for diabetes, there are still a multitude of challenges and barriers to full implementation. In the case of T1DM, the current division into T1a and T1b subtypes exists, however, the characterization of both is incomplete and does not have any substantial or concrete effect on the choice of patient management (Unnikrishnan et al., 2021). In addition, in order for cluster analysis to occur, the required data must be extracted from biomarker levels at the time of diagnosis. This is not standardized, and in most cases, these tests are not conducted routinely, which results in a data gap; extrapolating clusters in the context of this missing data would not be a reliable approach for precision diagnosis (Chung et al., 2020; Kleinberger & Pollin, 2015; Unnikrishnan et al., 2021).

As mentioned earlier, efforts to create clusters of T2DM have been made. However, there is a marked inability to replicate the clusters evenly, indicating that there may be differences based on variables such as geographic location, ethnicity, and other population characteristics (Unnikrishnan et al., 2021). Similar to T1DM, the clustering approach to precision diagnostics requires data on biomarkers which are not part of routine clinical testing; a further complication arises when one notes that the biomarkers are variable and can change over time and as the disease itself progresses.

Despite precision diagnostics being a reality in the context of monogenic diabetes, there are still a few barriers to implementing it in a standardized manner. With various options for testing modalities, accessibility is an issue when it comes to pricing and availability in countries, particularly LMICs (Unnikrishnan et al., 2021). This is without taking into account the need for provider knowledge; it is paramount that clinical workers be able to identify phenotypes with a high likelihood of being monogenic diabetes and accurately refer them for screening and further testing. (R et al., 2023)

Patient Barriers

Patient barriers are an important part of discussions centered around patient involvement. It is imperative that future research focuses on ameliorating the effects of these barriers and providing long-term solutions.

The major barriers that have been listed so far range from the patient's lack of will to accept that their condition is chronic and will likely require life-long care. This in turn manifests as an inability to adhere strictly to medication regimens and lifestyle changes until it reaches the point of no return, which is another hurdle that is frequently encountered in healthcare settings. While patients are interested in taking an active role in their care initially, it is preferable for them to be passive long-term and "have a pill given to them" (Brown et al., 2002). These barriers are strongly linked to how patients feel about their disease, which once again points to a need for better patient education and opportunities for counselling.

Other barriers that were identified include having co-morbidities and complications related to diabetes; patients often report a sense of helplessness. Cultural barriers, psychological barriers and financial/accessibility difficulties are all noteworthy challenges that should be analyzed and solutions are required for the same (Brown et al., 2002).

Future directions for the implementation of precision diagnostics (eventually on a larger scale) will require the meticulous use of advances in cost-effective technologies for genome sequencing and genetic testing. In the case of T1DM and T2DM, this will translate into greater opportunity for more accurate clustering with the ability to collect complete data outside of select research studies. These improvements will also enable increased access to screening individuals who fit one of the phenotypes likely to be monogenic in etiology (Kleinberger & Pollin, 2015).

Patient Engagement: The Importance of Patient Involvement and Education

Patient-centered medicine is at the forefront of healthcare approaches taken in recent decades. What this involves is offering patients the ability to be involved in their own care and to take a more active role in the decisions made regarding the management of their condition. This practice is currently more prevalent in Western nations (Lemmens et al., 2015). In a chronic disease such as diabetes, educating the patient on both their condition and the management required improves the outcome and has a positive effect on the individual as well. This is in part due to the fact that diabetes care involves a multitude of lifestyle changes as well as strict adherence to medication regimen; the knowledge and will of the patient are extremely important in this case (Ashrafzadeh & Hamdy, 2019).

Different Concepts and Levels of Patient Involvement

There are various terms such as patient education, involvement, and engagement that are often used interchangeably (Lemmens et al., 2015). These concepts encompass different parts of the process through which healthcare workers connect and discuss with the patient.

Patient education involves explaining diabetes' etiology, management and prognosis in clear and simple terms understandable by laymen. It is crucial that doctors do not use scientific terms that could scare the patients. Instead, empowering the patient and their families to take an active role in their care will enable them to achieve better outcomes and a good quality of life post-diagnosis (Mphasha et al., 2023). Patient engagement is the next step from education and imparting knowledge; here, the patient is invited to a discussion with the healthcare provider to evaluate and make decisions together for their management plan. Patient involvement, on the other hand, involves inviting the patient to discussions on a larger scale. This would create a space for them to offer their input in fields such as research and policies, which historically have been closed off to patients. (Lemmens et al., 2015)

There are also differing levels of patient involvement. A recent study classified these into five steps based on an adapted "Arnstein's ladder of citizen participation" for healthcare settings; these ranged from "passive participation (step one: being informed about decisions) to very active participation (step five: participating in formal decision-making)" (Lemmens et al., 2015, p.e044)

Overall, healthcare providers should aim to impart maximum understanding to patients not only regarding their condition but also the importance of adhering strictly to their care plan.

Current and Future Avenues for Patient Participation

The major avenues for patient involvement exist in three spheres: lifestyle changes, adherence to medication plans and monitoring of glucose levels and symptoms for potential complications. It is imperative that the patient takes accountability and works along with the healthcare provider to achieve the best outcome.

An interesting approach to this is the use of technological advancements to facilitate the process of shared-decision making. In particular, telemedicine would increase accessibility and the ability for the patient and healthcare provider to have more frequent check-ins and better keep track of the condition and its progression. Utilizing the convenience of telemedicine to the advantage of providing long-term diabetes is a valuable prospect and should be considered seriously when discussing future directions of personalised care. Another intriguing aspect is the ability to use digital tools for self-tracking that would motivate patients to stay on top of their lifestyle habits as well as provide a clear record of glucose levels from continuous monitoring (Ashrafzadeh & Hamdy, 2019).

Diabetes Mellitus and Precision Medicine

Diabetes mellitus is a significant worldwide health concern, requiring creative methods to enhance patient outcomes. Traditional standard management approaches have demonstrated constraints in effectively managing the diverse character of the disease. Applying artificial intelligence to diabetes is both possible and valuable for effectively managing data and creating tools and technologies for its management (Ellahham et al., 2020).

The Role of Advanced Treatment Modalities in Diabetes Management

Integrating connected devices into the health and wellbeing sector has led to the development of a digital ecosystem known as the Healthcare IoT, which is revolutionizing diabetes mellitus care. It has been noted that digital health, which includes linked devices, has the potential to significantly transform the healthcare business by providing widespread access to diagnosis, treatment, and prevention at a significantly reduced cost. This influence is particularly significant in the management of chronic diseases, such as diabetes (Roman et al., 2015). Various technological developments, such as linked glucometers, are readily available they can either serve as physical extensions of smartphones or they can be connected via Bluetooth. Some devices are often customized to the point that they may operate independently, although with their own internet connection. Moreover, specialized devices are specifically engineered to synchronize data from many glucometers that are not wirelessly accessing the internet (Heintzman, 2016). The data obtained from interconnected diabetes devices is very important as it can be gathered and merged with data from various other sources to serve numerous purposes in self-care and medical treatment. A healthcare recommendation system (HRS) employing machine learning has been utilized to forecast the likelihood of developing a disease, such as diabetes, by examining several aspects of a patient's lifestyle, physical health, mental health, and social network engagement. The predictive models developed to utilize big data analytics aim to forecast the likelihood of problems in individuals with diabetes. Several models have been created to forecast the progression of both chronic (such as retinal, cardiovascular, and renal) and acute (namely hypoglycaemia) issues associated with diabetes (Cichosz et al., 2016; Zou et al., 2018).

Artificial intelligence enables continuous, effortless remote monitoring of patient symptoms and biomarkers. In addition, digital media and online communities promote patient engagement in diabetes care. Optimizing diabetes care resources has been aided by recent technological developments. Improved glycemic management is the collective result of these thoughtful technical innovations.

Precision Medicine in Diabetes Treatment

Lifestyle variables exert a substantial influence on the onset of diabetes. Nevertheless, it is crucial to recognize that heredity also significantly influences an individual's susceptibility to this disease. Recently, there have been notable advancements in genetic research, which have yielded useful knowledge about the hereditary elements that increase the likelihood of developing diabetes. Acquiring a thorough comprehension of the genetic foundations of diabetes might have major ramifications for numerous facets, including risk prognosis, early identification, and the formulation of tailored treatment approaches (Htay et al., 2019; Pandit et al., 2023; Rescalli et al., 2022). It is generally agreed that precision medicine has the ability to better manage and treat chronic diseases like diabetes in all phases of care. In the case of diabetes in particular, its efficacy depends on incorporating both genetic and internal/external environmental factors that influence susceptibility, progression, and phenotype. After receiving a diabetes diagnosis, patients have access to a wide range of treatment interventions (Nathan et al., 2009). Over 400 genetic variants have been discovered in the quest to apply precision medicine to diabetes. However, further research is necessary to establish precision medicine as a sustainable approach for all forms of diabetes (Fitipaldi et al., 2018).

More comprehensive implementation of precision medicine in the treatment of diabetes has the potential to enhance outcomes that are tailored to individual patients and decrease costs. On the other hand, there are still obstacles preventing its widespread adoption. To enhance the capacity for implementing and enhancing diabetes detection methodologies among the general population, it is crucial to focus on addressing data gaps. Furthermore, as our knowledge of disease pathophysiology and environmental factors advances, it is necessary to streamline the process of detecting and classifying different forms of diabetes. The streamlining process would enhance the usage of healthcare providers to prevent misdiagnosis and minimize the delay in administering optimal treatments. Integrating advanced precision medicine approaches, such as diagnostic algorithms, can expand the use of these methods to enhance outcomes tailored to the specific needs of individuals living with diabetes (Galloway et al., 2022).

CONCLUSION

In conclusion, the integration of precision medicine and epigenetics stands as a transformative paradigm in the realm of personalized healthcare for non-communicable diseases, particularly diabetes. As we confront the escalating global burden of diabetes, a shift from conventional approaches to a precision-driven methodology emerges as a beacon of hope in tailoring interventions to the unique genetic and epigenetic profiles of individuals.

The advent of precision medicine in diabetes care allows for a granular understanding of an individual's genetic predisposition to the disease, facilitating early detection and personalized therapeutic strategies. By unraveling the intricate genetic markers associated with diabetes susceptibility, healthcare providers can proactively intervene, potentially preventing the onset or progression of the condition. This personalized approach not only enhances diagnostic accuracy but also ensures that interventions are precisely calibrated to address the specific needs and genetic variations of each patient.

Epigenetics further enriches this personalized healthcare landscape by shedding light on the dynamic interplay between genes and environmental factors. Understanding how epigenetic modifications influence gene expression in diabetes opens new avenues for targeted interventions and lifestyle modifications. This nuanced comprehension allows healthcare professionals to tailor their guidance to address individual risk factors and mitigate the impact of environmental influences on diabetes development and progression.

The amalgamation of precision medicine and epigenetics in diabetes care not only refines our understanding of the disease at a molecular level but also heralds a new era of patient-centric healthcare. By embracing these cutting-edge technologies, healthcare providers can move beyond traditional one-size-fits-all approaches, ushering in a future where diabetes management is highly personalized. This shift has the potential not only to improve clinical outcomes but also to enhance patient engagement, compliance, and overall quality of life.

As we navigate the complexities of non-communicable diseases, particularly diabetes, the commitment to advancing research, fostering interdisciplinary collaboration, and harnessing technological innovations becomes paramount. By collectively embracing and investing in the potential of personalized healthcare through precision medicine and epigenetics, we can aspire to revolutionize diabetes management, offering individuals tailored solutions that empower them to take control of their health and well-being.

REFERENCES

American Diabetes Association Professional Practice Committee; American Diabetes Association Professional Practice Committee. (2022). Prevention or delay of type 2 diabetes and associated comorbidities: Standards of Medical Care in Diabetes—2022. *Diabetes Care*, *45*(Supplement_1), S39–S45. doi:10.2337/dc22-S003 PMID:34964876

Ashrafzadeh, S., & Hamdy, O. (2019). Patient-Driven Diabetes Care of the Future in the Technology Era. *Cell Metabolism*, *29*(3), 564–575. doi:10.1016/j.cmet.2018.09.005 PMID:30269984

Bain, J. R., Stevens, R. D., Wenner, B. R., Ilkayeva, O., Muoio, D. M., & Newgard, C. B. (2009). Metabolomics Applied to Diabetes Research. *Diabetes*, *58*(11), 2429–2443. doi:10.2337/db09-0580 PMID:19875619

Banday, M. Z., Sameer, A. S., & Nissar, S. (2020). Pathophysiology of diabetes: An overview. *Avicenna Journal of Medicine*, *10*(04), 174–188. doi:10.4103/ajm.ajm_53_20 PMID:33437689

Baron, M., Veres, A., Wolock, S. L., Faust, A. L., Gaujoux, R., Vetere, A., Ryu, J. H., Wagner, B. K., Shen-Orr, S. S., Klein, A. M., Melton, D. A., & Yanai, I. (2016). A Single-Cell Transcriptomic Map of the Human and Mouse Pancreas Reveals Inter- and Intra-cell Population Structure. *Cell Systems*, *3*(4), 346–360.e4. doi:10.1016/j.cels.2016.08.011 PMID:27667365

Bintu, B., Mateo, L. J., Su, J.-H., Sinnott-Armstrong, N. A., Parker, M., Kinrot, S., Yamaya, K., Boettiger, A. N., & Zhuang, X. (2018). Super-resolution chromatin tracing reveals domains and cooperative interactions in single cells. *Science (New York, N.Y.)*, *362*(6413), eaau1783. doi:10.1126/science.aau1783

Bouchard, L., Thibault, S., Guay, S.-P., Santure, M., Monpetit, A., St-Pierre, J., Perron, P., & Brisson, D. (2010). Leptin gene epigenetic adaptation to impaired glucose metabolism during pregnancy. *Diabetes Care*, *33*(11), 2436–2441. doi:10.2337/dc10-1024 PMID:20724651

Brown, J. B., Harris, S. B., Webster-Bogaert, S., Wetmore, S., Faulds, C., & Stewart, M. (2002). The role of patient, physician and systemic factors in the management of type 2 diabetes mellitus. *Family Practice*, *19*(4), 344–349. doi:10.1093/fampra/19.4.344 PMID:12110552

Chung, W. K., Erion, K., Florez, J. C., Hattersley, A. T., Hivert, M.-F., Lee, C. G., McCarthy, M. I., Nolan, J. J., Norris, J. M., Pearson, E. R., Philipson, L., McElvaine, A. T., Cefalu, W. T., Rich, S. S., & Franks, P. W. (2020). Precision Medicine in Diabetes: A Consensus Report From the American Diabetes Association (ADA) and the European Association for the Study of Diabetes (EASD). *Diabetes Care*, *43*(7), 1617–1635. doi:10.2337/dci20-0022 PMID:32561617

Cichosz, S. L., Johansen, M. D., & Hejlesen, O. (2016). Toward big data analytics: Review of predictive models in management of diabetes and its complications. *Journal of Diabetes Science and Technology, 10*(1), 27–34. doi:10.1177/1932296815611680 PMID:26468133

Dayeh, T., Tuomi, T., Almgren, P., Perfilyev, A., Jansson, P.-A., de Mello, V. D., Pihlajamäki, J., Vaag, A., Groop, L., Nilsson, E., & Ling, C. (2016). DNA methylation of loci within ABCG1 and PHOSPHO1 in blood DNA is associated with future type 2 diabetes risk. *Epigenetics, 11*(7), 482–488. doi:10.1080/15592294.2016.1178418 PMID:27148772

Ellahham, S., Ellahham, N., & Simsekler, M. C. E. (2020). Application of artificial intelligence in the health care safety context: Opportunities and challenges. *American Journal of Medical Quality, 35*(4), 341–348. doi:10.1177/1062860619878515 PMID:31581790

Fitipaldi, H., McCarthy, M. I., Florez, J. C., & Franks, P. W. (2018). A global overview of precision medicine in type 2 diabetes. *Diabetes, 67*(10), 1911–1922. doi:10.2337/dbi17-0045 PMID:30237159

Galloway, N., Blanchette, J. E., Faizullabhoy, A., Mobley-Bukstein, W., Rinker, J., & Isaacs, D. (2022). Exploring the Role of Precision Medicine in Diabetes Diagnoses. *ADCES in Practice, 10*(1), 26–32. doi:10.1177/2633559X211060849

Ginsburg, G. S., & Phillips, K. A. (2018). Precision medicine: From science to value. *Health Affairs (Project Hope), 37*(5), 694–701. doi:10.1377/hlthaff.2017.1624 PMID:29733705

Heintzman, N. D. (2016). A digital ecosystem of diabetes data and technology: Services, systems, and tools enabled by wearables, sensors, and apps. *Journal of Diabetes Science and Technology, 10*(1), 35–41. doi:10.1177/1932296815622453 PMID:26685994

Holt, R. I. G., de Groot, M., & Golden, S. H. (2014). Groot M de, Golden SH. Diabetes and depression. *Current Diabetes Reports, 14*(6), 491. doi:10.1007/s11892-014-0491-3 PMID:24743941

Htay, T., Soe, K., Lopez-Perez, A., Doan, A. H., Romagosa, M. A., & Aung, K. (2019). Mortality and cardiovascular disease in type 1 and type 2 diabetes. *Current Cardiology Reports, 21*(6), 1–7. doi:10.1007/s11886-019-1133-9 PMID:31011838

Inzucchi, S., Bergenstal, R., Fonseca, V., Gregg, E., Mayer-Davis, B., Spollett, G., & Wender, R. (2010). Diagnosis and classification of diabetes mellitus. Diabetes care, 33, S62-S69. Anjana, R. M., Unnikrishnan, R., Deepa, M., Pradeepa, R., Tandon, N., Das, A. K., & Ghosh, S. (2023). *Metabolic non-communicable disease health report of India: the ICMR-INDIAB national cross-sectional study (ICMR-INDIAB-17).* The Lancet Diabetes & Endocrinology.

Ismail, L., Materwala, H., & Al Kaabi, J. (2021). Association of risk factors with type 2 diabetes: A systematic review. *Computational and Structural Biotechnology Journal, 19*, 1759–1785. doi:10.1016/j.csbj.2021.03.003 PMID:33897980

Jin, B., & Robertson, K. D. (2013). DNA Methyltransferases (DNMTs), DNA Damage Repair, and Cancer. *Advances in Experimental Medicine and Biology, 754*, 3–29. doi:10.1007/978-1-4419-9967-2_1 PMID:22956494

Jones, P. A., & Takai, D. (2001). The role of DNA methylation in mammalian epigenetics. *Science, 293*(5532), 1068–1070. doi:10.1126/science.1063852 PMID:11498573

Keating, S. T., van Diepen, J. A., Riksen, N. P., & El-Osta, A. (2018). Epigenetics in diabetic nephropathy, immunity and metabolism. *Diabetologia, 61*(1), 6–20. doi:10.1007/s00125-017-4490-1 PMID:29128937

Kleinberger, J. W., & Pollin, T. I. (2015). Personalized medicine in diabetes mellitus: Current opportunities and future prospects. *Annals of the New York Academy of Sciences, 1346*(1), 45–56. doi:10.1111/nyas.12757 PMID:25907167

Klonoff, D. C. (2008). Personalized Medicine for Diabetes. *Journal of Diabetes Science and Technology, 2*(3), 335–341. doi:10.1177/193229680800200301 PMID:19885196

Kong, L., Zhu, J., Han, W., Jiang, X., Xu, M., Zhao, Y., Dong, Q., Pang, Z., Guan, Q., Gao, L., Zhao, J., & Zhao, L. (2011). Significance of serum microRNAs in pre-diabetes and newly diagnosed type 2 diabetes: A clinical study. *Acta Diabetologica, 48*(1), 61–69. doi:10.1007/s00592-010-0226-0 PMID:20857148

Krueger, F., & Andrews, S. R. (2011). Bismark: A flexible aligner and methylation caller for Bisulfite-Seq applications. *Bioinformatics (Oxford, England), 27*(11), 1571–1572. doi:10.1093/bioinformatics/btr167 PMID:21493656

Kurdyukov, S., & Bullock, M. (2016). DNA Methylation Analysis: Choosing the Right Method. *Biology (Basel), 5*(1), 3. doi:10.3390/biology5010003 PMID:26751487

Lee, Y. Y., & Lin, J. L. (2009). The effects of trust in physician on self-efficacy, adherence and diabetes outcomes. *Social Science & Medicine, 68*(6), 1060–1068. doi:10.1016/j.socscimed.2008.12.033 PMID:19162386

Lemmens, L. C., de Bruin, S. R., Struijs, J. N., Rijken, M., Nijpels, G., & Baan, C. A. (2015). Patient involvement in diabetes care: Experiences in nine diabetes care groups. *International Journal of Integrated Care, 15*(4), e044. doi:10.5334/ijic.2207 PMID:27118961

Lillycrop, K. A., Slater-Jefferies, J. L., Hanson, M. A., Godfrey, K. M., Jackson, A. A., & Burdge, G. C. (2007). Induction of altered epigenetic regulation of the hepatic glucocorticoid receptor in the offspring of rats fed a protein-restricted diet during pregnancy suggests that reduced DNA methyltransferase-1 expression is involved in impaired DNA methylation and changes in histone modifications. *British Journal of Nutrition, 97*(6), 1064–1073. doi:10.1017/S000711450769196X PMID:17433129

Ling, C., & Rönn, T. (2019). Epigenetics in Human Obesity and Type 2 Diabetes. *Cell Metabolism, 29*(5), 1028–1044. doi:10.1016/j.cmet.2019.03.009 PMID:30982733

Mahajan, A., Taliun, D., Thurner, M., Robertson, N. R., Torres, J. M., Rayner, N. W., Payne, A. J., Steinthorsdottir, V., Scott, R. A., Grarup, N., Cook, J. P., Schmidt, E. M., Wuttke, M., Sarnowski, C., Mägi, R., Nano, J., Gieger, C., Trompet, S., Lecoeur, C., & McCarthy, M. I. (2018). Fine-mapping type 2 diabetes loci to single-variant resolution using high-density imputation and islet-specific epigenome maps. *Nature Genetics, 50*(11), 11. doi:10.1038/s41588-018-0241-6 PMID:30297969

Metzker, M. L. (2010). Sequencing technologies—The next generation. *Nature Reviews. Genetics, 11*(1), 31–46. doi:10.1038/nrg2626 PMID:19997069

Michalek, D. A., Onengut-Gumuscu, S., Repaske, D. R., & Rich, S. S. (2023). Precision Medicine in Type 1 Diabetes. *Journal of the Indian Institute of Science*, *103*(1), 335–351. doi:10.1007/s41745-023-00356-x PMID:37538198

Misra, S., Wagner, R., Ozkan, B., Schön, M., Sevilla-Gonzalez, M., Prystupa, K., Wang, C. C., Kreienkamp, R. J., Cromer, S. J., Rooney, M. R., Duan, D., Thuesen, A. C. B., Wallace, A. S., Leong, A., Deutsch, A. J., Andersen, M. K., Billings, L. K., Eckel, R. H., Sheu, W. H.-H., & Udler, M. S. (2023). Precision subclassification of type 2 diabetes: A systematic review. *Communications Medicine*, *3*(1), 1. doi:10.1038/s43856-023-00360-3 PMID:37798471

Morris, A. P., Voight, B. F., Teslovich, T. M., Ferreira, T., Segrè, A. V., Steinthorsdottir, V., Strawbridge, R. J., Khan, H., Grallert, H., Mahajan, A., Prokopenko, I., Kang, H. M., Dina, C., Esko, T., Fraser, R. M., Kanoni, S., Kumar, A., Lagou, V., Langenberg, C., & McCarthy, M. I. (2012). Large-scale association analysis provides insights into the genetic architecture and pathophysiology of type 2 diabetes. *Nature Genetics*, *44*(9), 981–990. doi:10.1038/ng.2383 PMID:22885922

Mphasha, M., Skaal, L., Mothiba, T., Ngoatle, C., & Hlahla, L. (2023). Primary health care–family partnership for better diabetes outcomes of patients: A systematic review. Journal of Endocrinology. *Jemdsa : the Journal of Endocrinology, Metabolism and Diabetes of South Africa (JEMDSA) : the Official Journal of the Society of Endocrinology, Metabolism and Diabetes of South Africa*, *28*(1), 1–6. doi:10.1080/16089677.2022.2140517

Naithani, N., Sinha, S., Misra, P., Vasudhevan, B., & Sahu, R. (2021). Precision medicine: Concept and tools. *Medical Journal, Armed Forces India*, *77*(3), 249–257. doi:10.1016/j.mjafi.2021.06.021 PMID:34305276

doi:10.1016/j.mjafi.2021.06.021

Nathan, D. M., Buse, J. B., Davidson, M. B., Ferrannini, E., Holman, R. R., Sherwin, R., & Zinman, B. (2009). Medical management of hyperglycemia in type 2 diabetes: a consensus algorithm for the initiation and adjustment of therapy: a consensus statement of the American Diabetes Association and the European Association for the Study of Diabetes. *Diabetes Care*, *32*(1), 193–203. doi:10.2337/dc08-9025 PMID:18945920

Nielsen, A. B., Gannik, D., Siersma, V., & de Fine Olivarius, N. (2011). The relationship between HbA1c level, symptoms and self-rated health in type 2 diabetic patients. *Scandinavian Journal of Primary Health Care*, *29*(3), 157–164. doi:10.3109/02813432.2011.585542 PMID:21707235

Nilsson, E., Jansson, P. A., Perfilyev, A., Volkov, P., Pedersen, M., Svensson, M. K., Poulsen, P., Ribel-Madsen, R., Pedersen, N. L., Almgren, P., Fadista, J., Rönn, T., Klarlund Pedersen, B., Scheele, C., Vaag, A., & Ling, C. (2014). Altered DNA methylation and differential expression of genes influencing metabolism and inflammation in adipose tissue from subjects with type 2 diabetes. *Diabetes*, *63*(9), 2962–2976. doi:10.2337/db13-1459 PMID:24812430

Olokoba, A. B., Obateru, O. A., & Olokoba, L. B. (2012). Type 2 diabetes mellitus: A review of current trends. *Oman Medical Journal*, *27*(4), 269–273. doi:10.5001/omj.2012.68 PMID:23071876

Pandit, S., Ho, A. C., & Yonekawa, Y. (2023). Recent advances in the management of proliferative diabetic retinopathy. *Current Opinion in Ophthalmology, 34*(3), 232–236. doi:10.1097/ICU.0000000000000946 PMID:36866849

Pearson, E. R. (2016). Personalized medicine in diabetes: The role of 'omics' and biomarkers. *Diabetic Medicine, 33*(6), 712–717. doi:10.1111/dme.13075 PMID:26802434

Powledge, T. M. (2003). Human genome project completed. *Genome Biology, 4*, spotlight-20030415-01. doi:10.1186/gb-spotlight-20030415-01

Poy, M. N., Eliasson, L., Krutzfeldt, J., Kuwajima, S., Ma, X., Macdonald, P. E., Pfeffer, S., Tuschl, T., Rajewsky, N., Rorsman, P., & Stoffel, M. (2004). A pancreatic islet-specific microRNA regulates insulin secretion. *Nature, 432*(7014), 226–230. doi:10.1038/nature03076 PMID:15538371

Ravnskjaer, K., Hogan, M. F., Lackey, D., Tora, L., Dent, S. Y. R., Olefsky, J., & Montminy, M. (2013). Glucagon regulates gluconeogenesis through KAT2B- and WDR5-mediated epigenetic effects. *The Journal of Clinical Investigation, 123*(10), 4318–4328. doi:10.1172/JCI69035 PMID:24051374

Reddy, M. A., Zhang, E., & Natarajan, R. (2015). Epigenetic mechanisms in diabetic complications and metabolic memory. *Diabetologia, 58*(3), 443–455. doi:10.1007/s00125-014-3462-y PMID:25481708

Rescalli, A., Varoni, E. M., Cellesi, F., & Cerveri, P. (2022). Analytical challenges in diabetes management: Towards glycated albumin point-of-care detection. *Biosensors (Basel), 12*(9), 687. doi:10.3390/bios12090687 PMID:36140073

Roden, D. M., McLeod, H. L., Relling, M. V., Williams, M. S., Mensah, G. A., Peterson, J. F., & Van Driest, S. L. (2019). Pharmacogenomics. *Lancet, 394*(10197), 521–532. doi:10.1016/S0140-6736(19)31276-0 PMID:31395440

Roman, D. H., Conlee, K. D., Abbott, I., Jones, R., Noble, A., Rich, N., & Costa, D. (2015). The digital revolution comes to US healthcare. *Internet of Things : Engineering Cyber Physical Human Systems, 5*, 1–54.

Rosenbloom, K. R., Armstrong, J., Barber, G. P., Casper, J., Clawson, H., Diekhans, M., Dreszer, T. R., Fujita, P. A., Guruvadoo, L., Haeussler, M., Harte, R. A., Heitner, S., Hickey, G., Hinrichs, A. S., Hubley, R., Karolchik, D., Learned, K., Lee, B. T., Li, C. H., & Kent, W. J. (2015). The UCSC Genome Browser database: 2015 update. *Nucleic Acids Research, 43*(Database issue), D670–D681. doi:10.1093/nar/gku1177 PMID:25428374

Sanghvi, V. R., Leibold, J., Mina, M., Mohan, P., Berishaj, M., Li, Z., Miele, M. M., Lailler, N., Zhao, C., de Stanchina, E., Viale, A., Akkari, L., Lowe, S. W., Ciriello, G., Hendrickson, R. C., & Wendel, H.-G. (2019). The Oncogenic Action of NRF2 Depends on De-glycation by Fructosamine-3-Kinase. *Cell, 178*(4), 807–819.e21. doi:10.1016/j.cell.2019.07.031 PMID:31398338

Sebastiani, G., Po, A., Miele, E., Ventriglia, G., Ceccarelli, E., Bugliani, M., Marselli, L., Marchetti, P., Gulino, A., Ferretti, E., & Dotta, F. (2015). MicroRNA-124a is hyperexpressed in type 2 diabetic human pancreatic islets and negatively regulates insulin secretion. *Acta Diabetologica, 52*(3), 523–530. doi:10.1007/s00592-014-0675-y PMID:25408296

Segerstolpe, Å., Palasantza, A., Eliasson, P., Andersson, E.-M., Andréasson, A.-C., Sun, X., Picelli, S., Sabirsh, A., Clausen, M., Bjursell, M. K., Smith, D. M., Kasper, M., Ämmälä, C., & Sandberg, R. (2016). Single-Cell Transcriptome Profiling of Human Pancreatic Islets in Health and Type 2 Diabetes. *Cell Metabolism*, *24*(4), 593–607. doi:10.1016/j.cmet.2016.08.020 PMID:27667667

Simpson, J. T., Workman, R. E., Zuzarte, P. C., David, M., Dursi, L. J., & Timp, W. (2017). Detecting DNA cytosine methylation using nanopore sequencing. *Nature Methods*, *14*(4), 407–410. doi:10.1038/nmeth.4184 PMID:28218898

Tajudeen, Y. O., & Shemishere, U. B. (2019). Role of epigenetics in aetiology and therapies for Type 1 Diabetes Mellitus: A narrative review. *Journal of Health and Social Sciences*. doi:10.19204/2019/rlfp12

The ENCODE Project Consortium. (2012). An integrated encyclopedia of DNA elements in the human genome. *Nature*, *489*(7414), 57–74. doi:10.1038/nature11247 PMID:22955616

Trivedi, H., Gray, L. J., Seidu, S., Davies, M. J., Charpentier, G., Lindblad, U., Kellner, C., Nolan, J., Pazderska, A., Rutten, G., Trento, M., & Khunti, K. (2017). Self-knowledge of HbA1c in people with type 2 diabetes mellitus and its association with glycaemic control. *Primary Care Diabetes*, *11*(5), 414–420. doi:10.1016/j.pcd.2017.03.011 PMID:28684227

Unnikrishnan, R., Radha, V., & Mohan, V. (2021). Challenges Involved in Incorporating Personalised Treatment Plan as Routine Care of Patients with Diabetes. *Pharmacogenomics and Personalized Medicine*, *14*, 327–333. doi:10.2147/PGPM.S271582 PMID:33758531

Villeneuve, L. M., Reddy, M. A., Lanting, L. L., Wang, M., Meng, L., & Natarajan, R. (2008). Epigenetic histone H3 lysine 9 methylation in metabolic memory and inflammatory phenotype of vascular smooth muscle cells in diabetes. *Proceedings of the National Academy of Sciences of the United States of America*, *105*(26), 9047–9052. doi:10.1073/pnas.0803623105 PMID:18579779

Yan, B., Yao, J., Liu, J.-Y., Li, X.-M., Wang, X.-Q., Li, Y.-J., Tao, Z.-F., Song, Y.-C., Chen, Q., & Jiang, Q. (2015). lncRNA-MIAT regulates microvascular dysfunction by functioning as a competing endogenous RNA. *Circulation Research*, *116*(7), 1143–1156. doi:10.1161/CIRCRESAHA.116.305510 PMID:25587098

Yang, B. T., Dayeh, T. A., Volkov, P. A., Kirkpatrick, C. L., Malmgren, S., Jing, X., Renström, E., Wollheim, C. B., Nitert, M. D., & Ling, C. (2012). Increased DNA methylation and decreased expression of PDX-1 in pancreatic islets from patients with type 2 diabetes. *Molecular Endocrinology (Baltimore, Md.)*, *26*(7), 1203–1212. doi:10.1210/me.2012-1004 PMID:22570331

Zhang, Y., Liu, T., Meyer, C. A., Eeckhoute, J., Johnson, D. S., Bernstein, B. E., Nusbaum, C., Myers, R. M., Brown, M., Li, W., & Liu, X. S. (2008). Model-based Analysis of ChIP-Seq (MACS). *Genome Biology*, *9*(9), R137. doi:10.1186/gb-2008-9-9-r137 PMID:18798982

Zhou, G., & Xia, J. (2018). OmicsNet: A web-based tool for creation and visual analysis of biological networks in 3D space. *Nucleic Acids Research*, *46*(W1), W514–W522. doi:10.1093/nar/gky510 PMID:29878180

Zou, Q., Qu, K., Luo, Y., Yin, D., Ju, Y., & Tang, H. (2018). Predicting diabetes mellitus with machine learning techniques. *Frontiers in Genetics*, *9*, 515. doi:10.3389/fgene.2018.00515 PMID:30459809

Chapter 15
Advancements in Precision Medicine:
Tailoring Healthcare to Individual Needs

Pranav Kumar Prabhakar
https://orcid.org/0000-0001-8130-1822
Lovely Professional University, India

ABSTRACT

The landscape of healthcare is undergoing a profound transformation with the emergence of precision medicine, a revolutionary approach that tailors medical interventions to the unique characteristics of each individual. This abstract provides a comprehensive overview of the advancements in precision medicine, exploring its foundations, applications in disease diagnosis and treatment, integration of big data and artificial intelligence, as well as the ethical considerations and challenges associated with this groundbreaking field. Precision medicine is fundamentally rooted in understanding the intricate interplay between an individual's genetic makeup, molecular profile, and environmental factors. Genome sequencing and the identification of genetic markers have become instrumental in predicting, preventing, and treating diseases on a personalized level, moving away from traditional population-based approaches. In the realm of oncology, precision medicine has redefined cancer care by targeting specific genetic mutations, enabling more effective and less toxic therapies.

INTRODUCTION TO PRECISION MEDICINE: A PARADIGM SHIFT IN HEALTHCARE

Precision medicine, also known as personalized medicine, marks a transformative paradigm in healthcare that endeavors to tailor medical treatments and interventions to the unique characteristics of individual patients. Precision medicine represents a revolutionary approach to healthcare that tailors medical treatments and interventions to the unique characteristics of each individual (Collins and Varmus, 2015). This innovative approach represents a departure from the traditional one-size-fits-all model, acknowledging the inherent variability among individuals in terms of genetic makeup, environmental exposures, and

DOI: 10.4018/979-8-3693-1922-2.ch015

lifestyle factors. The introduction to precision medicine encompasses its conceptual foundations, historical evolution, and the overarching goal of revolutionizing healthcare to meet each patient's specific needs (National Academies of Sciences, Engineering, and Medicine, 2018). In this chapter we explores the concept's evolution, starting from its historical roots to its contemporary significance in reshaping healthcare paradigms. Emphasis is placed on the foundational principles that underpin precision medicine, including a shift from a one-size-fits-all model to a more nuanced and individualized approach.

(i) *Conceptual Foundations:* At the core of precision medicine lies the recognition that no two individuals are identical, even when faced with similar health conditions. The traditional approach to medicine often treated patients based on generalized guidelines, irrespective of the considerable diversity in their genetic predispositions and responses to treatments. Precision medicine seeks to address this gap by considering the unique genetic, molecular, and environmental factors that contribute to an individual's health profile (Hamburg and Collins, 2010). The central tenet of precision medicine is the integration of diverse data sources, including genetic information, clinical data, lifestyle factors, and environmental exposures. By comprehensively understanding the intricacies of an individual's biology, healthcare providers can make more informed decisions about prevention, diagnosis, and treatment, thereby optimizing therapeutic outcomes.

(ii) *Historical Evolution:* The roots of precision medicine can be traced back to the completion of the Human Genome Project (HGP) in 2003. This monumental scientific endeavor, which involved mapping and sequencing the entire human genome, laid the foundation for unraveling the genetic basis of health and disease. The HGP not only provided valuable insights into the complexity of the human genome but also set the stage for the development of advanced technologies that enable the cost-effective sequencing of individual genomes (Schork, 2015). Following the HGP, precision medicine gained momentum as technological advancements, particularly in DNA sequencing, became more accessible. Next-Generation Sequencing (NGS) technologies emerged as a game-changer, enabling the rapid and cost-effective analysis of vast amounts of genetic information. This progress paved the way for the identification of genetic variants associated with diseases, offering a new level of understanding that could be translated into personalized medical interventions.

(iii) *Overarching Goals:* The overarching goal of precision medicine is to move beyond the constraints of a one-size-fits-all healthcare approach and usher in an era of individualized, targeted therapies. By embracing the principles of precision medicine, healthcare providers aim to optimize treatment efficacy, minimize adverse effects, and improve overall patient outcomes. This approach is particularly relevant in complex and multifactorial diseases where a nuanced understanding of individual variations is crucial for devising effective interventions. One of the key aspirations of precision medicine is to shift the emphasis from reactive to proactive healthcare (National Academies of Sciences, Engineering, and Medicine, 2018). Traditional medicine often addresses health issues after they manifest clinically. In contrast, precision medicine seeks to identify risk factors and susceptibilities at an earlier stage, enabling preemptive measures for disease prevention. This proactive stance aligns with the broader goals of public health, emphasizing the importance of personalized strategies for health promotion and disease prevention.

Moreover, precision medicine extends its reach to areas beyond diagnostics and therapeutics, encompassing lifestyle recommendations, nutrition plans, and behavioral interventions tailored to an individual's unique characteristics (Hamburg and Collins, 2010). This holistic approach acknowledges the interconnectedness of various factors influencing health and wellness, reinforcing the idea that personalized care extends beyond the prescription pad.

(iv) *Challenges and Considerations:* While precision medicine holds great promise, it is not without its challenges. One of the primary hurdles lies in the complexity of integrating vast datasets from diverse sources. The sheer volume of information generated by genomics, proteomics, metabolomics, and other omics technologies necessitates advanced computational tools and bioinformatics expertise for meaningful interpretation. Ethical considerations also come to the forefront in precision medicine. Issues such as privacy concerns, data security, and the potential for genetic discrimination raise important questions about the responsible implementation of precision medicine. Striking a balance between harnessing the power of genomic information and safeguarding individuals' privacy and autonomy is a critical aspect of navigating the ethical landscape.

Additionally, there are disparities in access to precision medicine, both on a global and individual level. Not all populations or individuals have equal access to advanced genetic testing or the latest therapeutic interventions. Addressing these disparities is essential to ensure that the benefits of precision medicine are equitably distributed, promoting health equity across diverse communities. The introduction to precision medicine signifies a profound shift in the way healthcare is conceptualized and delivered (Collins and Varmus, 2015; National Academies of Sciences, Engineering, and Medicine, 2018). As our understanding of genomics and molecular biology deepens, precision medicine holds the promise of revolutionizing healthcare, offering tailored solutions that consider the individuality of each patient. The journey from the completion of the Human Genome Project to the contemporary era of advanced omics technologies exemplifies the remarkable progress made in decoding the complexities of human biology. The overarching goals of precision medicine – individualization, optimization, and prevention – reflect a vision of healthcare that is not only more effective but also more patient-centered. However, challenges, including ethical considerations and disparities in access, underscore the importance of thoughtful implementation and ongoing efforts to ensure that the benefits of precision medicine are realized equitably. As precision medicine continues to evolve, its impact on healthcare is likely to be transformative, reshaping the landscape of medical practice and offering new hope for improved patient outcomes.

GENOMIC MEDICINE: UNRAVELING THE GENETIC CODE

The advent of genomic medicine has ushered in an era of unprecedented insight into the fundamental building blocks of life — the human genome. This field represents a pivotal component of precision medicine, aiming to decipher and utilize the information embedded within the genetic code for the betterment of healthcare (Jameson et al., 2015). In this exploration of genomic medicine, we will delve into the historical significance of the Human Genome Project, the revolutionary impact of next-generation sequencing technologies, and the application of genomic insights in fields such as pharmacogenomics.

(i) *Human Genome Project and its Impact:* The cornerstone of genomic medicine lies in the completion of the Human Genome Project (HGP) in 2003. This monumental scientific endeavor, involving an international collaboration of researchers, aimed to sequence and map the entire human genome, a complex structure comprised of approximately 3 billion base pairs. The successful completion of the HGP marked a historic achievement, providing a comprehensive reference for the genetic code that underlies human biology (Malin and Sweeney, 2004). The HGP had far-reaching implications, not only in terms of sheer scientific knowledge but also in reshaping our understanding of human health and disease.

It paved the way for identifying genes associated with various conditions, unraveling the genetic basis of hereditary diseases, and uncovering the genetic variations that contribute to individual differences in drug responses (Schork, 2015). The project laid the foundation for the subsequent advancements in genomic medicine, providing researchers and healthcare professionals with an invaluable resource to explore the intricacies of the human genome.

(ii) *Next-Generation Sequencing Technologies:* One of the most significant breakthroughs in genomic medicine is the development of Next-Generation Sequencing (NGS) technologies. Unlike traditional sequencing methods, which were time-consuming and expensive, NGS enables the rapid and cost-effective analysis of entire genomes, allowing researchers to decipher the genetic code with unprecedented speed and accuracy (Jameson et al., 2015). This technological leap has transformed genomics from a resource-intensive pursuit to a readily accessible tool for researchers and clinicians alike.

NGS technologies have not only expedited the sequencing process but have also enabled the analysis of individual variations within the genome. This capability is particularly crucial for identifying genetic mutations associated with diseases, elucidating the genetic basis of complex traits, and tailoring medical interventions to individual patients. The versatility of NGS has empowered the field of genomic medicine, opening new avenues for research and clinical applications.

(iii) *Genome-wide Association Studies (GWAS):* Genome-wide Association Studies (GWAS) represent a key methodology in genomic medicine for identifying genetic variants associated with specific diseases or traits. These large-scale studies analyze the genomes of thousands of individuals to identify common genetic variations that may be linked to a particular condition. GWAS have been instrumental in uncovering the genetic basis of a myriad of diseases, from common conditions like diabetes and cardiovascular diseases to more complex and rare disorders. The findings from GWAS have provided valuable insights into the polygenic nature of many diseases, highlighting the contribution of multiple genetic factors in concert with environmental influences (Jameson et al., 2015; Malin and Sweeney, 2004). This understanding is pivotal for developing targeted interventions and personalized treatment strategies. As the field of genomic medicine advances, the integration of GWAS data with other omics technologies further refines our understanding of the intricate interplay between genes, environment, and health outcomes.

(iv) *Pharmacogenomics: Tailoring Drug Therapies Based on Genetic Variations:* One of the most tangible applications of genomic medicine is witnessed in the realm of pharmacogenomics — the study of how an individual's genetic makeup influences their response to drugs. The variability in drug responses among individuals has long been recognized, and genomic medicine offers a novel approach to unraveling the underlying genetic factors. Pharmacogenomic insights enable healthcare providers to tailor drug therapies based on an individual's genetic profile, optimizing efficacy while minimizing adverse reactions (Gymrek et al., 2013). This personalized approach has the potential to revolutionize drug development, prescription practices, and patient outcomes. For example, certain genetic variations can affect the metabolism of drugs in the liver, influencing the efficacy and safety of medications. Understanding these genetic nuances allows for the customization of drug regimens, leading to more effective and safer treatments.

Moreover, pharmacogenomics has the potential to mitigate the trial-and-error approach often associated with medication prescription. By identifying genetic markers associated with drug response, healthcare providers can make informed decisions about the most suitable drugs and dosages for individual patients, thereby improving treatment outcomes and reducing the risk of adverse reactions.

Challenges and Ethical Considerations

While genomic medicine holds immense promise, it is not without its challenges and ethical considerations. The vast amount of data generated by genomic studies, especially with the advent of NGS technologies, poses challenges in terms of data storage, analysis, and interpretation. Advanced computational tools and bioinformatics expertise are crucial for making sense of the massive datasets generated by genomic studies, requiring ongoing advancements in these fields. Ethical considerations in genomic medicine are multifaceted. Privacy concerns surrounding the storage and sharing of genetic information, potential misuse of data, and the implications of genetic testing on individuals and their families necessitate robust ethical frameworks. Striking a balance between harnessing the power of genomics for medical advancement and safeguarding individuals' autonomy and privacy is an ongoing challenge that requires careful navigation.

Genomic medicine stands at the forefront of the precision medicine revolution, offering a profound understanding of the genetic code and its implications for human health. The completion of the Human Genome Project marked a historic milestone, providing the foundation for subsequent breakthroughs in genomics (Shringapure and Bustamante, 2015). Next-generation sequencing technologies have accelerated the pace of genomic research, making it more accessible and applicable in clinical settings. Genomic medicine's impact extends beyond the laboratory to real-world applications, particularly in the field of pharmacogenomics, where genetic insights are transforming drug development and prescription practices. However, challenges such as data management, analysis, and ethical considerations highlight the need for a thoughtful and multidisciplinary approach to advancing genomic medicine.

As genomic medicine continues to evolve, its integration with other omics technologies and clinical applications holds the promise of delivering personalized and targeted healthcare interventions. The unraveling of the genetic code represents not just a scientific achievement but a paradigm shift in how we understand and approach healthcare, paving the way for a future where medical treatments are as unique as the individuals they aim to help.

OMICS TECHNOLOGIES IN PRECISION MEDICINE: UNVEILING THE MOLECULAR LANDSCAPE

Precision medicine, a revolutionary approach to healthcare, relies on a comprehensive understanding of the individual's molecular landscape. Central to this understanding are various 'omics' technologies — powerful tools that delve into the intricacies of genomics, transcriptomics, proteomics, metabolomics, and microbiomics. This chapter explores how these omics technologies contribute to the precision medicine framework, unraveling the molecular complexity that underlies health and disease (Sanderson et al., 2017). From deciphering gene expression patterns to profiling proteins and uncovering metabolic signatures, omics technologies play a pivotal role in tailoring healthcare to individual needs.

(i) *Transcriptomics: Understanding Gene Expression:* At the heart of precision medicine is the ability to unravel the mysteries of gene expression, and transcriptomics serves as the key to deciphering this intricate code. Transcriptomics focuses on the study of RNA molecules, providing insights into which genes are actively transcribed in a particular cell or tissue. This information is crucial for understanding cellular function, identifying potential biomarkers for diseases, and uncovering novel therapeutic targets. Advancements in RNA sequencing technologies, such as RNA-Seq, have revolutionized transcriptomics by enabling the high-throughput analysis of entire transcriptomes (Jameson et al., 2015). This allows researchers and clinicians to quantify gene expression levels, identify alternative splicing events, and gain a holistic view of the transcriptional landscape. Transcriptomics is particularly valuable in precision oncology, where it aids in characterizing the molecular subtypes of cancers and guiding treatment decisions based on the unique gene expression profiles of individual tumors.

(ii) *Proteomics: Profiling Proteins for Personalized Medicine:* While genomics provides information about the genetic code, proteomics focuses on the study of proteins — the molecular machines that carry out the functions encoded by genes. Proteins are dynamic and diverse, and their activities play a crucial role in health and disease. Proteomics aims to decipher the identity, abundance, modifications, and interactions of proteins within a biological system, providing a deeper understanding of cellular processes. Mass spectrometry and high-throughput techniques have propelled proteomics into the precision medicine arena, enabling the simultaneous analysis of thousands of proteins (Shringapure and Bustamante, 2015). Proteomic profiling has been instrumental in identifying protein biomarkers associated with various diseases, facilitating early detection, prognosis, and treatment stratification. In cancer, for instance, proteomic analysis of tumor tissues can uncover specific protein signatures that guide targeted therapies, reflecting the heterogeneity of individual tumors.

(iii) *Metabolomics: Uncovering Metabolic Signatures:* Metabolomics focuses on the study of small molecules, known as metabolites, that are the end products of cellular processes. These molecules provide a snapshot of the metabolic state of a cell, tissue, or organism. Metabolomics aims to uncover metabolic signatures associated with health and disease, offering valuable insights into the functional consequences of genetic and environmental factors (Shringapure and Bustamante, 2015). Advancements in analytical techniques, such as mass spectrometry and nuclear magnetic resonance spectroscopy, have facilitated the high-throughput profiling of metabolites. Metabolomics has found applications in precision medicine by identifying metabolic signatures associated with specific diseases, predicting treatment responses, and uncovering metabolic pathways that can be targeted for therapeutic interventions. In conditions like metabolic disorders and certain cancers, metabolomics plays a pivotal role in unraveling the intricate metabolic alterations that characterize these diseases.

(iv) *Microbiomics: The Role of Microbiota in Personalized Healthcare:* The human microbiome, consisting of trillions of microorganisms residing in and on the body, has emerged as a critical player in health and disease. Microbiomics, focusing on the study of these microbial communities, provides a holistic view of the symbiotic relationship between humans and their microbiota. This field encompasses the study of bacteria, viruses, fungi, and other microorganisms, collectively influencing various aspects of human physiology. Advancements in DNA sequencing technologies have fueled microbiomics research, allowing for the characterization of microbial communities with unprecedented resolution. The gut microbiome, in particular, has garnered attention for its impact on metabolism, immune function, and even neurological health. Microbiome profiling holds promise in predicting disease susceptibility, personalizing dietary recommendations, and developing microbiota-targeted therapies (Sanderson et al., 2017). In precision medicine, understanding an individual's microbiome adds an additional layer of complexity to the personalized healthcare paradigm.

Data Integration and Analysis in Omics Technologies

The power of omics technologies lies not just in generating vast amounts of data but in the ability to integrate and analyze these diverse datasets comprehensively. The multi-omics approach involves combining data from genomics, transcriptomics, proteomics, metabolomics, and microbiomics to obtain a more holistic understanding of the molecular landscape. Bioinformatics and computational tools play a central role in processing and interpreting omics data. Advanced algorithms and machine learning techniques are employed to identify patterns, correlations, and potential biomarkers within these complex datasets. Integrating multi-omics data can uncover novel insights into disease mechanisms, guide treatment decisions, and enhance the predictive accuracy of precision medicine approaches.

Challenges and Future Directions

Despite the immense potential of omics technologies in precision medicine, several challenges need to be addressed. The sheer volume and complexity of data generated by these technologies pose challenges in terms of storage, processing, and interpretation. Developing robust analytical pipelines and computational methods is essential for translating raw omics data into actionable insights. Standardization of protocols and methodologies is another challenge in the omics field. Variability in experimental procedures and data processing methods can introduce biases and hinder the reproducibility of results. Collaborative efforts to establish standardized practices and share validated protocols are crucial for advancing the reliability and consistency of omics data.

Ethical considerations also come into play, especially concerning the privacy and security of individual genomic and health data. As omics technologies become more integrated into clinical practice, safeguarding patient privacy and ensuring responsible data stewardship are paramount. Looking ahead, the future of omics technologies in precision medicine holds exciting possibilities. Continued advancements in technology, particularly in the realm of single-cell omics, will provide even finer resolution in understanding cellular heterogeneity. Integrating omics data with other sources, such as electronic health records and environmental exposures, will further enhance the precision of personalized healthcare strategies.

Omics technologies represent a cornerstone in the edifice of precision medicine, providing a detailed and nuanced view of the molecular landscape. From genomics unraveling the genetic code to transcriptomics, proteomics, metabolomics, and microbiomics, these technologies contribute crucial layers of information that collectively shape the personalized healthcare paradigm. The ability to profile genes, transcripts, proteins, metabolites, and microbial communities on an individual level allows for a more precise understanding of health and disease. The integration of data from these omics disciplines enables a comprehensive approach to personalized medicine, where treatment decisions are tailored to the unique characteristics of each patient. While challenges persist, ongoing advancements in technology, analytical methods, and ethical frameworks position omics technologies at the forefront of future healthcare innovations. The journey from unraveling the genetic code to deciphering the intricacies of the entire molecular landscape exemplifies the transformative potential of omics technologies in reshaping the landscape of healthcare towards a more individualized, targeted, and effective approach.

DATA INTEGRATION AND ANALYSIS IN PRECISION MEDICINE

Precision medicine, a revolutionary paradigm in healthcare, relies on the integration and analysis of vast and diverse datasets to tailor medical interventions to the unique characteristics of individual patients. Central to this approach is the amalgamation of information from genomics, transcriptomics, proteomics, metabolomics, and clinical data, facilitated by advanced computational tools and bioinformatics (Green and Farahany, 2014). This chapter explores the intricacies of data integration and analysis in precision medicine, delving into the challenges, advancements, and the pivotal role these processes play in unlocking the potential of personalized healthcare.

(i) ***The Landscape of Data in Precision Medicine:*** Precision medicine operates at the intersection of diverse data streams, each providing a unique perspective on an individual's health. Genomic data reveals the genetic blueprint, transcriptomics elucidates gene expression patterns, proteomics unveils the protein landscape, metabolomics exposes the metabolic profile, and clinical data provides real-world information on patient health, treatment history, and outcomes. The challenge lies in integrating these disparate datasets to construct a comprehensive and cohesive portrait of an individual's health. This integration is essential for identifying patterns, correlations, and potential biomarkers that can inform personalized treatment strategies. Moreover, the integration of multi-omics data with clinical information enhances the precision and applicability of personalized interventions.

(ii) ***Bioinformatics and Computational Tools:*** Bioinformatics, a multidisciplinary field that combines biology, statistics, and computer science, plays a pivotal role in processing and interpreting the vast datasets generated by omics technologies. Advanced computational tools and algorithms are indispensable for handling the complexity and scale of genomic, transcriptomic, proteomic, metabolomic, and clinical data.

  a. *Genomic Data Analysis:* Genomic data analysis involves the identification of genetic variants, such as single nucleotide polymorphisms (SNPs) and structural variations, and their association with diseases or treatment responses. Bioinformatics tools, including variant callers and genome browsers, aid in annotating and interpreting genomic variations (Green and Farahany, 2014). The integration of genomic data with databases and resources allows for the identification of relevant pathways, functional annotations, and potential therapeutic targets.

  b. *Transcriptomic Data Analysis:* Transcriptomic data analysis focuses on understanding gene expression patterns and alternative splicing events. Differential gene expression analysis identifies genes that are upregulated or downregulated in specific conditions, offering insights into disease mechanisms or treatment responses (Sanderson et al., 2017). Clustering algorithms and pathway analysis tools help unravel the complex interplay of genes and pathways within biological systems.

c. *Proteomic Data Analysis:* Proteomic data analysis involves the identification and quantification of proteins, their post-translational modifications, and interactions (Green and Farahany, 2014). Mass spectrometry data analysis tools, such as MaxQuant and Proteome Discoverer, aid in processing raw data into interpretable results. Network analysis and pathway enrichment tools enable the identification of key protein networks and pathways associated with diseases or treatment outcomes.

d. *Metabolomic Data Analysis:* Metabolomic data analysis focuses on profiling small molecules, providing insights into metabolic pathways and alterations. Software tools like XCMS and MetaboAnalyst assist in peak detection, alignment, and statistical analysis of metabolomic data. Pathway analysis tools help link identified metabolites to relevant biological pathways, contributing to a holistic understanding of metabolic dysregulations in diseases (Evans, 2015).

e. *Clinical Data Integration:* Integrating omics data with clinical information is essential for translating molecular insights into actionable clinical strategies. Electronic health records (EHRs) and clinical databases provide a rich source of patient information, including demographics, medical history, and treatment outcomes. Integrative bioinformatics platforms enable the merging of omics and clinical data, facilitating a holistic approach to precision medicine (McGuire and Beskow, 2010).

(iii) ***Challenges in Data Integration:*** While the promise of precision medicine is enticing, data integration presents several challenges. One major hurdle is the sheer volume of data generated by omics technologies, especially with the increasing adoption of high-throughput sequencing and other advanced technologies. Integrating diverse data types, each with its unique characteristics and scale, requires sophisticated computational methodologies and robust infrastructure.

a. *Data Heterogeneity:* Omics data exhibit heterogeneity not only in terms of scale but also in experimental protocols, platforms, and data formats. Genomic, transcriptomic, proteomic, and metabolomic data may come from different laboratories using distinct technologies, making seamless integration challenging. Standardization efforts and the development of interoperable data formats are essential for overcoming this heterogeneity.

b. *Interdisciplinary Collaboration:* Effective data integration in precision medicine necessitates interdisciplinary collaboration between biologists, clinicians, bioinformaticians, and statisticians. Bridging the gap between diverse expertise areas is crucial for developing comprehensive and interpretable analytical pipelines. Collaborative platforms that facilitate communication and knowledge exchange among experts from different domains can enhance the success of data integration efforts.

c. *Data Quality and Reproducibility:* Ensuring the quality and reproducibility of omics data is paramount for reliable integration and analysis. Variability in experimental conditions, sample preparation, and data processing can introduce biases. Standardization of experimental procedures, rigorous quality control measures, and transparent reporting practices contribute to data reliability and reproducibility.

d. *Privacy and Ethical Concerns:* Patient privacy and ethical considerations become more pronounced when integrating clinical data with molecular information. Protecting sensitive health information while enabling data sharing for research purposes requires robust ethical frameworks and secure data-sharing platforms. Striking a balance between data accessibility and privacy safeguards is an ongoing challenge.

(iv) ***Advancements in Data Integration and Analysis:*** The dynamic landscape of precision medicine has prompted continuous advancements in data integration and analysis methodologies. From the development of sophisticated algorithms to the establishment of collaborative platforms, these innovations aim to enhance the accuracy, interpretability, and scalability of precision medicine approaches.

a. *Machine Learning and Artificial Intelligence:* Machine learning (ML) and artificial intelligence (AI) techniques have emerged as powerful tools for data integration and analysis in precision medicine. ML algorithms can identify complex patterns, predict patient outcomes, and uncover novel associations within integrated datasets. Deep learning approaches, such as neural networks, show promise in capturing intricate relationships within high-dimensional omics data.

b. *Network Analysis:* Network analysis provides a systems-level view of molecular interactions within biological systems. Integrating omics data into network models allows for the identification of key nodes, hubs, and modules associated with diseases or treatment responses. Network-based approaches facilitate the exploration of complex relationships among genes, proteins, and metabolites, contributing to a more holistic understanding of biological processes.

c. *Cloud Computing and Big Data Infrastructure:* The advent of cloud computing and big data infrastructure has significantly enhanced the scalability and accessibility of data integration and analysis in precision medicine. Cloud-based platforms enable researchers to store, process, and analyze large omics datasets without the need for extensive computational resources. This democratization of computational power fosters collaborative research and accelerates the pace of discoveries.

d. *Open Science and Data Sharing Initiatives:* Open science initiatives and data sharing platforms promote transparency, collaboration, and reproducibility in precision medicine research. Large-scale consortia, such as the Cancer Genome Atlas (TCGA) and the International Cancer Genome Consortium (ICGC), facilitate the sharing of omics and clinical data from diverse populations. These initiatives foster a global, collaborative approach to advancing precision medicine.

(v) ***Clinical Applications of Integrated Data:*** The integration and analysis of diverse datasets have tangible implications for clinical decision-making in precision medicine. From identifying potential therapeutic targets to predicting treatment responses, integrated data play a pivotal role in translating molecular insights into actionable clinical strategies.

a. *Identification of Therapeutic Targets:* Integrated analysis of omics data can uncover novel therapeutic targets for various diseases. By identifying genes, proteins, or metabolic pathways associated with disease progression, researchers can pinpoint vulnerabilities that may be targeted with specific drugs. This knowledge informs the development of targeted therapies tailored to the molecular characteristics of individual patients (Collins and Varmus, 2015).

b. *Prediction of Treatment Responses:* Integrating molecular and clinical data enhances the prediction of treatment responses and outcomes. Biomarker discovery through omics data analysis enables the identification of patients likely to respond to specific therapies or experience adverse reactions (Collins and Varmus, 2015). This personalized approach aids in treatment stratification, ensuring that patients receive interventions that are most likely to be effective for their unique molecular profile.

c. *Disease Subtyping and Classification:* Integrated omics data contribute to the classification and subtyping of diseases, particularly in the realm of oncology. Molecular subtyping based on genomic, transcriptomic, or proteomic profiles enables the categorization of tumors into distinct subgroups with unique characteristics (National Academies of Sciences, Engineering, and Medicine, 2018). This classification informs treatment decisions, allowing for more targeted and effective interventions tailored to the specific molecular features of each tumor.

d. *Prognostic and Predictive Biomarkers:* Integrated data analysis identifies prognostic and predictive biomarkers that aid in predicting disease outcomes and treatment responses. By correlating molecular signatures with clinical endpoints, researchers can uncover biomarkers that serve as indicators of disease progression, prognosis, or response to specific therapies (National Academies of Sciences, Engineering, and Medicine, 2018). These biomarkers guide clinicians in making informed decisions about patient management.

(vi) ***Future Directions and Challenges:*** The field of data integration and analysis in precision medicine continues to evolve, presenting both opportunities and challenges on the horizon. As technological advancements progress and the volume of available data increases, addressing these challenges becomes crucial for the continued success of precision medicine.

a. *Single-Cell Omics:* The advent of single-cell omics technologies offers the potential to unravel cellular heterogeneity at unprecedented resolution. Single-cell genomics, transcriptomics, and proteomics enable the study of individual cells within complex tissues, providing insights into cellular diversity and dynamics. Integrating single-cell data into precision medicine approaches holds promise for a more nuanced understanding of diseases and personalized interventions at the single-cell level (National Academies of Sciences, Engineering, and Medicine, 2018; Hamburg and Collins, 2010).

b. *Exposomics:* Integrating molecular data with environmental exposures, known as exposomics, represents the next frontier in precision medicine. Considering the impact of environmental factors, lifestyle, and external influences on an individual's health adds another layer of complexity to personalized healthcare. Incorporating exposomic data into precision medicine strategies enhances our understanding of gene-environment interactions and their contribution to health and disease (McGuire and Beskow, 2010).

c. *Patient-Generated Health Data:* The integration of patient-generated health data, including wearable device data, lifestyle information, and patient-reported outcomes, adds a real-world dimension to precision medicine. These data sources provide continuous monitoring of an individual's health, offering insights into daily activities, physiological parameters, and responses to interventions. Integrating such patient-generated data with molecular information enriches the holistic understanding of individual health trajectories.

d. *Ethical, Legal, and Social Implications (ELSI):* As the scope of precision medicine expands, ethical considerations regarding data privacy, consent, and equitable access become increasingly important. Addressing the ethical, legal, and social implications (ELSI) of data integration in precision medicine is essential for ensuring responsible research conduct and safeguarding patient rights. Collaborative efforts between researchers, policymakers, and ethicists are crucial for developing guidelines and frameworks that address emerging ELSI challenges.

Data integration and analysis form the backbone of precision medicine, shaping our ability to translate molecular insights into personalized healthcare strategies. From the complexities of integrating diverse omics datasets to the advancements in computational methodologies, this field is dynamic and multifaceted. The integration of genomics, transcriptomics, proteomics, metabolomics, and clinical data provides a holistic understanding of an individual's health, laying the foundation for personalized interventions that consider the unique molecular profile of each patient (National Academies of Sciences, Engineering, and Medicine, 2018; Jameson et al., 2015). Advancements in machine learning, network analysis, cloud computing, and open science initiatives contribute to the continuous progress in data integration and analysis methodologies. While challenges persist, the promise of precision medicine remains vast. The ongoing evolution of this field, with the incorporation of single-cell omics, exposomics, and patient-generated health data, holds the potential to further refine and expand our ability to deliver truly personalized healthcare. As we navigate the ethical considerations and ELSI implications, collaborative efforts between researchers, clinicians, policymakers, and the broader community are crucial for ensuring that precision medicine continues to advance responsibly and equitably, ultimately transforming the landscape of healthcare for the benefit of individuals worldwide.

CLINICAL APPLICATIONS OF PRECISION MEDICINE: TRANSFORMING HEALTHCARE FOR INDIVIDUALIZED TREATMENT

Precision medicine has emerged as a groundbreaking approach to healthcare, shifting the focus from traditional one-size-fits-all treatments to personalized interventions tailored to the unique characteristics of each patient. This paradigm is revolutionizing clinical practice across various medical domains, including oncology, cardiology, neurology, and infectious diseases. In this exploration of the clinical applications of precision medicine, we will delve into its impact on diagnosis, treatment selection, prognostication, and its potential to reshape the future of healthcare.

Oncology: Pioneering Precision Medicine Applications

One of the most significant success stories in precision medicine lies in the field of oncology. The heterogeneity of cancer, both across different types and within the same type, has made it an ideal candidate for personalized treatment strategies (Garraway and Lander, 2013). Genomic profiling of tumors has become a cornerstone in oncology, guiding treatment decisions based on the molecular characteristics of individual cancers.

(a) Targeted Therapies: Precision medicine has ushered in an era of targeted therapies designed to exploit specific molecular vulnerabilities in cancer cells. Targeted drugs, such as tyrosine kinase inhibitors and immune checkpoint inhibitors, are tailored to inhibit or enhance specific pathways involved in tumor growth. For example, the use of trastuzumab in HER2-positive breast cancer exemplifies how targeted therapies have improved treatment outcomes by directly addressing the molecular drivers of the disease.
(b) Biomarker-driven Treatment Decisions: Biomarkers play a pivotal role in guiding treatment decisions in precision oncology. Mutations in genes like EGFR, BRAF, and ALK serve as biomarkers that dictate the choice of targeted therapies in lung cancer. Additionally, the identification of mi-

crosatellite instability (MSI) or mismatch repair deficiency (dMMR) in colorectal cancer informs the use of immunotherapy, showcasing the importance of molecular profiling in tailoring treatment strategies (Dienstmann et al., 2017).

(c) Liquid Biopsies: Liquid biopsies, a non-invasive approach to analyze circulating tumor DNA (ctDNA), RNA, and proteins in the blood, have become valuable tools in precision oncology. These tests provide real-time information about tumor mutations, heterogeneity, and treatment responses. Liquid biopsies are particularly useful for monitoring minimal residual disease, predicting relapse, and adjusting treatment plans accordingly (Garraway and Lander, 2013).

(d) Resistance Mechanisms and Adaptive Therapies: Precision medicine has shed light on the mechanisms of treatment resistance, allowing for the development of adaptive therapeutic strategies. Understanding how tumors evolve and adapt to treatments at the molecular level enables clinicians to anticipate and counteract resistance mechanisms (Schwaederle et al., 2015). This knowledge has paved the way for combination therapies and treatment sequencing to enhance long-term treatment efficacy.

Cardiology: Tailoring Interventions for Cardiovascular Health

In cardiology, precision medicine is making significant strides in understanding the genetic and molecular underpinnings of cardiovascular diseases. From risk prediction to the selection of appropriate interventions, precision cardiology aims to optimize patient outcomes through individualized strategies.

(a) Pharmacogenomics in Cardiac Medications: Pharmacogenomics plays a crucial role in tailoring drug prescriptions for cardiovascular conditions. Genetic variations in drug-metabolizing enzymes and drug targets influence individual responses to medications (Mega and Sabatine, 2011). For instance, genetic testing can help identify patients who are poor metabolizers of clopidogrel, informing the choice of alternative antiplatelet therapies post-cardiac interventions.

(b) Genetic Risk Assessment: Precision medicine enables genetic risk assessment for cardiovascular diseases, allowing for early interventions and preventive measures. Understanding an individual's genetic predisposition to conditions like familial hypercholesterolemia or inherited arrhythmias allows for targeted screening, lifestyle modifications, and personalized management plans (Mega et al., 2009).

(c) Inherited Cardiomyopathies: The identification of genetic mutations associated with inherited cardiomyopathies has transformed the diagnosis and management of these conditions. Precision medicine guides risk stratification, allowing for targeted surveillance, early interventions, and family screening (Mega and Sabatine, 2011). Genetic testing in hypertrophic cardiomyopathy, for example, informs not only the diagnosis but also influences treatment decisions and family counseling.

(d) Personalized Antiplatelet Therapy: Precision medicine is influencing the choice of antiplatelet therapies based on individual genetic factors. Genetic variations in the CYP2C19 gene, for instance, impact the metabolism of clopidogrel. Tailoring antiplatelet therapy according to genetic profiles enhances treatment efficacy while minimizing the risk of adverse events (Maron et al., 2017).

Neurology: Navigating the Complex Landscape of Neurological Disorders

Precision medicine is increasingly becoming a guiding force in the field of neurology, where the intricate nature of neurological disorders demands a nuanced understanding of individual variability for accurate diagnosis and effective treatment.

(a) Genetic Testing in Neurodegenerative Diseases: Genetic testing plays a pivotal role in the diagnosis and risk assessment of neurodegenerative diseases. In conditions like Alzheimer's and Parkinson's diseases, identifying specific genetic mutations informs prognosis, allows for early intervention, and facilitates personalized treatment plans (Nalls et al., 2019). Genetic risk profiling is particularly valuable for individuals with a family history of neurodegenerative disorders.

(b) Pharmacogenomics in Epilepsy Treatment: Precision medicine in neurology extends to the treatment of epilepsy, where pharmacogenomics guides the selection of antiepileptic medications. Genetic variations influence drug metabolism and responses, aiding clinicians in choosing the most effective and well-tolerated medications for individual patients.

(c) Individualized Approaches in Neuropsychiatric Disorders: Precision medicine is influencing the management of neuropsychiatric disorders, tailoring treatment strategies based on genetic and molecular insights (Escott-Price et al., 2017). For example, pharmacogenetic testing can inform the selection of psychotropic medications, optimizing therapeutic outcomes while minimizing adverse effects in conditions like depression and bipolar disorder.

(d) Genetic Counseling and Familial Screening: The advent of precision medicine has brought genetic counseling to the forefront in neurology. Genetic testing and counseling are essential components for individuals with a family history of neurological disorders, providing insights into genetic risks, facilitating informed decision-making, and guiding family screening efforts (Moseley and Dhamija, 2019).

Infectious Diseases: A Personalized Approach to Pathogen Response

The application of precision medicine in infectious diseases revolves around understanding individual responses to pathogens, tailoring treatments based on host factors, and predicting susceptibility to infections.

(a) Host-Pathogen Interactions: Precision medicine in infectious diseases delves into the intricate interplay between host genetics and pathogen responses. Genetic variations in immune-related genes influence individual susceptibility, severity of infections, and responses to vaccines (Pulendran and Ahmed, 2011). Understanding these interactions allows for personalized interventions and vaccine strategies.

(b) Antimicrobial Stewardship: Precision medicine is optimizing antimicrobial therapy by considering individual factors such as drug metabolism, host immune response, and pathogen characteristics. Pharmacogenomics plays a role in tailoring antibiotic prescriptions, minimizing adverse reactions, and enhancing treatment efficacy while adhering to antimicrobial stewardship principles (Pulendran and Ahmed, 2011).

(c) Predictive Modeling of Outcomes: Precision medicine leverages predictive modeling to anticipate individual outcomes during infectious diseases. Factors such as genetic predisposition, comorbidities, and immune status contribute to the variability in responses to infections. Predictive models informed by precision medicine assist in risk stratification and guide personalized treatment plans (Hill, 2006; Lamsal et al., 2020).

(d) Vaccine Development and Individualized Strategies: Precision medicine contributes to the development of personalized vaccine strategies, considering individual variations in immune responses (Lamsal et al., 2020). Genetic factors influencing vaccine efficacy and adverse reactions are taken into account for tailoring vaccination schedules, especially in populations with diverse genetic backgrounds.

CHALLENGES AND FUTURE DIRECTIONS

While the clinical applications of precision medicine have demonstrated remarkable successes, several challenges persist, and ongoing efforts are needed to overcome them.

i. Data Integration and Interoperability: Integrating diverse datasets, including genomics, clinical information, and other omics data, remains a complex task. Achieving interoperability among different data sources, standardizing data formats, and implementing robust data-sharing platforms are essential for seamless integration and analysis.
ii. Ethical Considerations and Patient Privacy: As precision medicine relies heavily on genetic and molecular data, ethical considerations surrounding patient privacy, consent, and data security become paramount. Ensuring that individuals have control over their genomic information and implementing robust privacy safeguards are critical for maintaining public trust in precision medicine initiatives (Collins and Varmus, 2015; National Academies of Sciences, Engineering, and Medicine, 2018).
iii. Equity in Access and Implementation: Achieving equitable access to precision medicine is a challenge, considering the variability in resources, infrastructure, and expertise across different regions and healthcare systems. Efforts to bridge these disparities, enhance education and awareness, and promote inclusivity in precision medicine research are essential for its widespread implementation.
iv. Clinical Implementation and Physician Training: Integrating precision medicine into routine clinical practice requires a paradigm shift in physician training and healthcare infrastructure. Clinicians need to be proficient in interpreting genomic and molecular data, and healthcare systems must adapt to accommodate the complexities of personalized treatment plans.

CONCLUSION

Precision medicine has transformed from a concept to a reality, reshaping the landscape of clinical practice across various medical specialties. The applications in oncology, cardiology, neurology, and infectious diseases showcase the versatility and potential of personalized healthcare. As technology advances, challenges are addressed, and ethical frameworks evolve, the clinical applications of precision medicine will continue to expand, offering new avenues for improved patient outcomes and a paradigm shift toward truly individualized healthcare.

REFERENCES

Collins, F. S., & Varmus, H. (2015). A new initiative on precision medicine. *The New England Journal of Medicine, 372*(9), 793–795. doi:10.1056/NEJMp1500523 PMID:25635347

Dienstmann, R., Vermeulen, L., Guinney, J., Kopetz, S., Tejpar, S., & Tabernero, J. (2017). Consensus molecular subtypes and the evolution of precision medicine in colorectal cancer. *Nature Reviews. Cancer, 17*(2), 79–92. doi:10.1038/nrc.2016.126 PMID:28050011

Escott-Price, V., Shoai, M., Pither, R., Williams, J., & Hardy, J. (2017). Polygenic score prediction captures nearly all common genetic risk for Alzheimer's disease. *Neurobiology of Aging, 49*, 214.e7–214.e11. doi:10.1016/j.neurobiolaging.2016.07.018 PMID:27595457

Evans, B. J. (2015). Minimizing liability risks under the Genetic Information Nondiscrimination Act. *Genetics in Medicine, 17*(4), 297–305.

Garraway, L. A., & Lander, E. S. (2013). Lessons from the cancer genome. *Cell, 153*(1), 17–37. doi:10.1016/j.cell.2013.03.002 PMID:23540688

Green, R. C., & Farahany, N. A. (2014). Regulation: The FDA is overcautious on consumer genomics. *Nature, 505*(7483), 286–287. doi:10.1038/505286a PMID:24436984

Gymrek, M., McGuire, A. L., Golan, D., Halperin, E., & Erlich, Y. (2013). Identifying personal genomes by surname inference. *Science, 339*(6117), 321–324. doi:10.1126/science.1229566 PMID:23329047

Hamburg, M. A., & Collins, F. S. (2010). The path to personalized medicine. *The New England Journal of Medicine, 363*(4), 301–304. doi:10.1056/NEJMp1006304 PMID:20551152

Hill, A. V. (2006). Aspects of genetic susceptibility to human infectious diseases. *Annual Review of Genetics, 40*(1), 469–486. doi:10.1146/annurev.genet.40.110405.090546 PMID:17094741

Jameson, J. L., & Longo, D. L. (2015). Precision medicine—Personalized, problematic, and promising. *The New England Journal of Medicine, 372*(23), 2229–2234. doi:10.1056/NEJMsb1503104 PMID:26014593

Knoppers, B. M., Harris, J. R., Budin-Ljøsne, I., Dove, E. S., Ayme, S., Cambon-Thomsen, A., & Harris, H. (2014). A human rights approach to an international code of conduct for genomic and clinical data sharing. *Human Genetics, 133*(7), 895–903. doi:10.1007/s00439-014-1432-6 PMID:24573176

Lamsal, R., Higuchi, Y., & Umezawa, K. (2020). Applications of precision medicine in infectious diseases. *Advances in Experimental Medicine and Biology, 1255*, 153–174.

Malin, B. A., & Sweeney, L. (2004). How (not) to protect genomic data privacy in a distributed network: Using trail re-identification to evaluate and design anonymity protection systems. *Journal of Biomedical Informatics, 37*(3), 179–192. doi:10.1016/j.jbi.2004.04.005 PMID:15196482

Maron, B. A., Loscalzo, J., & Thiene, G. (2017). Moving beyond the alphabet soup of arrhythmias in cardiomyopathies: JACC Review Topic of the Week. *Journal of the American College of Cardiology, 70*(19), 2425–2434.

McGuire, A. L., & Beskow, L. M. (2010). Informed consent in genomics and genetic research. *Annual Review of Genomics and Human Genetics*, *11*(1), 361–381. doi:10.1146/annurev-genom-082509-141711 PMID:20477535

Mega, J. L., Close, S. L., Wiviott, S. D., Shen, L., Hockett, R. D., Brandt, J. T., Walker, J. R., Antman, E. M., Macias, W. L., Braunwald, E., & Sabatine, M. S. (2009). Cytochrome P450 genetic polymorphisms and the response to prasugrel: Relationship to pharmacokinetic, pharmacodynamic, and clinical outcomes. *Circulation*, *119*(19), 2553–2560. doi:10.1161/CIRCULATIONAHA.109.851949 PMID:19414633

Mega, J. L., & Sabatine, M. S. (2011). Antithrombotic therapy in patients with myocardial infarction. *Circulation Research*, *109*(10), 1060–1076.

Moseley, B., & Dhamija, R. (2019). Precision medicine in epilepsy. *Current Neurology and Neuroscience Reports*, *19*(11), 90. PMID:31720870

Nalls, M. A., Blauwendraat, C., Vallerga, C. L., Heilbron, K., Bandres-Ciga, S., Chang, D., Tan, M., Kia, D. A., Noyce, A. J., Xue, A., Bras, J., Young, E., von Coelln, R., Simón-Sánchez, J., Schulte, C., Sharma, M., Krohn, L., Pihlstrøm, L., Siitonen, A., & Zhang, F. (2019). Identification of novel risk loci, causal insights, and heritable risk for Parkinson's disease: A meta-analysis of genome-wide association studies. *Lancet Neurology*, *18*(12), 1091–1102. doi:10.1016/S1474-4422(19)30320-5 PMID:31701892

Pulendran, B., & Ahmed, R. (2011). Immunological mechanisms of vaccination. *Nature Immunology*, *12*(6), 509–517. doi:10.1038/ni.2039 PMID:21739679

Sanderson, S. C., Linderman, M. D., Kasarskis, A., Bashir, A., & Diaz, G. A. (2017). Implications of informing research participants of research results: An empirical study. *Genetics in Medicine*, *19*(6), 644–650.

Schork, N. J. (2015). Personalized medicine: Time for one-person trials. *Nature*, *520*(7549), 609–611. doi:10.1038/520609a PMID:25925459

Schwaederle, M., Zhao, M., Lee, J. J., Lazar, V., Leyland-Jones, B., & Schilsky, R. L. (2015). Association of biomarker-based treatment strategies with response rates and progression-free survival in refractory malignant neoplasms: A meta-analysis. *JAMA Oncology*, *1*(6), 870–879. PMID:27273579

Shringarpure, S. S., & Bustamante, C. D. (2015). Privacy risks from genomic data-sharing beacons. *American Journal of Human Genetics*, *97*(5), 631–646. doi:10.1016/j.ajhg.2015.09.010 PMID:26522470

The Global Alliance for Genomics and Health. (2013). A federated ecosystem for sharing genomic, clinical data. *Science*, *340*(6132), 1521–1522.

Vayena, E., Brownsword, R., Edwards, S. J., Greshake, B., Kahn, J. P., Ladher, N., Montgomery, J., O'Connor, D., O'Neill, O., Richards, M. P., Rid, A., Sheehan, M., Wicks, P., & Tasioulas, J. (2016). Research led by participants: A new social contract for a new kind of research. *Journal of Medical Ethics*, *42*(4), 216–219. doi:10.1136/medethics-2015-102663 PMID:25825527

National Academies of Sciences, Engineering, and Medicine. (2018). *Implementing and Evaluating Genomic Screening Programs in Health Care Systems: A Workshop*. National Academies Press.

Compilation of References

Abacha, A. B., & Zweigenbaum, P. (2015). MEANS: A medical question-answering system combining NLP techniques and semantic Web technologies. *Information Processing & Management, 51*(5), 570–594. doi:10.1016/j.ipm.2015.04.006

Abd-Alrazaq, A. A., Alajlani, M., Alalwan, A. A., Bewick, B. M., Gardner, P., & Househ, M. (2019). An overview of the features of chatbots in mental health: A scoping review. *International Journal of Medical Informatics, 132*, 103978. doi:10.1016/j.ijmedinf.2019.103978 PMID:31622850

Abdou, M. A. (2022). Literature review: Efficient deep neural networks techniques for medical image analysis. *Neural Computing & Applications, 34*(8), 5791–5812. doi:10.1007/s00521-022-06960-9

Abdullah, M. I., Raya, L., Norazman, M. A. A., & Suprihadi, U. (2022, July). Covid-19 patient health monitoring system using IoT. In *2022 IEEE 13th Control and System Graduate Research Colloquium (ICSGRC)* (pp. 155-158). IEEE. 10.1109/ICSGRC55096.2022.9845162

Abioye, S. O., Oyedele, L. O., Akanbi, L., Ajayi, A., Delgado, J. M. D., Bilal, M., & Ahmed, A. (2021). Artificial intelligence in the construction industry: A review of present status, opportunities and future challenges. *Journal of Building Engineering, 44*, 103299. doi:10.1016/j.jobe.2021.103299

Abouaitah, K., Bil, M., Pietrzykowska, E., Szałaj, U., Fudala, D., Woźniak, B., Nasiłowska, J., Swiderska-Sroda, A., Lojkowski, M., Sokołowska, B., Swieszkowski, W., & Lojkowski, W. (2021). Drug-releasing antibacterial coating made from nano-hydroxyapatite using the sonocoating method. *Nanomaterials (Basel, Switzerland), 11*(7), 1690. doi:10.3390/nano11071690 PMID:34203218

Abràmoff, M. D., Lavin, P. T., Birch, M., Shah, N., & Folk, J. C. (2018). Pivotal trial of an autonomous AI-based diagnostic system for detection of diabetic retinopathy in primary care offices. *NPJ Digital Medicine, 1*(1), 39. doi:10.1038/s41746-018-0040-6 PMID:31304320

Adeyemo, B. O., Biederman, J., Zafonte, R., Kagan, E., Spencer, T. J., Uchida, M., & Faraone, S. V. J. J. o. a. d. (2014). Mild traumatic brain injury and ADHD: a systematic review of the literature and meta-analysis. *Journal of Attention Disorders, 18*(7), 576-584.

Agbehadji, I. E., Awuzie, B. O., Ngowi, A. B., & Millham, R. C. (2020). Review of big data analytics, artificial intelligence and nature-inspired computing models towards accurate detection of COVID-19 pandemic cases and contact tracing. *International Journal of Environmental Research and Public Health, 17*(15), 5330. doi:10.3390/ijerph17155330 PMID:32722154

Aggarwal, N., Ahmed, M., Basu, S., Curtin, J. J., Evans, B. J., Matheny, M. E., Nundy, S., Sendak, M. P., Shachar, C., Shah, R. U., & Thadaney-Israni, S. (2020). (2020). Advancing Artificial Intelligence in Health Settings Outside the Hospital and Clinic. *NAM Perspectives*. doi:10.31478/202011f PMID:35291747

Agrawal, N., Mujwar, S., Goyal, A., & Gupta, J. K. (2021). Phytoestrogens as Potential Antiandrogenic Agents Against Prostate Cancer: An In Silico Analysis. *Letters in Drug Design & Discovery*, *19*(1), 69–78. doi:10.2174/1570180818666210813121431

Agrebi, S., & Larbi, A. (2020). Use of artificial intelligence in infectious diseases. In *Artificial intelligence in precision health* (pp. 415–438). Academic Press. doi:10.1016/B978-0-12-817133-2.00018-5

Ahmad, A., Garhwal, S., Ray, S. K., Kumar, G., Malebary, S. J., & Barukab, O. M. (2021). The number of confirmed cases of COVID-19 by using machine learning: Methods and challenges. *Archives of Computational Methods in Engineering*, *28*(4), 2645–2653. doi:10.1007/s11831-020-09472-8 PMID:32837183

Ahmadi, M. R., & Ameri, A. (2022). Skin Cancer Detection Based on Deep Learning. *Journal of Biomedical Physics & Engineering*, *12*(6), 559–568. doi:10.31661/jbpe.v0i0.2207-1517 PMID:36569567

Ahmed Kamal, M., Ismail, Z., Shehata, I. M., Djirar, S., Talbot, N. C., Ahmadzadeh, S., & Kaye, A. D. (2023). Telemedicine, e-health, and multi-agent systems for chronic pain management. *Clinics and Practice*, *13*(2), 470–482. doi:10.3390/clinpract13020042 PMID:36961067

Ahmed, I., Jeon, G., & Chehri, A. (2023). An IoT-enabled smart health care system for screening of COVID-19 with multi layers features fusion and selection. *Computing*, *105*(4), 743–760. doi:10.1007/s00607-021-00992-0

Ahmed, Z., Mohamed, K., Zeeshan, S., & Dong, X. (2020). Artificial intelligence with multi-functional machine learning platform development for better healthcare and precision medicine. *Database (Oxford)*, *2020*, baaa010. doi:10.1093/database/baaa010 PMID:32185396

Ahuja, A. S. (2019). The impact of artificial intelligence in medicine on the future role of the physician. *PeerJ*, *7*, e7702. doi:10.7717/peerj.7702 PMID:31592346

Akazawa, M., & Hashimoto, K. (2020). Artificial intelligence in ovarian cancer diagnosis. *Anticancer Research*, *40*(8), 4795–4800. doi:10.21873/anticanres.14482

Akhoon, N. (2021). Precision Medicine: A New Paradigm in Therapeutics. *International Journal of Preventive Medicine*, *12*(1), 12. PMID:34084309

Alamgir, A., Mousa, O., & Shah, Z. (2021). Artificial Intelligence in Predicting Cardiac Arrest: Scoping Review. *JMIR Medical Informatics*, *9*(12), e30798. doi:10.2196/30798 PMID:34927595

Alamo, T., Reina, D. G., Mammarella, M., & Abella, A. (2020). Covid-19: Open-data resources for monitoring, modeling, and forecasting the epidemic. *Electronics (Basel)*, *9*(5), 827. doi:10.3390/electronics9050827

AlDubayan, S. H., Conway, J. R., Camp, S. Y., Witkowski, L., Kofman, E., Reardon, B., Han, S., Moore, N., Elmarakeby, H., Salari, K., Choudhry, H., Al-Rubaish, A. M., Al-Sulaiman, A. A., Al-Ali, A. K., Taylor-Weiner, A., & Van Allen, E. M. (2020). Detection of Pathogenic Variants With Germline Genetic Testing Using Deep Learning vs Standard Methods in Patients With Prostate Cancer and Melanoma. *Journal of the American Medical Association*, *324*(19), 1957–1969. doi:10.1001/jama.2020.20457 PMID:33201204

Alnsour, Y., Hadidi, R., & Singh, N. (2019). Using data analytics to predict hospital mortality in sepsis patients. *International Journal of Healthcare Information Systems and Informatics*, *14*(3), 40–57. doi:10.4018/IJHISI.2019070104

Altun, E., Aydogdu, M. O., Chung, E., Ren, G., Homer-Vanniasinkam, S., & Edirisinghe, M. (2021). Metal-based nanoparticles for combating antibiotic resistance. *Applied Physics Reviews*, *8*(4), 041303. doi:10.1063/5.0060299

Amara, S. G., & Kuhar, M. (1993). Neurotransmitter transporters: recent progress. *Annual Review of Neuroscience*, *16*(1), 73-93.

Amendolara, A. B., Sant, D., Rotstein, H. G., & Fortune, E. (2023). LSTM-based recurrent neural network provides effective short term flu forecasting. *BMC Public Health*, *23*(1), 1788. doi:10.1186/s12889-023-16720-6 PMID:37710241

American Diabetes Association Professional Practice Committee; American Diabetes Association Professional Practice Committee. (2022). Prevention or delay of type 2 diabetes and associated comorbidities: Standards of Medical Care in Diabetes—2022. *Diabetes Care*, *45*(Supplement_1), S39–S45. doi:10.2337/dc22-S003 PMID:34964876

Aminizadeh, S., Heidari, A., Toumaj, S., Darbandi, M., Navimipour, N. J., Rezaei, M., Talebi, S., Azad, P., & Unal, M. (2023). The applications of machine learning techniques in medical data processing based on distributed computing and the Internet of Things. *Computer Methods and Programs in Biomedicine*, *241*, 107745. doi:10.1016/j.cmpb.2023.107745 PMID:37579550

Amit, D. J., & Amit, D. J. (1989). *Modeling brain function: The world of attractor neural networks*. Cambridge university press. doi:10.1017/CBO9780511623257

Andras, I., Mazzone, E., van Leeuwen, F. W. B., De Naeyer, G., van Oosterom, M. N., Beato, S., Buckle, T., O'Sullivan, S., van Leeuwen, P. J., Beulens, A., Crisan, N., D'Hondt, F., Schatteman, P., van Der Poel, H., Dell'Oglio, P., & Mottrie, A. (2020). Artificial intelligence and robotics: A combination that is changing the operating room. *World Journal of Urology*, *38*(10), 2359–2366. doi:10.1007/s00345-019-03037-6 PMID:31776737

Anggraini Ningrum, D. N., Li, Y. J., Hsu, C. Y., Solihuddin Muhtar, M., & Pandu Suhito, H. (2024). Artificial Intelligence Approach for Severe Dengue Early Warning System. *Studies in Health Technology and Informatics*, *310*, 881–885. doi:10.3233/SHTI231091 PMID:38269935

Arakaki, A. K., Skolnick, J., & McDonald, J. F. (2008). Marker metabolites can be therapeutic targets as well. *Nature*, *456*(7221), 443. doi:10.1038/456443c PMID:19037294

Archila, L. R., Smith, L., Sihvo, H. K., Koponen, V., Jenkins, S. M., O'Sullivan, D. M., & Moreira, R. K. (2023). Performance of an Artificial Intelligence Model for Recognition and Quantitation of Histologic Features of Eosinophilic Esophagitis on Biopsy Samples. *Modern Pathology*, *36*(10), 100285. doi:10.1016/j.modpat.2023.100285

Arciola, C. R., Campoccia, D., Speziale, P., Montanaro, L., & Costerton, J. W. (2012). Biofilm formation in *Staphylococcus* implant infections. A review of molecular mechanisms and implications for biofilm-resistant materials. *Biomaterials*, *33*(26), 5967–5982. doi:10.1016/j.biomaterials.2012.05.031 PMID:22695065

Armentano, R. L., & Kun, L. (2019). Leonardo da Vinci—the first bioengineer: Educational innovation to meet his desire for knowledge and promote his concept of interdisciplinarity. *Creative Education*, *10*(6), 1180–1191. doi:10.4236/ce.2019.106089

Armstrong, D. K., Alvarez, R. D., Bakkum-Gamez, J. N., Barroilhet, L., Behbakht, K., Berchuck, A., Chen, L., Cristea, M., DeRosa, M., Eisenhauer, E. L., Gershenson, D. M., Gray, H. J., Grisham, R., Hakam, A., Jain, A., Karam, A., Konecny, G. E., Leath, C. A., Liu, J., & Engh, A. M. (2021). Ovarian cancer, version 2.2020, NCCN clinical practice guidelines in oncology. *Journal of the National Comprehensive Cancer Network : JNCCN*, *19*(2), 191–226. doi:10.6004/jnccn.2021.0007

Artrith, N., & Urban, A. (2016). An implementation of artificial neural-network potentials for atomistic materials simulations: Performance for TiO2. *Computational Materials Science*, *114*, 135–150. doi:10.1016/j.commatsci.2015.11.047

Ashbaugh, A. G., Jiang, X., Zheng, J., Tsai, A. S., Kim, W. S., Thompson, J. M., Miller, R. J., Shahbazian, J. H., Wang, Y., Dillen, C. A., Ordonez, A. A., Chang, Y. S., Jain, S. K., Jones, L. C., Sterling, R. S., Mao, H. Q., & Miller, L. S. (2016). Polymeric nanofiber coating with tunable combinatorial antibiotic delivery prevents biofilm-associated infection *in vivo. Proceedings of the National Academy of Sciences of the United States of America, 113*(45), E6919–E6928. doi:10.1073/pnas.1613722113 PMID:27791154

Asherson, P., Brookes, K., Franke, B., Chen, W., Gill, M., Ebstein, R. P., & Eisenberg, J. J. A. J. o. P. (2007). Confirmation that a specific haplotype of the dopamine transporter gene is associated with combined-type ADHD. *American Journal of Psychiatry, 164*(4), 674-677.

Ashrafzadeh, S., & Hamdy, O. (2019). Patient-Driven Diabetes Care of the Future in the Technology Era. *Cell Metabolism, 29*(3), 564–575. doi:10.1016/j.cmet.2018.09.005 PMID:30269984

Ashton, J. J., Young, A., Johnson, M. J., & Beattie, R. M. (2023). Using machine learning to impact on long-term clinical care: Principles, challenges, and practicalities. *Pediatric Research, 93*(2), 324–333. doi:10.1038/s41390-022-02194-6 PMID:35906306

Askin, S., Burkhalter, D., Calado, G., & El Dakrouni, S. (2023). Artificial Intelligence Applied to clinical trials: Opportunities and challenges. *Health and Technology, 13*(2), 203–213. doi:10.1007/s12553-023-00738-2 PMID:36923325

Atkinson, A. J., Colburn, W. A., DeGruttola, V. G., DeMets, D. L., Downing, G. J., & Hoth, D. F. (2001). Biomarkers and surrogate endpoints: Preferred definitions and conceptual framework. *Clinical Pharmacology and Therapeutics, 69*(3), 89–95. doi:10.1067/mcp.2001.113989 PMID:11240971

Augier, F., Masbernat, O., & Guiraud, P. (2003). Slip velocity and drag law in a liquid-liquid homogeneous dispersed flow. *AIChE Journal. American Institute of Chemical Engineers, 49*(9), 2300–2316. doi:10.1002/aic.690490907

Ayano, G., Tsegay, L., Gizachew, Y., Necho, M., Yohannes, K., Abraha, M., & Alati, R. J. P. R. (2023). *Prevalence of attention deficit hyperactivity disorder in adults: Umbrella review of evidence generated across the globe.*

Azoulay, A. (2021). *Towards an ethics of artificial intelligence.* United Nations. [Report]

Baars, B. J. (2005). Global workspace theory of consciousness: Toward a cognitive neuroscience of human experience. *Progress in Brain Research, 150*, 45–53. doi:10.1016/S0079-6123(05)50004-9 PMID:16186014

Babu, M., & Snyder, M. (2023). Multi-omics profiling for health. *Molecular & Cellular Proteomics, 22*(6), 100561. doi:10.1016/j.mcpro.2023.100561 PMID:37119971

Back, L. H., Cho, Y. I., Crawford, D. W., & Cuffel, R. F. (1984). *Effect of mild atherosclerosis on flow resistance in a coronary artery casting of man.*

Baduge, S. K., Thilakarathna, S., Perera, J. S., Arashpour, M., Sharafi, P., Teodosio, B., Shringi, A., & Mendis, P. (2022). Artificial intelligence and smart vision for building and construction 4.0: Machine and deep learning methods and applications. *Automation in Construction, 141*, 104440. doi:10.1016/j.autcon.2022.104440

Bain, J. R., Stevens, R. D., Wenner, B. R., Ilkayeva, O., Muoio, D. M., & Newgard, C. B. (2009). Metabolomics Applied to Diabetes Research. *Diabetes, 58*(11), 2429–2443. doi:10.2337/db09-0580 PMID:19875619

Baker, S. G. (2003). The Central Role of Receiver Operating Characteristic (ROC) Curves in Evaluating Tests for the Early Detection of Cancer. *Journal of the National Cancer Institute, 95*(7), 511–515. doi:10.1093/jnci/95.7.511 PMID:12671018

Balagopal, P. B., de Ferranti, S. D., Cook, S., Daniels, S. R., Gidding, S. S., Hayman, L. L., McCrindle, B. W., Mietus-Snyder, M. L., & Steinberger, J. (2011). Nontraditional risk factors and biomarkers for cardiovascular disease: mechanistic, research, and clinical considerations for youth: a scientific statement from the American Heart Association. *Circulation, 123*(23), 2749–2769. doi:10.1161/CIR.0b013e31821c7c64 PMID:21555711

Ballyk, P. D., Steinman, D. A., & Ethier, C. R. (1994). Simulation of non-Newtonian blood flow in an end-to-side anastomosis. *Biorheology, 31*(5), 565–586. doi:10.3233/BIR-1994-31505 PMID:7833458

Baluprithviraj, K., Bharathi, K., Chendhuran, S., & Lokeshwaran, P. (2021). Artificial Intelligence based Smart Door with Face Mask Detection. *2021 International Conference on Artificial Intelligence and Smart Systems (ICAIS)* (pp. 543-548). IEEE. 10.1109/ICAIS50930.2021.9395807

Banday, M. Z., Sameer, A. S., & Nissar, S. (2020). Pathophysiology of diabetes: An overview. *Avicenna Journal of Medicine, 10*(04), 174–188. doi:10.4103/ajm.ajm_53_20 PMID:33437689

Bandyopadhyay, M. (2023). Development Paradigm of India as an Enabler for Practicing Science, Technology and Innovation Diplomacy. In *Science, Technology and Innovation Diplomacy in Developing Countries: Perceptions and Practice* (pp. 115–139). Springer Nature Singapore. doi:10.1007/978-981-19-6802-0_8

Banerjee, T. D., Middleton, F., & Faraone, S. V. J. A. p. (2007). Environmental risk factors for attention-deficit hyperactivity disorder. *Acta Paeddiatra, 96*(9), 1269-1274.

Bansal, M., Goyal, A., & Choudhary, A. (2022). A comparative analysis of K-nearest neighbor, genetic, support vector machine, decision tree, and long short term memory algorithms in machine learning. *Decision Analytics Journal, 3*, 100071. doi:10.1016/j.dajour.2022.100071

Baron, M., Veres, A., Wolock, S. L., Faust, A. L., Gaujoux, R., Vetere, A., Ryu, J. H., Wagner, B. K., Shen-Orr, S. S., Klein, A. M., Melton, D. A., & Yanai, I. (2016). A Single-Cell Transcriptomic Map of the Human and Mouse Pancreas Reveals Inter- and Intra-cell Population Structure. *Cell Systems, 3*(4), 346–360.e4. doi:10.1016/j.cels.2016.08.011 PMID:27667365

Barragán-Montero, A., Javaid, U., Valdés, G., Nguyen, D., Desbordes, P., Macq, B., Willems, S., Vandewinckele, L., Holmström, M., Löfman, F., Michiels, S., Souris, K., Sterpin, E., & Lee, J. A. (2021). Artificial intelligence and machine learning for medical imaging: A technology review. *Physica medica: PM: an international journal devoted to the applications of physics to medicine and biology: official journal of the Italian Association of Biomedical Physics (AIFB), 83*, 242–256. doi:10.1016/j.ejmp.2021.04.016

Barriga, A., Rutle, A., & Heldal, R. (2022). AI-powered model repair: An experience report—lessons learned, challenges, and opportunities. *Software & Systems Modeling, 21*(3), 1135–1157. doi:10.1007/s10270-022-00983-5

Başar, E. (2012). Brain function and oscillations: volume I: brain oscillations. Principles and approaches. Springer Science & Business Media.

Başar, E. (2012). Brain function and oscillations: volume ii: integrative brain function. Neurophysiology and cognitive processes. Springer Science & Business Media.

Basri, A. A., Zubair, M., Aziz, A. F., Ali, R. M., Tamagawa, M., & Ahmad, K. A. (2014, December). Computational fluid dynamics study of the aortic valve opening on hemodynamics characteristics. In *2014 IEEE Conference on Biomedical Engineering and Sciences (IECBES)* (pp. 99-102). IEEE. 10.1109/IECBES.2014.7047660

Baumgartner, R. (2021). Precision medicine and digital phenotyping: Digital medicine's way from more data to better health. *Big Data & Society, 8*(2), 20539517211066452. doi:10.1177/20539517211066452

Baumgartner, R., Arora, P., Bath, C., Burljaev, D., Ciereszko, K., Custers, B., Ding, J., Ernst, W., Fosch-Villaronga, E., Galanos, V., Gremsl, T., Hendl, T., Kropp, C., Lenk, C., Martin, P., Mbelu, S., Morais Dos Santos Bruss, S., Napiwodzka, K., Nowak, E., & Williams, R. (2023). Fair and equitable AI in biomedical research and healthcare: Social science perspectives. *Artificial Intelligence in Medicine*, *144*, 102658. doi:10.1016/j.artmed.2023.102658 PMID:37783540

Bee, S. L., Bustami, Y., Ul-Hamid, A., Lim, K., & Abdul Hamid, Z. A. (2021). Synthesis of silver nanoparticle-decorated hydroxyapatite nanocomposite with combined bioactivity and antibacterial properties. *Journal of Materials Science. Materials in Medicine*, *32*(9), 106. doi:10.1007/s10856-021-06590-y PMID:34426879

Behbahani, S. B., Kiridena, S. D., Wijayaratna, U. N., Taylor, C., Anker, J. N., & Tzeng, T. R. J. (2022). pH variation in medical implant biofilms: Causes, measurements, and its implications for antibiotic resistance. *Frontiers in Microbiology*, *13*(October), 1–28. doi:10.3389/fmicb.2022.1028560 PMID:36386694

Bell, B., Lin, J. J., Seidenberg, M., & Hermann, B. (2011). The neurobiology of cognitive disorders in temporal lobe epilepsy. *Nature Reviews. Neurology*, *7*(3), 154–164. doi:10.1038/nrneurol.2011.3 PMID:21304484

Bello, O., Holzmann, J., Yaqoob, T., & Teodoriu, C. (2015). Application of artificial intelligence methods in drilling system design and operations: A review of the state of the art. *Journal of Artificial Intelligence and Soft Computing Research*, *5*(2), 121–139. doi:10.1515/jaiscr-2015-0024

Bellugi, U., Bihrle, A., Neville, H., Doherty, S., & Jernigan, T. (2013). Language, cognition, and brain organization in a neurodevelopmental disorder. In *Developmental behavioral neuroscience* (pp. 201–232). Psychology Press.

Berg, K., Selbo, P. K., Weyergang, A., Dietze, A., Prasmickaite, L., Bonsted, A., Engesaeter, B. Ø., Angell-Petersen, E., Warloe, T., Frandsen, N., & Høgset, A. (2005). Porphyrin-related photosensitizers for cancer imaging and therapeutic applications. *Journal of Microscopy*, *218*(2), 133–147. doi:10.1111/j.1365-2818.2005.01471.x PMID:15857375

Berini, F., Orlandi, V., Gornati, R., Bernardini, G., & Marinelli, F. (2022). Nanoantibiotics to fight multidrug resistant infections by Gram-positive bacteria: Hope or reality? *Biotechnology Advances*, *57*, 107948. doi:10.1016/j.biotechadv.2022.107948 PMID:35337933

Bernal, G., Colombo, S., Al Ai Baky, M., & Casalegno, F. (2017, June). Safety++ designing IoT and wearable systems for industrial safety through a user centered design approach. In *Proceedings of the 10th international conference on pervasive technologies related to assistive environments* (pp. 163-170). IEEE. 10.1145/3056540.3056557

Berthier, B., Bouzerar, R., & Legallais, C.CBerthier. (2002). Blood flow patterns in an anatomically realistic coronary vessel: Influence of three different reconstruction methods. *Journal of Biomechanics*, *35*(10), 1347–1356. doi:10.1016/S0021-9290(02)00179-3 PMID:12231280

Bertini, I., Cacciatore, S., Jensen, B. V., Schou, J. V., Johansen, J. S., Kruhøffer, M., Luchinat, C., Nielsen, D. L., & Turano, P. (2012). Metabolomic NMR fingerprinting to identify and predict survival of patients with metastatic colorectal cancer. *Cancer Research*, *72*(1), 356–364. doi:10.1158/0008-5472.CAN-11-1543 PMID:22080567

Bessonov, N., Sequeira, A., Simakov, S., Vassilevskii, Y., & Volpert, V. (2016). Methods of blood flow modelling. *Mathematical Modelling of Natural Phenomena*, *11*(1), 1–25. doi:10.1051/mmnp/201611101

Best, J. R., Chiu, B. K., Hsu, C. L., Nagamatsu, L. S., & Liu-Ambrose, T. (2015). Long-term effects of resistance exercise training on cognition and brain volume in older women: Results from a randomized controlled trial. *Journal of the International Neuropsychological Society*, *21*(10), 745–756. doi:10.1017/S1355617715000673 PMID:26581787

Bhargava, A., & Bansal, A. (2021). Novel coronavirus (COVID-19) diagnosis using computer vision and artificial intelligence techniques: A review. *Multimedia Tools and Applications*, *80*(13), 19931–19946. doi:10.1007/s11042-021-10714-5 PMID:33686333

Bhattacharyya, S. (2021). Computational Fluid Dynamics-The Futuristic Innovation in Pharmaceutical Industry. *Indian Journal of Pharmaceutical Education & Research*, *55*(4).

Bherer, L., Erickson, K. I., & Liu-Ambrose, T. (2013). A review of the effects of physical activity and exercise on cognitive and brain functions in older adults. *Journal of Aging Research*, *2013*, 2013. doi:10.1155/2013/657508 PMID:24102028

Biesaga, M., Pyrzyńska, K., & Trojanowicz, M. (2000). Porphyrins in analytical chemistry. A review. *Talanta*, *51*(2), 209–224. doi:10.1016/S0039-9140(99)00291-X PMID:18967853

Billings, C., & Anderson, D. E. (2022). Role of Implantable Drug Delivery Devices with Dual Platform Capabilities in the Prevention and Treatment of Bacterial Osteomyelitis. *Bioengineering (Basel, Switzerland)*, *9*(2), 65. doi:10.3390/bioengineering9020065 PMID:35200418

Bintu, B., Mateo, L. J., Su, J.-H., Sinnott-Armstrong, N. A., Parker, M., Kinrot, S., Yamaya, K., Boettiger, A. N., & Zhuang, X. (2018). Super-resolution chromatin tracing reveals domains and cooperative interactions in single cells. *Science (New York, N.Y.)*, *362*(6413), eaau1783. doi:10.1126/science.aau1783

Biswal, B. B., Mennes, M., Zuo, X. N., Gohel, S., Kelly, C., Smith, S. M., & Milham, M. P. (2010). Toward discovery science of human brain function. *Proceedings of the National Academy of Sciences of the United States of America*, *107*(10), 4734–4739. doi:10.1073/pnas.0911855107 PMID:20176931

Blasco, B., Leroy, D., & Fidock, D. A. (2017). Antimalarial drug resistance: Linking Plasmodium falciparum parasite biology to the clinic. *Nature Medicine*, *23*(8), 917–928. doi:10.1038/nm.4381 PMID:28777791

Boehm, K. M., Aherne, E. A., Ellenson, L., Nikolovski, I., Alghamdi, M., Vázquez-García, I., Zamarin, D., Long Roche, K., Liu, Y., Patel, D., Aukerman, A., Pasha, A., Rose, D., Selenica, P., Causa Andrieu, P. I., Fong, C., Capanu, M., Reis-Filho, J. S., Vanguri, R., & Shah, S. P. (2022). Multimodal data integration using machine learning improves risk stratification of high-grade serous ovarian cancer. *Nature Cancer*, *3*(6), 723–733. doi:10.1038/s43018-022-00388-9

Bossuyt, P. M., Reitsma, J. B., Bruns, D. E., Gatsonis, C. A., Glasziou, P. P., Irwig, L. M., Lijmer, J. G., Moher, D., Rennie, D., & de Vet, H. C. W. (2003). Towards Complete and Accurate Reporting of Studies of Diagnostic Accuracy: The STARD Initiative. *Clinical Chemistry*, *49*(1), 1–6. doi:10.1373/49.1.1 PMID:12507953

Bouchard, L., Thibault, S., Guay, S.-P., Santure, M., Monpetit, A., St-Pierre, J., Perron, P., & Brisson, D. (2010). Leptin gene epigenetic adaptation to impaired glucose metabolism during pregnancy. *Diabetes Care*, *33*(11), 2436–2441. doi:10.2337/dc10-1024 PMID:20724651

Braun, J. M., Kahn, R. S., Froehlich, T., Auinger, P., & Lanphear, B. P. J. E. h. p. (2006). Exposures to environmental toxicants and attention deficit hyperactivity disorder in US children. *Environmental Health Perspective*, *114*(12), 1904-1909.

Breen, J., Allen, K., Zucker, K., Adusumilli, P., Scarsbrook, A., Hall, G., Orsi, N. M., & Ravikumar, N. (2023). Artificial intelligence in ovarian cancer histopathology: A systematic review. *NPJ Precision Oncology*, *7*(1), 83. doi:10.1038/s41698-023-00432-6

Brookes, K.-J., Mill, J., Guindalini, C., Curran, S., Xu, X., Knight, J., & Taylor, E. J. A. o. g. p. (2006). A common haplotype of the dopamine transporter gene associated with attention-deficit/hyperactivity disorder and interacting with maternal use of alcohol during pregnancy. *Arc. Gen. Psychology*, *63*(1), 74-81.

Brosset, P., Jain, A., Khemka, Y., Buvat, J., Thieullent, A., Khadikar, A., & Patsko, S. (2019). Scaling AI in manufacturing operations: A practitioners' perspective. Capgemini Research Institute.

Brot, H. F., & Mango, V. L. (2024). Artificial intelligence in breast ultrasound: Application in clinical practice. *Ultrasonography (Seoul, Korea)*, *43*(1), 3–14. doi:10.14366/usg.23116 PMID:38109894

Brown, J. B., Harris, S. B., Webster-Bogaert, S., Wetmore, S., Faulds, C., & Stewart, M. (2002). The role of patient, physician and systemic factors in the management of type 2 diabetes mellitus. *Family Practice, 19*(4), 344–349. doi:10.1093/fampra/19.4.344 PMID:12110552

Buchlak, Q. D., Esmaili, N., Leveque, J. C., Bennett, C., Farrokhi, F., & Piccardi, M. (2021). Machine learning applications to neuroimaging for glioma detection and classification: An artificial intelligence augmented systematic review. *Journal of clinical neuroscience: official journal of the Neurosurgical Society of Australasia, 89*, 177–198. doi:10.1016/j.jocn.2021.04.043

Bukłaho, P. A., Kiśluk, J., & Nikliński, J. (2023). Diagnostics and treatment of ovarian cancer in the era of precision medicine-opportunities and challenges. *Frontiers in Oncology, 13*, 1227657. doi:10.3389/fonc.2023.1227657

Butpheng, C., Yeh, K. H., & Xiong, H. (2020). Security and privacy in IoT-cloud-based e-health systems—A comprehensive review. *Symmetry, 12*(7), 1191. doi:10.3390/sym12071191

Cai, S., Xu, X., Yang, W., Chen, J., & Fang, X. (2019). Materials and Designs for Wearable Photodetectors. *Advanced Materials, 31*(18), 1808138. doi:10.1002/adma.201808138 PMID:30785644

Calabrese, G., Franco, D., Petralia, S., Monforte, F., Condorelli, G. G., Squarzoni, S., Traina, F., & Conoci, S. (2021). Dual-functional nano-functionalized titanium scaffolds to inhibit bacterial growth and enhance osteointegration. *Nanomaterials (Basel, Switzerland), 11*(10), 2634. doi:10.3390/nano11102634 PMID:34685075

Campero-Jurado, I., Márquez-Sánchez, S., Quintanar-Gómez, J., Rodríguez, S., & Corchado, J. M. (2020). Smart helmet 5.0 for industrial internet of things using artificial intelligence. *Sensors (Basel), 20*(21), 6241. doi:10.3390/s20216241 PMID:33139608

Candreva, A., De Nisco, G., Rizzini, M. L., D'Ascenzo, F., De Ferrari, G. M., Gallo, D., & Chiastra, C. (2022). Current and future applications of computational fluid dynamics in coronary artery disease. *Reviews in Cardiovascular Medicine, 23*(11), 377. doi:10.31083/j.rcm2311377

Cannon, Y. (2021). Closing the health justice gap: Access to justice in furtherance of health equity. *Colum. Hum. Rts. L. Rev., 53*, 517.

Capelli, C., Corsini, C., Biscarini, D., Ruffini, F., Migliavacca, F., Kocher, A., Laufer, G., Taylor, A. M., Schievano, S., Andreas, M., Burriesci, G., & Rath, C. (2017). Pledget-armed sutures affect the haemodynamic performance of biologic aortic valve substitutes: A preliminary experimental and computational study. *Cardiovascular Engineering and Technology, 8*(1), 17–29. doi:10.1007/s13239-016-0284-8 PMID:27873180

Caplin, J. D., & García, A. J. (2019). Implantable antimicrobial biomaterials for local drug delivery in bone infection models. *Acta Biomaterialia, 93*, 2–11. doi:10.1016/j.actbio.2019.01.015 PMID:30654212

Care, S. (2018). Stanford Partnership in AI-assisted care. *Stanford Medicine.* https://med.stanford.edu/pacresearch/research/senior-care.html

Castellanos, F. X., Lee, P. P., Sharp, W., Jeffries, N. O., Greenstein, D. K., Clasen, L. S., & Walter, J. M. J. J. (2002). Developmental trajectories of brain volume abnormalities in children and adolescents with attention-deficit/hyperactivity disorder. *JAMA, 288*(14), 1740-1748.

Cattaneo, G., Bartrés-Faz, D., Morris, T. P., Sánchez, J. S., Macià, D., Tarrero, C., Tormos, J. M., & Pascual-Leone, A. (2018). The Barcelona brain health initiative: A cohort study to define and promote determinants of brain health. *Frontiers in Aging Neuroscience, 10*, 321. doi:10.3389/fnagi.2018.00321 PMID:30405394

Centers for Disease Control and Prevention (CDC). (2021). *The Future of Work: Artificial Intelligence.* CDC. https://blogs.cdc.gov/niosh-science-blog/2021/05/24/ai-future-of-work/

Chadha, N. (2015). R. C. Gangwar, and Rajeev Bedi. "Current Challenges and Application of Speech Recognition Process using Natural Language Processing: A Survey.". *Int. J. Comput. Appl*, *131*(11), 28–31.

Chaichana, T., Sun, Z., & Jewkes, J. (2012). Computational fluid dynamics analysis of the effect of plaques in the left coronary artery. *Computational and Mathematical Methods in Medicine*, *2012*, 2012. doi:10.1155/2012/504367 PMID:22400051

Chakrabarti, M. S., & Ray, M. R. K. (2023). Artificial Intelligence And The Law. *Journal of Pharmaceutical Negative Results*, 87–95.

Chang, C.-H., Yu, C.-J., Du, J.-C., Chiou, H.-C., Chen, H.-C., Yang, W., & Mao, I.-F. J. E. R. (2018). The interactions among organophosphate pesticide exposure, oxidative stress, and genetic polymorphisms of dopamine receptor D4 increase the risk of attention deficit/hyperactivity disorder in children. *Environmental Resources, 160*, 339-346.

Chang, Z., Lichtenstein, P., Asherson, P. J., & Larsson, H. J. J. p. (2013). Developmental twin study of attention problems: high heritabilities throughout development. *JAMA, 70*(3), 311-318.

Charkhchi, P., Cybulski, C., Gronwald, J., Wong, F. O., Narod, S. A., & Akbari, M. R. (2020). CA125 and ovarian cancer: A comprehensive review. *Cancers (Basel), 12*(12), 3730. doi:10.3390/cancers12123730

Chatterjee, S., Chaudhuri, R., Thrassou, A., & Vrontis, D. (2022). Technology disruption in healthcare: Artificial intelligence application, challenges, and policy recommendations in India. *International Journal of Internet Marketing and Advertising*, *17*(3-4), 394–414. doi:10.1504/IJIMA.2022.126716

Chatterjee, S., & Kar, A. K. (2018). Regulation and governance of the Internet of Things in India. *Digital Policy. Regulation & Governance*, *20*(5), 399–412. doi:10.1108/DPRG-04-2018-0017

Check Hayden, E. (2014). The automated lab. *Nature*, *516*(7529), 131–132. doi:10.1038/516131a PMID:25471888

Chen, Q., Brikell, I., Lichtenstein, P., Serlachius, E., Kuja-Halkola, R., Sandin, S., & Psychiatry. (2017). Familial aggregation of attention-deficit/hyperactivity disorder. *Journal of Childhood Psychology, 58*(3), 231-239.

Chen, F., Xue, J., Zhou, L., Wu, S., & Chen, Z. (2011). Identification of serum biomarkers of hepatocarcinoma through liquid chromatography/mass spectrometry-based metabonomic method. *Analytical and Bioanalytical Chemistry*, *401*(6), 1899–1904. doi:10.1007/s00216-011-5245-3 PMID:21833635

Cheng, H., Xiong, W., Fang, Z., Guan, H., Wu, W., Li, Y., Zhang, Y., Alvarez, M. M., Gao, B., Huo, K., Xu, J., Xu, N., Zhang, C., Fu, J., Khademhosseini, A., & Li, F. (2016). Strontium (Sr) and silver (Ag) loaded nanotubular structures with combined osteoinductive and antimicrobial activities. *Acta Biomaterialia*, *31*, 388–400. doi:10.1016/j.actbio.2015.11.046 PMID:26612413

Chen, J. H., & Asch, S. M. (2017). Machine learning and prediction in medicine—Beyond the peak of inflated expectations. *The New England Journal of Medicine*, *376*(26), 2507–2509. doi:10.1056/NEJMp1702071 PMID:28657867

Chen, J., Shi, X., Zhu, Y., Chen, Y., Gao, M., Gao, H., Liu, L., Wang, L., Mao, C., & Wang, Y. (2020). On-demand storage and release of antimicrobial peptides using Pandora's box-like nanotubes gated with a bacterial infection-responsive polymer. *Theranostics*, *10*(1), 109–122. doi:10.7150/thno.38388 PMID:31903109

Chen, T., Xie, G., Wang, X., Fan, J., Qiu, Y., Zheng, X., Qi, X., Cao, Y., Su, M., Wang, X., Xu, L. X., Yen, Y., Liu, P., & Jia, W. (2011). Serum and urine metabolite profiling reveals potential biomarkers of human hepatocellular carcinoma. *Molecular & Cellular Proteomics*, *10*(7), M110.004945. Advance online publication. doi:10.1074/mcp.M110.004945 PMID:21518826

Chen, Z.-H., Lin, L., Wu, C.-F., Li, C.-F., Xu, R.-H., & Sun, Y. (2021). Artificial intelligence for assisting cancer diagnosis and treatment in the era of precision medicine. *Cancer Communications*, *41*(11), 1100–1115. doi:10.1002/cac2.12215 PMID:34613667

Chiao, J. Y. (Ed.). (2009). Cultural neuroscience: Cultural influences on brain function.

Chiao, J. Y., Li, S. C., Seligman, R., & Turner, R. (Eds.). (2016). *The Oxford handbook of cultural neuroscience*. Oxford University Press.

Chiao, J. Y., Li, S. C., Turner, R., & Lee-Tauler, S. Y. (2020). Cultural neuroscience and the research domain criteria: Implications for global mental health. *Neuroscience and Biobehavioral Reviews*, *116*, 109–119. doi:10.1016/j.neubiorev.2020.06.005 PMID:32540352

Chiao, J. Y., Li, S. C., Turner, R., Lee-Tauler, S. Y., & Pringle, B. A. (2017). Cultural neuroscience and global mental health: Addressing grand challenges. *Culture and Brain*, *5*(1), 4–13. doi:10.1007/s40167-016-0045-4 PMID:28642836

Choo, Y. J., & Chang, M. C. (2022). Use of Machine Learning in Stroke Rehabilitation: A Narrative Review. *Brain & Neurorehabilitation*, *15*(3), e26. doi:10.12786/bn.2022.15.e26 PMID:36742082

Chopra, V., Harley, K., Lahiff, M., & Eskenazi, B. J. E. r. (2014). Association between phthalates and attention deficit disorder and learning disability in US children, 6–15 years. *Environmental Resources, 128*, 64-69.

Chopra, D., Gulati, K., & Ivanovski, S. (2021). Understanding and optimizing the antibacterial functions of anodized nano-engineered titanium implants. *Acta Biomaterialia*, *127*, 80–101. doi:10.1016/j.actbio.2021.03.027 PMID:33744499

Chua, L. P., Song, G., Lim, T. M., & Zhou, T. (2006). Numerical analysis of the inner flow field of a biocentrifugal blood pump. *Artificial Organs*, *30*(6), 467–477. doi:10.1111/j.1525-1594.2006.00243.x PMID:16734599

Chung, W. K., Erion, K., Florez, J. C., Hattersley, A. T., Hivert, M.-F., Lee, C. G., McCarthy, M. I., Nolan, J. J., Norris, J. M., Pearson, E. R., Philipson, L., McElvaine, A. T., Cefalu, W. T., Rich, S. S., & Franks, P. W. (2020). Precision Medicine in Diabetes: A Consensus Report From the American Diabetes Association (ADA) and the European Association for the Study of Diabetes (EASD). *Diabetes Care*, *43*(7), 1617–1635. doi:10.2337/dci20-0022 PMID:32561617

Chung, W. W. S., Tariq, S., Mohandes, S. R., & Zayed, T. (2023). IoT-based application for construction site safety monitoring. *International Journal of Construction Management*, *23*(1), 58–74. doi:10.1080/15623599.2020.1847405

Cichosz, S. L., Johansen, M. D., & Hejlesen, O. (2016). Toward big data analytics: Review of predictive models in management of diabetes and its complications. *Journal of Diabetes Science and Technology*, *10*(1), 27–34. doi:10.1177/1932296815611680 PMID:26468133

Cirillo, D., & Valencia, A. (2019). Big data analytics for personalized medicine. *Current Opinion in Biotechnology*, *58*, 161–167. doi:10.1016/j.copbio.2019.03.004 PMID:30965188

Clarke, E. C., Fletcher, D. F., Stoodley, M. A., & Bilston, L. E. (2013). Computational fluid dynamics modelling of cerebrospinal fluid pressure in Chiari malformation and syringomyelia. *Journal of Biomechanics*, *46*(11), 1801–1809. doi:10.1016/j.jbiomech.2013.05.013 PMID:23769174

Clemente, A. S., Diniz, B. S., Nicolato, R., Kapczinski, F. P., Soares, J. C., Firmo, J. O., & Castro-Costa, É. J. B. J. o. P. (2015). Bipolar disorder prevalence: a systematic review and meta-analysis of the literature. *Braz Journal of Psychology, 37*, 155-161.

Clinical Trials Arena. (2024). *Deep 6 AI and Graticule develop clinical trial recruitment algorithm.* Clinical Trials Arena. https://www.clinicaltrialsarena.com/news/deep-6-ai-graticule-algorithm/?cf-view

Collins, F. S., & Varmus, H. (2015). A New Initiative on Precision Medicine. *The New England Journal of Medicine, 372*(9), 793–795. doi:10.1056/NEJMp1500523 PMID:25635347

Collins, P. Y., Patel, V., Joestl, S. S., March, D., Insel, T. R., Daar, A. S., & Walport, M. (2011). Grand challenges in global mental health. *Nature, 475*(7354), 27–30. doi:10.1038/475027a PMID:21734685

Collipp, P., Chen, S., Maitinsky, S. J. A. o. n., & metabolism. (1983). Manganese in infant formulas and learning disability. *Ann Nutr Metab, 27*(6), 488-494.

Comiter, M. (2019). Attacking artificial intelligence. *Belfer Center Paper, 8*, 2019–08.

Connaughton, A., Childs, A., Dylewski, S., & Sabesan, V. J. (2014). Biofilm disrupting technology for orthopedic implants: What's on the horizon? *Frontiers in Medicine, 1*(AUG), 1–4. doi:10.3389/fmed.2014.00022 PMID:25705632

Cook, N. R. (2007). Use and Misuse of the Receiver Operating Characteristic Curve in Risk Prediction. *Circulation, 115*(7), 928–935. doi:10.1161/CIRCULATIONAHA.106.672402 PMID:17309939

Corb Aron, R. A., Abid, A., Vesa, C. M., Nechifor, A. C., Behl, T., Ghitea, T. C., Munteanu, M. A., Fratila, O., Andronie-Cioara, F. L., Toma, M. M., & Bungau, S. (2021, March 17). Recognizing the benefits of pre-/probiotics in metabolic syndrome and type 2 diabetes mellitus considering the influence of akkermansia muciniphila as a key gut bacterium. *Microorganisms, 9*(3), 618. doi:10.3390/microorganisms9030618 PMID:33802777

Cox, D. R. (1972). Regression models and life tables. *Journal of the Royal Statistical Society. Series A (General), 34*, 187–220.

Cremin, C. J., Dash, S., & Huang, X. (2022). Big data: Historic advances and emerging trends in biomedical research. *Current Research in Biotechnology, 4*, 138–151. doi:10.1016/j.crbiot.2022.02.004

Czarnecki, F., Maguire, B. J., Harrell, M., Samo, D., McKinney, Z. J., Guidotti, T. L., & McLellan, R. K. (2024). Worker Health and Safety in Disaster Response. In *Ciottone's Disaster Medicine* (pp. 206–214). Elsevier. doi:10.1016/B978-0-323-80932-0.00033-1

D'Agostino, R. B. Sr, Grundy, S., Sullivan, L. M., & Wilson, P. (2001). Validation of the Framingham coronary heart disease prediction scores: Results of a multiple ethnic groups investigation. *Journal of the American Medical Association, 286*(2), 180–187. doi:10.1001/jama.286.2.180 PMID:11448281

D'Onofrio, B. M., Van Hulle, C. A., Waldman, I. D., Rodgers, J. L., Rathouz, P. J., & Lahey, B. B. J. A. o. g. p. (2007). Causal inferences regarding prenatal alcohol exposure and childhood externalizing problems. *Archive of General Psychiatry, 64*(11), 1296-1304.

D'Agati, E., Casarelli, L., Pitzianti, M. B., Pasini, A. J. P. i. N.-P., & Psychiatry, B. (2010). Overflow movements and white matter abnormalities in ADHD. *Prog Neuropsychopharmacol Biol Psychiatry, 34*(3), 441-445.

Dai, W., Yin, P., Chen, P., Kong, H., Luo, P., Xu, Z., Lu, X., & Xu, G. (2014). Study of urinary steroid hormone disorders: Difference between hepatocellular carcinoma in early stage and cirrhosis. *Analytical and Bioanalytical Chemistry, 406*(18), 4325–4335. doi:10.1007/s00216-014-7843-3 PMID:24817358

Damalas, C. A., & Koutroubas, S. D. (2016). Farmers' exposure to pesticides: Toxicity types and ways of prevention. *Toxics*, *4*(1), 1. doi:10.3390/toxics4010001 PMID:29051407

Damiani, C., Kalliatakis, G., Sreenivas, M., Al-Attar, M., Rose, J., Pudney, C., Lane, E. F., Cuzick, J., Montana, G., & Brentnall, A. R. (2023). Evaluation of an AI model to assess future breast cancer risk. *Radiology*, *307*(5), e222679. doi:10.1148/radiol.222679

Daoud, M. (2022). IoT Applications for Coronavirus Industry Protection. *Advances in Systems Analysis, Software Engineering, and High Performance Computing*, 225–237. doi:10.4018/978-1-7998-8367-8.ch012

Das, R., & Inuwa, M. M. (2023). A review on fog computing: issues, characteristics, challenges, and potential applications. *Telematics and Informatics Reports*, 100049.

Das, S., Garg, V., & Joseph, J. J. (2023, March). Advancements in e-Governance Initiatives: Digitalizing Healthcare in India. In *International Conference on Advances in IoT and Security with AI* (pp. 287-296). Singapore: Springer Nature Singapore. 10.1007/978-981-99-5085-0_28

Dauncey, M. J. (2009). New insights into nutrition and cognitive neuroscience: symposium on 'Early nutrition and later disease: current concepts, research and implications'. *The Proceedings of the Nutrition Society*, *68*(4), 408–415. doi:10.1017/S0029665109990188 PMID:19698201

Dave, P., Nambudiri, V., & Grant-Kels, J. M. (2023). The introduction of "Dr AI": What dermatologists should consider. *Journal of the American Academy of Dermatology*, *88*(6), 1401–1402. doi:10.1016/j.jaad.2022.01.014 PMID:35074441

Dayeh, T., Tuomi, T., Almgren, P., Perfilyev, A., Jansson, P.-A., de Mello, V. D., Pihlajamäki, J., Vaag, A., Groop, L., Nilsson, E., & Ling, C. (2016). DNA methylation of loci within ABCG1 and PHOSPHO1 in blood DNA is associated with future type 2 diabetes risk. *Epigenetics*, *11*(7), 482–488. doi:10.1080/15592294.2016.1178418 PMID:27148772

Debette, S., Wolf, P. A., Beiser, A., Au, R., Himali, J. J., Pikula, A., Auerbach, S., DeCarli, C., & Seshadri, S. (2009). Association of parental dementia with cognitive and brain MRI measures in middle-aged adults. *Neurology*, *73*(24), 2071–2078. doi:10.1212/WNL.0b013e3181c67833 PMID:20007524

Degu, A., Karimi, P. N., Opanga, S. A., & Nyamu, D. G. (2023). Survival outcomes among colorectal cancer patients at Kenyatta National Hospital: A retrospective cohort study. *Cancer Reports*, *6*(3), e1743. doi:10.1002/cnr2.1743

Dekker, I., De Jong, E. M., Schippers, M. C., De Bruijn-Smolders, M., Alexiou, A., & Giesbers, B. (2020). Optimizing Students' Mental Health and Academic Performance: AI-Enhanced Life Crafting. *Frontiers in Psychology*, *11*, 1063. doi:10.3389/fpsyg.2020.01063 PMID:32581935

Delipetrev, B., Tsinaraki, C., & Kostic, U. (2020). *Historical evolution of artificial intelligence.*

Demnitz, N., Zsoldos, E., Mahmood, A., Mackay, C. E., Kivimäki, M., Singh-Manoux, A., Dawes, H., Johansen-Berg, H., Ebmeier, K. P., & Sexton, C. E. (2017). Associations between mobility, cognition, and brain structure in healthy older adults. *Frontiers in Aging Neuroscience*, *9*, 155. doi:10.3389/fnagi.2017.00155 PMID:28588477

Deshpande, S. K., Hasegawa, R. B., Rabinowitz, A. R., Whyte, J., Roan, C. L., Tabatabaei, A., Baiocchi, M., Karlawish, J. H., Master, C. L., & Small, D. S. (2017). Association of playing high school football with cognition and mental health later in life. *JAMA Neurology*, *74*(8), 909–918. doi:10.1001/jamaneurol.2017.1317 PMID:28672325

Dewi, C., Fristiohady, A., Amalia, R., Khairul Ikram, N. K., Ibrahim, S., & Muchtaridi, M. (2022). Signaling pathways and natural compounds in triple-negative breast cancer cell line. *Molecules (Basel, Switzerland)*, *27*(12), 3661. doi:10.3390/molecules27123661

Dey, N. C. (2023). Innovative Digital Teaching and Learning Practices in Society in View of Nursing Education/Profession in India: A Comprehensive Review. *Profession in India: A Comprehensive Review.*

Dhar, Y., & Han, Y. (2020). Current developments in biofilm treatments: Wound and implant infections. *Engineered Regeneration, 1*, 64–75. doi:10.1016/j.engreg.2020.07.003

Dias, D., & Cunha, J. (2018). Wearable Health Devices—Vital Sign Monitoring, Systems and Technologies. *Sensors (Basel), 18*(8), 2414. Advance online publication. doi:10.3390/s18082414 PMID:30044415

Díaz-Rodríguez, N., Del Ser, J., Coeckelbergh, M., de Prado, M. L., Herrera-Viedma, E., & Herrera, F. (2023). Connecting the dots in trustworthy Artificial Intelligence: From AI principles, ethics, and key requirements to responsible AI systems and regulation. *Information Fusion, 99*, 101896. doi:10.1016/j.inffus.2023.101896

Dickinson, D., & Elvevåg, B. (2009). Genes, cognition and brain through a COMT lens. *Neuroscience, 164*(1), 72–87. doi:10.1016/j.neuroscience.2009.05.014 PMID:19446012

Dienstmann, R., Vermeulen, L., Guinney, J., Kopetz, S., Tejpar, S., & Tabernero, J. (2017). Consensus molecular subtypes and the evolution of precision medicine in colorectal cancer. *Nature Reviews. Cancer, 17*(2), 79–92. doi:10.1038/nrc.2016.126 PMID:28050011

DiMaio, S., Grizenko, N., Joober, R. J. J. o. P., & Neuroscience. (2003). Dopamine genes and attention-deficit hyperactivity disorder: a review. *Journal of Psychiatry and Neuroscience, 28*(1), 27-38.

Ding, Y., Hao, Y., Yuan, Z., Tao, B., Chen, M., Lin, C., Liu, P., & Cai, K. (2020). A dual-functional implant with an enzyme-responsive effect for bacterial infection therapy and tissue regeneration. *Biomaterials Science, 8*(7), 1840–1854. doi:10.1039/C9BM01924C PMID:31967110

Dlugatch, R., Georgieva, A., & Kerasidou, A. (2023). Trustworthy artificial intelligence and ethical design: Public perceptions of trustworthiness of an AI-based decision-support tool in the context of intrapartum care. *BMC Medical Ethics, 24*(1), 1–16. doi:10.1186/s12910-023-00917-w PMID:37340408

Dong, J., Fang, D., Zhang, L., Shan, Q., & Huang, Y. (2019). Gallium-doped titania nanotubes elicit anti-bacterial efficacy *in vivo* against *Escherichia coli* and *Staphylococcus aureus* biofilm. *Materialia, 5*, 100209. doi:10.1016/j.mtla.2019.100209

Donzelli, G., Carducci, A., Llopis-Gonzalez, A., Verani, M., Llopis-Morales, A., Cioni, L., & health, p. (2019). The association between lead and attention-deficit/hyperactivity disorder: a systematic review. *Int J Environ Res Public Health, 16*(3), 382.

Dopico, M., Gómez, A., De la Fuente, D., García, N., Rosillo, R., & Puche, J. (2016). A vision of industry 4.0 from an artificial intelligence point of view. In *Proceedings on the international conference on artificial intelligence (ICAI)* (p. 407). The Steering Committee of The World Congress in Computer Science, Computer Engineering and Applied Computing (WorldComp).

Dotoli, M., Fanti, M. P., Iacobellis, G., Martino, L., Moretti, A. M., & Ukovich, W. (2010, February). Modeling and management of a hospital department via petri nets. In *2010 IEEE Workshop on Health Care Management (WHCM)* (pp. 1-6). IEEE. 10.1109/WHCM.2010.5441248

Doukas, C., Pliakas, T., & Maglogiannis, I. (2010, August). Mobile healthcare information management utilizing Cloud Computing and Android OS. In *2010 Annual International Conference of the IEEE Engineering in Medicine and Biology* (pp. 1037-1040). IEEE. 10.1109/IEMBS.2010.5628061

Dubber, M. D., Pasquale, F., & Das, S. (Eds.). (2020). *The Oxford handbook of ethics of AI.* Oxford Handbooks. doi:10.1093/oxfordhb/9780190067397.001.0001

Dubey, N., & Vishwakarma, S. (2016). Cloud computing in healthcare. *Int. J. Curr. Trends Eng. Res*, *2*(5), 211–216.

Dwivedi, A., Mazumder, A., & Nasongkla, N. (2018). Layer-by-layer nanocoating of antibacterial niosome on orthopedic implant. *International Journal of Pharmaceutics*, *547*(1–2), 235–243. doi:10.1016/j.ijpharm.2018.05.075 PMID:29864515

Dwivedi, R., Mehrotra, D., & Chandra, S. (2022). Potential of Internet of Medical Things (IoMT) applications in building a smart healthcare system: A systematic review. *Journal of Oral Biology and Craniofacial Research*, *12*(2), 302–318. doi:10.1016/j.jobcr.2021.11.010 PMID:34926140

Ehrenstein, V., Kharrazi, H., & Lehmann, H. (2019) Obtaining Data From Electronic Health Records. In: *Gliklich RE, Leavy MB, Dreyer NA, editors. Tools and Technologies for Registry Interoperability, Registries for Evaluating Patient Outcomes: A User's Guide,* 3rd Edition, Addendum 2 [Internet]. Rockville (MD): Agency for Healthcare Research and Quality (US). https://www.ncbi.nlm.nih.gov/books/NBK551878/

Ellahham, S., Ellahham, N., & Simsekler, M. C. E. (2020). Application of artificial intelligence in the health care safety context: Opportunities and challenges. *American Journal of Medical Quality*, *35*(4), 341–348. doi:10.1177/1062860619878515 PMID:31581790

Elorriaga, M. Á., Neyro, J. L., Mieza, J., Cristóbal, I., & Llueca, A. (2021). Biomarkers in ovarian pathology: From screening to diagnosis. Review of the literature. *Journal of Personalized Medicine*, *11*(11), 1115. doi:10.3390/jpm11111115

Emond, V., Joyal, C., & Poissant, H. J. L. e. (2008). Structural and functional neuroanatomy of attention-deficit hyperactivity disorder (ADHD). *35*(2), 107-114.

Erickson, K. I., Hillman, C., Stillman, C. M., Ballard, R. M., Bloodgood, B., Conroy, D. E., & Powell, K. E. (2019). Physical activity, cognition, and brain outcomes: A review of the 2018 physical activity guidelines. *Medicine and Science in Sports and Exercise*, *51*(6), 1242–1251. doi:10.1249/MSS.0000000000001936 PMID:31095081

Erikson, K. M., Jones, B. C., Hess, E. J., Zhang, Q., Beard, J. L. J. P. b., & behavior. (2001). Iron deficiency decreases dopamine D1 and D2 receptors in rat brain. *69*(3-4), 409-418.

Ernst, M., Moolchan, E. T., Robinson, M. L. J. J. o. t. A. A. o. C., & Psychiatry, A. (2001). Behavioral and neural consequences of prenatal exposure to nicotine. *40*(6), 630-641.

Escott-Price, V., Shoai, M., Pither, R., Williams, J., & Hardy, J. (2017). Polygenic score prediction captures nearly all common genetic risk for Alzheimer's disease. *Neurobiology of Aging*, *49*, 214.e7–214.e11. doi:10.1016/j.neurobiolaging.2016.07.018 PMID:27595457

Esposito, C., El Ghoneimi, A., Yamataka, A., Rothenberg, S., Bailez, M., Ferro, M., Gamba, P., Castagnetti, M., Mattioli, G., Delagausie, P., Antoniou, D., Montupet, P., Marte, A., Saxena, A., Bertozzi, M., Philippe, P., Varlet, F., Lardy, H., Caldamone, A., & Schier, F. (2013). Work-related upper limb musculoskeletal disorders in paediatric laparoscopic surgery. A multicenter survey. *Journal of Pediatric Surgery*, *48*(8), 1750–1756. doi:10.1016/j.jpedsurg.2013.01.054 PMID:23932617

Eubanks, R. (2017). *Artificial Intelligence and the Healthcare Ecosystem–Part One.*

European Commission. (2020). On artificial intelligence – a European approach to excellence and trust. EC.

Evans, B. J. (2015). Minimizing liability risks under the Genetic Information Nondiscrimination Act. *Genetics in Medicine*, *17*(4), 297–305.

Fahle, S., Prinz, C., & Kuhlenkötter, B. (2020). Systematic review on machine learning (ML) methods for manufacturing processes–Identifying artificial intelligence (AI) methods for field application. *Procedia CIRP*, *93*, 413–418. doi:10.1016/j.procir.2020.04.109

Falzone, L., Scandurra, G., Lombardo, V., Gattuso, G., Lavoro, A., Distefano, A. B., Scibilia, G., & Scollo, P. (2021). A multidisciplinary approach remains the best strategy to improve and strengthen the management of ovarian cancer. *International Journal of Oncology*, *59*(1), 1–14. doi:10.3892/ijo.2021.5233

Fang, J., Wan, Y., Sun, Y., Sun, X., Qi, M., Cheng, S., Li, C., Zhou, Y., Xu, L., Dong, B., & Wang, L. (2022). Near-infrared-activated nanohybrid coating with black phosphorus/zinc oxide for efficient biofilm eradication against implant-associated infections. *Chemical Engineering Journal*, *435*(P1), 134935. doi:10.1016/j.cej.2022.134935

Farag, M. B., Karmonik, C., Rengier, F., Loebe, M., Karck, M., von Tengg-Kobligk, H., Ruhparwar, A., & Partovi, S. (2014). Review of recent results using computational fluid dynamics simulations in patients receiving mechanical assist devices for end-stage heart failure. *Methodist DeBakey Cardiovascular Journal*, *10*(3), 185. doi:10.14797/mdcj-10-3-185 PMID:25574347

Faraone, S. V., & Larsson, H. J. M. p. (2019). Genetics of attention deficit hyperactivity disorder. *24*(4), 562-575.

Farsad-Naeimi, A., Asjodi, F., Omidian, M., Askari, M., Nouri, M., Pizarro, A. B., & Daneshzad, E. J. C. t. i. m. (2020). Sugar consumption, sugar sweetened beverages and Attention Deficit Hyperactivity Disorder: A systematic review and meta-analysis. *53*, 102512.

Feldman, C. L., Ilegbusi, O. J., Hu, Z., Nesto, R., Waxman, S., & Stone, P. H. (2002). Determination of in vivo velocity and endothelial shear stress patterns with phasic flow in human coronary arteries: A methodology to predict progression of coronary atherosclerosis. *American Heart Journal*, *143*(6), 931–939. doi:10.1067/mhj.2002.123118 PMID:12075241

Ferziger, J. H., & Peri, C. M. (2002). *Computational methods for fluid dynamics*.

Filz, M. A., Langner, J. E. B., Herrmann, C., & Thiede, S. (2021). Data-driven failure mode and effect analysis (FMEA) to enhance maintenance planning. *Computers in Industry*, *129*, 103451. doi:10.1016/j.compind.2021.103451

Fioravanti, R., Kumar, K., Nakata, S., Chalamala, B., & Preger, Y. (2020). Predictive-maintenance practices: For operational safety of battery energy storage systems. *IEEE Power & Energy Magazine*, *18*(6), 86–97. doi:10.1109/MPE.2020.3014542

Fiorini, P., Goldberg, K. Y., Liu, Y., & Taylor, R. H. (2022). Concepts and Trends n Autonomy for Robot-Assisted Surgery. *Proceedings of the IEEE. Institute of Electrical and Electronics Engineers*, *110*(7), 993–1011. 10.1109/JPROC.2022.3176828

Fischer, H., Wittmann, D., Costa, A., Zhou, B., Joost, G., & Lukowicz, P. (2021). Masquare: A Functional Smart Mask Design for Health Monitoring. *Proceedings of the 2021 ACM International Symposium on Wearable Computers*. ACM. 10.1145/3460421.3478831

Fitipaldi, H., McCarthy, M. I., Florez, J. C., & Franks, P. W. (2018). A global overview of precision medicine in type 2 diabetes. *Diabetes*, *67*(10), 1911–1922. doi:10.2337/dbi17-0045 PMID:30237159

Fitriani, N., Sutrisna, B., & Pelupessy, N. U. (2020). Survival Analysis of Ovarian Cancer Patients based on Age at The Dr. Wahidin Sudirohusodo Makassar Hospital in 2014-2018. *Indian Journal of Public Health Research & Development*, *11*, 1397.

Fitt, A. D., & Gonzalez, G. (2006). Fluid mechanics of the human eye: Aqueous humour flow in the anterior chamber. *Bulletin of Mathematical Biology*, *68*(1), 53–71. doi:10.1007/s11538-005-9015-2 PMID:16794921

Flasarova, D., Urban, K., Strouhal, O., Klos, D., Lemstrova, R., Dvorak, P., Soucek, P., & Mohelnikova-Duchonova, B. (2023). DNA Repair Pathway in Ovarian Cancer Patients Treated with HIPEC. *International Journal of Molecular Sciences*, *24*(10), 8868. doi:10.3390/ijms24108868

Floridi, L., Cowls, J., Beltrametti, M., Chatila, R., Chazerand, P., Dignum, V., Luetge, C., Madelin, R., Pagallo, U., Rossi, F., Schafer, B., Valcke, P., & Vayena, E. (2018). Ai4people—an ethical framework for a good ai society: Opportunities, risks, principles, and recommendations. *Minds and Machines*, *28*(4), 689–707. doi:10.1007/s11023-018-9482-5 PMID:30930541

Flouris, A. D., & Cheung, S. S. (2006). Design and control optimization of microclimate liquid cooling systems underneath protective clothing. *Annals of Biomedical Engineering*, *34*(3), 359–372. doi:10.1007/s10439-005-9061-9 PMID:16463083

FontaineC.De BrakeleerS.TeugelsE.RenardV.Van den BulkH.VuylstekeP.De GrèveJ. (2022). Heterozygote germline mutations in Homologous Recombination core genes can predict for pathologic complete response in early triple negative breast cancer. doi:10.21203/rs.3.rs-1584056/v1

Fortune Business Insights. (2023). The global telemedicine market size was valued at $87.41 billion in 2022 & is projected to grow from $94.44 billion in 2023 to $286.22 billion by 2030. *Market research Report*. https://www.fortunebusinessinsights.com/industry-reports/telemedicine-market- 101067

Fosch-Villaronga, E. (2015). Creation of a care robot impact assessment. *Int J Soc Behav Edu Econ Bus Ind Eng*, *9*, 1913–1917.

Franke, B., Faraone, S., Asherson, P., Buitelaar, J., Bau, C., Ramos-Quiroga, J. A., & Haavik, J. J. M. p. (2012). The genetics of attention deficit/hyperactivity disorder in adults, a review. *17*(10), 960-987.

Franke, B., Michelini, G., Asherson, P., Banaschewski, T., Bilbow, A., Buitelaar, J. K., & Haavik, J. J. E. N. (2018). Live fast, die young? A review on the developmental trajectories of ADHD across the lifespan. *28*(10), 1059-1088.

Franke, B., Vasquez, A. A., Johansson, S., Hoogman, M., Romanos, J., Boreatti-Hümmer, A., & Casas, M. J. N. (2010). Multicenter analysis of the SLC6A3/DAT1 VNTR haplotype in persistent ADHD suggests differential involvement of the gene in childhood and persistent ADHD. *35*(3), 656-664.

Fraser, K. H., Taskin, M. E., Griffith, B. P., & Wu, Z. J. (2011). The use of computational fluid dynamics in the development of ventricular assist devices. *Medical Engineering & Physics*, *33*(3), 263–280. doi:10.1016/j.medengphy.2010.10.014 PMID:21075669

Freire, S. K., Panicker, S. S., Ruiz-Arenas, S., Rusák, Z., & Niforatos, E. (2022). A Cognitive Assistant for Operators: AI-Powered Knowledge Sharing on Complex Systems. *IEEE Pervasive Computing*, *22*(1), 50–58. doi:10.1109/MPRV.2022.3218600

Fukuda, M., Inamoto, K., Shibata, N., Ariji, Y., Yanashita, Y., Kutsuna, S., Nakata, K., Katsumata, A., Fujita, H., & Ariji, E. (2020). Evaluation of an artificial intelligence system for detecting vertical root fracture on panoramic radiography. *Oral Radiology*, *36*(4), 337–343. doi:10.1007/s11282-019-00409-x PMID:31535278

Furrer, D., Gregoire, J., Turcotte, S., Plante, M., Bachvarov, D., Trudel, D., Têtu, B., Douville, P., & Bairati, I. (2019). Performance of preoperative plasma tumor markers HE4 and CA125 in predicting ovarian cancer mortality in women with epithelial ovarian cancer. *PLoS One*, *14*(6), e0218621. doi:10.1371/journal.pone.0218621

Gaitskell, K., Hermon, C., Barnes, I., Pirie, K., Floud, S., Green, J., Beral, V., & Reeves, G. K.Million Women Study Collaborators. (2022). Ovarian cancer survival by stage, histotype, and pre-diagnostic lifestyle factors, in the prospective UK Million Women Study. *Cancer Epidemiology*, *76*, 102074. doi:10.1016/j.canep.2021.102074

Galloway, N., Blanchette, J. E., Faizullabhoy, A., Mobley-Bukstein, W., Rinker, J., & Isaacs, D. (2022). Exploring the Role of Precision Medicine in Diabetes Diagnoses. *ADCES in Practice*, *10*(1), 26–32. doi:10.1177/2633559X211060849

Gao, F., Li, G., Hu, R., & Okada, H. (2014). Computational Fluid Dynamic Analysis of Coronary Artery Stenting. *International Journal of Bioscience, Biochemistry, Bioinformatics*, *4*(3), 155–159. doi:10.7763/IJBBB.2014.V4.330

Gaona-Luviano, P., Medina-Gaona, L. A., & Magaña-Pérez, K. (2020). Epidemiology of ovarian cancer. *Chinese Clinical Oncology*, *9*(4), 47. doi:10.21037/cco-20-34

García-Herrero, S., Mariscal, M. A., García-Rodríguez, J., & Ritzel, D. O. (2012). Working conditions, psychological/physical symptoms and occupational accidents. Bayesian network models. *Safety Science*, *50*(9), 1760–1774. doi:10.1016/j.ssci.2012.04.005

García-Vázquez, J. P., Rodríguez, M. D., Tentori, M. E., Saldaña, D., Andrade, Á. G., & Espinoza, A. N. (2010). An agent-based architecture for developing activity-aware systems for assisting elderly. *Journal of Universal Computer Science*, *16*(12), 1500–1520.

Garnier, H., Murawski, M., Jastrzebski, T., Pawinska-Wasikowska, K., Balwierz, W., Sinacka, K., Gorecki, W., Izycka-Swieszewska, E., & Czauderna, P. (2021). Case report: cytoreductive surgery and hyperthermic intraperitoneal chemotherapy application in intraperitoneally disseminated inflammatory myofibroblastic tumor and in the youngest patient in the world: new indication and modification of technique. *Frontiers in Surgery*, *8*, 746700. doi:10.3389/fsurg.2021.746700

Garraway, L. A., & Lander, E. S. (2013). Lessons from the cancer genome. *Cell*, *153*(1), 17–37. doi:10.1016/j.cell.2013.03.002 PMID:23540688

Garrity, D. J., & Yusuf, S. A. (2021). A predictive decision-aid device to warn firefighters of catastrophic temperature increases using an AI-based time-series algorithm. *Safety Science*, *138*, 105237. doi:10.1016/j.ssci.2021.105237

Gavaskar, A., Rojas, D., & Videla, F. (2018). Nanotechnology: The scope and potential applications in orthopedic surgery. *European Journal of Orthopaedic Surgery & Traumatology*, *28*(7), 1257–1260. doi:10.1007/s00590-018-2193-z PMID:29602957

Geissbauer, R., Schrauf, S., Berttram, P., & Cheraghi, F. (2017). Digital Factories 2020: Shaping the future of manufacturing.

Ghaheri, A., Shoar, S., Naderan, M., & Hoseini, S. S. (2015). The applications of genetic algorithms in medicine. *Oman Medical Journal*, *30*(6), 406–416. doi:10.5001/omj.2015.82 PMID:26676060

Ghatak, B., Banerjee, S., Ali, S., Bandyopadhyay, R., Das, N., Mandal, D., & Tudu, B. (2020). Design of a Self-powered Smart Mask for COVID-19. arXiv: Medical Physics.

Ghilini, F., Fagali, N., Pissinis, D. E., Benítez, G., & Schilardi, P. L. (2021). Multifunctional Titanium Surfaces for Orthopedic Implants: Antimicrobial Activity and Enhanced Osseointegration. *ACS Applied Bio Materials*, *4*(8), 6451–6461. doi:10.1021/acsabm.1c00613 PMID:35006865

Gholamhosseini, L., Behmanesh, A., Nasiri, S., Ehsanzadeh, S. J., & Sadoughi, F. (2023). Cloud-Based Internet of Things in Healthcare Applications: A Systematic Literature Review. *Frontiers in Health Informatics*, *12*, 145. doi:10.30699/fhi.v12i0.451

Giddens, D. P., Zarins, C. K., & Glagov, S. (1990). *Response of arteries to near-wall fluid dynamic behavior.*

Gillan, C. M., & Rutledge, R. B. (2021). Smartphones and the neuroscience of mental health. *Annual Review of Neuroscience, 44*, 129-151.

Ginsburg, G. S., & Phillips, K. A. (2018). Precision medicine: From science to value. *Health Affairs (Project Hope)*, *37*(5), 694–701. doi:10.1377/hlthaff.2017.1624 PMID:29733705

Giordano, T. J., & McClean, M. (1975, October). Development of a Voice Projection System for Masked Firefighters. In *Audio Engineering Society Convention 52*. Audio Engineering Society.

Goar, V., Sharma, A., Yadav, N. S., Chowdhury, S., & Hu, Y. C. (2023). IOT-based smart mask protection against the waves of covid-19. *Journal of Ambient Intelligence and Humanized Computing*, *14*(8), 11153–11164. doi:10.1007/s12652-022-04395-7 PMID:36117515

Goff, D. C. Jr, Lloyd-Jones, D. M., Bennett, G., Coady, S., D'Agostino, R. B., Gibbons, R., Greenland, P., Lackland, D. T., Levy, D., O'Donnell, C. J., Robinson, J. G., Schwartz, J. S., Shero, S. T., Smith, S. C. Jr, Sorlie, P., Stone, N. J., & Wilson, P. W. F. (2014). 2013 ACC/AHA Guideline on the Assessment of Cardiovascular Risk: A Report of the American College of Cardiology/American Heart Association Task Force on Practice Guidelines. *Circulation*, *129*(25_suppl_2), S49–S73. doi:10.1161/01.cir.0000437741.48606.98 PMID:24222018

Golub, T. R., Slonim, D. K., Tamayo, P., Huard, C., Gaasenbeek, M., Mesirov, J. P., Coller, H., Loh, M. L., Downing, J. R., Caligiuri, M. A., Bloomfield, C. D., & Lander, E. S. (1999). Molecular classification of cancer: Class discovery and class prediction by gene expression monitoring. *Science*, *286*(5439), 531–537. doi:10.1126/science.286.5439.531 PMID:10521349

Gomes-Osman, J., Cabral, D. F., Morris, T. P., McInerney, K., Cahalin, L. P., Rundek, T., & Pascual-Leone, A. (2018). Exercise for cognitive brain health in aging: A systematic review for an evaluation of dose. *Neurology. Clinical Practice*, *8*(3), 257–265. doi:10.1212/CPJ.0000000000000460 PMID:30105166

Gondhowiardjo, S., Christina, N., Ganapati, N. P., Hawariy, S., Radityamurti, F., Jayalie, V. F., Octavianus, S., Prawira Putra, A., Sekarutami, S. M., Prajogi, G. B., Giselvania, A., Adham, M., Hamid, A. R. A. H., Widyastuti, E., Prabowo, Y., Aninditha, T., Purwoto, G., Aman, R. A., Siregar, T. P., & Agianda, F. (2021). Five-year cancer epidemiology at the national referral hospital: Hospital-based cancer registry data in Indonesia. *JCO Global Oncology*, *5*(1), 190–203. doi:10.1200/GO.20.00155

Gorbalenya, A., Baker, S., Baric, R., Groot, R., Drosten, C., Gulyaeva, A., Haagmans, B., Lauber, C., Leontovich, A., Neuman, B., Penzar, D., Perlman, S., Poon, L., Samborskiy, D., Sidorov, I., Sola, I., & Ziebuhr, J. (2020). The species Severe acute respiratory syndrome-related coronavirus: classifying 2019-nCoV and naming it SARS-CoV-2. *Nature Microbiology, 5,* 536 - 544. . doi:10.1038/s41564-020-0695-z

Gordon, E., Barnett, K. J., Cooper, N. J., Tran, N., & Williams, L. M. (2008). An" integrative neuroscience" platform: Application to profiles of negativity and positivity bias. *Journal of Integrative Neuroscience*, *7*(03), 345–366. doi:10.1142/S0219635208001927 PMID:18988297

Go, T., Kim, J. H., Byeon, H., & Lee, S. J. (2018). Machine learning-based in-line holographic sensing of unstained malaria-infected red blood cells. *Journal of Biophotonics*, *11*(9), e201800101. doi:10.1002/jbio.201800101 PMID:29676064

Goyal, A., Kuchana, M., & Ayyagari, K. P. R. (2020). Machine learning predicts live-birth occurrence before in-vitro fertilization treatment. *Scientific Reports*, *10*(1), 20925. doi:10.1038/s41598-020-76928-z PMID:33262383

Graham, S. F., Chevallier, O. P., Roberts, D., Hölscher, C., Elliott, C. T., & Green, B. D. (2013). Investigation of the human brain metabolome to identify potential markers for early diagnosis and therapeutic targets of Alzheimer's disease. *Analytical Chemistry*, *85*(3), 1803–1811. doi:10.1021/ac303163f PMID:23252551

Green, R. C., & Farahany, N. A. (2014). Regulation: The FDA is overcautious on consumer genomics. *Nature*, *505*(7483), 286–287. doi:10.1038/505286a PMID:24436984

Gries, A., Seekamp, A., Wrede, C., & Dodt, C. (2018). Training in clinical acute and emergency medicine–Supraspeciality in Germany: A concept for nationwide implementation! *Der Anaesthesist*, *67*(12), 895–900. doi:10.1007/s00101-018-0515-5 PMID:30511109

Grova, N., Schroeder, H., Olivier, J.-L., & Turner, J. D. J. I. j. o. g. (2019). Epigenetic and neurological impairments associated with early life exposure to persistent organic pollutants. *2019*.

Guerciotti, B., & Vergara, C. (2018). Computational comparison between Newtonian and non-Newtonian blood rheologies in stenotic vessels. *Biomedical Technology: Modeling, Experiments and Simulation*, 169-183.

Gumilang, T. J., & Hernawan, B. (2023). The Benefits of Physical Activity to Reduce Mortality in Postmenopausal Breast Cancer Patients: A Literature Review. *Indonesian Journal of Cancer*, *17*(1), 73–84. doi:10.33371/ijoc.v17i1.917

Gumusoglu-Acar, E., & Gunel, T. (2022). Ovarian Cancer Biomarkers. In *Advances in Diagnosis and Management of Ovarian Cancer* (pp. 27–42). Springer International Publishing. doi:10.1007/978-3-031-09169-8_3

Guo, G., Zhang, H., Shen, H., Zhu, C., He, R., Tang, J., Wang, Y., Jiang, X., Wang, J., Bu, W., & Zhang, X. (2020). Space-selective chemodynamic therapy of $CuFe_5O_8$ nanocubes for implant-related infections. *ACS Nano*, *14*(10), 13391–13405. doi:10.1021/acsnano.0c05255 PMID:32931252

Guo, K., Wu, M., Soo, Z., Yang, Y., Zhang, Y., Zhang, Q., Lin, H., Grosser, M., Venter, D., Zhang, G., & Lu, J. (2023). Artificial intelligence-driven biomedical genomics. *Knowledge-Based Systems*, *279*, 110937. doi:10.1016/j.knosys.2023.110937

Guo, Z., Shen, Y., Wan, S., Shang, W. L., & Yu, K. (2021). Hybrid intelligence-driven medical image recognition for remote patient diagnosis in internet of medical things. *IEEE Journal of Biomedical and Health Informatics*, *26*(12), 5817–5828. doi:10.1109/JBHI.2021.3139541 PMID:34971545

Gupta, M., Singh, N., & Verma, S. (2006). South Asians and Cardiovascular Risk: What Clinicians Should Know. *Circulation*, *113*(25), e924–e929. doi:10.1161/CIRCULATIONAHA.105.583815 PMID:16801466

Gupta, S., Johnson, E. M., Peacock, J. G., Jiang, L., McBee, M. P., Sneider, M. B., & Krupinski, E. A. (2020). Radiology, mobile devices, and internet of things (IoT). *Journal of Digital Imaging*, *33*(3), 735–746. doi:10.1007/s10278-019-00311-2 PMID:31898039

Gymrek, M., McGuire, A. L., Golan, D., Halperin, E., & Erlich, Y. (2013). Identifying personal genomes by surname inference. *Science*, *339*(6117), 321–324. doi:10.1126/science.1229566 PMID:23329047

Hamburg, M. A., & Collins, F. S. (2010). The path to personalized medicine. *The New England Journal of Medicine*, *363*(4), 301–304. doi:10.1056/NEJMp1006304 PMID:20551152

Hamilton, J. G., Garzon, M. G., Westerman, J. S., Shuk, E., Hay, J. L., Walters, C., Elkin, E., & … . (2019). "A tool, not a crutch": Patient perspectives about IBM Watson for oncology trained by Memorial Sloan Kettering. *Journal of Oncology Practice / American Society of Clinical Oncology*, *15*(4), e277–e288. doi:10.1200/JOP.18.00417 PMID:30689492

Hanley, J. A., & McNeil, B. J. (1982). The meaning and use of the area under a receiver operating characteristic (ROC) curve. *Radiology*, *143*(1), 29–36. doi:10.1148/radiology.143.1.7063747 PMID:7063747

Hanson, M. A. (2013). Green ergonomics: Challenges and opportunities. *Ergonomics*, *56*(3), 399–408. doi:10.1080/00140139.2012.751457 PMID:23384154

Haque, M. D. R., & Rubya, S. (2023). An Overview of Chatbot-Based Mobile Mental Health Apps: Insights From App Description and User Reviews. *JMIR mHealth and uHealth*, *11*, e44838. doi:10.2196/44838 PMID:37213181

Harrod, M., Weston, L. E., Gregory, L., Petersen, L., Mayer, J., Drews, F. A., & Krein, S. L. (2020). A qualitative study of factors affecting personal protective equipment use among health care personnel. *American Journal of Infection Control, 48*(4), 410–415. doi:10.1016/j.ajic.2019.08.031 PMID:31610895

Hasasneh, A., Hijazi, H., Talib, M. A., Afadar, Y., Nassif, A. B., & Nasir, Q. (2023). Wearable Devices and Explainable Unsupervised Learning for COVID-19 Detection and Monitoring. *Diagnostics (Basel), 13*(19), 3071. doi:10.3390/diagnostics13193071 PMID:37835814

Haskew, J., Rø, G., Turner, K., Kimanga, D., Sirengo, M., & Sharif, S. (2015). Implementation of a cloud-based electronic medical record to reduce gaps in the HIV treatment continuum in rural Kenya. *PLoS One, 10*(8), e0135361. doi:10.1371/journal.pone.0135361 PMID:26252212

Hassan, M., Awan, F. M., Naz, A., deAndrés-Galiana, E. J., Alvarez, O., Cernea, A., Fernández-Brillet, L., Fernández-Martínez, J. L., & Kloczkowski, A. (2022). Innovations in Genomics and Big Data Analytics for Personalized Medicine and Health Care: A Review. *International Journal of Molecular Sciences, 23*(9), 4645. doi:10.3390/ijms23094645 PMID:35563034

Haviland, J. A., Tonelli, M., Haughey, D. T., Porter, W. P., & Assadi-Porter, F. M. (2012). Novel diagnostics of metabolic dysfunction detected in breath and plasma by selective isotope-assisted labeling. *Metabolism: Clinical and Experimental, 61*(8), 1162–1170. doi:10.1016/j.metabol.2011.12.010 PMID:22304834

Hee, K., Cosa, A., Santhanam, N., Jannesari, M., Maros, M., & Ganslandt, T. (2022). Transfer learning for medical image classification: A literature review. *BMC Medical Imaging, 22*(1), 69. Advance online publication. doi:10.1186/s12880-022-00793-7 PMID:35418051

Heintzman, N. D. (2016). A digital ecosystem of diabetes data and technology: Services, systems, and tools enabled by wearables, sensors, and apps. *Journal of Diabetes Science and Technology, 10*(1), 35–41. doi:10.1177/1932296815622453 PMID:26685994

Hill, A. V. (2006). Aspects of genetic susceptibility to human infectious diseases. *Annual Review of Genetics, 40*(1), 469–486. doi:10.1146/annurev.genet.40.110405.090546 PMID:17094741

Hillman, C. H., Erickson, K. I., & Kramer, A. F. (2008). Be smart, exercise your heart: Exercise effects on brain and cognition. *Nature Reviews. Neuroscience, 9*(1), 58–65. doi:10.1038/nrn2298 PMID:18094706

Hiroto, S., Miyake, Y., & Shinokubo, H. (2017). Synthesis and functionalization of porphyrins through organometallic methodologies. *Chemical Reviews, 117*(4), 2910–3043. doi:10.1021/acs.chemrev.6b00427 PMID:27709907

Hlatky, M. A., Greenland, P., Arnett, D. K., Ballantyne, C. M., Criqui, M. H., Elkind, M. S., Go, A. S., Harrell, F. E. Jr, Hong, Y., Howard, B. V., Howard, V. J., Hsue, P. Y., Kramer, C. M., McConnell, J. P., Normand, S.-L. T., O'Donnell, C. J., Smith, S. C. Jr, & Wilson, P. W. F. (2009). Criteria for evaluation of novel markers of cardiovascular risk: A scientific statement from the American Heart Association. *Circulation, 119*(17), 2408–2416. doi:10.1161/CIRCULATIONAHA.109.192278 PMID:19364974

Hleg, A. (2019). *A Definition of AI: Main Capabilities and Disciplines–EU High-Level Expert Group on Artificial Intelligence.* [online] European Commission. https://www.aepd.es/sites/default/files/2019-09/ai-definition.pdf

Hoffman, S., & Podgurski, A. (2019). Artificial intelligence and discrimination in health care. *Yale J. Health Pol'y L. & Ethics, 19*, 1.

Holmes, E. A., Craske, M. G., & Graybiel, A. M. (2014). Psychological treatments: A call for mental-health science. *Nature, 511*(7509), 287–289. doi:10.1038/511287a PMID:25030152

Holmes, E., Loo, R. L., Stamler, J., Bictash, M., Yap, I. K. S., Chan, Q., Ebbels, T., De Iorio, M., Brown, I. J., Veselkov, K. A., Daviglus, M. L., Kesteloot, H., Ueshima, H., Zhao, L., Nicholson, J. K., & Elliott, P. (2008). Human metabolic phenotype diversity and its association with diet and blood pressure. *Nature*, *453*(7193), 396–400. doi:10.1038/nature06882 PMID:18425110

Holton, K. F., & Nigg, J. T. J. J. o. a. d. (2020). The association of lifestyle factors and ADHD in children. *24*(11), 1511-1520.

Holt, R. I. G., de Groot, M., & Golden, S. H. (2014). Groot M de, Golden SH. Diabetes and depression. *Current Diabetes Reports*, *14*(6), 491. doi:10.1007/s11892-014-0491-3 PMID:24743941

Hong, L., Liu, X., Tan, L., Cui, Z., Yang, X., Liang, Y., Li, Z., Zhu, S., Zheng, Y., Yeung, K. W. K., Jing, D., Zheng, D., Wang, X., & Wu, S. (2019). Rapid Biofilm Elimination on Bone Implants Using Near-Infrared-Activated Inorganic Semiconductor Heterostructures. *Advanced Healthcare Materials*, *8*(19), 1–11. doi:10.1002/adhm.201900835 PMID:31464096

Hong, Q., Huo, S., Tang, H., Qu, X., & Yue, B. (2021). Smart Nanomaterials for Treatment of Biofilm in Orthopedic Implants. *Frontiers in Bioengineering and Biotechnology*, *9*(September), 694635. doi:10.3389/fbioe.2021.694635 PMID:34589470

How, J. A., Jazaeri, A. A., Fellman, B., Daniels, M. S., Penn, S., Solimeno, C., Yuan, Y., Schmeler, K., Lanchbury, J. S., Timms, K., Lu, K. H., & Yates, M. S. (2021). Modification of homologous recombination deficiency score threshold and association with long-term survival in epithelial ovarian cancer. *Cancers (Basel)*, *13*(5), 946. doi:10.3390/cancers13050946

Htay, T., Soe, K., Lopez-Perez, A., Doan, A. H., Romagosa, M. A., & Aung, K. (2019). Mortality and cardiovascular disease in type 1 and type 2 diabetes. *Current Cardiology Reports*, *21*(6), 1–7. doi:10.1007/s11886-019-1133-9 PMID:31011838

Hu, B., & Koo, S. (2022). Development of wearable device designs for respiratory infection prevention. *International Journal of Clothing Science and Technology*, *34*(5), 764–781. doi:10.1108/IJCST-07-2021-0094

Hu, C. C., Chang, C. H., Chang, Y., Hsieh, J. H., & Ueng, S. W. N. (2020). Beneficial effect of TaON-Ag nanocomposite titanium on antibacterial capacity in orthopedic application. *International Journal of Nanomedicine*, *15*, 7889–7900. doi:10.2147/IJN.S264303 PMID:33116507

Huizink, A. C., Mulder, E. J. J. N., & Reviews, B. (2006). Maternal smoking, drinking or cannabis use during pregnancy and neurobehavioral and cognitive functioning in human offspring. *30*(1), 24-41.

Hu, J., Ding, Y., Tao, B., Yuan, Z., Yang, Y., Xu, K., Li, X., Liu, P., & Cai, K. (2022). Surface modification of titanium substrate via combining photothermal therapy and quorum-sensing-inhibition strategy for improving osseointegration and treating biofilm-associated bacterial infection. *Bioactive Materials*, *18*, 228–241. doi:10.1016/j.bioactmat.2022.03.011 PMID:35387171

Hung, T. K., Khalafvand, S. S., & Ng, E. Y. K. (2015). Fluid dynamic characteristics of systolic blood flow of the left ventricle. *Journal of Mechanics in Medicine and Biology*, *15*(03), 1550047. doi:10.1142/S0219519415500475

Hu, Y., Ruan, X., Lv, X., Xu, Y., Wang, W., Cai, Y., Ding, M., Dong, H., Shao, J., Yang, D., & Dong, X. (2022). Biofilm microenvironment-responsive nanoparticles for the treatment of bacterial infection. *Nano Today*, *46*, 101602. doi:10.1016/j.nantod.2022.101602

Hymel, P. A., Loeppke, R. R., Baase, C. M., Burton, W. N., Hartenbaum, N. P., Hudson, T. W., McLellan, R. K., Mueller, K. L., Roberts, M. A., Yarborough, C. M., Konicki, D. L., & Larson, P. W. (2011). Workplace health protection and promotion: A new pathway for a healthier—and safer—workforce. *Journal of Occupational and Environmental Medicine*, *53*(6), 695–702. doi:10.1097/JOM.0b013e31822005d0 PMID:21654443

Hyysalo, J., Dasanayake, S., Hannu, J., Schuss, C., Rajanen, M., Leppänen, T., Doermann, D., & Sauvola, J. (2022). Smart mask–Wearable IoT solution for improved protection and personal health. *Internet of Things : Engineering Cyber Physical Human Systems*, *18*, 100511. doi:10.1016/j.iot.2022.100511 PMID:37521492

Ibáñez, C., Simó, C., Martín-Álvarez, P. J., Kivipelto, M., Winblad, B., Cedazo-Mínguez, A., & Cifuentes, A. (2012). Toward a predictive model of Alzheimer's disease progression using capillary electrophoresis-mass spectrometry metabolomics. *Analytical Chemistry*, *84*(20), 8532–8540. doi:10.1021/ac301243k PMID:22967182

Ienca, M., Pollicino, O., Liguori, L., Stefanini, E., & Andorno, R. (Eds.). (2022). *The Cambridge handbook of information technology, life sciences and human rights*. Cambridge University Press. doi:10.1017/9781108775038

Imahori, H., Kashiwagi, Y., Endo, Y., Hanada, T., Nishimura, Y., Yamazaki, I., Araki, Y., Ito, O., & Fukuzumi, S. (2004). Structure and photophysical properties of porphyrin-modified metal nanoclusters with different chain lengths. *Langmuir*, *20*(1), 73–81. doi:10.1021/la035435p PMID:15745002

Im, H., Pathania, D., McFarland, P. J., Sohani, A. R., Degani, I., Allen, M., Coble, B., Kilcoyne, A., Hong, S., Rohrer, L., Abramson, J. S., Dryden-Peterson, S., Fexon, L., Pivovarov, M., Chabner, B., Lee, H., Castro, C. M., & Weissleder, R. (2018). Design and clinical validation of a point-of-care device for the diagnosis of lymphoma via contrast-enhanced microholography and machine learning. *Nature Biomedical Engineering*, *2*(9), 666–674. doi:10.1038/s41551-018-0265-3 PMID:30555750

Indelli, P. F., Ghirardelli, S., Iannotti, F., Indelli, A. M., & Pipino, G. (2021). Nanotechnology as an anti-infection strategy in periprosthetic joint infections (PJI). *Tropical Medicine and Infectious Disease*, *6*(2), 1–6. doi:10.3390/tropicalmed6020091 PMID:34071727

Iniesta, R. (2023). The human role to guarantee an ethical AI in healthcare: A five-facts approach. *AI and Ethics*, 1–13. doi:10.1007/s43681-023-00353-x

Inkster, B., Sarda, S., & Subramanian, V. (2018). An Empathy-Driven, Conversational Artificial Intelligence Agent (Wysa) for Digital Mental Well-Being: Real-World Data Evaluation Mixed-Methods Study. *JMIR mHealth and uHealth*, *6*(11), e12106. doi:10.2196/12106 PMID:30470676

Intravaia, J. T., Graham, T., Kim, H. S., Nanda, H. S., Kumbar, S. G., & Nukavarapu, S. P. (2023). Smart orthopedic biomaterials and implants. *Current Opinion in Biomedical Engineering*, *25*, 100439. doi:10.1016/j.cobme.2022.100439 PMID:36642994

Inzucchi, S., Bergenstal, R., Fonseca, V., Gregg, E., Mayer-Davis, B., Spollett, G., & Wender, R. (2010). Diagnosis and classification of diabetes mellitus. Diabetes care, 33, S62-S69. Anjana, R. M., Unnikrishnan, R., Deepa, M., Pradeepa, R., Tandon, N., Das, A. K., & Ghosh, S. (2023). *Metabolic non-communicable disease health report of India: the ICMR-INDIAB national cross-sectional study (ICMR-INDIAB-17).* The Lancet Diabetes & Endocrinology.

Iqbal, J., Jahangir, K., Mashkoor, Y., Sultana, N., Mehmood, D., Ashraf, M., Iqbal, A., & Hafeez, M. H. (2022). The future of artificial intelligence in neurosurgery: A narrative review. *Surgical Neurology International*, *13*, 13. doi:10.25259/SNI_877_2022 PMID:36447868

Islam, M. N., & Islam, A. N. (2020). A systematic review of the digital interventions for fighting COVID-19: The Bangladesh perspective. *IEEE Access : Practical Innovations, Open Solutions*, *8*, 114078–114087. doi:10.1109/ACCESS.2020.3002445 PMID:34192108

Ismail, L., Materwala, H., & Al Kaabi, J. (2021). Association of risk factors with type 2 diabetes: A systematic review. *Computational and Structural Biotechnology Journal*, *19*, 1759–1785. doi:10.1016/j.csbj.2021.03.003 PMID:33897980

Ivanoska-Dacikj, A., & Stachewicz, U. (2020). Smart textiles and wearable technologies – opportunities offered in the fight against pandemics in relation to current COVID-19 state. *Reviews on Advanced Materials Science*, *59*(1), 487–505. doi:10.1515/rams-2020-0048

Jackson, J. C., Mitchell, N., & Hopkins, R. O. (2009). Cognitive functioning, mental health, and quality of life in ICU survivors: An overview. *Critical Care Clinics*, *25*(3), 615–628. doi:10.1016/j.ccc.2009.04.005 PMID:19576534

Jaggessar, A., Shahali, H., Mathew, A., & Yarlagadda, P. K. D. V. (2017). Bio-mimicking nano and micro-structured surface fabrication for antibacterial properties in medical implants. *Journal of Nanobiotechnology*, *15*(1), 1–20. doi:10.1186/s12951-017-0306-1 PMID:28969628

Jain, P., Coogan, S. C., Subramanian, S. G., Crowley, M., Taylor, S., & Flannigan, M. D. (2020). A review of machine learning applications in wildfire science and management. *Environmental Reviews*, *28*(4), 478–505. doi:10.1139/er-2020-0019

Jameson, J. L., & Longo, D. L. (2015). Precision medicine—Personalized, problematic, and promising. *The New England Journal of Medicine*, *372*(23), 2229–2234. doi:10.1056/NEJMsb1503104 PMID:26014593

Jamuna, C., & Anburajan, M. (2011, April). Design of patient specific prosthetic aortic valve and to study its computational fluid dynamics. In *2011 3rd International Conference on Electronics Computer Technology* (Vol. 3, pp. 355-360). IEEE. 10.1109/ICECTECH.2011.5941771

Javaid, M., Haleem, A., Singh, R. P., & Suman, R. (2023). Sustaining the healthcare systems through the conceptual of biomedical engineering: A study with recent and future potentials. *Biomedical Technology*, *1*, 39–47. doi:10.1016/j.bmt.2022.11.004

Jha, D., Rauniyar, A., Srivastava, A., Hagos, D. H., Tomar, N. K., Sharma, V., Keles, E., Zhang, Z., Demir, U., Topcu, A., Yazidi, A., Haakegard, J. E., & Bagci, U. (2023). Ensuring Trustworthy Medical Artificial Intelligencethrough Ethical and Philosophical Principles. *arXiv preprint arXiv:2304.11530.*

Jha, S., & Topol, E. J. (2016). Adapting to artificial intelligence: Radiologists and pathologists as information specialists. *Journal of the American Medical Association*, *316*(22), 2353–2354. doi:10.1001/jama.2016.17438 PMID:27898975

Jia, Z., Xiu, P., Li, M., Xu, X., Shi, Y., Cheng, Y., Wei, S., Zheng, Y., Xi, T., Cai, H., & Liu, Z. (2016). Bioinspired anchoring AgNPs onto micro-nanoporous TiO_2 orthopedic coatings: Trap-killing of bacteria, surface-regulated osteoblast functions and host responses. *Biomaterials*, *75*, 203–222. doi:10.1016/j.biomaterials.2015.10.035 PMID:26513414

Jin, B., & Robertson, K. D. (2013). DNA Methyltransferases (DNMTs), DNA Damage Repair, and Cancer. *Advances in Experimental Medicine and Biology*, *754*, 3–29. doi:10.1007/978-1-4419-9967-2_1 PMID:22956494

JM, W. J. B. i. n., & aspects, b. (1988). Iron-dependent enzymes in the brain. 25-66.

Johnson, K. B., Wei, W.-Q., Weeraratne, D., Frisse, M. E., Misulis, K., Rhee, K., Zhao, J., & Snowdon, J. L. (2021). Precision medicine, AI, and the future of personalized health care. *Clinical and Translational Science*, *14*(1), 86–93. doi:10.1111/cts.12884 PMID:32961010

Jones, P. A., & Takai, D. (2001). The role of DNA methylation in mammalian epigenetics. *Science*, *293*(5532), 1068–1070. doi:10.1126/science.1063852 PMID:11498573

Jouffroy, R., Jost, D., & Prunet, B. (2020). Prehospital pulse oximetry: A red flag for early detection of silent hypoxemia in COVID-19 patients. *Critical Care (London, England)*, *24*(1), 313. doi:10.1186/s13054-020-03036-9 PMID:32513249

Junaid, S. B., Imam, A. A., Balogun, A. O., De Silva, L. C., Surakat, Y. A., Kumar, G., & Mahamad, S. (2022, October). Recent advancements in emerging technologies for healthcare management systems: A survey. [). MDPI.]. *Health Care, 10*(10), 1940. PMID:36292387

Jung, J., Kim, S.-H., Lee, H.-S., Choi, G. S., Jung, Y.-S., Ryu, D. H., Park, H.-S., & Hwang, G.-S. (2013). Serum metabolomics reveals pathways and biomarkers associated with asthma pathogenesis. *Clinical and Experimental Allergy, 43*(4), 425–433. doi:10.1111/cea.12089 PMID:23517038

Juyal, S., Sharma, S., & Shukla, A. S. (2021). Smart skin health monitoring using AI-enabled cloud-based IoT. *Materials Today: Proceedings, 46*, 10539–10545. doi:10.1016/j.matpr.2021.01.074

Kamath, P. S., Wiesner, R. H., Malinchoc, M., Kremers, W., Therneau, T. M., & Kosberg, C. L. (2001). A model to predict survival in patients with end-stage liver disease. *Hepatology (Baltimore, Md.), 33*(2), 464–470. doi:10.1053/jhep.2001.22172 PMID:11172350

Kanaga Suba Raja, S., Sandhiya, B., Kausalya, K., Ramya, B., Monish, S., & Sai Raja Gopal, R. (2023). IoT based Smart Door System for Monitoring Human Body Temperature and Face Mask Detection. *International Conference on Advances in Computing, Communication and Applied Informatics (ACCAI)*, Chennai, India. 10.1109/ACCAI58221.2023.10199273

Kanarek, R. B. J. N. r. (1994). Does sucrose or aspartame cause hyperactivity in children?, *52*(5), 173-175.

Kang, G., Liu, K., Hou, B., & Zhang, N. (2017). 3D multi-view convolutional neural networks for lung nodule classification. *PLoS One, 12*(11), e0188290. doi:10.1371/journal.pone.0188290 PMID:29145492

Kaplan, R. C., Aviles-Santa, M. L., Parrinello, C. M., Hanna, D. B., Jung, M., Castaneda, S. F., Hankinson, A. L., Isasi, C. R., Birnbaum-Weitzman, O., Kim, R. S., Daviglus, M. L., Talavera, G. A., Schneiderman, N., & Cai, J. (2014). Body Mass Index, Sex, and Cardiovascular Disease Risk Factors Among Hispanic/Latino Adults: Hispanic Community Health Study/Study of Latinos. *Journal of the American Heart Association, 3*(4), e000923. doi:10.1161/JAHA.114.000923 PMID:25008353

Kaptoge, S., Di Angelantonio, E., Pennells, L., Wood, A. M., White, I. R., & Gao, P. (2012). C-Reactive Protein, Fibrinogen, and Cardiovascular Disease Prediction. *The New England Journal of Medicine, 367*(14), 1310–1320. doi:10.1056/NEJMoa1107477 PMID:23034020

Karar, M. E., Alotaibi, B., & Alotaibi, M. (2022). Intelligent medical IoT-enabled automated microscopic image diagnosis of acute blood cancers. *Sensors (Basel), 22*(6), 2348. doi:10.3390/s22062348 PMID:35336523

Kareem, S. M., Al-Kadmy, I. M., Kazaal, S. S., Mohammed Ali, A. N., Aziz, S. N., Makharita, R. R., Algammal, A. M., Al-Rejaie, S., Behl, T., Batiha, G. E., El-Mokhtar, M. A., & Hetta, H. F. (2021, February 12). Detection of gyrA and parC mutations and prevalence of plasmid-mediated quinolone resistance genes in Klebsiella pneumoniae. *Infection and Drug Resistance, 14*, 555–563. doi:10.2147/IDR.S275852 PMID:33603418

Karim, N., Afroj, S., Lloyd, K., Oaten, L. C., Andreeva, D. V., Carr, C., Farmery, A. D., Kim, I.-D., & Novoselov, K. S. (2020). Sustainable personal protective clothing for healthcare applications: A review. *ACS Nano, 14*(10), 12313–12340. doi:10.1021/acsnano.0c05537 PMID:32866368

Kasai, K., Fukuda, M., Yahata, N., Morita, K., & Fujii, N. (2015). The future of real-world neuroscience: Imaging techniques to assess active brains in social environments. *Neuroscience Research, 90*, 65–71. doi:10.1016/j.neures.2014.11.007 PMID:25433093

Kaufman, D. R., Sheehan, B., Stetson, P., Bhatt, A. R., Field, A. I., Patel, C., & Maisel, J. M. (2016). Natural Language Processing-Enabled and Conventional Data Capture Methods for Input to Electronic Health Records: A Comparative Usability Study. *JMIR Medical Informatics, 4*(4), e35. doi:10.2196/medinform.5544 PMID:27793791

Kaur, A., Mujwar, S., & Adlakha, N. (2016). In-silico analysis of riboswitch of Nocardia farcinica for design of its inhibitors and pharmacophores. *International Journal of Computational Biology and Drug Design*, *9*(3), 9. doi:10.1504/IJCBDD.2016.078278

Kavitha, M. S., Gangadaran, P., Jackson, A., Venmathi, M. B. A., Kurita, T., & Ahn, B. C. (2022). Deep Neural Network Models for Colon Cancer Screening. Cancers (Basel).14(15):3707. *Published*, *2022*(Jul), 29. doi:10.3390/cancers14153707 PMID:35954370

Kc, K., Yin, Z., Wu, M., & Wu, Z. (2021). Evaluation of deep learning-based approaches for COVID-19 classification based on chest X-ray images. *Signal, Image and Video Processing*, *15*(5), 959–966. doi:10.1007/s11760-020-01820-2 PMID:33432267

Keating, D. P. (2004). Handbook of adolescent psychology: Vol. 45-84. *Cognitive and brain development*.

Keating, S. T., van Diepen, J. A., Riksen, N. P., & El-Osta, A. (2018). Epigenetics in diabetic nephropathy, immunity and metabolism. *Diabetologia*, *61*(1), 6–20. doi:10.1007/s00125-017-4490-1 PMID:29128937

Kelly, C. J., Karthikesalingam, A., Suleyman, M., Corrado, G., & King, D. (2019). Key challenges for delivering clinical impact with artificial intelligence. *BMC Medicine*, *17*(1), 1–9. doi:10.1186/s12916-019-1426-2 PMID:31665002

Kendal, E. (2020). Biological modification as prophylaxis: How extreme environments challenge the treatment/enhancement divide. *Human enhancements for space missions: Lunar, Martian, and future missions to the outer planets*, 35-46.

Kepenekian, V., Péron, J., You, B., Bonnefoy, I., Villeneuve, L., Alyami, M., Bakrin, N., Rousset, P., Benzerdjeb, N., & Glehen, O. (2022). Non-resectable malignant peritoneal mesothelioma treated with pressurized intraperitoneal aerosol chemotherapy (PIPAC) plus systemic chemotherapy could lead to secondary complete cytoreductive surgery: A cohort study. *Annals of Surgical Oncology*, *29*(3), 1–10. doi:10.1245/s10434-021-10983-2

Kessler, R. C., Adler, L. A., Barkley, R., Biederman, J., Conners, C. K., Faraone, S. V., & Spencer, T. J. B. p. (2005). Patterns and predictors of attention-deficit/hyperactivity disorder persistence into adulthood: results from the national comorbidity survey replication. *57*(11), 1442-1451.

Khalafvand, S. S., Ng, E. Y. K., Zhong, L., & Hung, T. K. (2017). Three-dimensional diastolic blood flow in the left ventricle. *Journal of Biomechanics*, *50*, 71–76. doi:10.1016/j.jbiomech.2016.11.032 PMID:27939353

Khatoon, Z., McTiernan, C. D., Suuronen, E. J., Mah, T. F., & Alarcon, E. I. (2018). Bacterial biofilm formation on implantable devices and approaches to its treatment and prevention. *Heliyon*, *4*(12), e01067. doi:10.1016/j.heliyon.2018.e01067 PMID:30619958

Khumalo, N. B., & Mnjama, N. (2019). The effect of eHealth information systems on health information management in hospitals in Bulawayo, Zimbabwe. *International Journal of Healthcare Information Systems and Informatics*, *14*(2), 17–27. doi:10.4018/IJHISI.2019040102

King, J., O'Neill, B., Ramsay, P., Linden, M. A., Darweish Medniuk, A., Outtrim, J., & Blackwood, B. (2019). Identifying patients' support needs following critical illness: A scoping review of the qualitative literature. *Critical Care*, *23*(1), 1–12. doi:10.1186/s13054-019-2441-6 PMID:31126335

Kirkpatrick, J., Pascanu, R., Rabinowitz, N., Veness, J., Desjardins, G., Rusu, A. A., Milan, K., Quan, J., Ramalho, T., Grabska-Barwinska, A., Hassabis, D., Clopath, C., Kumaran, D., & Hadsell, R. (2017). Overcoming catastrophic forgetting in neural networks. *Proceedings of the National Academy of Sciences of the United States of America*, *114*(13), 3521–3526. doi:10.1073/pnas.1611835114 PMID:28292907

Klein, M., Onnink, M., van Donkelaar, M., Wolfers, T., Harich, B., Shi, Y., & Reviews, B. (2017). Brain imaging genetics in ADHD and beyond–mapping pathways from gene to disorder at different levels of complexity. *80*, 115-155.

Kleinberger, J. W., & Pollin, T. I. (2015). Personalized medicine in diabetes mellitus: Current opportunities and future prospects. *Annals of the New York Academy of Sciences*, *1346*(1), 45–56. doi:10.1111/nyas.12757 PMID:25907167

Klonoff, D. C. (2008). Personalized Medicine for Diabetes. *Journal of Diabetes Science and Technology*, *2*(3), 335–341. doi:10.1177/193229680800200301 PMID:19885196

Knopik, V. S., Heath, A. C., Jacob, T., Slutske, W. S., Bucholz, K. K., Madden, P. A., & Martin, N. G. J. P. m. (2006). Maternal alcohol use disorder and offspring ADHD: disentangling genetic and environmental effects using a children-of-twins design. *36*(10), 1461-1471.

Knoppers, B. M., Harris, J. R., Budin-Ljøsne, I., Dove, E. S., Ayme, S., Cambon-Thomsen, A., & Harris, H. (2014). A human rights approach to an international code of conduct for genomic and clinical data sharing. *Human Genetics*, *133*(7), 895–903. doi:10.1007/s00439-014-1432-6 PMID:24573176

Kong, L., Zhu, J., Han, W., Jiang, X., Xu, M., Zhao, Y., Dong, Q., Pang, Z., Guan, Q., Gao, L., Zhao, J., & Zhao, L. (2011). Significance of serum microRNAs in pre-diabetes and newly diagnosed type 2 diabetes: A clinical study. *Acta Diabetologica*, *48*(1), 61–69. doi:10.1007/s00592-010-0226-0 PMID:20857148

Kong, X., Gao, P., Wang, J., Fang, Y., & Hwang, K. C. (2023). Advances of medical nanorobots for future cancer treatments. *Journal of Hematology & Oncology*, *16*(1), 74. doi:10.1186/s13045-023-01463-z PMID:37452423

Kontizas, E., Tastsoglou, S., Karamitros, T., Karayiannis, Y., Kollia, P., Hatzigeorgiou, A. G., & Sgouras, D. N. (2020). Impact of Helicobacter pylori infection and its major virulence factor CagA on DNA damage repair. *Microorganisms*, *8*(12), 2007. doi:10.3390/microorganisms8122007

Kopetz, H., & Steiner, W. (2022). Real-Time Communication. In *Real-time systems: Design principles for distributed embedded applications* (pp. 177–200). Springer International Publishing. doi:10.1007/978-3-031-11992-7_7

Krain, A. L., & Castellanos, F. X. J. C. p. r. (2006). Brain development and ADHD. *26*(4), 433-444.

Kramer, A. F., & Erickson, K. I. (2007). Capitalizing on cortical plasticity: Influence of physical activity on cognition and brain function. *Trends in Cognitive Sciences*, *11*(8), 342–348. doi:10.1016/j.tics.2007.06.009 PMID:17629545

Krishnan, G., Singh, S., Pathania, M., Gosavi, S., Abhishek, S., Parchani, A., & Dhar, M. (2023). Artificial intelligence in clinical medicine: Catalyzing a sustainable global healthcare paradigm. *Frontiers in Artificial Intelligence*, *6*, 6. doi:10.3389/frai.2023.1227091 PMID:37705603

Krizhevsky, A., Sutskever, I., & Hinton, G. E. (2012). Imagenet classification with deep convolutional neural networks. *Advances in Neural Information Processing Systems*, 25.

Kron, T., & Krishnan, P. (2019). Leonardo DaVinci's contributions to medical physics and biomedical engineering: Celebrating the life of a 'Polymath'. *Australasian Physical & Engineering Sciences in Medicine*, *42*(2), 403–405. doi:10.1007/s13246-019-00757-2 PMID:31049840

Krueger, F., & Andrews, S. R. (2011). Bismark: A flexible aligner and methylation caller for Bisulfite-Seq applications. *Bioinformatics (Oxford, England)*, *27*(11), 1571–1572. doi:10.1093/bioinformatics/btr167 PMID:21493656

Kumar, K. (2021). Comprehensive Method of Botnet Detection Using Machine Learning. [IJOSSP]. *International Journal of Open Source Software and Processes*, *12*(4), 37–61. doi:10.4018/IJOSSP.287613

Kumar, K., & Bhatnagar, V. (2021). Credit Card Fraud Transaction Detection System Using Neural Network-Based Sequence Classification Technique. [IJOSSP]. *International Journal of Open Source Software and Processes*, *12*(1), 21–40. doi:10.4018/IJOSSP.2021010102

Kumar, S., Behl, T., Sachdeva, M., Sehgal, A., Kumari, S., Kumar, A., Kaur, G., Yadav, H. N., & Bungau, S. (2021, January 1). Implicating the effect of ketogenic diet as a preventive measure to obesity and diabetes mellitus. *Life Sciences*, *264*, 118661. doi:10.1016/j.lfs.2020.118661 PMID:33121986

Kumar, Y., Koul, A., Singla, R., & Ijaz, M. F. (2023). Artificial intelligence in disease diagnosis: A systematic literature review, synthesizing framework and future research agenda. *Journal of Ambient Intelligence and Humanized Computing*, *14*(7), 8459–8486. doi:10.1007/s12652-021-03612-z PMID:35039756

Kurdyukov, S., & Bullock, M. (2016). DNA Methylation Analysis: Choosing the Right Method. *Biology (Basel)*, *5*(1), 3. doi:10.3390/biology5010003 PMID:26751487

Kwon, Y. H., Fingert, J. H., Kuehn, M. H., & Alward, W. L. (2009). Primary open-angle glaucoma. *The New England Journal of Medicine*, *360*(11), 1113–1124. doi:10.1056/NEJMra0804630 PMID:19279343

Kyriazis, M. (2015). Systems neuroscience in focus: From the human brain to the global brain? *Frontiers in Systems Neuroscience*, *9*, 7. doi:10.3389/fnsys.2015.00007 PMID:25705180

La Thangue, N. B., & Kerr, D. J. (2011). Predictive biomarkers: A paradigm shift towards personalized cancer medicine. *Nature Reviews. Clinical Oncology*, *8*(10), 587–596. doi:10.1038/nrclinonc.2011.121 PMID:21862978

Lacaze, P., Marquina, C., Tiller, J., Brotchie, A., Kang, Y. J., Merritt, M. A., Green, R. C., Watts, G. F., Nowak, K. J., Manchanda, R., Canfell, K., James, P., Winship, I., McNeil, J. J., & Ademi, Z. (2023). Combined population genomic screening for three high-risk conditions in Australia: A modelling study. *EClinicalMedicine*, *66*, 102297.. doi:10.1016/j.eclinm.2023.102297

Lai, C. T. A., Jiang, W., & Jackson, P. R. (2019). Internet of Things enabling condition-based maintenance in elevators service. *Journal of Quality in Maintenance Engineering*, *25*(4), 563–588. doi:10.1108/JQME-06-2018-0049

Lamia, A., & Fawaz, A. (2022). Detection of Pneumonia Infection by Using Deep Learning on a Mobile Platform. *Computational Intelligence and Neuroscience*, *7925668*, 1–9. doi:10.1155/2022/7925668 PMID:35942467

Lamsal, R., Higuchi, Y., & Umezawa, K. (2020). Applications of precision medicine in infectious diseases. *Advances in Experimental Medicine and Biology*, *1255*, 153–174.

Lancet, C.-D. G. o. t. P. G. C. J. T. (2013). Identification of risk loci with shared effects on five major psychiatric disorders: a genome-wide analysis. *381*(9875), 1371-1379.

Lange, K. W., Hauser, J., Lange, K. M., Makulska-Gertruda, E., Nakamura, Y., Reissmann, A., & Takeuchi, Y. J. C. p. r. (2017). The role of nutritional supplements in the treatment of ADHD: what the evidence says. *19*, 1-9.

Lange, K. W., Lange, K. M., Nakamura, Y., & Reissmann, A. J. C. N. R. (2023). Nutrition in the Management of ADHD: A Review of Recent Research. *12*(3), 383-394.

Lange, K. W., Nakamura, Y., & Reissmann, A. J. J. o. F. F. (2022). Diet and food in attention-deficit hyperactivity disorder. *2*(2), 112-118.

Lazaro, M., Lazaro, A., Villarino, R., & Girbau, D. (2021b). *Smart mask for temperature monitoring with LoRa backscattering communication.* 2021 6th International Conference on Smart and Sustainable Technologies (SpliTech), Bol and Split, Croatia. 10.23919/SpliTech52315.2021.9566327

Lazaro, M., Lázaro, A., Villarino, R., & Girbau, D. (2021a). Smart Face Mask with an Integrated Heat Flux Sensor for Fast and Remote People's Healthcare Monitoring. *Sensors (Basel)*, *21*(22), 7472. doi:10.3390/s21227472 PMID:34833547

Lee, H., Yang, H. L., Ryu, H. G., Jung, C. W., Cho, Y. J., Yoon, S. B., Yoon, H.-K., & Lee, H. C. (2023). Real-time machine learning model to predict in-hospital cardiac arrest using heart rate variability in ICU. *NPJ Digital Medicine*, *6*(1), 215. doi:10.1038/s41746-023-00960-2 PMID:37993540

Lee, J., Ni, J., Singh, J., Jiang, B., Azamfar, M., & Feng, J. (2020). Intelligent maintenance systems and predictive manufacturing. *Journal of Manufacturing Science and Engineering*, *142*(11), 110805. doi:10.1115/1.4047856

Lee, P., Kim, H., Kim, Y., Choi, W., Zitouni, M., Khandoker, A., Jelinek, H., Hadjileontiadis, L., Lee, U., & Jeong, Y. (2022). Beyond Pathogen Filtration: Possibility of Smart Masks as Wearable Devices for Personal and Group Health and Safety Management. *JMIR mHealth and uHealth*, *10*(6), e38614. doi:10.2196/38614 PMID:35679029

Lee, R., Wong, T. Y., & Sabanayagam, C. (2015). Epidemiology of diabetic retinopathy, diabetic macular edema and related vision loss. *Eye and Vision (London, England)*, *2*(1), 1–25. doi:10.1186/s40662-015-0026-2 PMID:26605370

Lee, Y. Y., & Lin, J. L. (2009). The effects of trust in physician on self-efficacy, adherence and diabetes outcomes. *Social Science & Medicine*, *68*(6), 1060–1068. doi:10.1016/j.socscimed.2008.12.033 PMID:19162386

Leigh, N., Solomon, D., Pletcher, E., Labow, D. M., Magge, D. R., Sarpel, U., & Golas, B. J. (2020). Is cytoreductive surgery and hyperthermic intraperitoneal chemotherapy indicated in hepatobiliary malignancies? *World Journal of Surgical Oncology*, *18*(1), 1–8. doi:10.1186/s12957-020-01898-5

Lemeshow, S. T. A. N., & Hosmer, D. W. Jr. (1982). A Review of Goodness of fit Statistics for use in the Development of Logistic Regression Models. *American Journal of Epidemiology*, *115*(1), 92–106. doi:10.1093/oxfordjournals.aje.a113284 PMID:7055134

Lemmens, L. C., de Bruin, S. R., Struijs, J. N., Rijken, M., Nijpels, G., & Baan, C. A. (2015). Patient involvement in diabetes care: Experiences in nine diabetes care groups. *International Journal of Integrated Care*, *15*(4), e044. doi:10.5334/ijic.2207 PMID:27118961

Letterie, G. (2021). Three ways of knowing: The integration of clinical expertise, evidence-based medicine, and artificial intelligence in assisted reproductive technologies. *Journal of Assisted Reproduction and Genetics*, *38*(7), 1617–1625. doi:10.1007/s10815-021-02159-4 PMID:33870475

Leung, T., & Vyas, D. (2014). Robotic surgery: Applications. *American Journal of Robotic Surgery*, *1*(1), 38–41. doi:10.1166/ajrs.2014.1004 PMID:26501128

Li, B. H., Zhang, L., & Chai, X. (2014). Smart Cloud Manufacturing (Cloud Manufacturing 2.0)-A New Paradigm and Approach of Smart Manufacturing. In ISPE CE 2014 Sep 23 (p. 26).

Liao, J., Li, X., Gan, Y., Han, S., Rong, P., Wang, W., Li, W., & Zhou, L. (2023). Artificial intelligence assists precision medicine in cancer treatment. *Frontiers in Oncology*, *12*, 998222. doi:10.3389/fonc.2022.998222 PMID:36686757

Liao, M.-S., & Scheiner, S. (2002). Electronic structure and bonding in metal porphyrins, metal = Fe, Co, Ni, Cu, Zn. *The Journal of Chemical Physics*, *117*(1), 205–219. doi:10.1063/1.1480872

Li, B. H., Hou, B. C., Yu, W. T., Lu, X. B., & Yang, C. W. (2017). Applications of artificial intelligence in intelligent manufacturing: A review. *Frontiers of Information Technology & Electronic Engineering*, *18*(1), 86–96. doi:10.1631/FITEE.1601885

Li, B., Ma, J., Wang, D., Liu, X., Li, H., Zhou, L., Liang, C., & Wang, H. (2019). Self-adjusting antibacterial properties of Ag-incorporated nanotubes on micro-nanostructured Ti surfaces. *Biomaterials Science*, *7*(10), 4075–4087. doi:10.1039/C9BM00862D PMID:31355387

Liberto, J. M., Chen, S. Y., Shih, I. M., Wang, T. H., Wang, T. L., & Pisanic, T. R. II. (2022). Current and emerging methods for ovarian cancer screening and diagnostics: A comprehensive review. *Cancers (Basel)*, *14*(12), 2885. doi:10.3390/cancers14122885

Li, J., Ma, Q., Chan, A. H., & Man, S. (2019). Health monitoring through wearable technologies for older adults: Smart wearables acceptance model. *Applied Ergonomics*, *75*, 162–169. doi:10.1016/j.apergo.2018.10.006 PMID:30509522

Li, J., Sun, H., Huang, Y., Wang, Y., Liu, Y., & Chen, X. (2019). Pathways and assays for DNA double-strand break repair by homologous recombination. *Acta Biochimica et Biophysica Sinica*, *51*(9), 879–889. doi:10.1093/abbs/gmz076

Li, J., Yin, J., Ramakrishna, S., & Ji, D. (2023, January 30). Smart Mask as Wearable for Post-Pandemic Personal Healthcare. *Biosensors*, *13*(2), 205. doi:10.3390/bios13020205 PMID:36831971

Lillycrop, K. A., Slater-Jefferies, J. L., Hanson, M. A., Godfrey, K. M., Jackson, A. A., & Burdge, G. C. (2007). Induction of altered epigenetic regulation of the hepatic glucocorticoid receptor in the offspring of rats fed a protein-restricted diet during pregnancy suggests that reduced DNA methyltransferase-1 expression is involved in impaired DNA methylation and changes in histone modifications. *British Journal of Nutrition*, *97*(6), 1064–1073. doi:10.1017/S000711450769196X PMID:17433129

Li, M., Zhang, D., Peng, F., Xie, J., Zhang, X., Qian, S., Zhang, Y., Liu, X., & Yu, B. (2022). Zinc-doped ferric oxyhydroxide nano-layer enhances the bactericidal activity and osseointegration of a magnesium alloy through augmenting the formation of neutrophil extracellular traps. *Acta Biomaterialia*, *152*, 575–592. doi:10.1016/j.actbio.2022.08.066 PMID:36070834

Lim, K. Y. H., Zheng, P., Chen, C. H., & Huang, L. (2020). A digital twin-enhanced system for engineering product family design and optimization. *Journal of Manufacturing Systems*, *57*, 82–93. doi:10.1016/j.jmsy.2020.08.011

Lindsey, R., Daluiski, A., Chopra, S., Lachapelle, A., Mozer, M., Sicular, S., Hanel, D., Gardner, M., Gupta, A., Hotchkiss, R., & Potter, H. (2018). Deep neural network improves fracture detection by clinicians. *Proceedings of the National Academy of Sciences of the United States of America*, *115*(45), 11591–11596. doi:10.1073/pnas.1806905115 PMID:30348771

Ling, C., & Rönn, T. (2019). Epigenetics in Human Obesity and Type 2 Diabetes. *Cell Metabolism*, *29*(5), 1028–1044. doi:10.1016/j.cmet.2019.03.009 PMID:30982733

Lin, H., Li, R., Liu, Z., Chen, J., Yang, Y., Chen, H., & Wu, X. (2019). Diagnostic efficacy and therapeutic decision-making capacity of an artificial intelligence platform for childhood cataracts in eye clinics: A multicentre randomized controlled trial. *EClinicalMedicine*, *9*, 52–59. doi:10.1016/j.eclinm.2019.03.001 PMID:31143882

Lin, L., Huang, Z., Gao, Y., Yan, X., Xing, J., & Hang, W. (2011). LC-MS based serum metabonomic analysis for renal cell carcinoma diagnosis, staging, and biomarker discovery. *Journal of Proteome Research*, *10*(3), 1396–1405. doi:10.1021/pr101161u PMID:21186845

Lin, M. H., Wang, Y. H., Kuo, C. H., Ou, S. F., Huang, P. Z., Song, T. Y., Chen, Y. C., Chen, S. T., Wu, C. H., Hsueh, Y. H., & Fan, F. Y. (2021). Hybrid ZnO/chitosan antimicrobial coatings with enhanced mechanical and bioactive properties for titanium implants. *Carbohydrate Polymers*, *257*, 117639. doi:10.1016/j.carbpol.2021.117639 PMID:33541664

Lin, X., Yang, S., Lai, K., Yang, H., Webster, T. J., & Yang, L. (2017). Orthopedic implant biomaterials with both osteogenic and anti-infection capacities and associated *in vivo* evaluation methods. *Nanomedicine; Nanotechnology, Biology, and Medicine*, *13*(1), 123–142. doi:10.1016/j.nano.2016.08.003 PMID:27553074

Lin, Z., Zhao, Y., Chu, P. K., Wang, L., Pan, H., Zheng, Y., Wu, S., Liu, X., Cheung, K. M. C., Wong, T., & Yeung, K. W. K. (2019). A functionalized TiO_2/Mg_2TiO_4 nano-layer on biodegradable magnesium implant enables superior bone-implant integration and bacterial disinfection. *Biomaterials, 219*, 119372. doi:10.1016/j.biomaterials.2019.119372 PMID:31362176

Liu, H. W., Wei, D. X., Deng, J. Z., Zhu, J. J., Xu, K., Hu, W. H., Xiao, S. H., & Zhou, Y. G. (2018). Combined antibacterial and osteogenic in situ effects of a bifunctional titanium alloy with nanoscale hydroxyapatite coating. *Artificial Cells, Nanomedicine and Biotechnology, 46*(sup3), S460–S470. doi:10.1080/21691401.2018.1499662

Liu, J., Liu, J., Attarilar, S., Wang, C., Tamaddon, M., Yang, C., Xie, K., Yao, J., Wang, L., Liu, C., & Tang, Y. (2020). Nano-Modified Titanium Implant Materials: A Way Toward Improved Antibacterial Properties. *Frontiers in Bioengineering and Biotechnology, 8*(November), 576969. doi:10.3389/fbioe.2020.576969 PMID:33330415

Liu, W., Li, J., Cheng, M., Wang, Q., Qian, Y., Yeung, K. W. K., Chu, P. K., & Zhang, X. (2019). A surface-engineered polyetheretherketone biomaterial implant with direct and immunoregulatory antibacterial activity against methicillin-resistant *Staphylococcus aureus. Biomaterials, 208*, 8–20. doi:10.1016/j.biomaterials.2019.04.008 PMID:30986611

Liu, X., Chen, H. H., Lin, Y. C., Nabilla, S. C., Liu, W. C., Wang, W. C., Shih, S. J., Li, Y., Lin, C. P., Zhao, G., & Chung, R. J. (2018). Composite polyelectrolyte multilayer and mesoporous bioactive glass nanoparticle coating on 316L stainless steel for controlled antibiotic release and biocompatibility. *Journal of Biomedical Nanotechnology, 14*(4), 725–735. doi:10.1166/jbn.2018.2531 PMID:31352946

Liu, Y., Kohlberger, T., Norouzi, M., Dahl, G. E., Smith, J. L., Mohtashamian, A., Olson, N., Peng, L. H., Hipp, J. D., & Stumpe, M. C. (2019). Artificial intelligence–based breast cancer nodal metastasis detection: Insights into the black box for pathologists. *Archives of Pathology & Laboratory Medicine, 143*(7), 859–868. doi:10.5858/arpa.2018-0147-OA PMID:30295070

Li, Y., Liu, X., Li, B., Zheng, Y., Han, Y., Chen, D. F., Yeung, K. W. K., Cui, Z., Liang, Y., Li, Z., Zhu, S., Wang, X., & Wu, S. (2020). Near-Infrared Light Triggered Phototherapy and Immunotherapy for Elimination of Methicillin-Resistant *Staphylococcus aureus* Biofilm Infection on Bone Implant. *ACS Nano, 14*(7), 8157–8170. doi:10.1021/acsnano.0c01486 PMID:32585104

Li, Ya., Yang, Q., Liu, S., Liu, X., Wu, Z., Wang, J., Zhou, M., Ren, M., Wan, C., Huo, W., Gao, Y., & Huang, X. (2022). A Self-Purifying Smart Mask Integrated With Metal–Organic Framework Membranes and Flexible Circuits. *Advanced Materials Interfaces, 10*(4), 2201895. doi:10.1002/admi.202201895

Llueca, A., Climent, M. T., Escrig, J., Carrasco, P., Serra, A., Gomez-Quiles, L., Játiva, R., Cebrian, G., Bosso, V., Villarin, A., Maiocchi, K., Delgado-Barriga, K., Rodrigo-Aliaga, M., Ruiz, N., Herrero, C., Frances, A., Beato, I., Ferrer, C., Aracil, J. P., & Adell, R. (2021). Validation of three predictive models for suboptimal cytoreductive surgery in advanced ovarian cancer. *Scientific Reports, 11*(1), 8111. doi:10.1038/s41598-021-86928-2

Lok, R., Karacali, H., Varol, A., Camli, U., & Yilmaz, E. (2022). Fabrication and characterization of resistance temperature detector by smart mask design. *International Journal of Advanced Manufacturing Technology, 122*(1), 147–158. doi:10.1007/s00170-022-09041-2

Loncar-Turukalo, T., Zdravevski, E., Machado da Silva, J., Chouvarda, I., & Trajkovik, V. (2019). Literature on Wearable Technology for Connected Health: Scoping Review of Research Trends, Advances, and Barriers. *Journal of Medical Internet Research, 21*(9), e14017. doi:10.2196/14017 PMID:31489843

Long, E., Lin, H., Liu, Z., Wu, X., Wang, L., Jiang, J., & Li, J. (2017). An artificial intelligence platform for the multihospital collaborative management of congenital cataracts. *Nature Biomedical Engineering, 1*(2), 0024.

Longest, P. W., Tian, G., Walenga, R. L., & Hindle, M. (2012). Comparing MDI and DPI aerosol deposition using in vitro experiments and a new stochastic individual path (SIP) model of the conducting airways. *Pharmaceutical Research, 29*(6), 1670–1688. doi:10.1007/s11095-012-0691-y PMID:22290350

Lotfali, H., & Meshkini, A. (2021). Synthesis and characterization of lysozyme-conjugated Ag.ZnO@HA nanocomposite: A redox and pH-responsive antimicrobial agent with photocatalytic activity. *Photodiagnosis and Photodynamic Therapy, 35*, 102418. doi:10.1016/j.pdpdt.2021.102418 PMID:34197967

Lott, J. A., Mitchell, L. C., Moeschberger, M. L., & Sutherland, D. E. (1992). Estimation of reference ranges: How many subjects are needed? *Clinical Chemistry, 38*(5), 648–650. doi:10.1093/clinchem/38.5.648 PMID:1582014

Lubans, D., Richards, J., Hillman, C., Faulkner, G., Beauchamp, M., Nilsson, M., Kelly, P., Smith, J., Raine, L., & Biddle, S. (2016). Physical activity for cognitive and mental health in youth: A systematic review of mechanisms. *Pediatrics, 138*(3), e20161642. doi:10.1542/peds.2016-1642 PMID:27542849

Lundberg, S. M., Nair, B., Vavilala, M. S., Horibe, M., Eisses, M. J., Adams, T., Liston, D. E., Low, D. K., Newman, S. F., Kim, J., & Lee, S. I. (2018). Explainable machine-learning predictions for the prevention of hypoxaemia during surgery. *Nature Biomedical Engineering, 2*(10), 749–760. doi:10.1038/s41551-018-0304-0 PMID:31001455

Mahajan, A., Taliun, D., Thurner, M., Robertson, N. R., Torres, J. M., Rayner, N. W., Payne, A. J., Steinthorsdottir, V., Scott, R. A., Grarup, N., Cook, J. P., Schmidt, E. M., Wuttke, M., Sarnowski, C., Mägi, R., Nano, J., Gieger, C., Trompet, S., Lecoeur, C., & McCarthy, M. I. (2018). Fine-mapping type 2 diabetes loci to single-variant resolution using high-density imputation and islet-specific epigenome maps. *Nature Genetics, 50*(11), 11. doi:10.1038/s41588-018-0241-6 PMID:30297969

Mahamuni-Badiger, P. P., Patil, P. M., Badiger, M. V., Patel, P. R., Thorat-Gadgil, B. S., Pandit, A., & Bohara, R. A. (2020). Biofilm formation to inhibition: Role of zinc oxide-based nanoparticles. *Materials Science and Engineering C, 108*, 110319. doi:10.1016/j.msec.2019.110319 PMID:31923962

Mahulea, C., Mahulea, L., García-Soriano, J. M., & Colom, J. M. (2014, October). Petri nets with resources for modeling primary healthcare systems. In *2014 18th International Conference on System Theory, Control and Computing (ICSTCC)* (pp. 639-644). IEEE. 10.1109/ICSTCC.2014.6982489

Majumder, S., Mondal, T., & Deen, M. J. (2017). Wearable Sensors for Remote Health Monitoring. *Sensors (Basel), 17*(1), 130. doi:10.3390/s17010130 PMID:28085085

Makris, N., Buka, S. L., Biederman, J., Papadimitriou, G. M., Hodge, S. M., Valera, E. M., & Caviness, V. S. J. C. c. (2008). Attention and executive systems abnormalities in adults with childhood ADHD: a DT-MRI study of connections. *18*(5), 1210-1220.

Maleki, R., Asadnia, M., & Razmjou, A. (2022). Artificial Intelligence-Based Material Discovery for Clean Energy Future. *Advanced Intelligent Systems, 4*(10), 2200073. doi:10.1002/aisy.202200073

Maleki, Z., Vali, M., Nikbakht, H. A., Hassanipour, S., Kouhi, A., Sedighi, S., Farokhi, R., & Ghaem, H. (2023). Survival rate of ovarian cancer in Asian countries: A systematic review and meta-analysis. *BMC Cancer, 23*(1), 1–11. doi:10.1186/s12885-023-11041-8

Malik, A. I., Zai, C. C., Abu, Z., Nowrouzi, B., & Beitchman, J. H. (2012). The role of oxytocin and oxytocin receptor gene variants in childhood-onset aggression. *Genes Brain & Behavior, 11*(5), 545–551. doi:10.1111/j.1601-183X.2012.00776.x PMID:22372486

Malin, B. A., & Sweeney, L. (2004). How (not) to protect genomic data privacy in a distributed network: Using trail re-identification to evaluate and design anonymity protection systems. *Journal of Biomedical Informatics*, *37*(3), 179–192. doi:10.1016/j.jbi.2004.04.005 PMID:15196482

Malottki, K., Biswas, M., Deeks, J. J., Riley, R. D., Craddock, C., Johnson, P., & Billingham, L. (2014). Stratified medicine in European Medicines Agency licensing: A systematic review of predictive biomarkers. *BMJ Open*, *4*(1), 4. doi:10.1136/bmjopen-2013-004188 PMID:24468721

Malve, M., Chandra, S., Lopez-Villalobos, J. L., Finol, E. A., Ginel, A., & Doblare, M. (2013). CFD analysis of the human airways under impedance-based boundary conditions: Application to healthy, diseased and stented trachea. *Computer Methods in Biomechanics and Biomedical Engineering*, *16*(2), 198–216. doi:10.1080/10255842.2011.615743 PMID:22220946

Mandal, B., Sarkar, S., & Sarkar, P. (2015). Theoretical Studies on Understanding the Feasibility of Porphyrin-Sensitized Graphene Quantum Dot Solar Cell. *The Journal of Physical Chemistry. C, Nanomaterials and Interfaces*, *119*(6), 3400–3407. doi:10.1021/jp511375a

Mapstone, M., Cheema, A. K., Fiandaca, M. S., Zhong, X., Mhyre, T. R., MacArthur, L. H., Hall, W. J., Fisher, S. G., Peterson, D. R., Haley, J. M., Nazar, M. D., Rich, S. A., Berlau, D. J., Peltz, C. B., Tan, M. T., Kawas, C. H., & Federoff, H. J. (2014). Plasma phospholipids identify antecedent memory impairment in older adults. *Nature Medicine*, *20*(4), 415–418. doi:10.1038/nm.3466 PMID:24608097

Ma, R., Vanstrum, E. B., Lee, R., Chen, J., & Hung, A. J. (2020). Machine learning in the optimization of robotics in the operative field. *Current Opinion in Urology*, *30*(6), 808–816. doi:10.1097/MOU.0000000000000816 PMID:32925312

MarcusG. (2018). Deep learning: A critical appraisal. arXiv. https://arxiv.org/abs/1801.00631

Mardonova, M., & Choi, Y. (2018). Review of wearable device technology and its applications to the mining industry. *Energies*, *11*(3), 547. doi:10.3390/en11030547

Markets and Markets. (2021). The global peripheral vascular devices market in terms of revenue was estimated to be worth $10.3 billion in 2021 and is poised to reach $13.4 billion by 2026, growing at a CAGR of 5.3% from 2021 to 2026. *Market research Report*. https://www.marketsandmarkets.com/Market-Reports/interventional-cardiology-peripheral-vascular-devices-market-790.html

Maron, B. A., Loscalzo, J., & Thiene, G. (2017). Moving beyond the alphabet soup of arrhythmias in cardiomyopathies: JACC Review Topic of the Week. *Journal of the American College of Cardiology*, *70*(19), 2425–2434.

Marr, B. (2022). How AI and machine learning will impact the future of healthcare. *Forbes*. https://www.forbes.com/sites/bernardmarr/2022/09/14/how-ai-and-machine-learning-will-impact-the-future-of-healthcare/?sh=3e9c257047e5

Martin, J., Walters, R. K., Demontis, D., Mattheisen, M., Lee, S. H., Robinson, E., & Lichtenstein, P. J. B. p. (2018). A genetic investigation of sex bias in the prevalence of attention-deficit/hyperactivity disorder. *83*(12), 1044-1053.

Masna, N. V. R., Kalavakonda, R. R., Dizon-Paradis, R. N., & Bhunia, S. (2021). Smart and Connected Mask for Protection beyond the Pandemic: (Invited Paper). *IEEE International Midwest Symposium on Circuits and Systems (MWSCAS)*. 10.1109/MWSCAS47672.2021.9531802

Mastronardi, C. A., Pillai, E., Pineda, D. A., Martinez, A. F., Lopera, F., Velez, J. I., & Acosta, M. T. J. M. p. (2016). Linkage and association analysis of ADHD endophenotypes in extended and multigenerational pedigrees from a genetic isolate. *21*(10), 1434-1440.

Matsas, A., Stefanoudakis, D., Troupis, T., Kontzoglou, K., Eleftheriades, M., Christopoulos, P., Panoskaltsis, T., Stamoula, E., & Iliopoulos, D. C. (2023). Tumor Markers and Their Diagnostic Significance in Ovarian Cancer. *Life (Basel, Switzerland)*, *13*(8), 1689. doi:10.3390/life13081689

Matsudaira, T. J. N., & health. (2007). Attention deficit disorders–drugs or nutrition?, *19*(1-2), 57-60.

Matsui, S., Terada, T., & Tsukamoto, M. (2017). Smart eye mask: eye-mask shaped sleep monitoring device. *Proceedings of the 2017 ACM International Joint Conference on Pervasive and Ubiquitous Computing and Proceedings of the 2017 ACM International Symposium on Wearable Computers*. ACM. 10.1145/3123024.3129264

McDermott, J. E., Wang, J., Mitchell, H., Webb-Robertson, B. J., Hafen, R., Ramey, J., & Rodland, K. D. (2013). Challenges in biomarker discovery: Combining expert insights with statistical analysis of complex omics data. *Expert Opinion on Medical Diagnostics*, *7*(1), 37–51. doi:10.1517/17530059.2012.718329 PMID:23335946

McDonough, I. M., Erwin, H. B., Sin, N. L., & Allen, R. S. (2022). Pet ownership is associated with greater cognitive and brain health in a cross-sectional sample across the adult lifespan. *Frontiers in Aging Neuroscience*, *14*, 953889. doi:10.3389/fnagi.2022.953889 PMID:36337704

McGuire, A. L., & Beskow, L. M. (2010). Informed consent in genomics and genetic research. *Annual Review of Genomics and Human Genetics*, *11*(1), 361–381. doi:10.1146/annurev-genom-082509-141711 PMID:20477535

McKinney, S. M., Sieniek, M., Godbole, V., Godwin, J., Antropova, N., Ashrafian, H., Back, T., Chesus, M., Corrado, G. S., Darzi, A., Etemadi, M., Garcia-Vicente, F., Gilbert, F. J., Halling-Brown, M., Hassabis, D., Jansen, S., Karthikesalingam, A., Kelly, C. J., King, D., & Shetty, S. (2020). International evaluation of an AI system for breast cancer screening. *Nature*, *577*(7788), 89–94. doi:10.1038/s41586-019-1799-6 PMID:31894144

McPherson, K. S., & Korzhnev, D. M. (2021). Targeting protein-protein interactions in the DNA damage response pathways for cancer chemotherapy. *RSC Chemical Biology*, *2*(4), 1167–1195. doi:10.1039/D1CB00101A

McShane, L. M., Altman, D. G., Sauerbrei, W., Taube, S. E., Gion, M., & Clark, G. M. (2005). Reporting Recommendations for Tumor Marker Prognostic Studies. *Journal of Clinical Oncology*, *23*(36), 9067–9072. doi:10.1200/JCO.2004.01.0454 PMID:16172462

Mega, J. L., Close, S. L., Wiviott, S. D., Shen, L., Hockett, R. D., Brandt, J. T., Walker, J. R., Antman, E. M., Macias, W. L., Braunwald, E., & Sabatine, M. S. (2009). Cytochrome P450 genetic polymorphisms and the response to prasugrel: Relationship to pharmacokinetic, pharmacodynamic, and clinical outcomes. *Circulation*, *119*(19), 2553–2560. doi:10.1161/CIRCULATIONAHA.109.851949 PMID:19414633

Mega, J. L., & Sabatine, M. S. (2011). Antithrombotic therapy in patients with myocardial infarction. *Circulation Research*, *109*(10), 1060–1076.

Mehrjou, B., Mo, S., Dehghan-Baniani, D., Wang, G., Qasim, A. M., & Chu, P. K. (2019). Antibacterial and Cytocompatible Nanoengineered Silk-Based Materials for Orthopedic Implants and Tissue Engineering. *ACS Applied Materials & Interfaces*, *11*(35), 31605–31614. doi:10.1021/acsami.9b09066 PMID:31385497

Mekid, S., & Chenaoua, K. (2023). IoT-Enabled Smart Mask for Monitoring Body Parameters and Location through Cloud. *Internet of Things : Engineering Cyber Physical Human Systems*, *22*, 100794. doi:10.1016/j.iot.2023.100794 PMID:37266184

Metzker, M. L. (2010). Sequencing technologies—The next generation. *Nature Reviews. Genetics*, *11*(1), 31–46. doi:10.1038/nrg2626 PMID:19997069

Michalek, D. A., Onengut-Gumuscu, S., Repaske, D. R., & Rich, S. S. (2023). Precision Medicine in Type 1 Diabetes. *Journal of the Indian Institute of Science*, *103*(1), 335–351. doi:10.1007/s41745-023-00356-x PMID:37538198

Mijwil, M. M., & Abttan, R. A. (2021). Utilisation of machine learning techniques in testing and training of different medical datasets. *Asian Journal of Computer and Information Systems*, *9*(4). doi:10.24203/ajcis.v9i4.6765

Milberger, S., Biederman, J., Faraone, S. V., & Jones, J. J. J. o. c. c. p. (1998). Further evidence of an association between maternal smoking during pregnancy and attention deficit hyperactivity disorder: findings from a high-risk sample of siblings. *27*(3), 352-358.

Milberger, S., Biederman, J., Faraone, S. V., Chen, L., & Jones, J. J. T. A. j. o. p. (1996). Is maternal smoking during pregnancy a risk factor for attention deficit hyperactivity disorder in children?, *153*(9), 1138-1142.

Miller, D. D., & Brown, E. W. (2018). Artificial intelligence in medical practice: The question to the answer? *The American Journal of Medicine*, *131*(2), 129–133. doi:10.1016/j.amjmed.2017.10.035 PMID:29126825

Minaam, D. S. A., & Abd-ELfattah, M. (2018). Smart drugs: Improving healthcare using smart pill box for medicine reminder and monitoring system. *Future Computing and Informatics Journal*, *3*(2), 443–456. doi:10.1016/j.fcij.2018.11.008

Min, J., Choi, K. Y., Dreaden, E. C., Padera, R. F., Braatz, R. D., Spector, M., & Hammond, P. T. (2016). Designer Dual Therapy Nanolayered Implant Coatings Eradicate Biofilms and Accelerate Bone Tissue Repair. *ACS Nano*, *10*(4), 4441–4450. doi:10.1021/acsnano.6b00087 PMID:26923427

Misra, S., Wagner, R., Ozkan, B., Schön, M., Sevilla-Gonzalez, M., Prystupa, K., Wang, C. C., Kreienkamp, R. J., Cromer, S. J., Rooney, M. R., Duan, D., Thuesen, A. C. B., Wallace, A. S., Leong, A., Deutsch, A. J., Andersen, M. K., Billings, L. K., Eckel, R. H., Sheu, W. H.-H., & Udler, M. S. (2023). Precision subclassification of type 2 diabetes: A systematic review. *Communications Medicine*, *3*(1), 1. doi:10.1038/s43856-023-00360-3 PMID:37798471

Mobley, R. K. (2002). *An introduction to predictive maintenance*. Elsevier.

Mohamadou, Y., Halidou, A., & Kapen, P. T. (2020). A review of mathematical modeling, artificial intelligence, and datasets used in the study, prediction, and management of COVID-19. *Applied Intelligence*, *50*(11), 3913–3925. doi:10.1007/s10489-020-01770-9 PMID:34764546

Mohammadzadeh, N., Gholamzadeh, M., Saeedi, S., & Rezayi, S. (2020). The application of wearable smart sensors for monitoring the vital signs of patients in epidemics: A systematic literature review. *Journal of Ambient Intelligence and Humanized Computing*, 1–15. PMID:33224305

Mohanta, B., Das, P., & Patnaik, S. (2019, May). Healthcare 5.0: A paradigm shift in digital healthcare system using artificial intelligence, IOT and 5G communication. In *2019 International Conference on Applied Machine Learning (ICAML)* (pp. 191-196). IEEE. 10.1109/ICAML48257.2019.00044

Momtazmanesh, S., Nowroozi, A., & Rezaei, N. (2022). Artificial intelligence in rheumatoid arthritis: current status and future perspectives: a state-of-the-art review. *Rheumatology and Therapy*, *9*(5), 1249–1304. doi:10.1007/s40744-022-00475-4 PMID:35849321

Morley, J., & Caio, C. V. (2020). The ethics of AI in health care: A mapping review. *Social Science & Medicine*, *260*, 113172. doi:10.1016/j.socscimed.2020.113172 PMID:32702587

Morris, A. P., Voight, B. F., Teslovich, T. M., Ferreira, T., Segrè, A. V., Steinthorsdottir, V., Strawbridge, R. J., Khan, H., Grallert, H., Mahajan, A., Prokopenko, I., Kang, H. M., Dina, C., Esko, T., Fraser, R. M., Kanoni, S., Kumar, A., Lagou, V., Langenberg, C., & McCarthy, M. I. (2012). Large-scale association analysis provides insights into the genetic architecture and pathophysiology of type 2 diabetes. *Nature Genetics*, *44*(9), 981–990. doi:10.1038/ng.2383 PMID:22885922

Morris, P. D., Narracott, A., von Tengg-Kobligk, H., Soto, D. A. S., Hsiao, S., Lungu, A., & Gunn, J. P. (2015). Computational fluid dynamics modelling in cardiovascular medicine. *Heart (British Cardiac Society)*. PMID:26512019

Morris, T. P., Tormos Muñoz, J. M., Cattaneo, G., Solana-Sánchez, J., Bartrés-Faz, D., & Pascual-Leone, A. (2019). Traumatic brain injury modifies the relationship between physical activity and global and cognitive health: Results from the barcelona brain health initiative. *Frontiers in Behavioral Neuroscience*, *13*, 135. doi:10.3389/fnbeh.2019.00135 PMID:31275124

Morrow, D. A., & Antman, E. M. (2009). Evaluation of High-Sensitivity Assays for Cardiac Troponin. *Clinical Chemistry*, *55*(1), 5–8. doi:10.1373/clinchem.2008.117218 PMID:19028812

Moseley, B., & Dhamija, R. (2019). Precision medicine in epilepsy. *Current Neurology and Neuroscience Reports*, *19*(11), 90. PMID:31720870

Mosselhy, D. A., Assad, M., Sironen, T., & Elbahri, M. (2021). Nanotheranostics: A possible solution for drug-resistant *Staphylococcus aureus* and their biofilms? *Nanomaterials (Basel, Switzerland)*, *11*(1), 1–36. doi:10.3390/nano11010082 PMID:33401760

Moure, J. S. F., Evangelopoulos, M., Colvill, K., Van Eps, J. L., & Tasciotti, E. (2017). Nanoantibiotics: A new paradigm for the treatment of surgical infection. *Nanomedicine (London)*, *12*(11), 1319–1334. doi:10.2217/nnm-2017-0401 PMID:28520517

Mozaffari, M. H., Li, Y., & Ko, Y. (2023). Real-time detection and forecast of flashovers by the visual room fire features using deep convolutional neural networks. *Journal of Building Engineering*, *64*, 105674. doi:10.1016/j.jobe.2022.105674

Mphasha, M., Skaal, L., Mothiba, T., Ngoatle, C., & Hlahla, L. (2023). Primary health care–family partnership for better diabetes outcomes of patients: A systematic review. Journal of Endocrinology. *Jemdsa : the Journal of Endocrinology, Metabolism and Diabetes of South Africa (JEMDSA) : the Official Journal of the Society of Endocrinology, Metabolism and Diabetes of South Africa*, *28*(1), 1–6. doi:10.1080/16089677.2022.2140517

Mujwar, S., Shah, K., Gupta, J. K. & Gour, A. (2022). *Docking Based Screening of Curcumin Derivatives: A Novel Approach in the Inhibition of Tubercular DHFR "Docking Based Screening of Curcumin*. Research Gate.

Mukhopadhyay, B. (2009). Undergraduate medical education in India and scope of pediatric surgery teaching. *Journal of Indian Association of Pediatric Surgeons*, *14*(3), 83. doi:10.4103/0971-9261.57697 PMID:20376246

Muratyan, A., Cheung, W., Dibbo, S. V., & Vhaduri, S. (2022, January). Opportunistic multi-modal user authentication for health-tracking IoT wearables. In *The Fifth International Conference on Safety and Security with IoT: SaSeIoT 2021* (pp. 1-18). Cham: Springer International Publishing.

Naithani, N., Sinha, S., Misra, P., Vasudhevan, B., & Sahu, R. (2021). Precision medicine: Concept and tools. *Medical Journal, Armed Forces India*, *77*(3), 249–257. doi:10.1016/j.mjafi.2021.06.021 PMID:34305276

Nakadi, I. E., Mélot, C., Closset, J., De Moor, V., Bétroune, K., Feron, P., Lingier, P., & Gelin, M. (2006). Evaluation of da Vinci Nissen fundoplication clinical results and cost minimization. *World Journal of Surgery*, *30*(6), 1050–1054. doi:10.1007/s00268-005-7950-6 PMID:16736336

Nalls, M. A., Blauwendraat, C., Vallerga, C. L., Heilbron, K., Bandres-Ciga, S., Chang, D., Tan, M., Kia, D. A., Noyce, A. J., Xue, A., Bras, J., Young, E., von Coelln, R., Simón-Sánchez, J., Schulte, C., Sharma, M., Krohn, L., Pihlstrøm, L., Siitonen, A., & Zhang, F. (2019). Identification of novel risk loci, causal insights, and heritable risk for Parkinson's disease: A meta-analysis of genome-wide association studies. *Lancet Neurology*, *18*(12), 1091–1102. doi:10.1016/S1474-4422(19)30320-5 PMID:31701892

Naskar, A., & Kim, K. S. (2023). Friends against the Foe: Synergistic Photothermal and Photodynamic Therapy against Bacterial Infections. *Pharmaceutics*, *15*(4), 1116. doi:10.3390/pharmaceutics15041116 PMID:37111601

Nasser, M., & Yusof, U. K. (2023). Deep Learning Based Methods for Breast Cancer Diagnosis: A Systematic Review and Future Direction. *Diagnostics (Basel)*, *13*(1), 161. doi:10.3390/diagnostics13010161 PMID:36611453

Nathan, D. M., Buse, J. B., Davidson, M. B., Ferrannini, E., Holman, R. R., Sherwin, R., & Zinman, B. (2009). Medical management of hyperglycemia in type 2 diabetes: a consensus algorithm for the initiation and adjustment of therapy: a consensus statement of the American Diabetes Association and the European Association for the Study of Diabetes. *Diabetes Care*, *32*(1), 193–203. doi:10.2337/dc08-9025 PMID:18945920

National Academies of Sciences, Engineering, and Medicine. (2018). *Implementing and Evaluating Genomic Screening Programs in Health Care Systems: A Workshop*. National Academies Press.

Navaratna, A. R., & Saxena, D. (2023). An Indian Approach to AI Policy: A Comparative Study Between Three Sectors. Handbook of Evidence Based Management Practices in Business, 440.

Nayak, R., Wang, L., & Padhye, R. (2015). Electronic textiles for military personnel. In *Electronic Textiles* (pp. 239–256). Woodhead Publishing. doi:10.1016/B978-0-08-100201-8.00012-6

Ndichu, S., Ban, T., Takahashi, T., & Inoue, D. (2023). AI-Assisted Security Alert Data Analysis with Imbalanced Learning Methods. *Applied Sciences (Basel, Switzerland)*, *13*(3), 1977. doi:10.3390/app13031977

Neale, B. M., Medland, S. E., Ripke, S., Asherson, P., Franke, B., Lesch, K.-P., & Psychiatry, A. (2010). Meta-analysis of genome-wide association studies of attention-deficit/hyperactivity disorder. *49*(9), 884-897.

Neuman, R. J., Lobos, E., Reich, W., Henderson, C. A., Sun, L.-W., & Todd, R. D. J. B. p. (2007). Prenatal smoking exposure and dopaminergic genotypes interact to cause a severe ADHD subtype. *61*(12), 1320-1328.

Ngarambe, J., Yun, G. Y., & Santamouris, M. (2020). The use of artificial intelligence (AI) methods in the prediction of thermal comfort in buildings: Energy implications of AI-based thermal comfort controls. *Energy and Building*, *211*, 109807. doi:10.1016/j.enbuild.2020.109807

Nicholson, J. K., & Wilson, I. D. (2003). Understanding 'global' systems biology: Metabonomics and the continuum of metabolism. *Nature Reviews. Drug Discovery*, *2*(8), 668–676. doi:10.1038/nrd1157 PMID:12904817

Nie, B., Huo, S., Qu, X., Guo, J., Liu, X., Hong, Q., Wang, Y., Yang, J., & Yue, B. (2022). Bone infection site targeting nanoparticle-antibiotics delivery vehicle to enhance treatment efficacy of orthopedic implant related infection. *Bioactive Materials*, *16*(February), 134–148. doi:10.1016/j.bioactmat.2022.02.003 PMID:35386313

Nielsen, A. B., Gannik, D., Siersma, V., & de Fine Olivarius, N. (2011). The relationship between HbA1c level, symptoms and self-rated health in type 2 diabetic patients. *Scandinavian Journal of Primary Health Care*, *29*(3), 157–164. doi:10.3109/02813432.2011.585542 PMID:21707235

Nigg, J., Nikolas, M., Burt, S. A. J. J. o. t. A. A. o. C., & Psychiatry, A. (2010). Measured gene-by-environment interaction in relation to attention-deficit/hyperactivity disorder. *49*(9), 863-873.

Nilsson, E., Jansson, P. A., Perfilyev, A., Volkov, P., Pedersen, M., Svensson, M. K., Poulsen, P., Ribel-Madsen, R., Pedersen, N. L., Almgren, P., Fadista, J., Rönn, T., Klarlund Pedersen, B., Scheele, C., Vaag, A., & Ling, C. (2014). Altered DNA methylation and differential expression of genes influencing metabolism and inflammation in adipose tissue from subjects with type 2 diabetes. *Diabetes*, *63*(9), 2962–2976. doi:10.2337/db13-1459 PMID:24812430

Nishijima, I., Oyadomari, S., Maedomari, S., Toma, R., Igei, C., Kobata, S., Koyama, J., Tomori, R., Kawamitsu, N., Yamamoto, Y., Tsuchida, M., Tokeshi, Y., Ikemura, R., Miyagi, K., Okiyama, K., & Iha, K. (2016). Use of a modified early warning score system to reduce the rate of in-hospital cardiac arrest. *Journal of Intensive Care*, *4*(1), 1–6. doi:10.1186/s40560-016-0134-7 PMID:26865981

Noh, J., & Kavuluru, R. (2018). Document Retrieval for Biomedical Question Answering with Neural Sentence Matching. *Proceedings of the ... International Conference on Machine Learning and Applications. International Conference on Machine Learning and Applications*. IEEE. 10.1109/ICMLA.2018.00036

Nowak, N., Kakade, P. P., & Annapragada, A. V. (2003). Computational fluid dynamics simulation of airflow and aerosol deposition in human lungs. *Annals of Biomedical Engineering*, *31*(4), 374–390. doi:10.1114/1.1560632 PMID:12723679

Ogdie, M. N., Fisher, S. E., Yang, M., Ishii, J., Francks, C., Loo, S. K., & Smalley, S. L. J. T. A. J. o. H. G. (2004). Attention deficit hyperactivity disorder: fine mapping supports linkage to 5p13, 6q12, 16p13, and 17p11. *75*(4), 661-668.

Okpala, I., Nnaji, C., & Karakhan, A. A. (2020). Utilizing emerging technologies for construction safety risk mitigation. *Practice Periodical on Structural Design and Construction*, *25*(2), 04020002. doi:10.1061/(ASCE)SC.1943-5576.0000468

Olatunji, G., Aderinto, N., Adefusi, T., Kokori, E., Akinmoju, O., Yusuf, I., Olusakin, T., & Muzammil, M. A. (2023). Efficacy of tumour-treating fields therapy in recurrent glioblastoma: A narrative review of current evidence. *Medicine*, *102*(48), e36421. doi:10.1097/MD.0000000000036421

Olokoba, A. B., Obateru, O. A., & Olokoba, L. B. (2012). Type 2 diabetes mellitus: A review of current trends. *Oman Medical Journal*, *27*(4), 269–273. doi:10.5001/omj.2012.68 PMID:23071876

Oner, O., Alkar, O. Y., & Oner, P. J. P. I. (2008). Relation of ferritin levels with symptom ratings and cognitive performance in children with attention deficit–hyperactivity disorder. *50*(1), 40-44.

Ooi, E. H., & Chiew, Y. S. (2023). *Model-Based Approaches in Biomedical Engineering*. IOP Publishing. doi:10.1088/978-0-7503-4016-8

Ooi, E. H., & Ng, E. Y. K. (2011). Effects of natural convection within the anterior chamber on the ocular heat transfer. *International Journal for Numerical Methods in Biomedical Engineering*, *27*(3), 408–423. doi:10.1002/cnm.1411

Ornoy, A. J. T. l. (2003). The impact of intrauterine exposure versus postnatal environment in neurodevelopmental toxicity: long-term neurobehavioral studies in children at risk for developmental disorders. *140*, 171-181.

Ornsø, K. B., Garcia-Lastra, J. M., & Thygesen, K. S. (2013). Computational screening of functionalized zinc porphyrins for dye sensitized solar cells. *Physical Chemistry Chemical Physics*, *15*(44), 19478–19486. doi:10.1039/c3cp54050b PMID:24129651

Oueida, S., Kotb, Y., Aloqaily, M., Jararweh, Y., & Baker, T. (2018). An edge computing based smart healthcare framework for resource management. *Sensors (Basel)*, *18*(12), 4307. doi:10.3390/s18124307 PMID:30563267

Ouyang, D., He, B., Ghorbani, A., Yuan, N., Ebinger, J., Langlotz, C. P., Heidenreich, P. A., Harrington, R. A., Liang, D. H., Ashley, E. A., & Zou, J. Y. (2020). Video-based AI for beat-to-beat assessment of cardiac function. *Nature*, *580*(7802), 252–256. doi:10.1038/s41586-020-2145-8 PMID:32269341

Ozaltin, O., Coskun, O., Yeniay, O., & Subasi, A. (2022). A Deep Learning Approach for Detecting Stroke from Brain CT Images Using OzNet. *Bioengineering (Basel, Switzerland)*, *9*(12), 783. doi:10.3390/bioengineering9120783 PMID:36550989

Pai, M. M., Ganiga, R., Pai, R. M., & Sinha, R. K. (2021). Standard electronic health record (EHR) framework for Indian healthcare system. *Health Services and Outcomes Research Methodology*, *21*(3), 339–362. doi:10.1007/s10742-020-00238-0

Pakkasjärvi, N., Luthra, T., & Anand, S. (2023). Artificial Intelligence in Surgical Learning. *Surgeries, 4*(1), 86–97. doi:10.3390/surgeries4010010

Panarello, A., Tapas, N., Merlino, G., Longo, F., & Puliafito, A. (2018). Blockchain and IoT Integration: A Systematic Survey. *Sensors (Basel), 18*(8), 2575. doi:10.3390/s18082575 PMID:30082633

Panch, T., Mattie, H., & Celi, L. A. (2019). The "inconvenient truth" about AI in healthcare. *NPJ Digital Medicine, 2*(1), 77. doi:10.1038/s41746-019-0155-4 PMID:31453372

Panda, G., Upadhyay, A. K., & Khandelwal, K. (2019). Artificial intelligence: A strategic disruption in public relations. *Journal of Creative Communications, 14*(3), 196–213. doi:10.1177/0973258619866585

Pandit, S., Ho, A. C., & Yonekawa, Y. (2023). Recent advances in the management of proliferative diabetic retinopathy. *Current Opinion in Ophthalmology, 34*(3), 232–236. doi:10.1097/ICU.0000000000000946 PMID:36866849

Pandya, S., Sur, A., & Solke, N. (2021, November 30). COVIDSAVIOR: A Novel Sensor-Fusion and Deep Learning Based Framework for Virus Outbreaks. *Frontiers in Public Health, 9*, 797808. doi:10.3389/fpubh.2021.797808 PMID:34917585

Pan, Y. (2016). Heading toward artificial intelligence 2.0. *Engineering (Beijing), 2*(4), 409–413. doi:10.1016/J.ENG.2016.04.018

Pan, Y., & Zhang, L. (2021). Roles of artificial intelligence in construction engineering and management: A critical review and future trends. *Automation in Construction, 122*, 103517. doi:10.1016/j.autcon.2020.103517

Paoletti, X., Lewsley, L. A., Daniele, G., Cook, A., Yanaihara, N., Tinker, A., Kristensen, G., Ottevanger, P. B., Aravantinos, G., Miller, A., Boere, I. A., Fruscio, R., Reyners, A. K. L., Pujade-Lauraine, E., Harkin, A., Pignata, S., Kagimura, T., Welch, S., Paul, J., & Glasspool, R. M. (2020). Assessment of progression-free survival as a surrogate end point of overall survival in first-line treatment of ovarian cancer: A systematic review and meta-analysis. *JAMA Network Open, 3*(1), e1918939–e1918939. doi:10.1001/jamanetworkopen.2019.18939

Papini, C., Palaniyappan, L., Kroll, J., Froudist-Walsh, S., Murray, R. M., & Nosarti, C. (2020). Altered cortical gyrification in adults who were born very preterm and its associations with cognition and mental health. *Biological Psychiatry: Cognitive Neuroscience and Neuroimaging, 5*(7), 640–650. doi:10.1016/j.bpsc.2020.01.006 PMID:32198001

Parry, C. M., & Aneja, U. (2020). *Artificial Intelligence for Healthcare: Insights from India*. Chatham House. https://www.chathamhouse.org/2020/07/artificial-intelligence-healthcare-insights-india/3-ai-healthcare-india-applications

Pasqual, E. M., Londero, A. P., Robella, M., Tonello, M., Sommariva, A., De Simone, M., Bacchetti, S., Baiocchi, G., Asero, S., Coccolini, F., De Cian, F., Guaglio, M., Cinquegrana, A., Cenzi, C., Scaringi, S., & Macrì, A. (2023). Repeated Cytoreduction Combined with Hyperthermic Intraperitoneal Chemotherapy (HIPEC) in Selected Patients Affected by Peritoneal Metastases: Italian PSM Oncoteam Evidence. *Cancers (Basel), 15*(3), 607. doi:10.3390/cancers15030607

Patel, S. R., Pundge, A., Zebanaaz, S., & Akther, N. (2023, May). An Analysis of Recent Developments in Healthcare Technology from the Standpoints of Artificial Intelligence, the Internet of Things, and Remote Health Monitoring. In *International Conference on Applications of Machine Intelligence and Data Analytics (ICAMIDA 2022)* (pp. 520-526). Atlantis Press. 10.2991/978-94-6463-136-4_45

Patel, V., Saxena, S., Lund, C., Thornicroft, G., Baingana, F., Bolton, P., Chisholm, D., Collins, P. Y., Cooper, J. L., Eaton, J., Herrman, H., Herzallah, M. M., Huang, Y., Jordans, M. J. D., Kleinman, A., Medina-Mora, M. E., Morgan, E., Niaz, U., Omigbodun, O., & UnÜtzer, J. Ü. (2018). The Lancet Commission on global mental health and sustainable development. *Lancet, 392*(10157), 1553–1598. doi:10.1016/S0140-6736(18)31612-X PMID:30314863

Patterson, A. D., Maurhofer, O., Beyoglu, D., Lanz, C., Krausz, K. W., Pabst, T., Gonzalez, F. J., Dufour, J.-F., & Idle, J. R. (2011). Aberrant lipid metabolism in hepatocellular carcinoma revealed by plasma metabolomics and lipid profiling. *Cancer Research, 71*(21), 6590–6600. doi:10.1158/0008-5472.CAN-11-0885 PMID:21900402

Paul, R. H., Gunstad, J., Cooper, N., Williams, L. M., Clark, C. R., Cohen, R. A., Lawrence, J. J., & Gordon, E. (2007). Cross-cultural assessment of neuropsychological performance and electrical brain function measures: Additional validation of an international brain database. *The International Journal of Neuroscience, 117*(4), 549–568. doi:10.1080/00207450600773665 PMID:17365135

Pauly, J. R., & Slotkin, T. A. J. A. P. (2008). Maternal tobacco smoking, nicotine replacement and neurobehavioural development. *97*(10), 1331-1337.

Paul, Y., Hickok, E., Sinha, A., Tiwari, U., Mohandas, S., Ray, S., & Bidare, P. M. (2018). *Artificial intelligence in the healthcare industry in India*. The Centre for Internet and Society.

Pawar, V., Topkar, H., & Srivastava, R. (2018). Chitosan nanoparticles and povidone iodine containing alginate gel for prevention and treatment of orthopedic implant associated infections. *International Journal of Biological Macromolecules, 115*, 1131–1141. doi:10.1016/j.ijbiomac.2018.04.166 PMID:29727653

Paz, C., Suárez, E., & Vence, J. (2017). CFD transient simulation of the cough clearance process using an Eulerian wall film model. *Computer Methods in Biomechanics and Biomedical Engineering, 20*(2), 142–152. doi:10.1080/10255842.2016.1206532 PMID:27380322

Pearson, E. R. (2016). Personalized medicine in diabetes: The role of 'omics' and biomarkers. *Diabetic Medicine, 33*(6), 712–717. doi:10.1111/dme.13075 PMID:26802434

Pee, L. G., Pan, S. L., & Cui, L. (2019). Artificial intelligence in healthcare robots: A social informatics study of knowledge embodiment. *Journal of the Association for Information Science and Technology, 70*(4), 351–369. doi:10.1002/asi.24145

Pencina, M. J., D'Agostino, R. B. Sr, D'Agostino, R. B. Jr, & Vasan, R. S. (2008). Evaluating the added predictive ability of a new marker: From area under the ROC curve to reclassification and beyond. *Statistics in Medicine, 27*(2), 157–172. doi:10.1002/sim.2929 PMID:17569110

Pepe, M. S. (2000). An Interpretation for the ROC Curve and Inference Using GLM Procedures. *Biometrics, 56*(2), 352–359. doi:10.1111/j.0006-341X.2000.00352.x PMID:10877289

Perera, C., Liu, C., Jayawardena, S., & Chen, M. (2015). A Survey on Internet of Things From Industrial Market Perspective. *IEEE Access : Practical Innovations, Open Solutions, 2*, 1660–1679. doi:10.1109/ACCESS.2015.2389854

Perlman, A., Sacks, R., & Barak, R. (2014). Hazard recognition and risk perception in construction. *Safety Science, 64*, 22–31. doi:10.1016/j.ssci.2013.11.019

Perl, R. M., Grimmer, R., Hepp, T., & Horger, M. S. (2021). Can a novel deep neural network improve the computer-aided detection of solid pulmonary nodules and the rate of false-positive findings in comparison to an established machine learning computer-aided detection? *Investigative Radiology, 56*(2), 103–108. doi:10.1097/RLI.0000000000000713 PMID:32796198

Philip, N. Y., Rodrigues, J. J., Wang, H., Fong, S. J., & Chen, J. (2021). Internet of Things for in-home health monitoring systems: Current advances, challenges and future directions. *IEEE Journal on Selected Areas in Communications, 39*(2), 300–310. doi:10.1109/JSAC.2020.3042421

Pires, T. d. O., da Silva, C. M. F. P., & de Assis, S. G. J. B. p. (2013). Association between family environment and attention deficit hyperactivity disorder in children–mothers' and teachers' views. *13*(1), 1-9.

Plathottam, S. J., Rzonca, A., Lakhnori, R., & Iloeje, C. O. (2023). A review of artificial intelligence applications in manufacturing operations. *Journal of Advanced Manufacturing and Processing*, *5*(3), 10159. doi:10.1002/amp2.10159

Polanczyk, G., De Lima, M. S., Horta, B. L., Biederman, J., & Rohde, L. A. J. A. j. o. p. (2007). The worldwide prevalence of ADHD: a systematic review and metaregression analysis. *164*(6), 942-948.

Pollock, L. (2023, August 22). *What Is an NLP Chatbot — And How Do NLP-Powered Bots Work?* Ultimate.ai. https://www.ultimate.ai/blog/ai-automation/how-nlp-text-based-chatbots-work

Poort, H., Fenton, A. T., Thompson, E., Dinardo, M. M., Liu, J. F., Arch, J. J., & Wright, A. A. (2021). Lived experiences of women reporting fatigue during PARP inhibitor maintenance treatment for advanced ovarian cancer: A qualitative study. *Gynecologic Oncology*, *160*(1), 227–233. doi:10.1016/j.ygyno.2020.10.034

Poplin, R., Varadarajan, A. V., Blumer, K., Liu, Y., McConnell, M. V., Corrado, G. S., Peng, L., & Webster, D. R. (2018). Prediction of cardiovascular risk factors from retinal fundus photographs via deep learning. *Nature Biomedical Engineering*, *2*(3), 158–164. doi:10.1038/s41551-018-0195-0 PMID:31015713

Powledge, T. M. (2003). Human genome project completed. *Genome Biology, 4*, spotlight-20030415-01. doi:10.1186/gb-spotlight-20030415-01

Powles, J., & Hodson, H. (2017). Google DeepMind and healthcare in an age of algorithms. *Health and Technology*, *7*(4), 351–367. doi:10.1007/s12553-017-0179-1 PMID:29308344

Poy, M. N., Eliasson, L., Krutzfeldt, J., Kuwajima, S., Ma, X., Macdonald, P. E., Pfeffer, S., Tuschl, T., Rajewsky, N., Rorsman, P., & Stoffel, M. (2004). A pancreatic islet-specific microRNA regulates insulin secretion. *Nature*, *432*(7014), 226–230. doi:10.1038/nature03076 PMID:15538371

Prabha, K. R., Nataraj, B., Swathi, P., Tharani, V. R., & Vinola, A. (2022, October). IoT Based Smart Mask for Health Monitoring. In *2022 3rd International Conference on Smart Electronics and Communication (ICOSEC)* (pp. 568-572). IEEE. 10.1109/ICOSEC54921.2022.9952121

Prabhakaran, P., Subbaiyan, A., Gopalakrishnan, D., Ramkumar, S., Veluswamy, S., Murugesan, D. K., ... Bhaskaran, P. (2022). Maintenance Methodologies Embraced for Railroad Systems: A Review. *Advances in Materials Science and Engineering*. Advance online publication. doi:10.1155/2022/7655245

Pradhan, P., Soni, N. K., Chaudhary, L., Mujwar, S., & Pardasani, K. R. (2015). In-silico prediction of riboswitches and design of their potent inhibitors for H1N1, H2N2 and H3N2 strains of influenza virus. *Biosciences Biotechnology Research Asia*, *12*(3), 2173–2186. doi:10.13005/bbra/1889

Prasenjit, C. (2021). Model for selecting a route for the transport of hazardous materials using a fuzzy logic system. *Vojnotehnički glasnik, 69*(2), 355-390.

Prosperi, M., Min, J. S., Bian, J., & Modave, F. (2018). Big data hurdles in precision medicine and precision public health. *BMC Medical Informatics and Decision Making*, *18*(1), 1–15. doi:10.1186/s12911-018-0719-2 PMID:30594159

Pulendran, B., & Ahmed, R. (2011). Immunological mechanisms of vaccination. *Nature Immunology*, *12*(6), 509–517. doi:10.1038/ni.2039 PMID:21739679

Puspitaningtyas, H., Espressivo, A., Hutajulu, S. H., Fuad, A., & Allsop, M. J. (2021). Mapping and visualization of cancer research in Indonesia: A scientometric analysis. *Cancer Control*, *28*, 10732748211053464. doi:10.1177/10732748211053464

Qing, Y., Cheng, L., Li, R., Liu, G., Zhang, Y., Tang, X., Wang, J., Liu, H., & Qin, Y. (2018). Potential antibacterial mechanism of silver nanoparticles and the optimization of orthopedic implants by advanced modification technologies. *International Journal of Nanomedicine*, *13*, 3311–3327. doi:10.2147/IJN.S165125 PMID:29892194

Qi, S., Ouyang, X., Wang, L., Peng, W., Wen, J., & Dai, Y. (2012). A pilot metabolic profiling study in serum of patients with chronic kidney disease based on 1H-NMR-spectroscopy. *Clinical and Translational Science*, *5*(5), 379–385. doi:10.1111/j.1752-8062.2012.00437.x PMID:23067349

Quan, K., Zhang, Z., Ren, Y., Busscher, H. J., van der Mei, H. C., & Peterson, B. W. (2021). On-demand pulling-off of magnetic nanoparticles from biomaterial surfaces through implant-associated infectious biofilms for enhanced antibiotic efficacy. *Materials Science and Engineering C*, *131*, 112526. doi:10.1016/j.msec.2021.112526 PMID:34857305

Quirós-Alcalá, L., Mehta, S., & Eskenazi, B. J. E. h. p. (2014). Pyrethroid pesticide exposure and parental report of learning disability and attention deficit/hyperactivity disorder in US children: NHANES 1999–2002. *122*(12), 1336-1342.

Radianti, J., Dokas, I., Boersma, K., Noori, N. S., Belbachir, N., & Stieglitz, S. (2019). *Enhancing disaster response for hazardous materials using emerging technologies: The role of AI and a research agenda*. In Engineering Applications of Neural Networks: 20th International Conference, EANN 2019, (pp. 368-376). Springer International Publishing.

Rahman, M. M., Manik, M. M. H., Islam, M. M., Mahmud, S., & Kim, J. H. (2020, September). An automated system to limit COVID-19 using facial mask detection in smart city network. In *2020 IEEE International IOT, Electronics and Mechatronics Conference (IEMTRONICS)* (pp. 1-5). IEEE.

Rahman, S., Sarker, S., Miraj, M. A., Nihal, R. A., Nadimul Haque, A. K., & Noman, A. A. (2021). Deep learning–driven automated detection of Covid-19 from radiography images: A comparative analysis. *Cognitive Computation*, 1–30. doi:10.1007/s12559-020-09779-5 PMID:33680209

Rajkomar, A., Oren, E., Chen, K., Dai, A. M., Hajaj, N., Hardt, M., Liu, P. J., Liu, X., Marcus, J., Sun, M., Sundberg, P., Yee, H., Zhang, K., Zhang, Y., Flores, G., Duggan, G. E., Irvine, J., Le, Q., Litsch, K., & Dean, J. (2018). Scalable and accurate deep learning with electronic health records. *NPJ Digital Medicine*, *1*(1), 18. doi:10.1038/s41746-018-0029-1 PMID:31304302

Raman, R. K., Dewang, Y., & Raghuwanshi, J. (2018). A review on applications of computational fluid dynamics. *International Journal of LNCT*, *2*(6), 137–143.

Rana, M., & Bhushan, M. (2023). Machine learning and deep learning approach for medical image analysis: Diagnosis to detection. *Multimedia Tools and Applications*, *82*(17), 26731–26769. doi:10.1007/s11042-022-14305-w PMID:36588765

Rao, M. V. V & Team DHR-ICMR AI Cell. (2023). Ethical guidelines for application of artificial intelligence in biomedical research and healthcare. *Indian Council of Medical Research*. https://main.icmr.nic.in/sites/default/files/upload_documents/Ethical_Guidelines_AI_Healthcare_2023.pdf

Raphel, J., Holodniy, M., Goodman, S. B., & Heilshorn, S. C. (2016). Multifunctional coatings to simultaneously promote osseointegration and prevent infection of orthopaedic implants. *Biomaterials*, *84*, 301–314. doi:10.1016/j.biomaterials.2016.01.016 PMID:26851394

Ravnskjaer, K., Hogan, M. F., Lackey, D., Tora, L., Dent, S. Y. R., Olefsky, J., & Montminy, M. (2013). Glucagon regulates gluconeogenesis through KAT2B- and WDR5-mediated epigenetic effects. *The Journal of Clinical Investigation*, *123*(10), 4318–4328. doi:10.1172/JCI69035 PMID:24051374

Razdan, S., & Sharma, S. (2022). Internet of medical things (IoMT): Overview, emerging technologies, and case studies. *IETE Technical Review*, *39*(4), 775–788. doi:10.1080/02564602.2021.1927863

Reddy, M. A., Zhang, E., & Natarajan, R. (2015). Epigenetic mechanisms in diabetic complications and metabolic memory. *Diabetologia*, *58*(3), 443–455. doi:10.1007/s00125-014-3462-y PMID:25481708

Reid, F. (2020). World Ovarian Cancer Coalition Atlas. 2020. *Global Trends in Incidence, Mortality, and Survival*, 1-42.

Rescalli, A., Varoni, E. M., Cellesi, F., & Cerveri, P. (2022). Analytical challenges in diabetes management: Towards glycated albumin point-of-care detection. *Biosensors (Basel)*, *12*(9), 687. doi:10.3390/bios12090687 PMID:36140073

Ricci, L., Maesa, D. D., Favenza, A., & Ferro, E. (2021). Blockchains for covid-19 contact tracing and vaccine support: A systematic review. *IEEE Access : Practical Innovations, Open Solutions*, *9*, 37936–37950. doi:10.1109/ACCESS.2021.3063152 PMID:34812382

Riek, L. D. (2016). Robotics technology in mental health care. In *Artificial intelligence in behavioral and mental health care* (pp. 185–203). Academic Press. doi:10.1016/B978-0-12-420248-1.00008-8

Rispoli, V. C., Nielsen, J. F., Nayak, K. S., & Carvalho, J. L. (2015). Computational fluid dynamics simulations of blood flow regularized by 3D phase contrast MRI. *Biomedical Engineering Online*, *14*(1), 1–23. doi:10.1186/s12938-015-0104-7 PMID:26611470

Robert, G., Donetto, S., & Williams, O. (2021). Co-designing healthcare services with patients. The Palgrave Handbook of Co-Production of Public Services and Outcomes, 313-333. Palgrave.

Robertson, S., Back, N., & Bartlett, J. D. (2016). Explaining match outcome in elite Australian Rules football using team performance indicators. *Journal of Sports Sciences*, *34*(7), 637–644. doi:10.1080/02640414.2015.1066026 PMID:26176890

Roden, D. M., McLeod, H. L., Relling, M. V., Williams, M. S., Mensah, G. A., Peterson, J. F., & Van Driest, S. L. (2019). Pharmacogenomics. *Lancet*, *394*(10197), 521–532. doi:10.1016/S0140-6736(19)31276-0 PMID:31395440

Rodger, J. A. (2012). Toward reducing failure risk in an integrated vehicle health maintenance system: A fuzzy multi-sensor data fusion Kalman filter approach for IVHMS. *Expert Systems with Applications*, *39*(10), 9821–9836. doi:10.1016/j.eswa.2012.02.171

Rodriguez, A., Olsen, J., Kotimaa, A., Kaakinen, M., Moilanen, I., Henriksen, T., & Psychiatry. (2009). Is prenatal alcohol exposure related to inattention and hyperactivity symptoms in children? Disentangling the effects of social adversity. *50*(9), 1073-1083.

Rodríguez, T. (2012). *Environmental pesticide exposure and neurobehavioral effects among children of Nicaraguan agricultural workers.* Acta Universitatis Upsaliensis, Saccaro, L., Schilliger, Z., Perroud, N., & Piguet, C. Inflammation, Anxiety, and Stress in Attention-Deficit/Hyperactivity Disorder. Biomedicines 2021, 9, 1313. In.

Rodríguez-Merchán, E. C., Davidson, D. J., & Liddle, A. D. (2021). Recent strategies to combat infections from biofilm-forming bacteria on orthopaedic implants. *International Journal of Molecular Sciences*, *22*(19), 10243. doi:10.3390/ijms221910243 PMID:34638591

Rolim, C. O., Koch, F. L., Westphall, C. B., Werner, J., Fracalossi, A., & Salvador, G. S. (2010, February). A cloud computing solution for patient's data collection in health care institutions. In *2010 Second International Conference on eHealth, Telemedicine, and Social Medicine* (pp. 95-99). IEEE. 10.1109/eTELEMED.2010.19

Roman, D. H., Conlee, K. D., Abbott, I., Jones, R., Noble, A., Rich, N., & Costa, D. (2015). The digital revolution comes to US healthcare. *Internet of Things : Engineering Cyber Physical Human Systems*, *5*, 1–54.

Romero-Brufau, S., Whitford, D., Johnson, M. G., Hickman, J., Morlan, B. W., Therneau, T., Naessens, J., & Huddleston, J. M. (2021). Using machine learning to improve the accuracy of patient deterioration predictions: Mayo Clinic Early Warning Score (MC-EWS). *Journal of the American Medical Informatics Association : JAMIA*, *28*(6), 1207–1215. doi:10.1093/jamia/ocaa347 PMID:33638343

Rong, G., Mendez, A., Assi, E. B., Zhao, B., & Sawan, M. (2020). Artificial intelligence in healthcare: Review and prediction case studies. *Engineering (Beijing)*, *6*(3), 291–301. doi:10.1016/j.eng.2019.08.015

Rosenbloom, K. R., Armstrong, J., Barber, G. P., Casper, J., Clawson, H., Diekhans, M., Dreszer, T. R., Fujita, P. A., Guruvadoo, L., Haeussler, M., Harte, R. A., Heitner, S., Hickey, G., Hinrichs, A. S., Hubley, R., Karolchik, D., Learned, K., Lee, B. T., Li, C. H., & Kent, W. J. (2015). The UCSC Genome Browser database: 2015 update. *Nucleic Acids Research, 43*(Database issue), D670–D681. doi:10.1093/nar/gku1177 PMID:25428374

Rossell, S. L., Coakes, J., Shapleske, J., Woodruff, P. W. R., & David, A. S. (2003). Insight: Its relationship with cognitive function, brain volume and symptoms in schizophrenia. *Psychological Medicine, 33*(1), 111–119. doi:10.1017/S0033291702006803 PMID:12537042

Rothan, H., & Byrareddy, S. (2020). The epidemiology and pathogenesis of coronavirus disease (COVID-19) outbreak. *Journal of Autoimmunity, 109*, 102433–102433. doi:10.1016/j.jaut.2020.102433 PMID:32113704

Roy, S., Das, A., Vernekar, M., Mandal, S., & Chatterjee, N. (2022). Understanding the correlation between metabolic regulator SIRT1 and exosomes with CA-125 in ovarian cancer: A clinicopathological study. *BioMed Research International, 2022*, 1–16. doi:10.1155/2022/5346091

Ruiz-Aguilar, J. J., Turias, I., González-Enrique, J., Urda, D., & Elizondo, D. (2021). A permutation entropy-based EMD-ANN forecasting ensemble approach for wind speed prediction. *Neural Computing & Applications, 33*(7), 2369–2391. doi:10.1007/s00521-020-05141-w

Rupp, M., Tkatchenko, A., Müller, K.-R., & von Lilienfeld, O. A. (2012). Fast and accurate modeling of molecular atomization energies with machine learning. *Physical Review Letters, 108*(5), 058301. doi:10.1103/PhysRevLett.108.058301 PMID:22400967

Sadybekov, A. V., & Katritch, V. (2023). Computational approaches streamlining drug discovery. *Nature, 616*(7958), 673–685. doi:10.1038/s41586-023-05905-z PMID:37100941

Saha, T. (2018). *Artificial intelligence based healthcare startups in India.*

Sahney, R., & Sharma, M. (2018). Electronic health records: A general overview. *Current Medicine Research and Practice, 8*(2), 67–70. doi:10.1016/j.cmrp.2018.03.004

Sahoo, J., Sarkhel, S., Mukherjee, N., & Jaiswal, A. (2022). Nanomaterial-Based Antimicrobial Coating for Biomedical Implants: New Age Solution for Biofilm-Associated Infections. *ACS Omega, 7*(50), 45962–45980. doi:10.1021/acsomega.2c06211 PMID:36570317

Sahoo, S. K., & Goswami, S. S. (2023). A comprehensive review of multiple criteria decision-making (MCDM) Methods: Advancements, applications, and future directions. *Decision Making Advances, 1*(1), 25–48. doi:10.31181/dma1120237

Said, A. M., Yahyaoui, A., & Abdellatif, T. (2021). Efficient anomaly detection for smart hospital IoT systems. *Sensors (Basel), 21*(4), 1026. doi:10.3390/s21041026 PMID:33546169

Saidi, A., Gauvin, C., Ladhari, S., & Nguyen-Tri, P. (2021). Advanced functional materials for intelligent thermoregulation in personal protective equipment. *Polymers, 13*(21), 3711. doi:10.3390/polym13213711 PMID:34771268

Saini, K., & Raj, P. (2022). Edge platforms, frameworks and applications. []. Elsevier.] *Advances in Computers, 127*, 237–258. doi:10.1016/bs.adcom.2022.02.005

Samuel, S. (2023). Artificial intelligence in disaster prediction and risk reduction. *Applied Intelligence in Human-Computer Interaction*, 83-95.

Sanderson, S. C., Linderman, M. D., Kasarskis, A., Bashir, A., & Diaz, G. A. (2017). Implications of informing research participants of research results: An empirical study. *Genetics in Medicine, 19*(6), 644–650.

Sanghvi, V. R., Leibold, J., Mina, M., Mohan, P., Berishaj, M., Li, Z., Miele, M. M., Lailler, N., Zhao, C., de Stanchina, E., Viale, A., Akkari, L., Lowe, S. W., Ciriello, G., Hendrickson, R. C., & Wendel, H.-G. (2019). The Oncogenic Action of NRF2 Depends on De-glycation by Fructosamine-3-Kinase. *Cell*, *178*(4), 807–819.e21. doi:10.1016/j.cell.2019.07.031 PMID:31398338

San, H., Paresoglou, M., Minneboo, M., van Hengel, I. A. J., Yilmaz, A., Gonzalez-Garcia, Y., Fluit, A. C., Hagedoorn, P. L., Fratila-Apachitei, L. E., Apachitei, I., & Zadpoor, A. A. (2022). Fighting Antibiotic-Resistant Bacterial Infections by Surface Biofunctionalization of 3D-Printed Porous Titanium Implants with Reduced Graphene Oxide and Silver Nanoparticles. *International Journal of Molecular Sciences*, *23*(16), 9204. doi:10.3390/ijms23169204 PMID:36012467

Sarker, I. H. (2021). Deep Learning: A Comprehensive Overview on Techniques, Taxonomy, Applications and Research Directions. *SN Computer Science*, *2*(6), 420. doi:10.1007/s42979-021-00815-1 PMID:34426802

Sarker, I. H. (2021). Machine Learning: Algorithms, Real-World Applications and Research Directions. *SN Computer Science*, *2*(3), 160. doi:10.1007/s42979-021-00592-x PMID:33778771

Sarrouti, M., & Ouatik, E. (2017). A machine learning-based method for question type classification in biomedical question answering. *Methods of Information in Medicine*, *56*(3), 209–216. doi:10.3414/ME16-01-0116 PMID:28361158

Schachar, R. A. (2006). The mechanism of accommodation and presbyopia. *International Ophthalmology Clinics*, *46*(3), 39–61. doi:10.1097/00004397-200604630-00006 PMID:16929224

Schork, N. J. (2015). Personalized medicine: Time for one-person trials. *Nature*, *520*(7549), 609–611. doi:10.1038/520609a PMID:25925459

Schumann, G., Andreassen, O. A., Banaschewski, T., Calhoun, V. D., Clinton, N., Desrivieres, S., Brandlistuen, R. E., Feng, J., Hese, S., Hitchen, E., Hoffmann, P., Jia, T., Jirsa, V., Marquand, A. F., Nees, F., Nöthen, M. M., Novarino, G., Polemiti, E., Ralser, M., & Ogoh, G. (2023). Addressing Global Environmental Challenges to Mental Health Using Population Neuroscience: A Review. *JAMA Psychiatry*, *80*(10), 1066. doi:10.1001/jamapsychiatry.2023.2996 PMID:37610741

Schwaederle, M., Zhao, M., Lee, J. J., Lazar, V., Leyland-Jones, B., & Schilsky, R. L. (2015). Association of biomarker-based treatment strategies with response rates and progression-free survival in refractory malignant neoplasms: A meta-analysis. *JAMA Oncology*, *1*(6), 870–879. PMID:27273579

Schwalbe, N., & Wahl, B. (2020). Artificial intelligence and the future of global health. *Lancet*, *395*(10236), 1579–1586. doi:10.1016/S0140-6736(20)30226-9 PMID:32416782

Sebastiani, G., Po, A., Miele, E., Ventriglia, G., Ceccarelli, E., Bugliani, M., Marselli, L., Marchetti, P., Gulino, A., Ferretti, E., & Dotta, F. (2015). MicroRNA-124a is hyperexpressed in type 2 diabetic human pancreatic islets and negatively regulates insulin secretion. *Acta Diabetologica*, *52*(3), 523–530. doi:10.1007/s00592-014-0675-y PMID:25408296

Segerstolpe, Å., Palasantza, A., Eliasson, P., Andersson, E.-M., Andréasson, A.-C., Sun, X., Picelli, S., Sabirsh, A., Clausen, M., Bjursell, M. K., Smith, D. M., Kasper, M., Ämmälä, C., & Sandberg, R. (2016). Single-Cell Transcriptome Profiling of Human Pancreatic Islets in Health and Type 2 Diabetes. *Cell Metabolism*, *24*(4), 593–607. doi:10.1016/j.cmet.2016.08.020 PMID:27667667

Selvadass, S., Paul, J. J., & Bella Mary, I. (2022). IoT-Enabled smart mask to detect COVID19 outbreak. *Health and Technology*, *12*(5), 1025–1036. doi:10.1007/s12553-022-00695-2 PMID:36000088

Serag, A., Ion-Margineanu, A., Qureshi, H., McMillan, R., Saint Martin, M.-J., Diamond, J., O'Reilly, P., & Hamilton, P. (2019). Translational AI and deep learning in diagnostic pathology. *Frontiers in Medicine*, *6*, 185. doi:10.3389/fmed.2019.00185 PMID:31632973

Serrano, D. R., Kara, A., Yuste, I., Luciano, F. C., Ongoren, B., Anaya, B. J., & Lalatsa, A. (2023). 3D printing technologies in personalized medicine, nanomedicines, and biopharmaceuticals. *Pharmaceutics*, *15*(2), 313. doi:10.3390/pharmaceutics15020313 PMID:36839636

Šerý, O., Paclt, I., Drtílková, I., Theiner, P., Kopečková, M., Zvolský, P., & Functions, B. (2015). A 40-bp VNTR polymorphism in the 3′-untranslated region of DAT1/SLC6A3 is associated with ADHD but not with alcoholism. *11*(1), 1-8.

Sforza, D. M., Putman, C. M., & Cebral, J. R. (2012). Computational fluid dynamics in brain aneurysms. *International Journal for Numerical Methods in Biomedical Engineering*, *28*(6-7), 801–808. doi:10.1002/cnm.1481 PMID:25364852

Shademan, A., Decker, R. S., Opfermann, J. D., Leonard, S., Krieger, A., & Kim, P. C. (2016). Supervised autonomous robotic soft tissue surgery. *Science Translational Medicine*, *8*(337), 337ra64. doi:10.1126/scitranslmed.aad9398 PMID:27147588

Shafranskaya, D., Chori, A., & Korobeynikov, A. (2021). Graph-based approaches significantly improve the recovery of antibiotic resistance genes from complex metagenomic datasets. *Frontiers in Microbiology*, *12*, 714836. doi:10.3389/fmicb.2021.714836 PMID:34690959

Shah, H., & Srivastava, A. (2014). Signature provisions in the amended Indian Information Technology Act 2000: Legislative chaos. *Common Law World Review*, *43*(3), 208–230. doi:10.1350/clwr.2014.43.3.0271

Shah, L., Dandawate, V., & Gutam, S. (2016). *The changing landscape and future of Open Access in India*. Not Available.

Shao, Y., Zhu, B., & Zheng, R. (2014). Development of urinary pseudotargeted LC-MS-based metabolomics method and its application in hepatocellular carcinoma biomarker discovery. *Journal of Proteome Research*. doi:10.1021/pr500973d PMID:25483141

Sharada, K., Kiran, P. U., Katipelli, R., Maddi, Y., & Rajendar, P. (2021). Investigation Of Compressible Fluid Flow Between Spinning Rotor Cascades In Elevated Aerodynamics Using Numerical Techniques. *Turkish Online Journal of Qualitative Inquiry, 12*(10)

Sharma, A., Singh, A., Gupta, V., & Arya, S. (2022). Advancements and future prospects of wearable sensing technology for healthcare applications. *Sensors & Diagnostics*, *1*(3), 387–404. doi:10.1039/D2SD00005A

Sharma, A., & Singh, B. (2022). Measuring Impact of E-commerce on Small Scale Business: A Systematic Review. *Journal of Corporate Governance and International Business Law*, *5*(1).

Sharma, M., Luthra, S., Joshi, S., & Kumar, A. (2022). Implementing challenges of artificial intelligence: Evidence from public manufacturing sector of an emerging economy. *Government Information Quarterly*, *39*(4), 101624. doi:10.1016/j.giq.2021.101624

Shastry, K. A., & Shastry, A. (2023). An integrated deep learning and natural language processing approach for continuous remote monitoring in digital health. *Decision Analytics Journal*, *8*, 100301. doi:10.1016/j.dajour.2023.100301

Shaw, P., & Rabin, C. J. C. p. r. (2009). New insights into attention-deficit/hyperactivity disorder using structural neuroimaging. *11*(5), 393-398.

Shaw, P., Eckstrand, K., Sharp, W., Blumenthal, J., Lerch, J., Greenstein, D., & Rapoport, J. J. P. o. t. n. a. o. s. (2007). Attention-deficit/hyperactivity disorder is characterized by a delay in cortical maturation. *104*(49), 19649-19654.

Shaw, P., Gornick, M., Lerch, J., Addington, A., Seal, J., Greenstein, D., & Castellanos, F. X. J. A. o. g. p. (2007). Polymorphisms of the dopamine D4 receptor, clinical outcome, and cortical structure in attention-deficit/hyperactivity disorder. *64*(8), 921-931.

Shaw, P., Lerch, J., Greenstein, D., Sharp, W., Clasen, L., Evans, A., & Rapoport, J. J. A. o. g. p. (2006). Longitudinal mapping of cortical thickness and clinical outcome in children and adolescents with attention-deficit/hyperactivity disorder. *63*(5), 540-549.

Shen, X., Zhang, F., Li, K., Qin, C., Ma, P., Dai, L., & Cai, K. (2016). Cecropin B loaded TiO_2 nanotubes coated with hyaluronidase sensitive multilayers for reducing bacterial adhesion. *Materials & Design*, *92*, 1007–1017. doi:10.1016/j.matdes.2015.12.126

Shibeshi, S. S., & Collins, W. E. (2005). The rheology of blood flow in a branched arterial system. *Applied Rheology (Lappersdorf, Germany)*, *15*(6), 398–405. doi:10.1515/arh-2005-0020 PMID:16932804

Shigley, J. E., Mischke, C. R., & Brown, T. H. Jr. (2004). *Standard handbook of machine design*. McGraw-Hill Education.

Shneiderman, B. (2020). Bridging the gap between ethics and practice: Guidelines for reliable, safe, and trustworthy human-centered AI systems. [TiiS]. *ACM Transactions on Interactive Intelligent Systems*, *10*(4), 1–31. doi:10.1145/3419764

Shringarpure, S. S., & Bustamante, C. D. (2015). Privacy risks from genomic data-sharing beacons. *American Journal of Human Genetics*, *97*(5), 631–646. doi:10.1016/j.ajhg.2015.09.010 PMID:26522470

Shuaib, A., Arian, H., & Shuaib, A. (2020). The increasing role of artificial intelligence in health care: Will robots replace doctors in the future? *International Journal of General Medicine*, *13*, 891–896. doi:10.2147/IJGM.S268093 PMID:33116781

Shu, X., & Ye, Y. (2023). Knowledge Discovery: Methods from data mining and machine learning. *Social Science Research*, *110*, 102817. doi:10.1016/j.ssresearch.2022.102817

Si, A., Das, S., & Kar, S. (2021). Picture fuzzy set-based decision-making approach using Dempster-Shafer theory of evidence and grey relation analysis and its application in COVID-19 medicine selection. *Soft Computing*, 1–15. doi:10.1007/s00500-021-05909-9

Sibley, M. H., Mitchell, J. T., & Becker, S. P. J. T. L. P. (2016). Method of adult diagnosis influences estimated persistence of childhood ADHD: a systematic review of longitudinal studies. *3*(12), 1157-1165.

Sikkandar, M. Y., Sudharsan, N. M., Begum, S. S., & Ng, E. Y. K. (2019). Computational Fluid Dynamics: A Technique to Solve Complex Biomedical Engineering Problems-A Review. *WSEAS Transactions on Biology and Biomedicine*, *16*, 121–137.

Silk, T. J., Vance, A., Rinehart, N., Bradshaw, J. L., & Cunnington, R. J. H. b. m. (2009). White-matter abnormalities in attention deficit hyperactivity disorder: A diffusion tensor imaging study. *30*(9), 2757-2765.

Simonite, T. (2018). Google's AI eye doctor gets ready to go to work in India. *Wired Magazine.*

Simpson, J. T., Workman, R. E., Zuzarte, P. C., David, M., Dursi, L. J., & Timp, W. (2017). Detecting DNA cytosine methylation using nanopore sequencing. *Nature Methods*, *14*(4), 407–410. doi:10.1038/nmeth.4184 PMID:28218898

Singh, B. (2023). Blockchain Technology in Renovating Healthcare: Legal and Future Perspectives. In Revolutionizing Healthcare Through Artificial Intelligence and Internet of Things Applications (pp. 177-186). IGI Global.

Singh, A. K., Kumar, S., & Singh, V. P. (2014, March 4). Optimization of parameters using conductive powder in dielectric for EDM of super Co 605 with multiple quality characteristics. *Materials and Manufacturing Processes*, *29*(3), 267–273. doi:10.1080/10426914.2013.864397

Singh, B. (2019). Profiling Public Healthcare: A Comparative Analysis Based on the Multidimensional Healthcare Management and Legal Approach. *Indian Journal of Health and Medical Law*, *2*(2), 1–5.

Singh, B. (2020). Global Science and Jurisprudential Approach Concerning Healthcare and Illness. *Indian Journal of Health and Medical Law*, *3*(1), 7–13.

Singh, B. (2022). COVID-19 Pandemic and Public Healthcare: Endless Downward Spiral or Solution via Rapid Legal and Health Services Implementation with Patient Monitoring Program. *Justice and Law Bulletin*, *1*(1), 1–7.

Singh, B. (2022). Relevance of Agriculture-Nutrition Linkage for Human Healthcare: A Conceptual Legal Framework of Implication and Pathways. *Justice and Law Bulletin*, *1*(1), 44–49.

Singh, B. (2022). Understanding Legal Frameworks Concerning Transgender Healthcare in the Age of Dynamism. *Electronic Journal of Social and Strategic Studies*, *3*(1), 56–65. doi:10.47362/EJSSS.2022.3104

Singh, B. (2023). Federated Learning for Envision Future Trajectory Smart Transport System for Climate Preservation and Smart Green Planet: Insights into Global Governance and SDG-9 (Industry, Innovation and Infrastructure). *National Journal of Environmental Law*, *6*(2), 6–17.

Singh, B. (2024). Legal Dynamics Lensing Metaverse Crafted for Videogame Industry and E-Sports: Phenomenological Exploration Catalyst Complexity and Future. *Journal of Intellectual Property Rights Law*, *7*(1), 8–14.

Singh, D., Vignat, J., Lorenzoni, V., Eslahi, M., Ginsburg, O., Lauby-Secretan, B., Arbyn, M., Basu, P., Bray, F., & Vaccarella, S. (2023). Global estimates of incidence and mortality of cervical cancer in 2020: A baseline analysis of the WHO Global Cervical Cancer Elimination Initiative. *The Lancet. Global Health*, *11*(2), e197–e206. doi:10.1016/S2214-109X(22)00501-0

Sinha, S. (2023). State of IoT 2023: Number of connected IoT devices growing 16% to 16.7 billion globally. *Iot Analytics*. https://iot-analytics.com/number-connected-iot-devices/

Sirois, E., Wang, Q., & Sun, W. (2011). Fluid simulation of a transcatheter aortic valve deployment into a patient-specific aortic root. *Cardiovascular Engineering and Technology*, *2*(3), 186–195. doi:10.1007/s13239-011-0037-7

Sisinni, E., Saifullah, A., Han, S., Jennehag, U., & Gidlund, M. (2018). Industrial Internet of Things: Challenges, Opportunities, and Directions. *IEEE Transactions on Industrial Informatics*, *14*(11), 4724–4734. doi:10.1109/TII.2018.2852491

Skłodowski, K., Chmielewska-Deptuła, S. J., Piktel, E., Wolak, P., Wollny, T., & Bucki, R. (2023). Metallic Nanosystems in the Development of Antimicrobial Strategies with High Antimicrobial Activity and High Biocompatibility. *International Journal of Molecular Sciences*, *24*(3), 2104. doi:10.3390/ijms24032104 PMID:36768426

Small, G. W., Lee, J., Kaufman, A., Jalil, J., Siddarth, P., Gaddipati, H., & Bookheimer, S. Y. (2022). Brain health consequences of digital technology use. *Dialogues in Clinical Neuroscience*. PMID:32699518

Solberg, H. E. (1987). The concept of reference values. *Journal of Clinical Chemistry and Clinical Biochemistry. Zeitschrift fur Klinische Chemie und Klinische Biochemie*, *25*, 337–342. PMID:3612033

Song, P., Zha, M., Yang, Q., Zhang, Y., Li, X., & Rudan, I. J. J. o. g. h. (2021). The prevalence of adult attention-deficit hyperactivity disorder: A global systematic review and meta-analysis. *11*.

Stahl, B. C., Andreou, A., Brey, P., Hatzakis, T., Kirichenko, A., Macnish, K., Shaelou, S. L., Patel, A., Ryan, M., & Wright, D. (2021). Artificial intelligence for human flourishing–Beyond principles for machine learning. *Journal of Business Research*, *124*, 374–388. doi:10.1016/j.jbusres.2020.11.030

Stangl, M., Maoz, S. L., & Suthana, N. (2023). Mobile cognition: Imaging the human brain in the 'real world'. *Nature Reviews. Neuroscience*, *24*(6), 1–16. doi:10.1038/s41583-023-00692-y PMID:37046077

Stein, D. J., He, Y., Phillips, A., Sahakian, B. J., Williams, J., & Patel, V. (2015). Global mental health and neuroscience: Potential synergies. *The Lancet. Psychiatry*, *2*(2), 178–185. doi:10.1016/S2215-0366(15)00014-0 PMID:26359754

Stevens, L. J., Kuczek, T., Burgess, J. R., Hurt, E., & Arnold, L. E. J. C. p. (2011). Dietary sensitivities and ADHD symptoms: thirty-five years of research. *50*(4), 279-293.

Stewart, C., Ralyea, C., & Lockwood, S. (2019, April). Ovarian cancer: an integrated review. In Seminars in oncology nursing (Vol. 35, No. 2, pp. 151-156). WB Saunders. https://doi.org/ doi:10.1016/j.soncn.2019.02.001

Steyerberg, E. W., Vickers, A. J., Cook, N. R., Gerds, T., Gonen, M., Obuchowski, N., Pencina, M. J., & Kattan, M. W. (2010). Assessing the performance of prediction models: A framework for traditional and novel measures. *Epidemiology (Cambridge, Mass.)*, *21*(1), 128–138. doi:10.1097/EDE.0b013e3181c30fb2 PMID:20010215

Subbiah, V. (2023). The next generation of evidence-based medicine. *Nature Medicine*, *29*(1), 49–58. doi:10.1038/s41591-022-02160-z PMID:36646803

Sudlow, C., Gallacher, J., Allen, N., Beral, V., Burton, P., Danesh, J., Downey, P., Elliott, P., Green, J., Landray, M., Liu, B., Matthews, P., Ong, G., Pell, J., Silman, A., Young, A., Sprosen, T., Peakman, T., & Collins, R. (2015). UK biobank: An open access resource for identifying the causes of a wide range of complex diseases of middle and old age. *PLoS Medicine*, *12*(3), e1001779. doi:10.1371/journal.pmed.1001779 PMID:25826379

Su, E. P., Justin, D. F., Pratt, C. R., Sarin, V. K., Nguyen, V. S., Oh, S., & Jin, S. (2018). Effects of titanium nanotubes on the osseointegration, cell differentiation, mineralisation and antibacterial properties of orthopaedic implant surfaces. *The Bone & Joint Journal*, *100B*(1), 9–16. doi:10.1302/0301-620X.100B1.BJJ-2017-0551.R1 PMID:29292334

Sukuma, R., Athavale, M. M., Makhijani, V. B., & Przekwas, A. J. (1996). Application of computational fluid dynamics techniques to blood pumps. *Artificial Organs*, *20*(5), 529–533. doi:10.1111/j.1525-1594.1996.tb04475.x PMID:28868695

Sullivan, P. F., Daly, M. J., & O'donovan, M. J. N. R. G. (2012). Genetic architectures of psychiatric disorders: the emerging picture and its implications. *13*(8), 537-551.

Sultan, M. B., Mansberger, S. L., & Lee, P. P. (2009). Understanding the importance of IOP variables in glaucoma: A systematic review. *Survey of Ophthalmology*, *54*(6), 643–662. doi:10.1016/j.survophthal.2009.05.001 PMID:19665744

Sun, H., Yuan, F., Shen, X., Xiong, G., & Wu, J. J. M. n. (2014). Role of COMT in ADHD: a systematic meta-analysis. *49*, 251-261.

Sunderman, F. W. Jr. (1975). Current Concepts of "Normal Values," "Reference Values," and "Discrimination Values" in Clinical Chemistry. *Clinical Chemistry*, *21*(13), 1873–1877. doi:10.1093/clinchem/21.13.1873 PMID:1192578

Sun, T. Q. (2021). Adopting artificial intelligence in public healthcare: The effect of social power and learning algorithms. *International Journal of Environmental Research and Public Health*, *18*(23), 12682. doi:10.3390/ijerph182312682 PMID:34886404

Sun, Z., & Xu, L. (2014). Computational fluid dynamics in coronary artery disease. *Computerized Medical Imaging and Graphics*, *38*(8), 651–663. doi:10.1016/j.compmedimag.2014.09.002 PMID:25262321

Suo, J., Liu, Y., Wu, C., Chen, M., Huang, Q., Liu, Y., Yao, K., Chen, Y., Pan, Q., Chang, X., Leung, A., Chan, H., Zhang, G., Yang, Z., Daoud, W., Li, X., Roy, V., Shen, J., Yu, X., & Li, W. (2022). Wide-Bandwidth Nanocomposite-Sensor Integrated Smart Mask for Tracking Multiphase Respiratory Activities (Adv. Sci. 31/2022). *Advanced Science (Weinheim, Baden-Wurttemberg, Germany)*, *9*(31), 2270200. Advance online publication. doi:10.1002/advs.202270200

Sushila, S. D., & Talekar, S. C. (2023). Recent Developments of Protective Textiles 6. in India. *Innovation and Recent Advancement in Fashion and Textile*, 63.

Su, Z., Kong, L., Dai, Y., Tang, J., Mei, J., Qian, Z., Ma, Y., Li, Q., Ju, S., Wang, J., Fan, W., & Zhu, C. (2022). Bioresponsive nano-antibacterials for H_2S-sensitized hyperthermia and immunomodulation against refractory implant–related infections. *Science Advances*, *8*(14), 1–20. doi:10.1126/sciadv.abn1701 PMID:35394829

Suzuki, K., Minei, L. J., Johnson, E. E. J. A. j. o. o., & gynecology. (1980). Effect of nicotine upon uterine blood flow in the pregnant rhesus monkey. *136*(8), 1009-1013.

Sweetman, B., & Linninger, A. A. (2011). Cerebrospinal fluid flow dynamics in the central nervous system. *Annals of Biomedical Engineering*, *39*(1), 484–496. doi:10.1007/s10439-010-0141-0 PMID:20737291

Tajik, P., Zwinderman, A. H., Mol, B. W., & Bossuyt, P. M. (2013). Trial Designs for Personalizing Cancer Care: A Systematic Review and Classification. [PubMed] [Google Scholar]. *Clinical Cancer Research*, *19*(17), 4578–4588. doi:10.1158/1078-0432.CCR-12-3722 PMID:23788580

Tajudeen, Y. O., & Shemishere, U. B. (2019). Role of epigenetics in aetiology and therapies for Type 1 Diabetes Mellitus: A narrative review. *Journal of Health and Social Sciences*. doi:10.19204/2019/rlfp12

Talamini, M. A., Chapman, S., Horgan, S., & Melvin, W. S. (2003). A prospective analysis of 211 robotic-assisted surgical procedures. *Surgical Endoscopy*, *17*(10), 1521–1524. doi:10.1007/s00464-002-8853-3 PMID:12915974

Tamm, E. R., & Fuchshofer, R. (2007). What increases outflow resistance in primary open-angle glaucoma? *Survey of Ophthalmology*, *52*(6), S101–S104. doi:10.1016/j.survophthal.2007.08.002 PMID:17998032

Tamminga, C. A. (2014). Approaching human neuroscience for disease understanding. *World Psychiatry; Official Journal of the World Psychiatric Association (WPA)*, *13*(1), 41–43. doi:10.1002/wps.20099 PMID:24497244

Tan, F. P. P., Xu, X. Y., Torii, R., Wood, N. B., Delahunty, N., Mullen, M., Moat, N., & Mohiaddin, R. (2012). Comparison of aortic flow patterns before and after transcatheter aortic valve implantation. *Cardiovascular Engineering and Technology*, *3*(1), 123–135. doi:10.1007/s13239-011-0073-3

Tang, X., Dai, J., Sun, H., Nabanita, S., Petr, S., Wang, D., Cheng, Q., & Wei, J. (2019). Mechanical Strength, Surface Properties, Cytocompatibility and Antibacterial Activity of Nano Zinc-Magnesium Silicate/Polyetheretherketone Biocomposites. *Journal of Nanoscience and Nanotechnology*, *19*(12), 7615–7623. doi:10.1166/jnn.2019.16727 PMID:31196268

Tan, Y., Yin, P., Tang, L., Xing, W., Huang, Q., Cao, D., Zhao, X., Wang, W., Lu, X., Xu, Z., Wang, H., & Xu, G. (2012). Metabolomics study of stepwise hepatocarcinogenesis from the model rats to patients: Potential biomarkers effective for small hepatocellular carcinoma diagnosis. *Molecular & Cellular Proteomics*, *11*(2), M111.010694. doi:10.1074/mcp.M111.010694 PMID:22084000

Tao, B., Yi, W., Qin, X., Wu, J., Li, K., Guo, A., Hao, J., & Chen, L. (2023). Improvement of antibacterial, anti-inflammatory, and osteogenic properties of OGP loaded Co-MOF coating on titanium implants for advanced osseointegration. *Journal of Materials Science and Technology*, *146*, 131–144. doi:10.1016/j.jmst.2022.11.013

Tenori, L., Oakman, C., Claudino, W. M., Bernini, P., Cappadona, S., Nepi, S., Biganzoli, L., Arbushites, M. C., Luchinat, C., Bertini, I., & Di Leo, A. (2012). Exploration of serum metabolomic profiles and outcomes in women with metastatic breast cancer: A Pilot Study. *Molecular Oncology*, *6*(4), 437–444. doi:10.1016/j.molonc.2012.05.003 PMID:22687601

Tham, S., Koh, F. H., Teo, E. K., Lin, C. L., & Foo, F. J. (2023). Knowledge, perceptions and behaviours of endoscopists towards the use of artificial intelligence-aided colonoscopy. *Surgical Endoscopy*, *37*(10), 7395–7400. doi:10.1007/s00464-023-10412-3 PMID:37670191

Thanoon, M. A., Zulkifley, M. A., Mohd Zainuri, M. A. A., & Abdani, S. R. (2023). A Review of Deep Learning Techniques for Lung Cancer Screening and Diagnosis Based on CT Images. *Diagnostics (Basel), 13*(16), 2617. doi:10.3390/diagnostics13162617 PMID:37627876

Thapa, K., Khan, H., Singh, T. G., & Kaur, A. (2021, September). Traumatic brain injury: Mechanistic insight on pathophysiology and potential therapeutic targets. *Journal of Molecular Neuroscience, 71*(9), 1725–1742. doi:10.1007/s12031-021-01841-7 PMID:33956297

Thapar, A., Cooper, M., Eyre, O., Langley, K. J. J. o. C. P., & Psychiatry. (2013). Practitioner review: what have we learnt about the causes of ADHD?, *54*(1), 3-16.

The ENCODE Project Consortium. (2012). An integrated encyclopedia of DNA elements in the human genome. *Nature, 489*(7414), 57–74. doi:10.1038/nature11247 PMID:22955616

The Global Alliance for Genomics and Health. (2013). A federated ecosystem for sharing genomic, clinical data. *Science, 340*(6132), 1521–1522.

The Hindu Bureau. (2023). Monitoring app and AI-based model for predictive analysis of dengue to be launched. *The Hindu.* https://www.thehindu.com/news/cities/bangalore/app-and-ai-based-model-for- predictive-analysis-of-dengue-to-be-launched/article67281225.ece

Thompson, P. M., Jahanshad, N., Ching, C. R., Salminen, L. E., Thomopoulos, S. I., Bright, J., & Sønderby, I. E. (2020). ENIGMA and global neuroscience: A decade of large-scale studies of the brain in health and disease across more than 40 countries. *Translational Psychiatry, 10*(1), 100. doi:10.1038/s41398-020-0705-1 PMID:32198361

Thygesen, K., Alpert, J. S., & White, H. D. (2007). TASK FORCE MEMBERS: Chairpersons: Kristian Thygesen (Denmark), Biomarker Group: Allan S. Jaffe CU. Universal Definition of Myocardial Infarction. *Circulation, 116*(22), 2634–2653. doi:10.1161/CIRCULATIONAHA.107.187397 PMID:17951284

Tian, G., & Longest, P. W. (2010). *Development of a CFD boundary condition to model transient vapor absorption in the respiratory airways.*

Tian, G., Longest, P. W., Su, G., Walenga, R. L., & Hindle, M. (2011). Development of a stochastic individual path (SIP) model for predicting the tracheobronchial deposition of pharmaceutical aerosols: Effects of transient inhalation and sampling the airways. *Journal of Aerosol Science, 42*(11), 781–799. doi:10.1016/j.jaerosci.2011.07.005

Tian, S., Yang, W., Le Grange, J. M., Wang, P., Huang, W., & Ye, Z. (2019). Smart healthcare: Making medical care more intelligent. *Global Health Journal (Amsterdam, Netherlands), 3*(3), 62–65. doi:10.1016/j.glohj.2019.07.001

Tian, S., Zhang, Y., Feng, Y., Elsagan, N., Ko, Y., Mozaffari, M. H., Xi, D. D., & Lee, C. G. (2023). Time series classification, augmentation and artificial-intelligence-enabled software for emergency response in freight transportation fires. *Expert Systems with Applications, 233*, 120914. doi:10.1016/j.eswa.2023.120914

Tomczak, K., Czerwińska, P., & Wiznerowicz, M. (2015). Review The Cancer Genome Atlas (TCGA): an immeasurable source of knowledge. *Contemporary Oncology/Współczesna Onkologia*, 68-77.

Tomita, M., & Kami, K. (2012). Systems biology, metabolomics, and cancer metabolism. *Science, 336*(6084), 990–991. doi:10.1126/science.1223066 PMID:22628644

Tong, W., Hussain, A., Bo, W. X., & Maharjan, S. (2019). Artificial intelligence for vehicle-to-everything: A survey. *IEEE Access : Practical Innovations, Open Solutions, 7*, 10823–10843. doi:10.1109/ACCESS.2019.2891073

Tran, P. A., O'brien-Simpson, N., Palmer, J. A., Bock, N., Reynolds, E. C., Webster, T. J., Deva, A., Morrison, W. A., & O'connor, A. J. (2019). Selenium nanoparticles as anti-infective implant coatings for trauma orthopedics against methicillin-resistant *Staphylococcus aureus* and *epidermidis: In vitro* and *in vivo* assessment. *International Journal of Nanomedicine, 14*, 4613–4624. doi:10.2147/IJN.S197737 PMID:31308651

Trelis, A., Lago, V., Padilla, P., Martínez, R. P., Ripollés, V. B., Matute, L., & Domingo, S. (2023). Optimal cytoreduction: Is a CT's picture worth a surgeon's word? *Surgical Oncology, 101948*, 101948. Advance online publication. doi:10.1016/j.suronc.2023.101948

Trivedi, H., Gray, L. J., Seidu, S., Davies, M. J., Charpentier, G., Lindblad, U., Kellner, C., Nolan, J., Pazderska, A., Rutten, G., Trento, M., & Khunti, K. (2017). Self-knowledge of HbA1c in people with type 2 diabetes mellitus and its association with glycaemic control. *Primary Care Diabetes, 11*(5), 414–420. doi:10.1016/j.pcd.2017.03.011 PMID:28684227

Tsamados, A., Aggarwal, N., Cowls, J., Morley, J., Roberts, H., Taddeo, M., & Floridi, L. (2021). *The ethics of algorithms: key problems and solutions.* Ethics, Governance, and Policies in Artificial Intelligence.

Tsutsui, H., Maeda, T., Min, J. Z., Inagaki, S., Higashi, T., Kagawa, Y., & Toyo'oka, T. (2011). Biomarker discovery in biological specimens (plasma, hair, liver and kidney) of diabetic mice based upon metabolite profiling using ultra-performance liquid chromatography with electrospray ionization time-of-flight mass spectrometry. *Clinica Chimica Acta, 412*(11-12), 861–872. doi:10.1016/j.cca.2010.12.023 PMID:21185819

Turing, A. M. (2009). *Computing machinery and intelligence.* Springer Netherlands. doi:10.1007/978-1-4020-6710-5_3

Turing, D. (2015). *Prof: Alan Turing Decoded.* Pitkin.

Uchida, M., Spencer, T. J., Faraone, S. V., & Biederman, J. J. J. o. a. d. (2018). Adult outcome of ADHD: an overview of results from the MGH longitudinal family studies of pediatrically and psychiatrically referred youth with and without ADHD of both sexes. *22*(6), 523-534.

Ukil, A., Bandyoapdhyay, S., Puri, C., & Pal, A. (2016, March). IoT healthcare analytics: The importance of anomaly detection. In *2016 IEEE 30th international conference on advanced information networking and applications (AINA)* (pp. 994-997). IEEE.

Ulhaq, A., Born, J., Khan, A., Gomes, D. P., Chakraborty, S., & Paul, M. (2020). COVID-19 control by computer vision approaches: A survey. *IEEE Access : Practical Innovations, Open Solutions, 8*, 179437–179456. doi:10.1109/ACCESS.2020.3027685 PMID:34812357

Um, Y. H., Hong, S.-C., Jeong, J.-H. J. C. P., & Neuroscience. (2017). Sleep problems as predictors in attention-deficit hyperactivity disorder: causal mechanisms, consequences and treatment. *15*(1), 9.

Unnikrishnan, R., Radha, V., & Mohan, V. (2021). Challenges Involved in Incorporating Personalised Treatment Plan as Routine Care of Patients with Diabetes. *Pharmacogenomics and Personalized Medicine, 14*, 327–333. doi:10.2147/PGPM.S271582 PMID:33758531

Van Cott, H. P., & Kinkade, R. G. (Eds.). (1972). *Human engineering guide to equipment design.* Department of Defense.

van Hengel, I. A. J., Riool, M., Fratila-Apachitei, L. E., Witte-Bouma, J., Farrell, E., Zadpoor, A. A., Zaat, S. A. J., & Apachitei, I. (2017). Selective laser melting porous metallic implants with immobilized silver nanoparticles kill and prevent biofilm formation by methicillin-resistant *Staphylococcus aureus. Biomaterials, 140*, 1–15. doi:10.1016/j.biomaterials.2017.02.030 PMID:28622569

Vasan, R. S. (2006). Biomarkers of Cardiovascular Disease: Molecular Basis and Practical Considerations. *Circulation, 113*(19), 2335–2362. doi:10.1161/CIRCULATIONAHA.104.482570 PMID:16702488

Vayena, E., Brownsword, R., Edwards, S. J., Greshake, B., Kahn, J. P., Ladher, N., Montgomery, J., O'Connor, D., O'Neill, O., Richards, M. P., Rid, A., Sheehan, M., Wicks, P., & Tasioulas, J. (2016). Research led by participants: A new social contract for a new kind of research. *Journal of Medical Ethics*, *42*(4), 216–219. doi:10.1136/medethics-2015-102663 PMID:25825527

Vijayan, V., Connolly, J. P., Condell, J., McKelvey, N., & Gardiner, P. (2021). Review of Wearable Devices and Data Collection Considerations for Connected Health. *Sensors (Basel)*, *21*(16), 5589. doi:10.3390/s21165589 PMID:34451032

Villeneuve, L. M., Reddy, M. A., Lanting, L. L., Wang, M., Meng, L., & Natarajan, R. (2008). Epigenetic histone H3 lysine 9 methylation in metabolic memory and inflammatory phenotype of vascular smooth muscle cells in diabetes. *Proceedings of the National Academy of Sciences of the United States of America*, *105*(26), 9047–9052. doi:10.1073/pnas.0803623105 PMID:18579779

Vora, L. K., Gholap, A. D., Jetha, K., Thakur, R. R. S., Solanki, H. K., & Chavda, V. P. (2023). Artificial Intelligence in Pharmaceutical Technology and Drug Delivery Design. *Pharmaceutics*, *15*(7), 1916. doi:10.3390/pharmaceutics15071916 PMID:37514102

Vouk, K., Hevir, N., Ribič-Pucelj, M., Haarpaintner, G., Scherb, H., Osredkar, J., Möller, G., Prehn, C., Rižner, T. L., & Adamski, J. (2012). Discovery of phosphatidylcholines and sphingomyelins as biomarkers for ovarian endometriosis. *Human Reproduction (Oxford, England)*, *27*(10), 2955–2965. doi:10.1093/humrep/des152 PMID:22859507

Vrana, J., & Singh, R. (2021). NDE 4.0—A design thinking perspective. *Journal of Nondestructive Evaluation*, *40*(1), 8. doi:10.1007/s10921-020-00735-9 PMID:33424070

Wade, A. N. G. I. (2000). Derivation versus validation. *Archives of Disease in Childhood*, *83*(6), 459–460. doi:10.1136/adc.83.6.459 PMID:11087274

Walenga, R. L., Tian, G., & Worth Longest, P. (2013). Development of characteristic upper tracheobronchial airway models for testing pharmaceutical aerosol delivery. *Journal of Biomechanical Engineering*, *135*(9), 091010. doi:10.1115/1.4024630 PMID:23722698

Walsh, R. (2011). Lifestyle and mental health. *The American Psychologist*, *66*(7), 579–592. doi:10.1037/a0021769 PMID:21244124

Wang, B., Chen, D., Chen, Y., Hu, Z., Cao, M., Xie, Q., Chen, Y., Xu, J., Zheng, S., & Li, L. (2012). Metabonomic profiles discriminate hepatocellular carcinoma from liver cirrhosis by ultraperformance liquid chromatography-mass spectrometry. *Journal of Proteome Research*, *11*(2), 1217–1227. doi:10.1021/pr2009252 PMID:22200553

Wang, B., Wu, Z., Lan, J., Li, Y., Xie, L., Huang, X., Zhang, A., Qiao, H., Chang, X., Lin, H., Zhang, H., Li, T., & Huang, Y. (2021). Surface modification of titanium implants by silk fibroin/Ag co-functionalized strontium titanate nanotubes for inhibition of bacterial-associated infection and enhancement of *in vivo* osseointegration. *Surface and Coatings Technology*, *405*, 126700. doi:10.1016/j.surfcoat.2020.126700

Wang, H., Jia, S., Li, Z., Duan, Y., Tao, G., & Zhao, Z. (2022). A Comprehensive Review of Artificial Intelligence in Prevention and Treatment of COVID-19 Pandemic. *Frontiers in Genetics*, *13*, 845305. doi:10.3389/fgene.2022.845305 PMID:35559010

Wang, J., Li, C., Han, S., Sarkar, S., & Zhou, X. (2017). Predictive maintenance based on event-log analysis: A case study. *IBM Journal of Research and Development*, *61*(1), 11–121. doi:10.1147/JRD.2017.2648298

Wang, J., Wang, L., Pan, J., Zhao, J., Tang, J., Jiang, D., Hu, P., Jia, W., & Shi, J. (2021). Magneto-Based Synergetic Therapy for Implant-Associated Infections via Biofilm Disruption and Innate Immunity Regulation. *Advancement of Science*, *8*(6), 1–12. doi:10.1002/advs.202004010 PMID:33747747

Wang, J., Zhou, H., Guo, G., Tan, J., Wang, Q., Tang, J., Liu, W., Shen, H., Li, J., & Zhang, X. (2017). Enhanced anti-infective efficacy of ZnO nanoreservoirs through a combination of intrinsic anti-biofilm activity and reinforced innate defense. *ACS Applied Materials & Interfaces*, *9*(39), 33609–33623. doi:10.1021/acsami.7b08864 PMID:28884578

Wang, L. (2022). Deep Learning Techniques to Diagnose Lung Cancer. *Cancers (Basel)*, *14*(22), 5569. doi:10.3390/cancers14225569 PMID:36428662

Wang, M., & Tang, T. (2019). Surface treatment strategies to combat implant-related infection from the beginning. *Journal of Orthopaedic Translation*, *17*, 42–54. doi:10.1016/j.jot.2018.09.001 PMID:31194031

Wang, M., Yang, B., Liu, Y., Yang, Y., Ji, H., & Yang, C. (2023). Emerging infectious disease surveillance using a hierarchical diagnosis model and the Knox algorithm. *Scientific Reports*, *13*(1), 19836. doi:10.1038/s41598-023-47010-1 PMID:37963966

Wang, T. J. (2011). Assessing the Role of Circulating, Genetic, and Imaging Biomarkers in Cardiovascular Risk Prediction. *Circulation*, *123*(5), 551–565. doi:10.1161/CIRCULATIONAHA.109.912568 PMID:21300963

Wang, T. J., Gona, P., Larson, M. G., Tofler, G. H., Levy, D., Newton-Cheh, C., Jacques, P. F., Rifai, N., Selhub, J., Robins, S. J., Benjamin, E. J., D'Agostino, R. B., & Vasan, R. S. (2006). Multiple Biomarkers for the Prediction of First Major Cardiovascular Events and Death. *The New England Journal of Medicine*, *355*(25), 2631–2639. doi:10.1056/NEJMoa055373 PMID:17182988

Wang, T. J., Larson, M. G., Vasan, R. S., Cheng, S., Rhee, E. P., McCabe, E., Lewis, G. D., Fox, C. S., Jacques, P. F., Fernandez, C., O'Donnell, C. J., Carr, S. A., Mootha, V. K., Florez, J. C., Souza, A., Melander, O., Clish, C. B., & Gerszten, R. E. (2011). Metabolite profiles and the risk of developing diabetes. *Nature Medicine*, *17*(4), 448–453. doi:10.1038/nm.2307 PMID:21423183

Wang, W. H., & Hsu, W. S. (2023). Integrating artificial intelligence and wearable IoT system in long-term care environments. *Sensors (Basel)*, *23*(13), 5913. doi:10.3390/s23135913 PMID:37447763

Wang, W., Cheng, X., Liao, J., Lin, Z., Chen, L., Liu, D., Zhang, T., Li, L., Lu, Y., & Xia, H. (2019). Synergistic Photothermal and Photodynamic Therapy for Effective Implant-Related Bacterial Infection Elimination and Biofilm Disruption Using Cu_9S_8 Nanoparticles. *ACS Biomaterials Science & Engineering*, *5*(11), 6243–6253. doi:10.1021/acsbiomaterials.9b01280 PMID:33405531

Wang, Y., Teng, W., Zhang, Z., Zhou, X., Ye, Y., Lin, P., Liu, A., Wu, Y., Li, B., Zhang, C., Yang, X., Li, W., Yu, X., Gou, Z., & Ye, Z. (2021). A trilogy antimicrobial strategy for multiple infections of orthopedic implants throughout their life cycle. *Bioactive Materials*, *6*(7), 1853–1866. doi:10.1016/j.bioactmat.2020.11.030 PMID:33336116

Wang, Y., Tetko, I. V., Hall, M. A., Frank, E., Facius, A., Mayer, K. F., & Mewes, H. W. (2005). Gene selection from microarray data for cancer classification—A machine learning approach. *Computational Biology and Chemistry*, *29*(1), 37–46. doi:10.1016/j.compbiolchem.2004.11.001 PMID:15680584

Wang, Y., Zhang, M., Chen, Y., & Qu, H. (2022). Signed neuron with memory: Towards simple, accurate and high-efficient ann-snn conversion. In *International Joint Conference on Artificial Intelligence*. IEEE. 10.24963/ijcai.2022/347

Warburton, P. E., & Sebra, R. P. (2023). Long-Read DNA Sequencing: Recent Advances and Remaining Challenges. *Annual Review of Genomics and Human Genetics*, *24*(1), 24. doi:10.1146/annurev-genom-101722-103045 PMID:37075062

Ware, J. H. (2006). The Limitations of Risk Factors as Prognostic Tools. *The New England Journal of Medicine*, *355*(25), 2615–2617. doi:10.1056/NEJMp068249 PMID:17182986

Wei, H., Song, X., Liu, P., Liu, X., Yan, X., & Yu, L. (2022). Antimicrobial coating strategy to prevent orthopaedic device-related infections: Recent advances and future perspectives. *Biomaterials Advances*, *135*, 212739. doi:10.1016/j.bioadv.2022.212739 PMID:35929213

Wei, W., Visweswaran, S., & Cooper, G. F. (2011). The application of naive Bayes model averaging to predict Alzheimer's disease from genome-wide data. *Journal of the American Medical Informatics Association : JAMIA*, *18*(4), 370–375. doi:10.1136/amiajnl-2011-000101 PMID:21672907

Wendell, D. C., Samyn, M. M., Cava, J. R., Krolikowski, M. M., & LaDisa, J. F. Jr. (2016). The impact of cardiac motion on aortic valve flow used in computational simulations of the thoracic aorta. *Journal of Biomechanical Engineering*, *138*(9), 091001. doi:10.1115/1.4033964 PMID:27367143

Whitaker, R. (2011). *Anatomy of an epidemic: Magic bullets, psychiatric drugs, and the astonishing rise of mental illness in America*. Crown.

White House. (2016). *Precision Medicine.* White House. https://www.whitehouse.gov/precision-medicine

Wickramasinghe, S., Ju, M., Milbrandt, N. B., Tsai, Y. H., Navarreto-Lugo, M., Visperas, A., Klika, A., Barsoum, W., Higuera-Rueda, C. A., & Samia, A. C. S. (2020). Photoactivated Gold Nanorod Hydrogel Composite Containing D-Amino Acids for the Complete Eradication of Bacterial Biofilms on Metal Alloy Implant Materials. *ACS Applied Nano Materials*, *3*(6), 5862–5873. doi:10.1021/acsanm.0c01018

Wilens, T. E., & Spencer, T. J. J. P. m. (2010). Understanding attention-deficit/hyperactivity disorder from childhood to adulthood. *122*(5), 97.

Williams, R. (2019). *European perspectives on the anticipatory governance of AI. AI Governance: A Year in Review: Observations of 50 Global Experts.* Institute for Science of Science.

Wilson, E. B., & Sudan, R. (2013). The evolution of robotic bariatric surgery. *World Journal of Surgery*, *37*(12), 2756–2760. doi:10.1007/s00268-013-2125-3 PMID:23778837

Winarto, H., Welladatika, A., Habiburrahman, M., Purwoto, G., Kusuma, F., Utami, T. W., & Nuryanto, K. H. (2022). Overall survival and related factors of advanced-stage epithelial ovarian cancer patients underwent debulking surgery in Jakarta, Indonesia: A single-center experience. *Open Access Macedonian Journal of Medical Sciences*, *10*(B), 265-280. https://doi.org/ doi:10.3889/oamjms.2022.8296

Wirtz, J., Hofmeister, J., Chew, P., & Ding, X. (2023). Digital Service Technologies, Service Robots, AI, and the Strategic Pathways to Cost-Effective Service Excellence. *Service Industries Journal*, *43*(15-16), 1173–1196. doi:10.1080/02642069.2023.2226596

Wölfel, R., Corman, V., Guggemos, W., Seilmaier, M., Zange, S., Müller, M., Niemeyer, D., Jones, T., Vollmar, P., Rothe, C., Hoelscher, M., Bleicker, T., Brünink, S., Schneider, J., Ehmann, R., Zwirglmaier, K., Drosten, C., & Wendtner, C. (2020). Virological assessment of hospitalized patients with COVID-2019. *Nature*, *581*(7809), 465–469. doi:10.1038/s41586-020-2196-x PMID:32235945

Wong, J., Murray Horwitz, M., Zhou, L., & Toh, S. (2018). Using machine learning to identify health outcomes from electronic health record data. *Current Epidemiology Reports*, *5*(4), 331–342. doi:10.1007/s40471-018-0165-9 PMID:30555773

Woo, H. M., Kim, K. M., Choi, M. H., Jung, B. H., Lee, J., Kong, G., Nam, S. J., Kim, S., Bai, S. W., & Chung, B. C. (2009). Mass spectrometry based metabolomic approaches in urinary biomarker study of women's cancers. *Clinica Chimica Acta*, *400*(1-2), 63–69. doi:10.1016/j.cca.2008.10.014 PMID:19010317

Worden, K., Tsialiamanis, G., Cross, E. J., & Rogers, T. J. (2023). Artificial neural networks. In *Machine Learning in Modeling and Simulation: Methods and Applications* (pp. 85–119). Springer International Publishing. doi:10.1007/978-3-031-36644-4_2

World Health Organization. (2001). [Mental health: new understanding, new hope.]. *World Health Report*. WHO.

World Health Organization. (2021). *Ethics and governance of artificial intelligence for health: WHO guidance*. WHO.

Wright, J. D. (2017). Robotic-assisted surgery: Balancing evidence and implementation. *Journal of the American Medical Association, 318*(16), 1545–1547. doi:10.1001/jama.2017.13696 PMID:29067404

Wu, J., Xiao, H., Sun, H., Zou, L., & Zhu, L.-Q. J. M. n. (2012). Role of dopamine receptors in ADHD: a systematic meta-analysis. *45*, 605-620.

Wu, L., Zhang, J., Zhou, W., An, P., Shen, L., Liu, J., Jiang, X., Huang, X., Mu, G., Wan, X., Lv, X., Gao, J., Cui, N., Hu, S., Chen, Y., Hu, X., Li, J., Chen, D., Gong, D., & Yu, H. G. (2019). Randomised controlled trial of WISENSE, a real-time quality improving system for monitoring blind spots during esophagogastroduodenoscopy. *Gut*, *68*(12), 2161–2169. Advance online publication. doi:10.1136/gutjnl-2018-317366 PMID:30858305

Wu, Q., Zhang, Y. D., Tao, W., & Amin, M. G. (2015). Radar-based fall detection based on Doppler time–frequency signatures for assisted living. *IET Radar, Sonar & Navigation*, *9*(2), 164–172. doi:10.1049/iet-rsn.2014.0250

Wu, Z., Chan, B., Low, J., Chu, J. J. H., Hey, H. W. D., & Tay, A. (2022). Microbial resistance to nanotechnologies: An important but understudied consideration using antimicrobial nanotechnologies in orthopaedic implants. *Bioactive Materials*, *16*, 249–270. doi:10.1016/j.bioactmat.2022.02.014 PMID:35415290

Xiang, Y., Liu, X., Mao, C., Liu, X., Cui, Z., Yang, X., Yeung, K. W. K., Zheng, Y., & Wu, S. (2018). Infection-prevention on Ti implants by controlled drug release from folic acid/ZnO quantum dots sealed titania nanotubes. *Materials Science and Engineering C*, *85*, 214–224. doi:10.1016/j.msec.2017.12.034 PMID:29407150

Xiao, J. F., Varghese, R. S., Zhou, B., Nezami Ranjbar, M. R., Zhao, Y., Tsai, T.-H., Di Poto, C., Wang, J., Goerlitz, D., Luo, Y., Cheema, A. K., Sarhan, N., Soliman, H., Tadesse, M. G., Ziada, D. H., & Ressom, H. W. (2012). LC-MS based serum metabolomics for identification of hepatocellular carcinoma biomarkers in Egyptian cohort. *Journal of Proteome Research*, *11*(12), 5914–5923. doi:10.1021/pr300673x PMID:23078175

Xu, B., Jia, T., Macare, C., Banaschewski, T., Bokde, A. L., Bromberg, U., & Psychiatry, A. (2017). Impact of a common genetic variation associated with putamen volume on neural mechanisms of attention-deficit/hyperactivity disorder. *56*(5), 436-444. e434.

Yamashita, R., Nishio, M., Do, R. K. G., & Togashi, K. (2018). Convolutional neural networks: An overview and application in radiology. *Insights Into Imaging*, *9*(4), 611–629. doi:10.1007/s13244-018-0639-9 PMID:29934920

Yan, B., Yao, J., Liu, J.-Y., Li, X.-M., Wang, X.-Q., Li, Y.-J., Tao, Z.-F., Song, Y.-C., Chen, Q., & Jiang, Q. (2015). lncRNA-MIAT regulates microvascular dysfunction by functioning as a competing endogenous RNA. *Circulation Research*, *116*(7), 1143–1156. doi:10.1161/CIRCRESAHA.116.305510 PMID:25587098

Yang, B. T., Dayeh, T. A., Volkov, P. A., Kirkpatrick, C. L., Malmgren, S., Jing, X., Renström, E., Wollheim, C. B., Nitert, M. D., & Ling, C. (2012). Increased DNA methylation and decreased expression of PDX-1 in pancreatic islets from patients with type 2 diabetes. *Molecular Endocrinology (Baltimore, Md.)*, *26*(7), 1203–1212. doi:10.1210/me.2012-1004 PMID:22570331

Yang, Y., Liu, L., Luo, H., Zhang, D., Lei, S., & Zhou, K. (2019). Dual-Purpose Magnesium-Incorporated Titanium Nanotubes for Combating Bacterial Infection and Ameliorating Osteolysis to Realize Better Osseointegration. *ACS Biomaterials Science & Engineering*, *5*(10), 5368–5383. doi:10.1021/acsbiomaterials.9b00938 PMID:33464078

Yan, J., Xia, D., Zhou, W., Li, Y., Xiong, P., Li, Q., Wang, P., Li, M., Zheng, Y., & Cheng, Y. (2020). pH-responsive silk fibroin-based CuO/Ag micro/nano coating endows polyetheretherketone with synergistic antibacterial ability, osteogenesis, and angiogenesis. *Acta Biomaterialia*, *115*, 220–234. doi:10.1016/j.actbio.2020.07.062 PMID:32777292

Yao, L., Zhang, H., Zhang, M., Chen, X., Zhang, J., Huang, J., & Zhang, L. (2021). Application of artificial intelligence in renal disease. *Clinical eHealth, 4*, 54-61.

Yao, C., Zhu, M., Han, X., Xu, Q., Dai, M., Nie, T., & Liu, X. (2021). A Bone-Targeting Enoxacin Delivery System to Eradicate *Staphylococcus aureus*-Related Implantation Infections and Bone Loss. *Frontiers in Bioengineering and Biotechnology*, *9*(November), 1–17. doi:10.3389/fbioe.2021.749910 PMID:34869262

Ye, D., Liu, Z., Sun, M., Shi, B., Zhao, P., Wu, H., Yu, H., Yang, S., Wu, X., Guo, Q., Chen, Q., Yin, Y., Zhang, H., Shi, T., Wang, L., Fu, Q., Yang, W., & Huang, L. (2020, April). Mastering complex control in MOBA games with deep reinforcement learning. *Proceedings of the AAAI Conference on Artificial Intelligence*, *34*(04), 6672–6679. doi:10.1609/aaai.v34i04.6144

Yedukondalu, M., & Ravikanth, M. (2011). Core-modified porphyrin based assemblies. *Coordination Chemistry Reviews*, *255*(5-6), 547–573. doi:10.1016/j.ccr.2010.10.039

Ye, J. (2021). Advancing mental health and psychological support for health care workers using digital technologies and platforms. *JMIR Formative Research*, *5*(6), e22075. doi:10.2196/22075 PMID:34106874

Ye, Z., Ling, Y., Yang, M., Xu, Y., Zhu, L., Yan, Z., & Chen, P. (2022). A Breathable, Reusable, and Zero-Power Smart Face Mask for Wireless Cough and Mask-Wearing Monitoring. *ACS Nano*, *16*(4), 5874–5884. doi:10.1021/acsnano.1c11041 PMID:35298138

Yiallourou, T. I., Kröger, J. R., Stergiopulos, N., Maintz, D., Martin, B. A., & Bunck, A. C. (2012). Comparison of 4D phase-contrast MRI flow measurements to computational fluid dynamics simulations of cerebrospinal fluid motion in the cervical spine. *PLoS One*, *7*(12), e52284. doi:10.1371/journal.pone.0052284 PMID:23284970

Yolton, K., Cornelius, M., Ornoy, A., McGough, J., Makris, S., Schantz, S. J. N., & teratology. (2014). Exposure to neurotoxicants and the development of attention deficit hyperactivity disorder and its related behaviors in childhood. *44*, 30-45.

Yoon, H. J., Jeong, Y. J., Kang, H., Jeong, J. E., & Kang, D.-Y. (2019). Medical image analysis using artificial intelligence. *Progress in Medical Physics*, *30*(2), 49–58. doi:10.14316/pmp.2019.30.2.49

Yoshida, M., Hatano, N., Nishiumi, S., Irino, Y., Izumi, Y., Takenawa, T., & Azuma, T. (2012). Diagnosis of gastroenterological diseases by metabolome analysis using gas chromatography-mass spectrometry. *Journal of Gastroenterology*, *47*(1), 9–20. doi:10.1007/s00535-011-0493-8 PMID:22041921

Yoshimasu, K., Kiyohara, C., Takemura, S., & Nakai, K. J. N. (2014). A meta-analysis of the evidence on the impact of prenatal and early infancy exposures to mercury on autism and attention deficit/hyperactivity disorder in the childhood. *44*, 121-131.

Yuan, M., Breitkopf, S. B., Yang, X., & Asara, J. M. (2012). A positive/negative ion-switching, targeted mass spectrometry-based metabolomics platform for bodily fluids, cells, and fresh and fixed tissue. *Nature Protocols*, *7*(5), 872–881. doi:10.1038/nprot.2012.024 PMID:22498707

Yuan, Z., Tao, B., He, Y., Mu, C., Liu, G., Zhang, J., Liao, Q., Liu, P., & Cai, K. (2019). Remote eradication of biofilm on titanium implant via near-infrared light triggered photothermal/photodynamic therapy strategy. *Biomaterials*, *223*, 119479. doi:10.1016/j.biomaterials.2019.119479 PMID:31520885

Yu, K. H., Fitzpatrick, M. R., Pappas, L., Chan, W., Kung, J., & Snyder, M. (2018). Omics AnalySIs System for PRecision Oncology (OASISPRO): A web-based omics analysis tool for clinical phenotype prediction. *Bioinformatics (Oxford, England)*, *34*(2), 319–320. doi:10.1093/bioinformatics/btx572 PMID:28968749

Yu, K. H., Levine, D. A., Zhang, H., Chan, D. W., Zhang, Z., & Snyder, M. (2016). Predicting Ovarian Cancer Patients' Clinical Response to Platinum-Based Chemotherapy by Their Tumor Proteomic Signatures. *Journal of Proteome Research*, *15*(8), 2455–2465. doi:10.1021/acs.jproteome.5b01129 PMID:27312948

Yu, K. H., & Snyder, M. (2016). Omics profiling in precision oncology. *Molecular & Cellular Proteomics*, *15*(8), 2525–2536. doi:10.1074/mcp.O116.059253 PMID:27099341

Yurttas, C., Überrück, L., Nadiradze, G., Königsrainer, A., & Horvath, P. (2022). Limitations of laparoscopy to assess the peritoneal cancer index and eligibility for cytoreductive surgery with HIPEC in peritoneal metastasis. *Langenbeck's Archives of Surgery*, *407*(4), 1667–1675. doi:10.1007/s00423-022-02455-2

Zachow, S., Steinmann, A., Hildebrandt, T., & Heppt, W. (2007). Understanding nasal airflow via CFD simulation and visualization. *Proc. Comp Aid Surg Around the Head*, 173-176.

Zaidi, M. Y., Lee, R. M., Gamboa, A. C., Speegle, S., Cloyd, J. M., Kimbrough, C., Grotz, T., Leiting, J., Fournier, K., Lee, A. J., Dineen, S., Dessureault, S., Kelly, K. J., Kotha, N. V., Clarke, C., Gamblin, T. C., Patel, S. H., Lee, T. C., Hendrix, R. J., & Staley, C. A. III. (2020). Preoperative risk score for predicting incomplete cytoreduction: A 12-institution study from the US HIPEC collaborative. *Annals of Surgical Oncology*, *27*(1), 156–164. doi:10.1245/s10434-019-07626-y

Zeng, Q., He, Y., Dong, S., Zhao, X., Chen, Z., Song, Z., Chang, G., Yang, F., & Wang, Y. (2014). Optimal cut-off values of BMI, waist circumference and waist:height ratio for defining obesity in Chinese adults. *British Journal of Nutrition*, *112*(10), 1735–1744. doi:10.1017/S0007114514002657 PMID:25300318

Zhang, H.-B., Wang, Y.-F., Li, J., Wang, B., & Yang, L. J. Z. e. k. z. z. C. J. o. P. (2005). Association between dopamine beta hydroxylase gene and attention deficit hyperactivity disorder complicated with disruptive behavior disorder. *43*(1), 26-30.

Zhang, A., Sun, H., Han, Y., Yuan, Y., Wang, P., Song, G., Yuan, X., Zhang, M., Xie, N., & Wang, X. (2012). Exploratory urinary metabolic biomarkers and pathways using UPLC-Q-TOF-HDMS coupled with pattern recognition approach. *Analyst*, *137*(18), 4200–4208. doi:10.1039/c2an35780a PMID:22852134

Zhang, A., Sun, H., & Wang, X. (2012). Saliva metabolomics opens door to biomarker discovery, disease diagnosis, and treatment. *Applied Biochemistry and Biotechnology*, *168*(6), 1718–1727. doi:10.1007/s12010-012-9891-5 PMID:22971835

Zhang, A., Sun, H., Wu, X., & Wang, X. (2012). Urine metabolomics. *Clinica Chimica Acta*, *414*, 65–69. doi:10.1016/j.cca.2012.08.016 PMID:22971357

Zhang, A., Sun, H., Yan, G., Wang, P., Han, Y., & Wang, X. (2014). Metabolomics in diagnosis and biomarker discovery of colorectal cancer. *Cancer Letters*, *345*(1), 17–20. doi:10.1016/j.canlet.2013.11.011 PMID:24333717

Zhang, B., Zheng, Y., Qiu, Y., Yang, L., Zhang, R., Liao, Y., & Xu, J. (2022). Overview of smart masks and research on new technology. *Wearable Technologies*, *2*(2), 46. doi:10.54517/wt.v2i2.1648

Zhang, G., Yang, Y., Shi, J., Yao, X., Chen, W., Wei, X., Zhang, X., & Chu, P. K. (2021). Near-infrared light II - assisted rapid biofilm elimination platform for bone implants at mild temperature. *Biomaterials*, *269*, 120634. doi:10.1016/j.biomaterials.2020.120634 PMID:33421708

Zhang, J., & Zhang, Z. M. (2023). Ethics and governance of trustworthy medical artificial intelligence. *BMC Medical Informatics and Decision Making*, *23*(1), 7. doi:10.1186/s12911-023-02103-9 PMID:36639799

Zhang, Q., Zhou, C., Xiong, N., Qin, Y., Li, X., & Huang, S. (2015). Multimodel-based incident prediction and risk assessment in dynamic cybersecurity protection for industrial control systems. *IEEE Transactions on Systems, Man, and Cybernetics. Systems*, *46*(10), 1429–1444. doi:10.1109/TSMC.2015.2503399

Zhang, S., Vaida, J., Parenti, J., Lindsey, B. A., Xing, M., & Li, B. (2021). Programmed Multidrug Delivery Based on Bio-Inspired Capsule-Integrated Nanocoatings for Infected Bone Defect Treatment. *ACS Applied Materials & Interfaces*, *13*(10), 12454–12462. doi:10.1021/acsami.0c20332 PMID:33683872

Zhang, T., Wu, X., Yin, M., Fan, L., Zhang, H., Zhao, F., Zhang, W., Ke, C., Zhang, G., Hou, Y., Zhou, X., Lou, G., & Li, K. (2012). Discrimination between malignant and benign ovarian tumors by plasma metabolomic profiling using ultra performance liquid chromatography/mass spectrometry. *Clinica Chimica Acta*, *413*(9-10), 861–868. doi:10.1016/j.cca.2012.01.026 PMID:22309680

Zhang, W., Yang, D., & Wang, H. (2019). Data-driven methods for predictive maintenance of industrial equipment: A survey. *IEEE Systems Journal*, *13*(3), 2213–2227. doi:10.1109/JSYST.2019.2905565

Zhang, Y., Liu, T., Meyer, C. A., Eeckhoute, J., Johnson, D. S., Bernstein, B. E., Nusbaum, C., Myers, R. M., Brown, M., Li, W., & Liu, X. S. (2008). Model-based Analysis of ChIP-Seq (MACS). *Genome Biology*, *9*(9), R137. doi:10.1186/gb-2008-9-9-r137 PMID:18798982

Zhang, Z., Wen, F., Sun, Z., Guo, X., He, T., & Lee, C. (2022). Artificial intelligence-enabled sensing technologies in the 5G/internet of things era: From virtual reality/augmented reality to the digital twin. *Advanced Intelligent Systems*, *4*(7), 2100228. doi:10.1002/aisy.202100228

Zhao, L., Ferraro, P., & Shorten, R. (2022). A DLT enabled smart mask system to enable social compliance. ArXiv, abs/2205.13256. https://doi.org//arXiv.2205.13256. doi:10.48550

Zhao, J., & Gómez Fariñas, B. (2023). Artificial intelligence and sustainable decisions. *European Business Organization Law Review*, *24*(1), 1–39. doi:10.1007/s40804-022-00262-2

Zhao, S., Wei, C., Tang, H., Ding, H., Han, B., Chen, S., Song, X., Gu, Q., Zhang, Y., Liu, W., & Wang, J. (2021). Elevated DNA polymerase delta 1 expression correlates with tumor progression and immunosuppressive tumor microenvironment in hepatocellular carcinoma. *Frontiers in Oncology*, *11*, 736363. doi:10.3389/fonc.2021.736363

Zhao, Y. Y. (2013). Metabolomics in chronic kidney disease. *Clinica Chimica Acta*, *422*, 59–69. doi:10.1016/j.cca.2013.03.033 PMID:23570820

Zheng, G., Ye, B., Zheng, Y., Xiong, Z., Xia, R., Qiu, P., Tao, J., & Chen, L. (2019). The effects of exercise on the structure of cognitive related brain regions: A meta-analysis of functional neuroimaging data. *The International Journal of Neuroscience*, *129*(4), 406–415. doi:10.1080/00207454.2018.1508135 PMID:30073877

Zhou, G., & Xia, J. (2018). OmicsNet: A web-based tool for creation and visual analysis of biological networks in 3D space. *Nucleic Acids Research*, *46*(W1), W514–W522. doi:10.1093/nar/gky510 PMID:29878180

Zhou, W., Bai, T., Wang, L., Cheng, Y., Xia, D., Yu, S., & Zheng, Y. (2023). Biomimetic AgNPs@antimicrobial peptide/silk fibroin coating for infection-trigger antibacterial capability and enhanced osseointegration. *Bioactive Materials*, *20*(5), 64–80. doi:10.1016/j.bioactmat.2022.05.015 PMID:35633877

Zhou, W., Jia, Z., Xiong, P., Yan, J., Li, Y., Li, M., Cheng, Y., & Zheng, Y. (2017). Bioinspired and Biomimetic AgNPs/Gentamicin-Embedded Silk Fibroin Coatings for Robust Antibacterial and Osteogenetic Applications. *ACS Applied Materials & Interfaces*, *9*(31), 25830–25846. doi:10.1021/acsami.7b06757 PMID:28731325

Zhuhadar, L. P., & Lytras, M. D. (2023). The Application of AutoML Techniques in Diabetes Diagnosis: Current Approaches, Performance, and Future Directions. *Sustainability (Basel)*, *15*(18), 13484. doi:10.3390/su151813484

Zou, F., Jiang, J., Lv, F., Xia, X., & Ma, X. (2020). Preparation of antibacterial and osteoconductive 3D-printed PLGA/Cu(I)@ZIF-8 nanocomposite scaffolds for infected bone repair. *Journal of Nanobiotechnology*, *18*(1), 1–14. doi:10.1186/s12951-020-00594-6 PMID:32103765

Zou, Q., Qu, K., Luo, Y., Yin, D., Ju, Y., & Tang, H. (2018). Predicting diabetes mellitus with machine learning techniques. *Frontiers in Genetics*, *9*, 515. doi:10.3389/fgene.2018.00515 PMID:30459809

About the Contributors

Pranav Kumar Prabhakar is currently working as a Professor and Head at the Research and Development Cell, Lovely Professional University, Punjab, India. He is one amongst the World's Top 2% Scientists (list published by Stanford University, USA, 2021, 2022, and 2023). He has completed his PhD in Biotechnology from IIT Madras. His main area of research interest focuses Elucidate Molecular Mechanism and Strategies for Oral Insulin Delivery, Mimic the signaling pathways in Metabolic Disorders (diabetes) with Natural Products. He is member of Royal Society of Chemistry, and Asia-Pacific Chemical, Biological& Environmental Engineering Society. He is also serving as editorial board member and reviewer for many reputed national and International Journals. Dr. Pranav received honors includes recipient of travel grant towards attending ATTD 2009 in Greece from Indian Institute of Technology Madras & Council for scientific and Industrial research (CSIR) and approved by Department of Science and Technology (DST). He has published 85+ research articles in journals, authored 3 books, and 14 book chapters. He also delivered 9 oral and poster presentations in scientific meeting.

***

Ishwar Basavan is an MBBS graduate, currently employed as an intern at KLE Prabhakar Kore Hospital and Research Centre,Belagavi, Karnataka.

Madhura Bodkhe has always been curious to learn about emerging fields in chemical engineering. With her study in chemical engineering, she was able to explore a vast array of fundamental topics such as heat transfer, mass transfer, chemical reaction engineering, thermodynamics, process control, and dynamics. This broadened her understanding and greatly expanded her prospects. She has worked on the "Phosphoric acid-based thermochemically activated ground nut shell carbon for removal of acid red 33 dye" project during her Polytechnic years, and won first place in a national conference run by the University Institute of Chemical Technology Jalgaon for it. For a month, she was a trainee at Smaart Pharmaceuticals in Jalgaon.

Rekha C. M. has completed B.E(Electrical and Electronics Engineering) and M.E(Power and Energy Systems) in GVIT and UVCE from Bangalore University, Karnataka, India. Currently working as Assistant Professor in the Department of Electrical and Electronics, Acharya Institute of Technology, Bengaluru, Karnataka, India. Her current research interests include Power systems, Distributed generation, Renewables, and heuristic algorithms. She is a life member of the Indian Society for Technical Education (ISTE).

Tejas Chalke is pursuing a Bachelor of Technology (B.Tech.) in chemical engineering at the prestigious University Institute of Technology in Jalgaon. Throughout this time at this esteemed institution, he has demonstrated a keen interest in science and technology, consistently seeking to expand his knowledge and skills in the field of chemical engineering.

Ajaygiri K. Goswami has completed Masters and Doctorate in Chemical engineering and technology and presently working as Head Pharmaceutical Technology, KBCNorth Maharashtra University, Jalgaon. He has presented more than fifty research articles in International Conferences and also published research papers more than hundred in National and international reputed Journals. He is working in various academic bodies of the University like BOS,BOG and Coordinators for IIC, Startup, KCIL cell etc. He has more than 25 years Teaching and research experience in Engineering College and University Departments and is research area are Pharmaceutical technology, Chemical engineering, Oil technology, Bio fuel, Nano catalyst etc.

Vishvesh Joshi is currently serving as V.P(R&D) at Chartwell Pharmaceuticals, NY, USA. Dr Joshi has 19 years of Pharmaceutical Industry experience with over 10 years in leading cross functional CMC teams. From 2019-23, he served as V.P(R&D) of Alembic Labs, USA. From 2013-18 he served as Director(R&D&BD), Phoenix Pharmaceuticals. From 2005-13 he advanced through various R&D managerial positions in Dr Reddy's, Piramal Healthcare and in Phoenix Pharmaceutical, USA. He also worked as a consultant to various pharmaceutical companies. He has contributed to the development of platform technologies in controlled and extended-release dosage forms and approval of more than 50 "ANDAs". His expertise is in developing dosage forms like oral solids, liquids, pediatric formulations and injectables in the form of Liposomes, SLNP and nanocomplex. Dr Joshi Holds a Ph.D in Pharmaceutical Technology from the Institute of Chemical Technology, Mumbai. He has authored and co-authored several research publications and book chapters.

Paranjeet Kaur is a researcher in the field of Medicinal Chemistry, Organic Chemistry, Drug Discovery and Design, In-Silico Drug Designing, Molecular Biology and Instrumental Analysis with main focus in discovery of small molecules for the management of Diabetes and Cancer, Synthetic organic chemistry encompassing asymmetric synthesis, Heterocyclic chemistry. Dr. Kaur has significant research achievements in the field of pharmacy, with a published patents and multiple publications in peer reviewed journals of high repute with a high impact factors (total published papers: 40 & 7 Book Chapters). Dr. Kaur worked as Research Scientist under Department of science and technology funded project with total budget 25.77 lakhs (INR), at LPU, Punjab, India (DST No: SB/FT/CS-087/2014) from 2015 till 2018. In addition to this, Dr. Kaur received University Honor Award of appreciation in Academics (Ph.D, 2015-2018 and M.Pharm, 2013-2015). Apart from research, Dr. Kaur has enriched experience of teaching in field of pharmaceutical Sciences and has held teaching position in various esteemed institutions in India. Currently, she is serving as Assistant Professor in Chitkara College of Pharmacy, Chitkara University. Dr. Kaur has been awarded a "Bharat Siksha Gaurav Puruskar Award" and "Certificate of Excellence" in teaching by KTK Outstanding Achiever and Education foundation and Helium Productions, 2022. She has also received "Young scientist Awards" In several National and International conferences. Dr. Kaur, has been invited as Guest speaker at "18th international conference on Targeted Drug delivery and Drug designing during December 06-08, 2017 at Dallas, Texas, USA"; "5th International conference on Diabetes, Hypertension and Metabolic syndrome" held during July

18-19th 2022, Melbourne. Also served as Presenting Author in prestigious national and international conferences. She is a member of Indian Pharmacy Graduate Association, IPGA. Dr. Kaur has an h-index: 11; i10-index = 12 and citations: 336.

Manju Khari is a Professor at Jawaharlal Nehru University, New Delhi, prior to the university she worked with Netaji Subhas University of Technology, East Campus, Under Govt. Of NCT Delhi. Her PhD is in Computer Science and Engineering from the National Institute of Technology Patna and She received her master's degree in Information Security from Guru Gobind Singh Indraprastha University, Delhi, India. She has 100+ published papers in refereed National/International Journals and Conferences (viz. IEEE, ACM, Springer, Inderscience, and Elsevier), 10 book chapters in a Springer, CRC press, IGI Global, Auerbach. She is also co-author of two books published by NCERT of XI and XII and co-editor of 10 edited books. She has also organized 05 International conference sessions, 03 Faculty Development Programmes, 02 workshops, and 01 industrial meet her experience. She delivered an expert talk, was a guest lecturer at International Conferences and was a member of the reviewer/technical program committee at various International Conferences.

Rita Komalasari is a lecturer at YARSI University. Her current work is focused on Transformative Insights: Harnessing Artificial Intelligence for Enhanced Ovarian Cancer Prediction and Prognosis.

Sunil J. Kulkarni has done his PhD in Chemical Engineering with focus on adsorption for wastewater treatment. He has done his Masters in chemical engineering from Tatyasaheb Kore Institute of Engineering and Technology, Warananagar. He has worked as an Assistant Professor in Chemical Engineering Department of Datta Meghe College of Engineering, Airoli, Navi Mumbai, India for 24 years. He has 25 years of teaching and research experience. Dr. Sunil is involved with different academic activities across the universities in Maharashtra. His area of investigation in doctoral studies is phenol and heavy metal removal from waste water. He has also worked on kinetics and isotherms of various sorption operations. His area of research includes adsorption, clean technology and environmental biotechnology. He is on the reviewer board of many international journals and reviewed many international papers. He is also governing body member, promotional editor and executive member of many reputed international journals. Currently he is working as Associate Professor in chemical engineering and and Dean Research and Development at GIT, Lavel, India.

Animesh Laha, an Assistant General Manager at Welspun India Limited, specializes in Material, fibrous structures, nonwoven structures, fibre orientation, fibrous composites and polymer composites. He obtained his Ph.D. in Textile and Fibre Engineering from IIT Delhi, where a wide range of fibrous structures for soft body armour development was explored. His doctoral research primarily focused on the development of soft body armour using textile weave structures treated with shear thickening fluid. With more than 9 years of experience in research and development, Dr. Laha has authored more than 15 research papers, 4 book chapters and delivered presentations at over 15 conferences of National and International repute. His primary interest involves advancing the field of fibrous material design for high-end applications. His current research interest includes sustainable textiles, technical textiles with natural materials and polymer composites.

G. Muthugurunathan works as an Assistant Professor in Madanapalle Institute of Technology & Science, Madanapalle, AP. He has completed B.E. in Computer Science and Engineering in 2011. He has completed in M.E. in Computer Science & Engineering in 2014. His research area involves Machine Learning and IoT.

Raghu N. has Ten years of experience in Academic and Research. He received the Ph.D. (2020) in Electronics Engineering from Jain University, and B.E. in Electronics and Telecommunications Engineering from VTU University, Karnataka. His research interests were in RF Communication and Image Processing. He has published his research findings in various Scopus indexed Journals and International conferences. He is a member of various professional bodies.

Priyanka Prabhu received her M.Pharm in Drug Delivery Technology from Mumbai University in 2010 and her Ph.D. in Pharmaceutics from Institute of Chemical Technology in 2017. In 2016 she was hired as an Assistant Professor, Pharmaceutical Sciences by Shobhaben Pratapbhai Patel School of Pharmacy & Technology Management at Shri Vile Parle Kelavani Mandal 's (SVKM's) Narsee Monjee Institute of Management Studies (NMIMS), Mumbai. She received the 'Ambuja Cement Best Ph.D. (Tech.) Thesis Award' and 'Institute of Chemical Technology Golden Jubilee Innovative Ph.D. (Tech.) Thesis Award' for Ph.D. thesis, and 'Third Prize for Preclinical Oral Presentation' at the Eighth International Annual Conference on "Translational Clinical Pharmacology Research in Drug Development". Dr. Priyanka has 26 publications to her credit. Her research interests include development of nanocarriers for treatment of infections and inflammatory disorders.

Sabyasachi Pramanik is a professional IEEE member. He obtained a PhD in Computer Science and Engineering from Sri Satya Sai University of Technology and Medical Sciences, Bhopal, India. Presently, he is an Associate Professor, Department of Computer Science and Engineering, Haldia Institute of Technology, India. He has many publications in various reputed international conferences, journals, and book chapters (Indexed by SCIE, Scopus, ESCI, etc). He is doing research in the fields of Artificial Intelligence, Data Privacy, Cybersecurity, Network Security, and Machine Learning. He also serves on the editorial boards of several international journals. He is a reviewer of journal articles from IEEE, Springer, Elsevier, Inderscience, IET and IGI Global. He has reviewed many conference papers, has been a keynote speaker, session chair, and technical program committee member at many international conferences. He has authored a book on Wireless Sensor Network. He has edited 8 books from IGI Global, CRC Press, Springer and Wiley Publications.

Rupayan Roy, an Assistant Professor at the National Institute of Fashion Technology in Kannur, India, specializes in fibrous structures, nonwoven structures, fibre orientation, and fibrous composites. He obtained his Ph.D. in Textile and Fibre Engineering from IIT Delhi, where a wide range of fibrous structures for various applications was explored. His doctoral research primarily focused on the development of highly efficient air filter media utilizing needle-punched nonwovens. With 3 years of experience in academics, Dr. Roy has authored more than 25 research papers and delivered presentations at over 10 conferences of International repute. His primary interest involves advancing the field of fibrous material design for high-end applications. His current research interest includes technical textiles with natural materials.

Gorthy Shanmukha Rama is a detail-oriented individual with sound knowledge of dentistry and clinical skills. Ability to understand and quickly learn study Protocols and Concepts. Knowledge of basics of clinical research and designs. I look forward to opportunities to prove my abilities by accepting challenges through continuous learning and commitment to become an asset to the organization.

Ashish Shukla is working as Professor, Dept. Of Biotechnology, RBS Engineering Technical Campus Bichpuri Agra.

N.L. Swathi is a writer and research enthusiast specializing in medicine, Advanced technologies, and Diabetes. I've contributed as a reviewer, ensuring quality research dissemination. Additionally, I've served as a hands-on research assistant, gaining expertise in experimental design and data analysis. My passion lies in bridging academia with wider audiences and communicating complex scientific concepts effectively.

Index

www.ingramcontent.com/pod-product-compliance
Lightning Source LLC
LaVergne TN
LVHW060301011026
849161LV00011B/2749